Globalization: Effects on Fisheries Resources

Globalization is a multidimensional issue, and its impacts on world resources cross and integrate environmental, economic, political, and cultural boundaries. Over the last few decades, the push towards globalization has brought a new dimension in which managers of fisheries and water resources will need to operate, both at the local and global level of governance. In order to address effectively the future sustainability of these resources, it is critical to understand the driving factors of globalization and their effect on fisheries ecosystems and the people who depend on these resources for their cultural and societal well-being. This book discusses the social and political changes affecting fisheries, the changes to ecological processes due to direct and indirect impacts of globalization, the changing nature of the goods and services that fisheries ecosystems are able to provide, and the resultant changes in markets and economic assessment of our fishery resources.

WILLIAM W. TAYLOR is a University Distinguished Professor and Chairperson in the Department of Fisheries and Wildlife at Michigan State University.

MICHAEL G. SCHECHTER is Professor and Chairperson of the International Relations Field in James Madison College at Michigan State University.

LOIS G. WOLFSON is a Specialist in the Department of Fisheries and Wildlife and the Institute of Water Research at Michigan State University.

Drawing by Katrina Mueller, Michigan State University.

Globalization:

Effects on Fisheries Resources

Edited by
WILLIAM W. TAYLOR
MICHAEL G. SCHECHTER
LOIS G. WOLFSON
Michigan State University

CAMBRIDGE UNIVERSITY PRESS
Cambridge, New York, Melbourne, Madrid, Cape Town, Singapore, São Paulo

Cambridge University Press
The Edinburgh Building, Cambridge CB2 8RU, UK

Published in the United States of America by Cambridge University Press, New York

www.cambridge.org
Information on this title: www.cambridge.org/9780521875936

© Cambridge University Press 2007

This publication is in copyright. Subject to statutory exception and to the provisions of relevant collective licensing agreements, no reproduction of any part may take place without the written permission of Cambridge University Press.

First published 2007

Printed in the United Kingdom at the University Press, Cambridge

A catalog record for this publication is available from the British Library

ISBN 978-0-521-87593-6 hardback

Cambridge University Press has no responsibility for the persistence or accuracy of URLs for external or third-party internet websites referred to in this publication, and does not guarantee that any content on such websites is, or will remain, accurate or appropriate.

Contents

List of contributors	*page* ix
About the editors	xiv
Foreword	
JIANGUO LIU	xvii
Preface	xix
Acknowledgments	xxiii

Introduction
Globalization and fisheries: a necessarily interdisciplinary inquiry
JOHN ROOD AND MICHAEL G. SCHECHTER 1

Part I
Impacts of globalization on fisheries and aquatic habitats 19

1 Globalization: implications for fish, fisheries, and their management
WILLIAM W. TAYLOR, NANCY J. LEONARD, JUD F. KRATZER, CHRIS GODDARD, AND PATRICIA STEWART 21

2 Fisheries globalization: fair trade or piracy?
JACKIE ALDER AND REG WATSON 47

3 Effects of globalization on freshwater systems and strategies for conservation
JESSICA SEARES, KATHERINE SMITH, ELIZABETH ANDERSON, AND CATHERINE PRINGLE 75

4	Globalization effects on water quality: monitoring the impact on and control of waterborne disease JOAN B. ROSE AND STEPHANIE L. MOLLOY	92
5	Health challenges to aquatic animals in the globalization era MOHAMED FAISAL	120
6	Globalization, biological invasions, and ecosystem changes in North America's Great Lakes KRISTEN T. HOLECK, EDWARD L. MILLS, AND HUGH J. MACISAAC	156

Part II
Case studies of globalization and fisheries resources   183

7	Possible contributions of globalization in creating and addressing sea horse conservation problems A. C. J. VINCENT, A. D. MARSDEN, AND U. R. SUMAILA	185
8	Wronging rights and righting wrongs: some lessons on community viability from the colonial era in the Pacific KENNETH RUDDLE	215
9	Cooperation and conflict between large- and small-scale fisheries: a Southeast Asian example RICHARD B. POLLNAC	229
10	Response of Alaskan fishermen to aquaculture and the salmon crisis ROSAMOND NAYLOR, JOSH EAGLE, AND WHITNEY SMITH	244
11	Tilapia: a fish with global reach JOSEPH J. MOLNAR AND WILLIAM H. DANIELS	269
12	The influence of globalization on the sustainability of North Pacific salmon fisheries WILLIAM W. TAYLOR AND NANCY J. LEONARD	291

Part III
Governance and multilevel management systems   303

13 Great Lakes fisheries as a bellwether of global governance
    GRANT FOLLAND AND MICHAEL G. SCHECHTER   305

14 Ecosystem-based insights on northwest Atlantic fisheries
    in an age of globalization
    DEAN BAVINGTON AND JAMES KAY   331

15 "Fishy" food laws
    P. VINCENT HEGARTY   364

Part IV
Ethical, economic, and policy implications   383

16 The intersection of global trade, social networks, and
    fisheries
    KENNETH A. FRANK, KATRINA MUELLER,
    ANN KRAUSE, WILLIAM W. TAYLOR, AND
    NANCY J. LEONARD   385

17 Fishing for consumers: market-driven factors
    affecting the sustainability of the fish and
    seafood supply chain
    H. CHRISTOPHER PETERSON AND KARL FRONC   424

18 Globalization and worth of fishery resources in an
    integrated market-based system
    WILLIAM KNUDSON AND H. CHRISTOPHER PETERSON   453

19 Can transgenic fish save fisheries?
    REBECCA M. BRATSPIES   468

20 Contributing to fisheries sustainability
    through the adoption of a broader ethical
    approach
    TRACY DOBSON AND HENRY A. REGIER   499

**Part V
Conclusions and recommendations** 525

21 Globalization and fisheries: recommendations for policy
    and management
    TRACY L. KOLB AND WILLIAM W. TAYLOR 527

    Index 545

The color plates are situated between pages 74 and 75

Contributors

Jackie Alder
Fisheries Centre, The University of British Columbia, 2202 Main Mall Road, Vancouver, BC, V6T 1Z4 Canada

Elizabeth Anderson
Global Water for Sustainability Program, Department of Environmental Studies, Florida International University, 11200 SW 8th Street, ECS 347, Miami, FL 33199, USA

Dean Bavington
27 Hatcher Street, St. John's, NL, A1B 1Z3 Canada

Rebecca M. Bratspies
CUNY School of Law, 62-21 Main Street, Flushing, NY 11367, USA

William H. Daniels
International Center for Aquaculture and Aquatic Environments, Department of Fisheries and Allied Aquacultures, Auburn University, Auburn, AL 36849, USA

Tracy Dobson
Department of Fisheries and Wildlife, Michigan State University, 13 Natural Resource Building, East Lansing, MI 48824, USA

Josh Eagle
University of South Carolina School of Law, 701 Main Street, Columbia, SC 29208, USA

Mohamed Faisal
Departments of Pathobiology and Diagnostic Investigation and Fisheries and Wildlife, Michigan State University, S-110 Plant Biology Building, East Lansing, MI 48824, USA

Grant Folland
School of Law, University of Chicago, 1307 E 60th Street No. 307, Chicago, IL 60637, USA

Kenneth A. Frank
Measurement and Quantitative Methods Counseling, Educational Psychology and Special Education and Fisheries and Wildlife, Michigan State University, 460 Erickson Hall, East Lansing, MI 48824, USA

Karl Fronc
The Product Center for Agriculture and Natural Resources, Michigan State University, 4700 S. Hagadorn Road, Suite 210, East Lansing, MI 48823, USA

P. Vincent Hegarty
Institute for Food Laws and Regulations, Michigan State University, Agriculture Hall, East Lansing, MI 48824, USA

Kristen T. Holeck
Department of Natural Resources, Cornell University Biological Field Station, 900 Shackelton Point Road, Bridgeport, NY 13030, USA

James Kay
Department of Environment and Resource Studies, Faculty of Environmental Studies, University of Waterloo, 200 University Avenue West, Waterloo, ON, N2L 3G1 Canada

William Knudson
Department of Agricultural Economics, Michigan State University, Agriculture Hall, East Lansing, MI 48824, USA

Tracy L. Kolb
Department of Fisheries and Wildlife, Michigan State University, 13 Natural Resource Building, East Lansing, MI 48824, USA

Jud F. Kratzer
Department of Fisheries and Wildlife, Michigan State University,
13 Natural Resource Building, East Lansing, MI 48824, USA

Ann Krause
Center for Systems Integration and Sustainability, Michigan State University, 13 Natural Resources Building, East Lansing, MI 48824, USA

Nancy J. Leonard
Great Lakes Fishery Commission, 2100 Commonwealth Boulevard, Suite 100, Ann Arbor, MI 48105, USA

Hugh J. MacIsaac
Great Lakes Institute for Environmental Research, University of Windsor, Windsor, ON, N9B 3P4 Canada

A. D. Marsden
Fisheries Economics Research Unit and the Sea Around Us Project, Fisheries Centre, The University of British Columbia, 2202 Main Mall Road, Vancouver, BC, V6T 1Z4 Canada

Edward L. Mills
Department of Natural Resources, Cornell University Biological Field Station, 900 Shackelton Point Road, Bridgeport, NY 13030, USA

Stephanie L. Molloy
Department of Fisheries and Wildlife, Michigan State University, 13 Natural Resource Building, East Lansing, MI 48824, USA

Joseph J. Molnar
International Center for Aquaculture and Aquatic Environments, Department of Agricultural Economics and Rural Sociology, Auburn University, Auburn, AL 36849, USA

Katrina Mueller
Department of Fisheries and Wildlife, Michigan State University, 13 Natural Resource Building, East Lansing, MI 48824, USA

Rosamond Naylor
Center for Environmental Science and Policy, Stanford University, Encina Hall E418, Stanford, CA 94305, USA

H. Christopher Peterson
Consumer-Responsive Agriculture, Department of Agricultural Economics, Michigan State University, Agriculture Hall, East Lansing, MI 48824, USA

Richard B. Pollnac
Departments of Anthropology and Marine Affairs and the Coastal Resources Center, University of Rhode Island, Kingston, RI 02881, USA

Catherine Pringle
Institute of Ecology, University of Georgia, Ecology Building, Athens, GA 30602, USA

Henry A. Regier
Institute of Environmental Studies, University of Toronto, and Department of Fisheries and Wildlife, Michigan State University, 10 Ernst Street, Elmira, ON, N3B 1K5 Canada

John Rood
Graduate Division of Social Sciences, University of Chicago, 5330 S. Blackstone Avenue, Chicago, IL 60615, USA

Joan B. Rose
Department of Fisheries and Wildlife, Michigan State University, 13 Natural Resource Building, East Lansing, MI 48824, USA

Kenneth Ruddle
Asahigaoka-cho 7-22-511, Ashiya-shi, Hyogo-ken, Japan 659-0012

Michael G. Schechter
James Madison College, Michigan State University, South Case Hall, East Lansing, MI 48825, USA

Jessica Seares
Department of Environmental Studies, Emory University, Math & Science Center, Suite E510, 400 Dowman Drive, Atlanta, GA 30322, USA

Katherine Smith
Institute of Ecology, University of Georgia, Ecology Building, Athens, GA 30602, USA

Whitney Smith
Center for Environmental Science and Policy, Stanford University, Stanford, CA 94305, USA

Patricia Stewart
Michigan Department of Natural Resources, Box 30028, Lansing, MI 48909, USA

U. R. Sumaila
Fisheries Economics Research Unit and the Sea Around Us Project, Fisheries Centre, The University of British Columbia, 2202 Main Mall Road, Vancouver, BC, V6T 1Z4 Canada

William W. Taylor
Department of Fisheries and Wildlife, Michigan State University, 13 Natural Resource Building, East Lansing, MI 48824, USA

Amanda C. J. Vincent
Project Seahorse, Fisheries Centre, The University of British Columbia, 2202 Main Mall Road, Vancouver, BC, V6T 1Z4 Canada

Reg Watson
Fisheries Centre, University of British Columbia, 2202 Lower Mall Road, Vancouver, BC, V6T 1Z4 Canada

About the editors

Dr. William W. Taylor is Chairperson of the Department of Fisheries and Wildlife at Michigan State University. He is immediate past President of the American Fisheries Society and is a Fellow of the American Institute of Fishery Research Biologists. His research has centered on understanding the biotic and abiotic factors that regulate the dynamics and yield of fishes, particularly in the Great Lakes, and how this understanding can be used to more effectively manage these species. He is an internationally recognized expert in fisheries ecology, population dynamics, and Great Lakes fisheries management. He has authored numerous articles and has edited a book on *Great Lakes Fisheries Policy and Management: A Binational Perspective* and served as the Series Editor for *Fisheries*. He has received numerous honors and awards, including a recent Distinguished Faculty Award at Michigan State University.

Dr. Michael G. Schechter is Professor and Chairperson of the international relations major in James Madison College, Michigan State University. Professor Schechter, who earned his Ph.D. in political science at Columbia University, specializes in international law, global governance, and globalization. His recent book publications include *The Historical Dictionary of International Organizations; Innovation in Multilateralism; Future Multilateralism: The Political and Social Framework; The Revival of Civil Society: Global and Comparative Perspectives; Rethinking Globalization(s): From Corporate Transnationalism to Local Interventions; United Nations-Sponsored World Conferences: Focus on Implementation and Follow-Up; The Political Economy of a Plural World: Power, Morals and Civilizations*; and *United Nations Global Conferences*. His awards at Michigan State University include a Distinguished Faculty Award and a Teacher-Scholar Award. He is a past President of the International Studies Association, Midwest Region.

Dr. Lois G. Wolfson is a Specialist in the Department of Fisheries and Wildlife and Institute of Water Research at Michigan State University. She is the State Extension Water Quality Coordinator and serves on the Regional Great Lakes Water Quality Leadership Team, a consortium of the six Great Lakes states in Environmental Protection Agency Region 5 which provides regional and national leadership in the water quality arena. Her applied research interests are in the area of lake and watershed management and nuisance aquatic species effects on native biota. She is a past President of the Michigan Chapter of the North American Lake Management Society. She is co-editor of a book on groundwater contamination and former editor of a monthly newsletter on water issues.

Foreword

Globalization is an increasingly important driving force that connects various corners of the world. Although the phenomenon of globalization (worldwide economic, cultural, political, and technological linkages) is not new, the magnitude and speed of today's globalization are unprecedented. Since the latter part of the twentieth century, consumption of natural resources in distant locations has created an accelerating demand on local resources. The modern era of globalization has profoundly affected fisheries resources, one of the most important sources of protein for humans, in many ways. Numerous fisheries scientists, policy-makers, managers, producers, and consumers are deeply concerned about the impact of globalization on future fisheries from local to global levels.

This timely book fills an important void by including insightful and comprehensive papers from eminent scholars documenting the intricate relationships between globalization and fisheries. By elegantly integrating theory and case studies, this book offers a clear overview and detailed analyses of the socioeconomic and ecological consequences of globalization. The authors also critically analyze relevant governance and multilevel management systems, and provide fresh perspectives on ethical and socioeconomic dimensions of globalization in the context of fisheries management.

As the authors vividly illustrate, globalization is a double-edged sword. It threatens fisheries by reducing aquatic ecosystem productivity and diminishing fish stocks. Despite the use of more efficient technologies (e.g., vessels and fishing gear) to increase fish production, global capture fisheries have been decreasing rapidly in the past several decades. Furthermore, international trade of fish products drastically increases waterborne diseases, poses great health challenges to aquatic animals, introduces numerous invasive species, displaces local communities, threatens small-scale fishers, and increases risks to food security. On the other hand, globalization can produce positive outcomes through facilitating transfer of remediation techniques for aquatic ecosystems and vulnerable fish stocks, capital transactions, and sharing research and management information. For instance, many hail aquaculture as an alternative to capture fisheries, as its production is more controllable and can reduce pressure on wild fish stocks. However, aquaculture can also pollute aquatic ecosystems, destroy habitat of wild fish stocks, release exotic species, spread diseases, raise concerns about potential risks of transgenic fish, and compromise the livelihood of people who have depended on wild fish stocks for generations. Thus, the effects of globalization are sometimes ambiguous and require more extensive and intensive research.

Sustainable fisheries require integrating ethics and ecological responsibility of producers and consumers into decision-making processes and management practices. More innovative technical measures (e.g., eco-labeling and Web-based relational databases) are needed for consumers to choose ecologically sustainable fish products and shape fish markets. As the authors have forcefully argued, new ways of thinking, fair and equitable trade, more effective government policy, coordinated management across juridical boundaries, strictly enforced regulations, frequent monitoring, internalizing environmental costs, eliminating perverse subsidies, and multilevel governance for global fish stocks are also critical to steer fisheries into a sustainable future.

The impact of this book will reach far beyond fisheries. Although the book focuses on the effects on globalization on fisheries, its interdisciplinary approach and remarkable insights have important implications for sustainable management of other common resources (e.g., wildlife, forests) and the environment in the new age of globalization.

Jianguo Liu
Rachel Carson Chair in Ecological Sustainability and Director,
Center for Systems Integration and Sustainability,
Michigan State University

Preface

While the word globalization has been extensively used in recent times, the processes of exchanging information, goods, and culture between people and nations have been occurring since the dawn of human existence. Globalization, as defined by Held and McGrew[1] refers to "the expanding scale, growing magnitude, speeding up and deepening impact of interregional flows and patterns of social interaction ... [as well as] to a shift or transformation in the scale of human social organization that links regions and continents." What is novel about globalization today is the accelerating rate at which these exchanges are occurring. All segments of society and all societies are experiencing the impacts of globalization with the effects being viewed either as positive or negative depending on the value systems used to evaluate them.

The world's fisheries resources and fisheries-dependent communities have long been impacted by globalization. There is, however, a significant deficit in analyzing and understanding the influences of globalization acting on these systems. These analyses would allow for insightful policy reform needed to ensure the sustainability of these coupled human and natural resources systems. Our lack of such knowledge and policy integration have been clearly demonstrated by the collapse of a number of socially and economically important fisheries over the past several decades. This deficit in our understanding, however, is beginning to change as fisheries policy-makers and managers are realizing that globalization is impacting fisheries in numerous

---

[1] Held, David and Anthony McGrew. 2000. The great debate: an introduction. In *The Global Transformations Reader: An Introduction to the Globalization Debate*, ed. David Held and Anthony McGrew. Cambridge, UK: Polity Press, pp. 3–4.

ways at all levels of governance, and in view of this are taking action toward understanding and ameliorating the impacts of globalization on these resources.

Both human and fish populations have benefited from advances in communications, transportation, and governance systems due to globalization. Many of the changes arising from these advances have positively impacted our abilities to locate freshwater and marine fish species, whether for food, recreational fishing, or ecosystem restoration purposes. Additionally, these changes have enhanced our abilities to harvest, process, and transport fish and fish products, thereby greatly improving the quality and diversity of fish products available to humans throughout the world. With the technological improvements spurred by globalization, we have also improved our ability to raise fish in highly managed aquaculture operations thus permitting food security to be realized in many regions of the world. Lastly, globalization has facilitated our understanding of the entire fisheries supply chain, from its productivity to its marketing and consumption. This knowledge is facilitating our examination and management of fisheries in a holistic ecosystem-based approach, as well as incorporating both the biological and social factors within our governance decisions and their implementations.

The advances and benefits that are associated with globalization and fisheries resources have not been without controversy, and this resounds throughout the book. Some of the disputes that occur among and within nations include: who has rights to a fishery, an increasingly contentious issue as fishing nations are harvesting distant stocks that were formerly fished by locally based communities; loss of genetic diversity of fish stocks resulting from aquaculture escapees and inappropriate stocking protocols; habitat degradation and pollution related to our alterations of both aquatic and terrestrial ecosystems; and homogenization of cultures which, among many aspects, can threaten the existence of traditional fishing cultures. Effectively addressing these "new" challenges in fisheries management has created the need for a more agile governance approach of local and transboundary fisheries. Fortunately, globalization also provides us with the necessary tools to move towards developing approaches that can better solve these emerging challenges and should ultimately result in improved conditions for fish and their aquatic ecosystems. For instance, the ability to access expertise from across the world provides practitioners with better knowledge on, as well as possible solutions to, perplexing fisheries problems.

The path to successfully managing globally influenced fisheries resources is not easy, as is clearly illustrated in numerous chapters in this book. We believe that a key to facilitating the sustainability of our fish and aquatic resources in the future will depend on having an informed public that are cognizant of the impacts that their desires and demands are having on local as well as distant fishing communities and fisheries ecosystems. Thus, much like our ability to share information that improves the health of humans throughout the world, such exchanges and public viewing of the impact of globalization on our fisheries resources are needed to assure that more sustainable fisheries policies and practices evolve in the future. As we learn from each other we will hopefully develop culturally sensitive best management practices for sustainable fisheries. It is equally important that fisheries policy-makers and professionals understand the good, the bad, and the ugly of globalization agents acting on local communities, the economies involved, and the environmental factors of concern. Such understanding would ensure that we meet the challenges and realize the opportunities put forth by globalization of our fisheries and their ecosystems. Knowledge is power, and we hope that the sharing of our experiences on a large number of fisheries will facilitate the process of more sustainable use and management of the world's magnificent fisheries.

It was for these reasons that we chose to host a symposium at an annual meeting of the American Fisheries Society which attempted to depict the impact of globalization on coupled human and fisheries resources systems. We assembled a collection of case studies that provide guidance for designing ecologically and socially sustainable systems in the face of accelerating globalization. This book provides such an overview, and it is our hope that this information will enable an attentive public, fisheries professionals, and policy-makers to make better decisions in regard to the management of local and global fisheries in the years to come.

This book is written for a very diverse audience, including fisheries and aquatic ecologists, natural resource managers, governmental agency personnel, conservation biologists, social scientists, non-governmental organizations, policy-makers, graduate students, and advanced undergraduate students. Each chapter provides a world or regional view of one of the many facets of globalization, including its relationship to and impact on fisheries resources and their management. An overall theme throughout this book is to address the social and political changes affecting globalized fisheries and provide recommendations for their future sustainability. The book contains a number of case studies that focus on the relationship of changing markets and international trade agreements on fisheries resources. These studies also assess and evaluate the management of these systems and the factors affecting their sustainability. We also provide information on declining biodiversity as a result of overexploitation, habitat destruction, infectious diseases, and exotic species introductions, and how these factors impact fisheries, aquatic resources, and human health. The book also includes information on management and governance challenges using examples and lessons learned to illustrate ecosystem-based approaches to monitoring, managing, and governing globalized fisheries systems. As such, this book can serve as a resource for graduate and undergraduate courses, including courses related to fishery management, international political economy, global governance, environmental politics, globalization, natural resource policy and management, and conservation biology.

We were fortunate that more than 85 ecologists, social scientists, and natural resource managers enthusiastically participated in this book endeavor, either as contributors or as reviewers. To ensure the highest-quality information possible and the appropriate coverage of perspectives from both academic fields and policy agencies, two to three experts from academic institutions or management agencies and policy-makers reviewed each chapter. Thus, it is fair to say that the completion of this book is an excellent example of close collaboration between academics, fisheries and aquatic professionals, and policy-makers. We hope that this teamwork will continue and that this book will help to cement the bond between these groups on a worldwide basis. Ultimately, by doing so, we can better manage the world's fisheries resources in a sustainable manner.

William W. Taylor
Michael G. Schechter
Lois G. Wolfson

Acknowledgments

This book results from individuals in multiple disciplines and career paths in academia, governance entities, and management communities that have been involved in providing sustainable solutions for local and global fisheries. These individuals inspired passion among the team as each relied upon the energy and knowledge of others to make this book as complete as possible. Their cooperation, patience, and support in this endeavor are very much appreciated. We are also grateful to the following who shared their time and ideas, and provided excellent and insightful comments while reviewing manuscripts for this book: Jay Austin, Conner Bailey, Ted Batterson, S. Tamer Cavusgil, Patrick John Christie, Joseph Cullon, Tracy Dobson, Roger Eberhardt, David Ehrenfeld, Paola Ferreri, George Francis, Marc Gaden, Leon Gordenker, Jack Helle, Jeffrey Johnson, Gunnar Knapp, W. Andy Knight, Ann Krause, John Kurien, Robert Lackey, Linda Lobao, Andy Loftus, Donald MacDonald, Diane Mollenkopf, Shauna Oh, Rosemary Ommer, Pierre Payment, Rob Pennock, Ken Poff, Bryan Ritchie, Henry Regier, Yvonne Sadovy, Gary Sakagawa, Paul Seelbach, Paulette Stenzel, Merrit Turetsky, Ewen Todd, Marylynn Yates, and Jonas Zoninsein.

This book grew out of a symposium held during an American Fisheries Society Annual Meeting in Quebec City, Quebec. We wish to thank the Canadian Embassy and Government of Canada for supporting the symposium and subsequent publication of this book through their Canadian Studies Conference Grant Program. A special thank you goes to Dennis Moore, Canadian Consulate General in Detroit, Michigan for his invaluable assistance. Additional intellectual and financial support for this project was provided by the Great Lakes Fishery Commission and the Canadian Studies Center at Michigan State University. We also acknowledge the tremendous efforts of the American Fisheries Society conference organizers that contributed to the success of our symposium.

We thank Katrina Mueller of Michigan State University who drew the frontispiece illustration. Several people were also involved in the preparation of this book. We wish to thank Leslie Johnson for her editorial assistance, as well as Rachael Loucks, Kerry Waco, Katrina Mueller, Cecilia Lewis, Tracy Kolb, and Brittany Denison for their assistance with the index. Finally, we would like to express our sincere appreciation to the staff at Cambridge University Press, particularly Alison Evans, Clare Georgy, Anna Hodson, and Jeanette Alfoldi for their support and assistance.

The co-editors were assisted by the exceptional support of their families, colleagues, and friends who provided the needed intellectual and emotional support that is required to accomplish such an undertaking. Specifically, we would like to acknowledge Evelyn Taylor, Sir Thomas Huxley, Ilene Schechter, Jianguo (Jack) Liu, Nancy Leonard, Daniel Taylor, Tom Bedell, Jim Martin, John Robertson, Denny Grinold, Tom Coon, Pat Stewart, Stephanie Smith, Kris Lynch, Paola Ferreri, and Jack Laurie. Special thanks are extended to Alan Crowden who first encouraged us to undertake this project and publish with Cambridge University Press.

JOHN ROOD AND MICHAEL G. SCHECHTER

Introduction

Globalization and fisheries – a necessarily interdisciplinary inquiry

Solutions to the key problems of the twenty-first century require interdisciplinary inquiry. Globalization's impact on fisheries, including overfishing, is no exception. This conviction is the foundation for the American Fisheries Society symposium on globalization and fisheries out of which this book arose. Though the language of globalization has rarely been used by fisheries scholars or practitioners, this unique volume readily provides evidence that they had the empirical data and could write detailed case studies about globalization that are almost entirely lacking in the current volumes on that subject being published by social scientists. Moreover, as Folland and Schechter's chapter on global governance suggests many of the key concepts in that field were also preshadowed by works by fisheries scholars, but in materials rarely consulted by social scientists.

In this introductory chapter, we will introduce the concept of globalization and also begin to show how the study of fisheries in this volume can provide insights into many of the key questions animating globalization studies today, such as: Is globalization really anything new? What are the drivers of globalization? What role has the technology revolution played in accelerating the current era of globalization? What are the consequences of globalization, including who benefits and who loses from globalization? What roles do various international actors (e.g., states, intergovernmental organizations, non-governmental organizations) play in the governance of globalization in general and the management of fisheries more specifically? Why do scholars study globalization in general and the relationship between globalization and fisheries in particular?

Globalization: Effects on Fisheries Resources, ed. William W. Taylor, Michael G. Schechter, and Lois G. Wolfson. Published by Cambridge University Press. © Cambridge University Press 2007.

WHAT IS GLOBALIZATION?

Over the past several years, "globalization" has gone from international economics and business jargon to a worldwide buzzword.[1] Still, the definition and the concept of globalization are nothing if not amorphous and contentious. There are nearly as many definitions of globalization as there are authors who discuss globalization. In fact, a number of scholars have urged us to think in terms of globalizations (plural) rather than globalization: "The move from the singular to the plural is deliberate and implies deep skepticism of the idea that there can ever be a single theory or interpretation of globalization." This call has been made in the hope of widening "the debate on globalization beyond the definition of the processes as simply economic, or even worse, as about 'free trade' and liberalization" (Gills 2004). By 2007, the scope of globalization studies has been broadened. That is, though some books are still written from a single disciplinary perspective – particularly economics – there is widespread understanding that globalization refers to complex multidimensional processes. However, it is also true that most authors still privilege one disciplinary angle or another, treating "its debates as authoritative without awareness or acknowledgement of their partial status" (Pieterse 2004:15). Although many of the chapters in this volume demonstrate this proclivity as well, it is largely ameliorated by the diverse disciplinary backgrounds and professional experiences of the authors. Still, one must take care to decipher how each author thinks about globalization. Though it is true that there is no consensual definition of globalization (Pieterse 2004), and, indeed, that it is probably best to think of globalization as multifaceted and constantly in flux, there are several clear common threads in thinking about globalization. These are present in Held and McGrew's definition:

> Simply put, globalization denotes the expanding scale, growing magnitude, speeding up and deepening impact of interregional flows and patterns of social interaction. It refers to a shift or transformation in the scale of human social organization that links regions and continents. (Held and McGrew 2000)

Globalization has reduced the importance of distance through increased communication and transportation technology, while giving

---

[1] Globalization has been referred to as "the most over-used term in the current political lexicon" (Bromley 1996).

disparate peoples reasons to connect with one another to solve common problems and engage in trade. The deepening of interregional flows and the thickening of networks also suggests that people in different places are now more important and more linked to one another than ever before. These increased linkages are amply demonstrated in the governance of common fisheries resources as discussed by Folland and Schechter (Chapter 13) in this volume, who use the Great Lakes as their prime example.

IS GLOBALIZATION NEW?

A recurring question in the globalization literature is whether globalization is a new phenomenon or whether it is a consistently evolving process. If globalization is thought of, very generally, as interconnectedness driven most especially by trade and technology, then the world has, with a few exceptions, been getting more globalized for several hundred years. As technology has progressed, the world has been perceived as getting progressively smaller. The world known to Columbus was very large indeed; the "discovery" of a new world indicated that what was not known was at least as important as what was known. The advent of locomotives substantially shortened travel times throughout the world; airplanes did the same. Indeed, until the outbreak of the world wars, global capital flows consistently increased as technology made communication and transportation easier. After World War II, the value of trade again increased dramatically, along with the well-known explosion in international collaboration, intergovernmental organizations such as the United Nations and the European Union, and increased flows of people and ideas (Murphy 1994). Thus, though globalization is surely not a wholly new phenomenon (even if the term is relatively new), it is clear that the post-World War II era, the present era of globalization (what Pieterse refers to as "contemporary accelerated globalization"), differs from previous eras of globalization (including that prior to World War I) in several respects (Pieterse 2004:16).

First, the scope and speed of globalization have advanced so much. The speed of communication, for example, has consistently been growing for years. The days of messages sent by horseback faded to memory with the development of the telegraph and then the telephone. In the current age of globalization, communication has become so fast and so inexpensive as to provide no practical barriers to the exchange of information. Even 25 years ago, intercontinental phone calls could be a substantial business expense; such communication is available for pennies today. At present, even the land telephone is becoming obsolete because the Internet has created the ability to send full-motion video and sound instantaneously.

Second, trade in goods has also increased remarkably; it now makes solid economic sense for Volkswagen to make cars in Venezuela and for Nike to make shoes in Vietnam rather than to make them locally (Enloe 2000). Transport costs have become so low as to make production of goods almost exclusively a question of inputs (labor and raw material). International trade has grown substantially. In 1900, foreign exchange trading was measured in the millions of dollars, according to the Bank for International Settlements; in 1998, trade equaled $650 billion per day, and by 2004, trade was at $1.8 trillion per day. Of course, there are still limits. Transport costs will always be higher between Kansas City and Singapore than between Kansas City and St. Louis. In addition, capital flows will likely never be completely open, as Adam Smith lamented hundreds of years ago: politicians, responding to domestic pressures, will never allow it. The continuing need for the World Trade Organization (WTO) to arbitrate disputes suggests that many economies, even the largest, try to cheat the free trade system for domestic economic and political reasons. Shaffer details one of the most contentious cases, the United States shrimp–turtle case in which the WTO found that the United States applied its ban on shrimp imports discriminatorily; it had provided countries in the Western Hemisphere – mainly the Caribbean – technical and financial assistance and longer transition periods for their fishers to start using turtle excluder devices than it had granted for those in the four Asian countries (India, Malaysia, Pakistan, and Thailand) that had filed the complaint with the WTO (Shaffer 2005). On balance, however, markets and capital flows are significantly more open today to the point that a restriction on trade is seen as the exception and is likely to be defeated by the WTO.

Third, it has been argued that the new (1980s on) era of globalization differs because it involves a much greater magnitude of people and states. Hobsbawm (1975:50) details the extraordinary increase in trade volume from 1840 to 1975 and concludes that "the value of exchanges between the most industrialized economy and the most remote or backward regions of the world had increased sixfold." The Silk Road once connected two distant powers (Rome and China); today the financial centers of the world are connected not just to other financial centers but to markets throughout the world (Germain 1997). The current phase of globalization provides the opportunity for previously closed and economically less privileged societies to realize significant gains by opening themselves to global markets, but, as will be discussed later, it also raises questions about social justice.

Finally, globalization today is different because of the scope of sameness and standardization that has accompanied the current round of globalization. Hobsbawm writes, "There is a substantial difference between the process as we experience it today and that in the previous century. What is most striking about it in the later twentieth century is an international standardization which goes far beyond the purely economic and technological" (Hobsbawm 1975:65). This standardization closely follows several of the key drivers and dimensions of globalization that will later be discussed. The rising international norm toward a free market has converted all but the staunchest holdouts of closed or controlled economies. Technology now spreads quickly around the world, even in states with populations that have very limited access to communication lines, such as China. Global ethical norms, such as a ban on landmines, spread quickly through an environment in which non-governmental organizations (NGOs) and grassroots activists can use inexpensive and rapid communications technology to coordinate their strategies to put great pressure on government leaders. Similar impacts are being made by intergovernmental organizations, such as the International Standardization Organization (ISO), which promulgates a wide variety of global standards; the World Intellectual Property Organization (WIPO), which works to standardize norms relating to intellectual property; and the International Whaling Commission, which seeks to conserve whale stocks.

The current age of globalization has also made it possible to share cultural traditions and practices, an opportunity taken up by many around the world. Though it would be premature to say that the world is becoming homogeneous, the rate of international music, film, and culture exchange is certainly increasing. One must only consider the worldwide popularity of Coca Cola and Pepsi Cola or the wide availability of ethnic cuisine around the world to understand that trends and fashions are becoming globalized and "glocalized" along with trade.[2]

---

[2] "Glocalization" is a term that was invented to emphasize that the globalization of a product is more likely to succeed when the product or service is adapted

For both fisheries practitioners and managers, the current globalization era is also a new process or, at least, has taken a new form as a consequence of its acceleration. Fisheries management concerns reflect that, indeed, the world is becoming both smaller and more interconnected. The increased prevalence of long-distance fleets from developing states (see Alder and Watson, Chapter 2) and the dangers associated with the introduction of foreign species (see Holeck, Mills, and MacIsaac, Chapter 6) demonstrate that, in the current round of globalization, even a once localized activity has become global in nature.[3] One is no longer tied to fishing grounds located near or around a port; increased refrigeration, ship-based freezing and processing technology enable long-distance catches to travel globally. The establishment of exclusive economic zones has done little to limit the use of local fisheries resources by distant countries, as developing states often felt compelled to sell their local fishing rights to distant fleets in return for hard currency (Alder and Watson, Chapter 2). Globalization in fisheries resources is indeed new in the sense of globalization's greater influence on fisheries stock and consequently requires new governance approaches and policy solutions.

specifically to each locality or culture it is marketed in. The term combines the word "globalization" with "localization." The term began in the field of business, and was subsequently adopted by cultural sociologists. Others refer to the phenomenon as "hybridization."

[3] The importance of exotic species on the sustainability of the Great Lakes fishery is also undeniable (IJC–GLFC 1990; Mills *et al*. 1993), including the impacts observed from the establishment of sea lamprey, alewife, and more recently the zebra mussel (Fetterolf 1980; Brandt *et al*. 1987; Mills *et al*. 1993). The success of exotics in displacing native species in both the terrestrial and aquatic environment makes it certain that these aquatic invasive species will continue to have a detrimental impact on the Great Lakes fishery. The correlation reported between increasing human activities in the Great Lakes, such as transoceanic shipping and canal construction, and the increase in exotic species strongly suggests that as globalization increases, so will the number of exotic species introductions unless action is taken to deter these unwanted invaders (Mills *et al*. 1993; Ricciardi and Rasmussen 1998). Currently, Great Lakes fisheries managers and all stakeholders in the Great Lakes are facing another threat to the Great Lakes ecosystem, the Asian carp. Asian carp, imported to the southern United States to function as biocontrols in catfish aquaculture, escaped and have become established in the Mississippi River Basin (Rasmussen n.d.). These voracious carp are rapidly swimming upstream toward the Chicago Sanitary and Shipping Canal, a canal constructed to connect the Mississippi River with the Great Lakes.

WHAT ARE THE CHIEF DRIVERS OF GLOBALIZATION?

The increasing interconnectedness of the world can hardly be disputed. Even today, at the height of global connectedness, new projects for political integration and economic development are being created and implemented at a fantastic pace. Two questions of interest that aid in our understanding of the globalization process are: what factors drove globalization in the past, and what factors will cause globalization to continue to be a pressing policy issue for the foreseeable future?

The simplest response to these questions is the first driver we will discuss, economic integration. Globalization has long been thought to have been driven by the will to acquire new markets and gain access to new resources. Current multinational corporations are only contemporary versions of global actors such as the Dutch East India Company. However, today's much more numerous and globally dispersed multinational corporations have the benefit of the actions taken by the Bretton Woods institutions (the World Bank and the International Monetary Fund [IMF]) and, perhaps especially, the General Agreements on Tariffs and Trade (GATT).

In 1944, the leaders of the allied powers met at Bretton Woods, New Hampshire, to discuss the makeup of the global economic system. They saw as a causal agent of World War II the inward economic turn of many of the most powerful states, that is, neo-mercantilist policies often referred to as "beggar-thy-neighbor" policies. Meeting toward the end of that war, the leaders believed they had a rare and valuable opportunity to recreate international economic procedures as they saw fit. Believing in the chance for an economic perpetual peace, the leaders sought to make states dependent on one another to the extent that international organizations could accomplish such a lofty goal. Economic globalization, though likely inevitable, was greatly facilitated by leaders who held economic integration as a political good. They strongly believed that economic connectivity demanded economic openness. Thus states would be required to leave behind antiquated notions of mercantilism and embrace the free exchange of products, material, and capital. A principal means of accomplishing this goal was the development in 1947 of the GATT, a treaty system creating formal rules stipulating increased openness in trade relations and a relaxation of formal trade barriers, especially tariffs. The GATT was adopted after the U.S. government rejected the proposed third Bretton Woods institution, the International Trade Organization. Under the GATT, members would allow third-party arbitration to resolve formal disputes among the signatory states. GATT proved to be a success. Along with the Bretton Woods institutions, it substantially pressured states to relax protections and lower trade barriers. In 1995, GATT was subsumed under the WTO. The WTO has a permanent secretariat, formal identity as an organization, and a quicker and more binding system of arbitration than had the GATT (Shaffer 2005). Although the WTO has been widely credited with contributing to substantial increases in gains from trade, it has been criticized for allowing subsidies, including those in fisheries, to exist in the developed world and distort the real cost of doing business. For example, Vincent, Marsden, and Sumaila (Chapter 7) argue that if the real cost of harvesting sea horses were to become apparent, the quantity taken would decrease substantially.[4] And Alder and Watson (Chapter 2) contend that the majority of governments of the developing world have, partly under pressure from the international financial institutions, opened their markets completely, surrendering fishing rights for a fraction of their market worth. Moreover, Seares, Smith, Anderson, and Pringle (Chapter 3) contend that much opposition to the WTO comes from those in the environmental movement who are suspicious of its potential for punishing states that seek greater environmental considerations – e.g., for allegedly favoring free trade and open markets over ecological concerns.

Consequently, the impacts of economically driven globalization on fisheries are manyfold, including pressure on the growth of the fish market from a local to a more global consumer base. In previous eras of globalization, goods from distant lands were seen as extravagant; today, middle classes around the world have included high-value fish such as salmon in their diets. Additionally, the increase in the number of supermarkets worldwide has created a demand for a stable fish source of constant quality for consumption by an audience that is unfamiliar with the inconsistencies of shopping in an open fish market (Knudson and Peterson, Chapter 18). These modifications in the way that people think about and shop for fish suggest that worldwide demand for fish will be met through greater use of distant fisheries and aquaculture. In this book, Naylor, Eagle, and Smith (Chapter 10) demonstrate the ways that developments in Europe and Asia in the evolution of aquaculture and increased standardization have adversely affected the market in wild Alaskan salmon fisheries. Moreover,

[4] Fishing subsidies, including the role of the WTO regarding them, have become a widely discussed and disputed topic (Schrank 2003).

Vincent, Marsden, and Sumaila (Chapter 7) demonstrate that, as cultures (in this case, Chinese) expand globally, their particular cultural needs go with them – for example, greatly expanding global demand for sea horses for medicinal use.

Globalization would be impossible without the development and spread of technology, the second major driver that we will discuss. It's clear that the previous rounds of globalization were greatly aided (or even caused) by advances in technology, whether it be the steam engine or the telegraph. Today, quick airplane travel, satellite-based communications, and the Internet have revolutionized the ways in which huge numbers of people conduct their lives. Technology has largely made possible the interconnectedness of economy, culture, and ideas. The most essential technological development has likely been the development of the Internet, which has allowed for the unprecedented pace of current communication; large data sets can now be sent easily across oceans or continents. Likewise, those seeking technical assistance can now acquire considerable knowledge virtually. Of course, the speed of traditional transportation continues to increase, but the speed of communications and computing technologies has increased exponentially.

Technology has been as crucial to fisheries resources as it has been to other areas of globalization. Technology has produced distant fleets, armed with fishing technology (e.g., drift nets), that have been recognized as having potentially negative consequences for fisheries resources. Conversely, Taylor, Leonard, Kratzer, Goddard, and Stewart (Chapter 1) suggest that technological development has greatly aided monitoring by fisheries managers. Additionally, the development of technologies such as turtle excluder devices provides a check on some of the most wasteful and destructive fishing practices. Technology, along with the globalization of demand, is the source of both problems for fisheries stocks and solutions to those problems.

A third major driver of globalization, resulting from the growing interconnectedness of peoples, is more perceptual or ideational, resulting in a greater awareness of truly global problems: Barbara Ward's "Spaceship Earth" phenomenon (Ward 1966). These problems include humanitarian tragedy, international health problems such as AIDS, overfishing, the threat of nuclear holocaust, and the problem of environmental destruction and conservation. One hundred years ago, industrial development was seen as a symbol of strength and economic vitality. Today, building a new factory is just as likely to elicit a negative reaction from environmentalists and locals who insist "not in my backyard." As scientists grew more aware of the problems that industrial development poses to the environment, it became clear that environmental damage crossed all borders. Still, creating global norms and, even more so, binding universal treaties for environmental protection has posed an incredible challenge. Countries that had strong environmental protections were bound to lose out economically to states that had fewer protections, at least in the near term. Even states with few restrictions had a powerful incentive to maintain or to lower their protections to stay competitive with neighbors doing the same thing, the classic "race to the bottom." Efforts to stave off the environmental problems of habitat destruction and loss of vertebrate biodiversity have followed a similar global course in the development of international treaties such as the Convention on International Trade in Endangered Species (CITES) (Vincent, Marsden, and Sumaila, Chapter 7). Though CITES has had more success than many other environmental treaties, including Kyoto, its provisions are a constant source of tension as states try to decrease protections on native species while shifting the burden of protection to others.

Such problems have been at the core of energizing an additional driver of globalization: global civil society. Among the most prominent organized global civil society actors are NGOs, which draw their membership from around the world and push hard at all levels of government for change, including ways to increase the benefits and decrease the negative consequences of globalization. Moreover, NGOs are proving to be one of the central organizing features of the new era of global governance – for example, they monitor and publicize the (more frequently negative) impacts of globalization. It would be premature to stipulate that globalization indicates a turn toward global government, but it does indicate a turn toward global governance. Different situations involve and, indeed, seem to require the participation of different actors and stakeholders. Folland and Schechter (Chapter 13) demonstrate the ways in which NGOs have long been influential in governance of the Great Lakes and demonstrate how NGOs can be especially successful at "agenda setting." One of the consequences of the increased governance role of non-state actors, including NGOs and multinational corporations, is a debate over the most efficient and democratic level of governance. In this context, Folland and Schechter discuss the concept of subsidiarity as it applies to fisheries – that is, the belief that decisions are best made at the lowest (closest to the people) possible level of governance.

WHAT ARE THE CONSEQUENCES OF GLOBALIZATION?

One of the most controversial issues relating to the impact of globalization is its tendency to produce unequal results between developed and economically less developed countries and within individual countries. In spite of this tendency, Dollar and Kraay (2004) argue forcefully that globalization is indeed good for the poor. They argue that economic inequality between states peaked in 1975, and also that states that have accepted globalization and opened their markets do much better than those that do not. Within countries, they argue that economic inequality results chiefly from social and economic policies within those countries and is not the fault of growing openness in trade and capital markets, which many, of course, equate with globalization. Others point to the continued persistence of poverty and large gaps in education and access to technology between the rich and the poor. Though it is clear that it would be beneficial to create social and economic arrangements that would be more beneficial to many of the poor, the question of whether globalization per se is responsible for the gaps between the rich and the poor remains a contested issue (Lambright and van de Walle 2000).

The crux of this debate clearly resonates in the fisheries literature in this volume. For example, one recurring concern is the tendency of globalized fisheries to divert an essential source of protein away from those in economically less developed countries, a variant of the much better publicized cash crop phenomenon, wherein priority is given to the production of crops to satisfy the external market rather than food crops for the local population (Mittelman and Pasha 1997:19). As fisheries in economically less developed countries focus more on high-dollar catches for export, they focus less on low-value catch eaten in villages (Alder and Watson, Chapter 2). There seem, however, to be a few solutions to this conundrum.

First, economically less developed countries must create a system that charges realistic prices for access to waters in their exclusive economic zones. Governments of economically less developed countries must take advantage of scarcity of their resources and demand comparative prices. Higher prices for access rights will tend to decrease fisheries operations, leaving more room (and a larger stock) for local fishing efforts. Second, economically less developed states must focus on becoming a larger part of the production and distribution of fisheries products rather than a supplier of raw biomass. This will certainly take a measure of political will, but it is possible that international organizations can come to their aid (e.g., with studies arguing about the harmful effects of direct and indirect fishery subsidies). Another possible alternative lies in creating consumer demand, in the developed world, for fairer trade practices toward those in economically less developed countries. Peterson and Fronc (Chapter 17) label this process "eco-labeling," although their efforts focus chiefly on creating a consumer preference for capture tuna and not aquaculture products. Related, Taylor *et al.* (Chapter 1) discuss how U.S. citizens agitated for and eventually won worldwide regulations for the capture and sale of dolphin-safe tuna. If civil society in the developed world would view the conditions of trade in economically less developed countries as a central concern and then convey that concern to their governments as well as to NGOs, positive social change is possible.

Another related contentious issue relates to the impact of increased global interconnectedness on the traditional power of the state, and related, the empowerment of other actors in the international system, including multinational corporations, NGOs, intergovernmental organizations, and individuals, a phenomenon most pronounced in relation to human rights but also evident in the communications, accounting, and insurance fields (Strange 1996).

In the age of globalization, the sovereign state system is breaking down in two ways. First, transnational organizations such as the European Union (EU) and the WTO sap state power with the approval of the state. To have effective organization at the international level, states must agree to lose some control of their affairs, a point made concrete in Hegarty's chapter on the regulation of fish and fishery products (Chapter 15). Second, and more controversially, growing global concern for human rights issues, for example, has created powerful international incentives for international intervention in the inner workings of states. This movement can be traced both to better international exposure of human rights violations by NGOs and intergovernmental organizations, and better communication and organization on behalf of human rights organizations. States no longer have the power to treat their citizens as they want. This powerful limitation on the power of the state has become of central international importance in places such as the former Yugoslavia and Rwanda. Some suggest that an aggressive application of the precautionary principle (e.g., in relation to overfishing) would have analogous consequences for traditional understandings of state sovereignty – i.e., leading to an erosion of state power.

The relative weakening of the state has also involved the empowerment of a relatively new set of actors on the international stage:

individuals. The rise of the global media and the greater demands for accountability have resulted in a turn toward transparency in government affairs. This allows for a level of citizen interest that would have been unknown when state affairs were more secretive. Additionally, more open global communication has allowed citizens to form opinions on a wide variety of subjects. Matthews writes, "Widely accessible and affordable technology has broken government's monopoly on the collection and management of large amounts of information and deprived governments of the deference they enjoyed because of it" (Matthews 1997). Individuals now have the power to conduct their own research and reach conclusions that are often at odds with the will of the state. This is true of individuals deciding what sort of fish to catch, or to buy (or boycott) or, for chefs, whether to limit their menus to "good fish" such as tilapia, as discussed by Molnar and Daniels (Chapter 11).

The impact of globalization on the changing role of the state in the international system and the increased empowerment of non-state actors are also related to the evolution of global governance. As the chapters by Bavington and Kay (Chapter 14), Folland and Schechter (Chapter 13), Bratspies (Chapter 19), and Hegarty (Chapter 15) show, although the state remains the primary actor in the international system, viewing the state as a unitary actor is misleading. Moreover, their studies evidence the increasing roles of intergovernmental organizations, including not only the United Nations (UN) but also the fishery commissions, in the creation of norms, regulations, and, in some cases, international laws. Oftentimes these are found to be reflective of the concerns and pressure exerted by the ever proliferating NGOs, whose access to the Internet has facilitated and decreased the cost of policy coordination and information dissemination.

One of the most widely noted consequences of globalization is what is often referred to as resistance to globalization or globalization backlash.[5] The forms it takes are multiple, from the highly publicized street demonstrations against the WTO in Seattle to the less well-known slow food movements in Italy and France (Mittelman 2000). Pollnac's case study in this book (Chapter 9) provides an example of resistance to capitalism and modernity. It takes the form of Southeast Asian fishers' economically inefficient distribution methods

---

[5] Robertson suggests that it makes sense to see these countervailing trends as an integral part of globalization, a perspective he recognizes as contestable (Robertson 1992:174–5).

underscoring that the community is "too bound by notions of fairness to succumb to Western notions of economic efficiency." However, many localities would have difficulty making such sacrifices, preferring instead merely to sell off fishing rights or, at best, to find employment in a regional processing plant. That is, they can't effectively resist globalization. However, Frank, Mueller, Krause, Taylor, and Leonard (Chapter 16) suggest that the closer local ties remain, the better localities can resist outside pressure to surrender old ways for efficient (and also exploitative) practices.

CONCLUSIONS

Globalization, we have argued, is made up of several interlocking dimensions. Reviewing a few will reveal the ways in which globalization is omnipresent in the world around us and how the insights of fishery scholars augment our knowledge of each.

The most obvious and most often mentioned dimension of globalization is economic globalization. The interconnectedness of markets and some of the causes and effects thereof have already been mentioned. The concern we wish to develop in this context is the tendency to externalize environmental damage while internalizing economic benefit, as in the environmental exploitation of poorer countries by richer ones. The decrease in transportation and communication costs and the lengthening of the supply chain have made it possible for environmentally damaging processes to be done in economically less developed states, with the high value-added product sold by economically developed countries (Sachs 2000). This is problematic for several reasons. While one group feasts on salmon, another is left with the environmental disaster that results from aquaculture often inadequately funded or poorly managed.

A second major dimension of globalization is the tendency toward the spread of ideas and beliefs, what some refer to as ideational globalization. A major driver of this process is the increased speed of communication technology already discussed. Ideas are spread through the global media, including print and 24-hour television news. More importantly, the Internet has provided a forum for like-minded people to come together and commiserate or debate. When a large faction of organized individuals pressures governments, they may be compelled to listen. A fine example of this is the push toward dolphin-safe tuna fishing (detailed by Taylor *et al.*, Chapter 1), in which grassroots activism led to government action. The interconnected nature of the fishing industry then led to the creation of an international limit on bycatch. Ideas and beliefs are no doubt important in their own regard, but when magnified with the power of instant global communication, they lay the groundwork for potentially powerful grassroots movements.

Perhaps the most contentious dimension of globalization is the push toward greater cultural exchange and, some would say, accelerating homogenization. The advances in mass communication have made it possible for matters of cultural preferences to be easily transmitted between people; increased economic interconnectedness makes this exchange profitable to investors, what some term the "commodification of culture." There are two widely divergent perspectives: one argues that cultural exchange is one of cultural liberation; the other argues that this exchange is an impetus for cultural decay, cultural homogenization, or even cultural imperialism. The closed nature of many of the world's cultures proves distasteful to some of its citizens, who feel confined in traditional ways. To them, the era of globalization has been a revolution in the most positive sense. The model of the spread of ideas for the cultural imperialist is the global corporation; the model for those rejoicing in choice is the Internet. The Internet gives individuals unprecedented access to others who share their intellectual and social proclivities (Crane 2002). Though the broader globalization literature suggests that there are many opportunities for hybridization and free cultural choice, many suggest that the purely economic exchanges that most often govern global fisheries leave little room for gaining what is good from other cultures. The difficulty is often framed in terms of Westernization. Several authors in this volume, for example, have suggested that traditional fishing practices were both more ecologically friendly and less stressful on cultural networks. Specifically, Pollnac (Chapter 9) demonstrates the "moral economy" approach in Southeast Asian fishing villages, in which an economically inefficient amount of each catch is distributed around the community.

We conclude this chapter by making explicit why we believe there has been such an upsurge in studies of globalization and global governance, including those related to fisheries.

In one of the most quoted sentences in contemporary international relations scholarship, Robert W. Cox wrote, "Theory is always *for* someone and *for* some purpose." He went on to note, "All theories have a perspective. Perspectives derive from a position in time and space, specifically social and political time and space" (Cox 1996:87).

Globalization theory is no exception. It aims at understanding and explaining the evolving world order, with a goal of ensuring that the direction it pursues is one that makes the future better than the past. And that students of globalization widely differ on the means for achieving their goal, most want to improve the lives of *all* – rich and poor, rural and urban, fishers, fish managers, consumers, and corporate executives of major fishery-related corporations. Students of global governance disagree on the details, but all are animated by a belief that a different governance structure is needed to achieve such a goal. Fittingly, this volume concludes with the most explicitly normative of chapters, those by Bratspies (Chapter 19) and by Dobson and Regier (Chapter 20). And fittingly as well, though their goals seem similar, the means for achieving them are not identical.

Acknowledgments

We appreciate the suggestions for revision of an earlier version made by Tracy Kolb, Nancy Leonard, William Taylor, and Lois Wolfson. We also appreciate the financial support provided by Michigan State University's Institute for Public Policy and Social Research (IPPSR).

References

Brandt, S. B., Mason, D. M., MacNell, D. B., Coates, T., and Gannon, J. E. 1987. Predation by alewives on larvae of yellow perch in Lake Ontario. *Transactions of the American Fishery Society* **116**: 641–645.

Bromley, S. 1996. Feature article. *New Political Economy* **1**: 120.

Cox, R. W. 1996. Social forces, states, and world orders: beyond international relations theory. In *Approaches to World Order*, eds. R. W. Cox and T. J. Sinclair. Cambridge, UK: Cambridge University Press, pp. 85–123.

Crane, D. 2002. Culture and globalization: theoretical models and emerging trends. In *Global Culture: Media, Arts, Policy, and Globalization*, eds. D. Crane, N. Kawashima, and K. Kawasaki. New York: Routledge, pp. 1–28.

Dollar, D. and Kraay, A. 2004. Growth is good for the poor. In *The Globalization Reader, 2nd edn*, eds. F. J. Lechner and J. Boli. Oxford, UK: Blackwell, pp. 178–179.

Enloe, C. 2000. Daughters and generals in the politics of the globalized sneaker. In *Rethinking Globalization(s): From Corporate Transnationalism to Local Interventions*, eds. P. S. Aulakh and M. G. Schechter. London: Macmillan, pp. 238–246.

Fetterolf, C. M., Jr. 1980. Why a Great Lakes Fishery Commission and Why a Sea Lamprey International Symposium. *Canadian Journal of Fisheries and Aquatic Science* **37**: 1588–1593.

Germain, R. D. 1997. *The International Organization of Credit: States and Global Finance in the World Economy*. Cambridge, UK: Cambridge University Press.

Gills, B. K. 2004. Editorial: The turning of the tide. *Globalizations* **1**: 1–2.

Held, D. and McGrew, A. 2000. The great debate: an introduction. In *The Global Transformations Reader: An Introduction to the Globalization Debate*, eds. D. Held and A. McGrew. Cambridge, UK: Polity Press, pp. 3-4.

Hobsbawm, E. J. 1975. *The Age of Capital 1848-1875*. London: Weidenfeld and Nicholson.

IJC-GLFC (International Joint Commission-Great Lakes Fishery Commission). 1990. *Exotic Species and the Shipping Industry: The Great Lakes-St. Lawrence Ecosystem at Risk*. Ottawa, Ontario: International Joint Commission; Ann Arbor, MI: Great Lakes Fishery Commission.

Lambright, G. and van de Walle, N. 2000. Globalization in the developing world: state capacity, social fragmentation and embeddedness. In *Rethinking Globalization(s): From Corporate Transnationalism to Local Interventions*, eds. P. S. Aulakh and M. G. Schechter. London: Macmillan, pp. 123-152.

Matthews, J. 1997. Power shift. *Foreign Affairs* **76**: 50-66.

Mills, E. L., Leach, J., Carlton, J. T., and Secor, C. L. 1993. Exotic species in the Great Lakes: a history of biotic crises and anthropogenic introductions, *Journal of Great Lakes Research* **1**: 1-54.

Mittelman, J. H. 2000. *The Globalization Syndrome: Transformation and Resistance*. Princeton, NJ: Princeton University Press.

Mittelman, J. H. and Pasha, M. K. 1997. *Out from Underdevelopment Revisited: Changing Global Structures and the Remaking of the Third World*. New York: St. Martin's Press.

Murphy, C. N. 1994. *International Organization and Industrial Change: Global Governance since 1850*. New York: Oxford University Press.

Pieterse, J. N. 2004. *Globalization and Culture: Global Mélange*. Lanham, MD: Rowman & Littlefield.

Rasmussen, J. L. n.d. *The Cal-Sag and Chicago Sanitary and Ship Canal: A Perspective on the Spread and Control of Selected Aquatic Nuisance Fish Species*. Rock Island, IL: U.S. Fish and Wildlife Service. Available online at wwwaux.cerc.cr.usgs.gov/MICRA/

Ricciardi, A. and Rasmussen, J. B. 1998. Predicting the identity and impact of future biological invaders: a priority for aquatic resource management. *Canadian Journal of Fisheries and Aquatic Sciences* **55**: 1759-1765.

Robertson, R. 1992. *Globalization: Social Theory and Global Culture*. London: Sage.

Sachs, W. 2000. Globalization and sustainability. In *The Globalization Reader, 2nd edn*, eds. F. J. Lechner and J. Boli. Oxford, UK: Blackwell, pp. 398-409.

Schrank, W. E. 2003. *Introducing Fisheries Subsidies*, FAO Fisheries Technical Paper 432. Rome: Food and Agriculture Organization of the United Nations.

Shaffer, G. 2005. Power, governance, and the WTO: a comparative institutional approach. In *Power in Global Governance*, eds. M. Barnett and R. Duvall. Cambridge, UK: Cambridge University Press, pp. 140-155.

Strange, S. 1996. *The Retreat of the State: The Diffusion of Power in the World Economy*. Cambridge, UK: Cambridge University Press.

Ward, B. 1966. *Spaceship Earth*. New York: Columbia University Press.

Part I Impacts of globalization on fisheries and aquatic habitats

WILLIAM W. TAYLOR, NANCY J. LEONARD, JUD F. KRATZER, CHRIS GODDARD, AND PATRICIA STEWART

1

Globalization: implications for fish, fisheries, and their management

People, nations, and ecosystems are becoming more integrated as the exchange of goods and services among countries and ecosystems is occurring at an ever-increasing rate. Transportation and communication systems over the past century have enhanced this integration, which is resulting in a highly interdependent world community. This phenomenon, referred to as globalization, has significantly affected the world's environmental and social systems (see Alder and Watson, Chapter 2; Rose and Molloy, Chapter 4; Ruddle, Chapter 8; Frank *et al.*, Chapter 16), and has captured the attention of the public and professionals from a wide spectrum of disciplines. Globalization is defined as "the widening, deepening and speeding up of worldwide interconnectedness in all aspects of contemporary social life, from the cultural to the criminal, the financial to the spiritual" (Held *et al.* 1999:2).

The attention given to recent issues related to globalization, e.g., cultural and economic influences, may lead many to think that globalization is an entirely new trend in human history. Globalization, however, is hardly new and can be traced back to the time when individuals and communities began interacting and exchanging goods with one another (Lentner 2000; Simmons and Oudraat 2001). What is new is the accelerated rate of these interactions and exchanges beginning during the latter part of the twentieth century. This acceleration is related to the significantly improved communication and transportation systems that have allowed for the rapid transfer of goods, services, and knowledge throughout the world. This rapid exchange has also been the impetus for improving methods for worldwide trade, which have resulted in increasingly opened borders among economic sectors, and thereby further facilitated exchanges among nations (Friedman 2000). In addition to the more accessible international market (Ihonvbere 1996), the reduction in impediments to global communication and

Globalization: Effects on Fisheries Resources, ed. William W. Taylor, Michael G. Schechter, and Lois G. Wolfson. Published by Cambridge University Press. © Cambridge University Press 2007.

trade is also facilitating the worldwide exchange of knowledge and technology. Emerging from this growing exchange among cultures and nations is the increased harmonization and homogenization of our technology, businesses, knowledge, cultures, and societal values.

The integration of societies is a result of the processes of globalization as well as a driver facilitating greater globalization. Harmonization and homogenization provide societies with commonalities that further enable and enhance global interactions among nations. Two of the products of this integration are the almost seamless communication and transportation networks that underpin the rate and magnitude of the acceleration of globalization. In particular, the World Wide Web, an important part of today's communication sector, is allowing us to exchange information more rapidly and effectively than ever before with more individuals and across larger geographic distances. These advances in communication have made it possible for a remote village to connect with the rest of the world through the use of satellite technology. Additionally, the transportation network, including railroads, shipping lanes, canals, and airports, is providing the world with more reliable and expeditious transportation and allowing greater numbers of people and cargo to traverse large geographic distances faster than ever before. The concurrent development of highly efficient and coordinated transportation networks and communication systems has played an important role in the exchange of goods and ideas between individuals, communities, and nations. This, in turn, has spurred governance systems related to commerce to become more integrated to increase the effectiveness of international trade. The result of these advances and integration has been the ability rapidly to deliver products, such as live or fresh fish products, to almost anywhere in the global marketplace.

Though it has many positive effects, globalization is also having negative impacts on individuals and their environment. As globalization facilitates exchanges and increases the speed of both interactions and transfer of goods among distant people, unwanted hitchhikers, such as human and animal diseases, are being transported rapidly across geographic and societal borders. Recent examples of these unintentional exchanges include the threat to both humans and livestock from avian flu in poultry, mad cow disease in cattle, and the severe acute respiratory syndrome (SARS) epidemic in humans (Centers for Disease Control and Prevention 2006; World Health Organization 2006). Natural systems are also affected by this increased exchange, which has facilitated the introduction and spread of non-native species that are drastically altering many ecosystems. The rapid exchange of live organisms between regional communities and the ease with which these exchanges can translate from the local to the global scale create a need to maintain a constant watch on international affairs to minimize emerging threats to the socioeconomic and natural ecosystems.

In this chapter, we introduce the impact that the intensification of globalization has on the aquatic ecosystems, fisheries, and their management. We provide a brief overview of how globalization, through fishing pressure, alterations of aquatic ecosystems, and the rapid expansion of the aquaculture industry, is affecting fish stocks. Additionally, we examine how the influence of globalization on natural resources is requiring a change in the approach used to manage fisheries to assure that the challenges arising from this growing influence are successfully met. We conclude with a look at the future of fisheries management in this increasingly globalized world.

GLOBALIZATION OF FISHERIES

Natural resources such as fisheries stocks are affected by the global integration of nations' economies, industries, policies, and cultures. Globalization's impacts on these resources have been associated with technological improvements. Some of these new technologies related to fishing allow for greater accessibility to fish populations that are located farther from fishing ports and also to those fish stocks that reside at great depth (Thiele 2001). While globalization has increased the fishing industry's ability to harvest, store, process, and market fish products, this phenomenon has also been associated with the degradation of aquatic ecosystems. This degradation is related to human disturbance of the natural ecosystem, which often results in increased sediment, nutrient, and contaminant loads being introduced into waterways, such as those from agricultural, residential, and industrial discharges. Additionally, the introduction of exotic species through the construction of canals and ballast water discharge has also led to the destruction of critical fish habitat (Mills *et al.* 1993; Cohen 1999; McClanahan 2002). These impacts, in turn, affect the biological diversity, productivity, abundance, and health of the fishery ecosystem.

Importantly, globalization also provides us with the knowledge, socioeconomic incentives, and technological abilities to restore aquatic ecosystems. For instance, globalization has facilitated the growth of an effective and efficient aquaculture industry which, while providing the protein needs of an increasing human population, enables fisheries managers to reduce the wild harvest catch, thereby allowing for the rehabilitation of depleted stocks (e.g., Pacific salmon; see Taylor and Leonard, Chapter 12). In the following sections, we will elaborate on the impact of globalization on fisheries ecosystems in four main areas: fishing pressure, aquatic ecosystem degradation, fish habitat restoration, and aquaculture. These areas, the majority of factors that influence fisheries, are developed further in other chapters of this book.

Fishing pressure

Recreational fisheries have benefited from the advancements in communication and transportation technologies associated with globalization. These benefits are primarily related to the increased ease of accessing previously unknown or inaccessible fishing sites by a greater number of anglers. For instance, the World Wide Web is an accessible source of information that anglers can use to find information on location, access points and fishing conditions on local or distant fishing sites (e.g., Recreational Boating and Fishing Foundation, Maumee Bait and Tackle Shop[1]). The ease of obtaining fishing information may prove especially valuable in assisting anglers in accessing fishery resources distant from their homes or to novice anglers learning about local fishing spots. The use of this communication network was documented by a target market focus group conducted on "avid American anglers," which indicated that there is a tendency for a portion of sport fishers, especially tournament anglers, to utilize the Web when searching for information on sport fishing (Responsive Management National Office 1999:36–37). Although the exact level of Web usage by American sport anglers was undetermined in the 1999 report from the Responsive Management National Office, it was observed to be low in a 1999 survey of all fishing license holders in New Hampshire (Salz and Loomis 2001:11). We expect Web usage to increase as more people gain access to and familiarity with the Web and discover the wealth of fishing-related information available on these sites. A side effect of this availability of information is the apparent continuing increase, in the number of individuals wanting access

[1] Recreational Boating and Fishing Foundation, 601 N. Fairfax St., Suite 140, Alexandria, VA 22314, www.rbff.org/; Maumee Bait and Tackle Shop, 104 E. Wayne St., Maumee, OH, 43537, www.maumeetackle.net/

to the same resource. In some situations, this increasing number of both local and non-local residents interested in using a specific aquatic resource has provided a context for conflict between various stakeholder groups. This increasing and frequently conflicting demand on the aquatic resource has forced managers to contend with the needs of the local fishery and its habitat while balancing the demands of these stakeholder groups in determining how these resources should be managed and utilized, e.g., fishing, boating, swimming (Jones 1996; Miranda 2001).

In addition to the progress made in the communication networks, the dramatic improvements in the world's transportation networks (roads, aircraft, and ships) is facilitating the expansion of fishing opportunities by providing anglers with relatively easy and inexpensive access to once isolated fishing locations worldwide. This improved transportation system, combined with the general increase in available leisure time, increased disposable income, and improved fishing facilities and access capabilities has resulted in greater participation and improved economic health of the recreational and charter fishing industries in the United States (Dawson et al. 1989; Lichtkoppler and Hushak 1993). The strength of this industry is evidenced by the magnitude of the U.S. recreational marine and freshwater fishing industry. During 2001, the U.S. recreational fishing industry was estimated to have generated more than $41 billion from anglers' direct expenditures on traveling to fishing locations, lodging, food, fishing equipment, licenses, boats, and the hiring of guides (Southwick Associates 2002). An important benefit of this increase in recreational users is the increasing number of individuals who are invested in and concerned with the health of aquatic ecosystems and thus participating in sustainable stewardship practices.

The commercial fishing industry has also benefited from globalization and associated technological improvements. The invention of the factory trawling operation is an example of advancement in fishing gear and boats that relied on the innovations of globalization for its development and profitability (Warner 1983; Harris 1998; Kennelly and Broadhurst 2002; Cole 2003). The fishing technological innovations facilitated by globalization are allowing for the location and capture of larger numbers of fish per unit of effort, processing fish with onboard filleting machines, and preserving fish using freezer technology until the trawler returns to its port. This results in larger amounts of fish harvested and the ability to fish in areas distant from the fishing vessel's home port, thus opening new fishing grounds when the near-port fisheries become less available and unprofitable. In concert with these on-board fishing innovations, the advancement in communication technology provides owners of fishing industries the ability to identify consumers residing across a larger geographical expanse and to transport fish products to these widespread customers. This broader consumer base also serves to provide the needed market demand for purchasing the substantially larger fish catch harvested by the improved trawlers. The increase in fishers' abilities to harvest and market fish products locally and globally has led to an increase in fishing pressure on commercially targeted fish stocks as fishers seek to maximize their harvest per unit effort while minimizing costs. In 1998, commercial operators worldwide provided nearly 40 percent of total fish harvest to the international marketplace (Le Sann 1998). However, consumer demand and fishers' abilities to meet this increasing demand have led to unsustainable pressure on many fish stocks. An estimated 10 percent of the main commercial stocks or species are being significantly depleted, 18 percent are overexploited, and 47 percent of the stocks are fully exploited (FAO Fisheries Department 2002).

Aquatic ecosystem degradation

Globalization is contributing to the worldwide degradation of aquatic ecosystems. This degradation is taking place in numerous ways, both directly and indirectly. Two of the main impacts on fishery ecosystems are the increasing demand for fresh water by local residents and distant consumers, which has the potential to decrease both the quantity and quality of the fresh water available to the natural system (Barlow 2001; Gleick et al. 2002), and the changes in ecosystem composition that occur with the introduction and establishment of aquatic exotic species (Mills et al. 1993).

The increasing demands for the world's limited fresh water by all sectors of society are significantly affecting fisheries productivity. Though the exact percentage of use varies by region, agriculture usually ranks high among users of fresh water (about 70 percent), followed by industrial (about 20 percent) and household users (about 10 percent), with the amount and quality of the water returning to the natural system after being used varying, depending on the activity (Serageldin 1999; Cosgrove and Rijsberman 2000). Noticeably absent from the user categories listed above is water needed for ecosystem services. These needs are of paramount importance to ecosystem integrity and fisheries productivity, yet generally little consideration is given to the needs of fish, other aquatic organism, and their aquatic habitats (Gleick et al. 2002). Additionally, pressures placed on fresh water resources by society have led to alterations of landscapes and waterways that severely affect aquatic ecosystems. The development of watersheds to support human activities (i.e., residential development, agricultural and industrial activities) has resulted in increased sediment loading in waterways and anthropogenic pollutants that reduce both the quantity and quality of the fresh water resource. Thus societal consumption of fresh water has led to the degradation of aquatic ecosystems and their embedded fish habitats. These degradations thereby decrease the productivity and health of fish populations (Lake Michigan Federation 2002; Tanner and Tody 2002), as evidenced by the need for restrictive harvest regulations and the establishment of fish consumption advisories (see Environmental Protection Agency 2006).

An extreme example of how the water demands of society can severely affect the biota and health of aquatic ecosystems when its needs are ignored is the Aral Sea. The Aral Sea, bordered by Kazakhstan and Uzbekistan, has been severely affected by water withdrawal and water quality degradation through human activities. Diversion of 94 percent of the water from the Amu Darya and Syr Darya, tributaries to the Aral Sea, led to a substantial reduction in volume (75 percent) of the Aral Sea between 1960 and 1995. This resulted in the extinction of 20 of its 24 endemic fish populations, and the loss of 44 000 tonnes of fish harvest annually and of 60 000 jobs locally (Abramovitz 1996). The impacts of decreased water quantity and quality on the Aral Sea's aquatic ecosystem strongly illustrate the link between a healthy and sustainable fishery and water security, a story repeated worldwide (e.g., CALFED Bay-Delta Program in California).

Introduction and successful establishments of aquatic exotic species are a significant threat to the health of aquatic ecosystems. The rate of introductions, as detected in the Great Lakes (Mills *et al.* 1993), appears to have increased with the rapid expansion and recent technological advances in transportation systems that have facilitated trade and consequently the movement of live organisms. The transportation of these organisms can occur either intentionally, such as live cargo for the pet trade or the aquaculture industry, or unintentionally, such as species found within a ship's ballast water. Intentional official introductions of exotic species as a food source, for increased recreational fishing opportunity, or for biological control of exotic species, are taking place throughout the world.[2] Species such as the common carp (*Cyprinus carpio*), brown trout (*Salmo trutta*), chinook salmon (*Oncorhynchus tshawytscha*), and coho salmon (*O. kisutch*) were intentionally introduced to inland U.S. waters to provide a new food source, or recreational fishing opportunity, or to control exotic species. Brown trout and common carp were transported in 1881 and 1877, respectively, from Europe to North America to improve fisheries (Nielson 1999). Chinook and coho salmon were brought from Oregon to the Great Lakes in 1964–65 to create a recreational fishery (Nielson 1999) that would also prey on the problematic exotic alewives (*Alosa pseudoharengus*) (Tanner and Tody 2002).

Though a number of these official intentional introductions have had significant societal benefits, many also have detrimental impacts on the native ecosystem and their species (Mills et al. 1993; Manchester and Bullock 2000). The common carp, introduced as a food source in the Great Lakes, alters native aquatic habitats by uprooting vegetation, increasing water turbidity, degrading fish habitat, and reducing aquatic productivity (Mills et al. 1993). The introduction of a non-native salmonid in the Colorado River has resulted in hybridization of the native Colorado River cutthroat trout (*Oncorhynchus clarki pleuriticus*), decreasing the abundance of this native species (Novinger and Rahel 2003). Introduction of the chinook salmon in Lake Ontario for increased recreational fishing (Nielson 1999; Tanner and Tody 2002) seems to be having a negative impact on the reproductive success of the native Atlantic salmon (*Salmo salar*) by allegedly delaying the onset of spawning, and increasing activity and mortality levels of the native salmonids (Scott et al. 2003). Many fisheries management activities now focus on reducing or eliminating these intentional introductions to rehabilitate native species. An example is the elimination of rainbow trout (*Oncorhynchus mykiss*) and other non-native salmonids from rivers in the Great Smoky Mountains National Park, USA, to restore native brook trout (*Salvelinus fontinalis*) populations (Kulp and Moore 2000).

Vectors of unintentional introduction include releases from the live fish food market, the aquaculture industry (e.g., Asian carp,

---

[2] Official introductions refer to introductions of exotic species that were coordinated and implemented by a governmental organization. Thus introductions that were intentionally implemented by an individual citizen without the sanction of the responsible government agency, such as release of a pet fish, are not included under the "official" introduction category but instead are included under the "unintentional" category.

Cyprinidae family), the exotic aquarium trade (e.g., goldfish, *Carassius auratus*, and snakehead, *Channa* spp.), live bait fish release and ballast water from shipping vessels (e.g., zebra and quagga mussels, *Dreissena* spp.; round goby, *Neogobius melanostomus*; see Mills et al. 1993; Koel et al. 2000; Florida Fish and Wildlife Conservation Commission 2001; USGS 2003a, b, c). In addition to these vectors, the extensive infrastructure of canals and locks constructed to support the shipping industry has contributed to the spread of exotics by joining separated bodies of waters. Well-known examples in the Great Lakes are the Erie Barge Canal and Welland Canal, which facilitated the expansion of alewife and sea lamprey populations into the Great Lakes (Scott and Crossman 1998). Overall, unintentional introduction of exotic species has generally affected native species through competition for food and habitat, predation, and parasitism (Marsden and Jude 1995; USGS 2002).

Rehabilitation of fish habitat

Globalization, conversely, also provides the information, finances, and tools necessary to restore degraded aquatic ecosystems. The communication networks arising with globalization have provided scientists and managers with the ability to share technological advances and knowledge for the protection and rehabilitation of fish and their habitats. An example of how international collaboration and sharing of scientific knowledge among all levels of governance is improving a degraded aquatic ecosystem is the North American Laurentian Great Lakes. Great Lakes water quality and shoreline habitat were severely degraded by human activities during the early twentieth century (Beeton et al. 1999). The governments of the United States and Canada, in response to this water security concern, agreed to establish the International Joint Commission (IJC) through the signing of the 1909 Boundary Waters Treaty. This treaty provides the principles and mechanisms to resolve and prevent disputes about water quantity and quality along the boundary between Canada and the United States, including the Great Lakes (IJC 1989). Furthermore, in 1978, both nations signed the Great Lakes Water Quality Agreement, which protects water quality in the upper Great Lakes (Superior, Huron, and Michigan) and restores and improves water quality in the lower Great Lakes (Erie and Ontario), thus enhancing the two countries' resolution to protect the integrity of the Great Lakes water resource. A key factor for the success of these agreements was the empowerment of individuals and institutions at the local, regional, national, and international levels which resulted in the necessary changes for attaining these goals. Moreover, these improvements in water quality were facilitated with the aid of scientific and technological improvements from around the globe that allowed for such innovations as the development and installation of water treatment plants to eliminate or reduce raw sewage from being directly released into the Great Lakes. The culmination of all these factors, international agreements, public empowerment, and science and technological advancements was the successful outcome of these binational agreements in making the Great Lakes ecosystem healthier than it was during the early twentieth century (Beeton et al. 1999).

It is important to recognize that the ability to restore degraded aquatic ecosystems depends on economic resources and the political will of people at all levels of governance. In the United States and Canada, the push for restoring the Great Lakes ecosystem occurred during the mid-twentieth century. This was a time period characterized by a robust and growing economy and aquatic ecosystem degradation that was becoming increasingly visible. For example, when the Cuyahoga River in Ohio caught on fire, the event was widely communicated through the print and broadcast media, thereby galvanizing public environmental sentiment (Hummer 2001; BBC-h2g2 2004). This economic wealth overlapping with increasing political and public support as well as the rise of environmental non-governmental organizations (NGOs) that represented the concern of various public segments to improve the health of the environment allowed for the passing of many foundational environmental acts at various levels of governance. Examples of these are the 1948 Federal Water Pollution Control Act, revised in 1972 as the Clean Water Act, and the 1973 U.S. Endangered Species Act, which created policy that focused on the rehabilitation of degraded ecosystems. On an international level, societal sentiment to protect and restore ecosystems and their fauna has been growing, as exhibited through the signing of international environmental agreements such as the 1971 Convention on Wetlands of International Importance Especially as Waterfowl Habitat (Ramsar Convention), the 1992 Convention on Biological Diversity, and the 1973 Convention to Regulate International Trade in Endangered Species of Flora and Fauna. Thus, globalization has the potential to provide the economic, political, and social incentives at local and global levels of governance to restore and enhance fish stocks and their habitats throughout the world.

Aquaculture

The knowledge and techniques used in producing fish in aquaculture systems, as well as the ability to transfer fish stocks among facilities, have been greatly enhanced by the influence of globalization. Globalization has permitted the sharing of new technologies that have increased the ability to grow, process, market, and sell an increasing variety of fish products throughout the world. These new technologies include advances in transportation systems, hatchery design, water treatment, diet formulation, brood stock composition, and improved diagnosis and treatment of fish health problems.

These technological improvements, combined with the ability to process and transport fish worldwide, are providing fisheries managers with an important source of fish to use in management. Aquaculture-reared fish can be used for creating and/or improving recreational and commercial fisheries, rehabilitating wild stocks, and assisting in the preservation of threatened and endangered species (Heidinger 1999; Rahel *et al.* 1999). For instance, in the North American Laurentian Great Lakes, aquaculture is providing the means for mitigating the loss of fisheries related to ecosystem degradations. This includes the ability to introduce new species, such as salmon (*Oncorhynchus* spp.), which created a valuable recreational fishery; to assist in the rehabilitation of lake trout (*Salvelinus namaycush*) by supplementing natural populations with aquaculture-raised individuals; and to preserve lake sturgeon (*Acipenser fulvescens*) stocks. Similar use of aquaculture-reared fish can be observed worldwide (Hansen 1999; Heidinger 1999; Rahel *et al.* 1999).

The development of the aquaculture industry, facilitated by these technological advancements, is also providing society with an alternative source of fish protein for human consumption (see Knudson and Peterson, Chapter 18). The production capacity of the aquaculture industry has exponentially increased during the past decade (Anonymous 2003a, b) (Fig. 1.1) and has become an important, reliable and affordable year-round source of fish products for both local and global markets. The "Blue Revolution" (Anonymous 2003a, b) resulted in a substantial growth of the aquaculture industry, with the industry augmenting its contribution to the worldwide fish market from 5 percent in 1985 to 33 percent within a 15-year period (Robbins 2001). Although the aquaculture industry has provided an increasing proportion of the protein source to the world's citizens and has helped reduce the harvest of selected wild stocks, e.g., Pacific salmon (see Taylor and

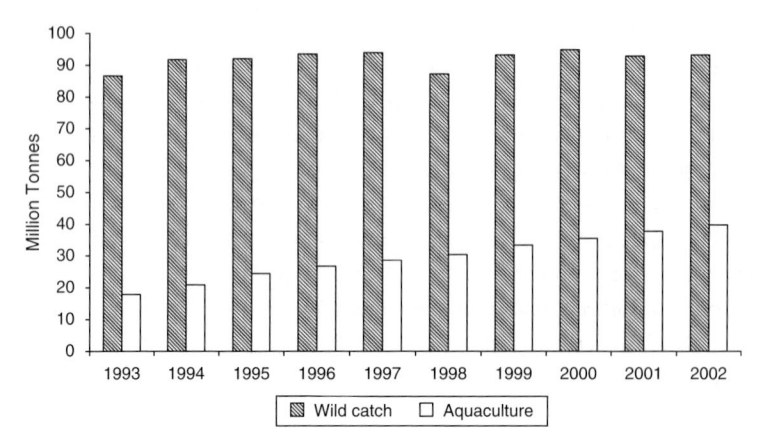

Figure 1.1 World aquaculture and commercial wild catches, 1993–2002. (*Source:* NMFS 2001, 2002, 2004.)

Leonard, Chapter 12), it also has been associated with negative impacts on aquatic ecosystems and related fisheries. These include overharvest of fish species used for fish feed in aquaculture operations, habitat destruction, hybridization between wild fish and genetically modified fish, disease and parasite outbreaks, and water quality degradation (Robbins 2001; Alaska Department of Fish and Game 2002; see Rose and Molloy, Chapter 4; Molnar and Daniels, Chapter 11; Knudson and Peterson, Chapter 18; Bratspies, Chapter 19).

FISHERIES MANAGERS AND GLOBALIZATION

Fishers harvesting oceanic fish stocks located outside of their country's exclusive economic zone (EEZ) have long been aware of the impact of globalization on fisheries. This awareness can be observed in the response actions they selected to implement with the aim of protecting the fishery from global influence. The cod fishery in the northern Atlantic Ocean and the associated response of fishers and governments to the global influence acting on this fishery are one illustration of this long-held knowledge. The northern Atlantic cod fishery experienced a rapid destruction following the advent of more advanced fishing vessels, improved harvesting equipment, increased number of nations and fishing vessels targeting the same fish stock, and mismanagement (Warner 1983; Parsons and Beckett 1997; Roy 1997; Harris 1998; Wright 2001; Kennelly and Broadhurst 2002). Recognizing the threat of this growing global fleet and its increasing efficiency at harvesting large numbers of cod per unit of effort, Canada extended its 2-mile exclusive economic zone to 200 miles in 1977 (Roy 1997). This response was aimed at excluding foreign vessels from harvesting cod stocks located near the Canadian coast, thereby limiting the influence of foreign fishing vessels on the fishery. However, the response was too late, thus ineffective because the combined pressures of the fishery prior to 1977 had already substantially reduced the cod population. The continued and intensified fishing by both Canadian fishers within the 200-mile EEZ and foreign fishing nations outside the EEZ led to the collapse of the North Atlantic cod fishery in 1992 (Roy 1997).

Unlike oceanic fisheries, most inland fisheries management, whether managing stocks that are shared or wholly owned by one nation, has had limited consideration for the influence of globalization on fish stocks. This limited consideration may partially be due to the less conspicuous aspects of globalization acting on inland fisheries. Examples of these less conspicuous aspects include the exchange of fish products and fisheries technology between nations and increased access to fisheries resources located in deeper and more distant waters. The governance structure for inland fisheries management also contributed to the minimal attention given to global factors acting on the local environment and fisheries. The responsibilities of most inland fisheries managers were designed to focus on local fish stocks and local activities affecting these stocks and their habitat. In the absence of acknowledgment of the global influence on fish stocks, the management strategies of local fisheries managers have consisted of localized action such as banning gear, establishing catch regulations on total number and the size of fish harvested, stocking fish and, in some cases, establishing no-fishing areas (Noble and Jones 1993). Therefore, management decisions infrequently considered the effect of external factors on local fisheries such as evaluating the demand from foreign nations for local fisheries, or how local management decisions may affect fisheries located outside of their jurisdictions. This approach is changing rapidly, partially because ecosystems and the goods and services they produce are being more visibly affected by decisions from outside the local communities' boundaries, and because researchers are undertaking more globally oriented studies (see Vincent *et al.*, Chapter 7; Pollnac, Chapter 9; Folland and Schechter, Chapter 13) as fisheries professionals' recognition of the external influences acting on local fish stocks is heightened.

Recognition that impacts on inland and oceanic fisheries ecosystems arise from a variety of sources that originate from local and external communities is compelling fisheries managers to incorporate these influences into their management plans. This necessary expansion of focus, especially for inland fisheries, results in a need to alter fisheries management philosophy and practice to address officially the global characteristics of these impacts. For instance, the establishment of multijurisdictional fisheries management organizations, such as the Great Lakes Fishery Commission[3] and MICRA,[4] provides a forum for managers to address external factors acting on their shared resources to ensure their sustainability. If managers fail to incorporate within their fisheries management plans all local and global factors affecting a fisheries resource, their plans will likely result in an unsustainable fishery and ecosystem degradation (see Alder and Watson, Chapter 2). This will culminate in the classic scenario of Garrett Hardin's "Tragedy of the Commons" (1968), whereby many stakeholders are exploiting the same ecosystem and maximizing their individual gain without assessing the combined impact on the ecosystem.

Implications of global stakeholders on fisheries management

As mentioned previously, the number and types of stakeholders with interest in a given aquatic ecosystem are increasing as globalization facilitates access to these resources. This change is affecting fisheries management most notably with respect to the role of the public and the incorporation of multijurisdictional governance tools and forums. Globalization is providing the public with the tools to become more efficiently organized through the development of NGOs that can effectively engage and influence fisheries management agencies. Additionally, by facilitating cooperation among fisheries management organizations and their stakeholders, globalization is increasing the likelihood that all interested parties are included in the management discussions of shared stocks through multijurisdictional policies and related organizations.

The public is also increasing its ability to influence fisheries management as changes associated with globalization facilitate the public's abilities to play a significant role in determining the policy actions of management agencies. Improvements in communication,

---

[3] Great Lakes Fishery Commission, 2100 Commonwealth Boulevard, Suite 100, Ann Arbor, MI 48105, www.glfc.org

[4] Mississippi Interstate Cooperative Resource Association (MICRA), P.O. Box 774, Betterndorf, IA 52722, www.aux.cerc.cr.usgs.gov/MICRA/

especially the development of the Web, have made it easier for people to communicate with groups that share their interest. This has augmented the likelihood of providing a unified voice for their interests and concerns to be transmitted to the agency and the public at large. For instance, a recreational angler concerned with protecting salmonids and their habitats can now more easily locate and join organized stakeholder groups that represent the angler's interests, such as Trout Unlimited.[5] By having a unified voice, fishers can ensure that their personal interests are considered in the development and implementation of management plans that are complimentary with their interests. The Whale and Dolphin Conservation Society[6] is an example of a society that successfully advocated for the decrease of dolphin bycatch in the tuna fishery. The public pressure created by the society resulted in changes in tuna harvest methods, the creation of dolphin-safe tuna labels (Hunter *et al.* 2002), and empowerment of consumers in influencing harvest techniques used by national and international fishing fleets (see Peterson and Fronc, Chapter 17). The effectiveness of the Whale and Dolphin Conservation Society in achieving its goal of reducing dolphin bycatch mortality by generating the needed financial and political support was greatly facilitated by the tools of globalization.

In addition to the increasingly organized role and engagement of the public in resource management, fisheries managers are facing new challenges as improvements in the transportation network are allowing more and distant fishers access to local stocks. This expansion in the number and diversity of fishery stakeholders increases the complexity of fisheries managers' responsibilities. In the past, managers worked within small, easily defined jurisdictions and with few stakeholder types, such as local residents interested in angling and localized water use. Today, fisheries and aquatic managers generally have to contend with demands from diverse stakeholder groups that lie across geopolitical boundaries. Thus, the expectations and demands of these stakeholders now influence fisheries management from the local level to the regional, national, and international communities (Warner 1983; Pollock *et al.* 1994), and require that fisheries managers contend with this growing global pressure on their fish communities and habitats.

[5] Trout Unlimited, 1300 N. 17th St. Suite 500, Arlington, VA 22209, www.tu.org/index.asp#
[6] The Whale and Dolphin Conservation Society, Brookfield House, 38 St. Paul Street, Chippenham, Wiltshire, UK, www.wdcs.org

Approaches for handling these cross-jurisdictional pressures include a large repertoire of agreements, conventions, and treaties that influence the local and global management of fish populations with respect to the timing, locations, methods, and quantities of a species that can be harvested. For instance, the signatory countries to the 1995 United Nations Straddling and Highly Migratory Fish Stocks Convention (United Nations 1982) agreed to harvest selected fish stocks in a sustainable manner promoting long-term conservation of these species. Commissions are another means by which to construct policy for sustainable fisheries. These work toward coalescing diverse stakeholders' values into a common management plan that provides optimal benefits for all while protecting the fish stocks (Cole 2003). These commissions are engaging in cross-jurisdictional activities at the local, regional, national, and global levels of governance. They are composed of representatives from multiple jurisdictions and, generally, they aid in facilitating internationally based cooperative efforts to protect, rehabilitate, or enhance shared fisheries. Illustrations of fisheries management organizations attempting to manage oceanic fisheries influenced by globalization include international fishery commissions such as the Commission for the Conservation of Southern Bluefin Tuna, the Inter-American Tropical Tuna Commission, the Great Lakes Fishery Commission, and the International Whaling Commission.[7]

Agreements, treaties, conventions, and organizations, however, are sometimes limited in their abilities to assure proper management of shared stocks. For instance, Radonski (1991) summarized the apparent failure of the contracting parties of the International Commission of the Conservation of Atlantic Tuna (ICCAT) to meet their responsibility to maintain tuna populations at a level that would allow the maximum sustainable catch for human use. Despite data that indicated that harvest reductions were necessary, these were never implemented. By 1998, the ICCAT parties acknowledged the problem of overexploitation and developed a concerted 20-year rebuilding program for the western Atlantic bluefin tunas (SCRS 2005). Another example of a multinational cooperative effort to manage a common fishery and its industry has been the evolution of the 1970 Common Measures in Europe into the 1983 Common Fishery Policy (Symes 1997; EC-Fisheries 1998; Song 1998). The Common Measures created a common fishery market, assured equal access to fishing grounds, and provided structural

---

[7] International Whaling Commission, The Red House, 135 Station Road, Impington, Cambridge, Cambridgeshire, UK, www.iwcoffice.org

improvements for fishing fleets and onshore installations for all members of the European Community. The Common Fishery Policy (CFP) contains similar aspects to the Common Measures but is expanded to include conservation of fish stocks through regulations on harvest and gear types, allowance of the European Community to represent its members in international fishing agreements and organizations, and the establishment of a means to assure that CFP rules are respected and enforced by member states. The success of the CFP in attaining its conservation goals while balancing the socioeconomic components of the fishery is controversial. Its greatest success has been its ability to negotiate third-party fishing agreements for its member states while making modest progress in balancing harvest and fish availability (Laurec and Armstrong 1997; Symes 1997; Song 1998; Hatcher and Robinson 1999). A multijurisdictional strategic plan that is successful in unifying two nations and their states, provinces, and tribal authorities is the Great Lakes Joint Strategic Plan for Fisheries Management (GLFC) (Dochoda 1999). This plan has helped increase the coordination of fisheries management agencies throughout the Great Lakes Basin, resulting in the enhancement of fisheries productivity and societal benefits.

Use of communication networks and technologies in fisheries

With globalization, fisheries managers and conservation officers are finding themselves responsible for monitoring fish stocks and enforcing fisheries regulations over an expanding territory. Fisheries professionals, therefore, need the skills and information to participate effectively and efficiently in cross-jurisdictional management activities. One of the requirements for successful cooperative management is the ability to communicate ideas freely and broadly and to have access to common information sources. The technological and communication advancements over the past century have provided the means for facilitating this level of information exchange, enhancing coordination and enabling fisheries agencies to become more effective at monitoring, managing shared fish stocks, and enforcing fisheries regulations.

The development of important tools for sampling, analyzing, monitoring, reporting, and enforcing fisheries regulations is facilitating interjurisdictional management. For instance, scientists are now better equipped to study fish movement with the use of telemetry equipment using radio or sonic tags (USGS 2000). Fish can be tracked from land, boat, airplane, or satellite. Computer technology has allowed for the development of global positioning systems (GPS) and geographic information systems (GIS), which aid fisheries professionals in navigating to specific sites, recording exact locations, studying wild fish populations, rapidly analyzing data, and detecting status and trends of fish abundance (USGS 2000). This information can be electronically transferred to an easily accessible database, where it is rapidly processed for real-time use in management and enforcement (e.g., Peru's anchovetta commercial fleet).

Enforcement efforts have been greatly enhanced because of innovations that allow for detection and monitoring of fishing vessels. These technologies include use of aircraft, satellite imagery, and telecommunication that survey and monitor where a fishing boat is located and when it is deploying its fishing gear (Pollock et al. 1994; Keus 1997). The liberation of trade in goods and ideas also has set in motion many cooperative ventures between local fisheries professionals and their enforcement colleagues worldwide. This liberation of communication pathways is enabling law enforcement officers to identify needs for combined enforcement operations and overcome political barriers to cooperate on cross-jurisdictional enforcement activities, e.g., through informal agreements or more formal memorandums of understanding. Enforcement of fisheries regulations along the Canada–U.S. border in Lake Erie provides an example of such a combined enforcement effort, in which officers from the U.S. Coast Guard, the U.S. Fish and Wildlife Service, the Ohio Department of Natural Resources, the Michigan Department of Natural Resources, the Pennsylvania Fish and Boat Commission, the New York State Department of Environmental Conservation, and the Ontario Ministry of Natural Resources joined together to monitor for illegal commercial fishing incursions along this international border (Kirshman and Leonard 2003).

CULTURES AND ETHICS

Increasingly, globalization is making us aware of the diversity of values held by various stakeholders in relation to management approaches or valuation of fisheries (see Bratspies, Chapter 19; Dobson and Regier, Chapter 20). Fisheries management in this globalized era is influenced by the cultural and ethical differences of nations that have converged as fishers and consumers from around the globe and interact in relation to the demand and supply of fish products. In some instances, this convergence of fishers and consumers results in disagreements and conflicts over issues such as which species to fish, which management and harvest techniques to apply, and what levels of bycatch mortalities are acceptable. Many of the conflicts arising from these differences are being addressed either through lengthy negotiations or more formally through the passing of law. Several examples of fishery-related conflicts among nations exist, including those related to the dolphin bycatch of the tuna fishery (Constance and Bonanno 1999), sea turtle (Sands 2000) and sea horse (see Vincent *et al.*, Chapter 7) bycatch in shrimp harvest, and acceptable fishing gear.

The bycatch mortality of dolphins associated with the tuna harvest in the eastern tropical Pacific was deemed ethically unacceptable by portions of the U.S. public, thus creating controversy among the countries engaged in tuna harvest. Initially, this public pressure prompted the U.S. government to pass the Marine Mammal Protection Act of 1972 and the U.S. Dolphin Protection Consumer Information Act in 1990. The latter identified and labeled dolphin-safe tuna products and prohibited the import of non-dolphin-safe tuna into the U.S. market (Constance and Bonanno 1999). This course of action was negatively viewed by nations such as Venezuela and Mexico, which argued that this restriction was a violation of the General Agreement on Tariffs and Trade (GATT) because the restriction unfairly discriminated against tuna harvested by fishers from Mexico and Venezuela. Eventually, the obligation to assure the long-term sustainability of dolphins and other associated marine species as stipulated under the 1982 United Nations Convention on the Law of the Sea led to the 1995 Panama Declaration, which established a cap on annual allowable dolphin mortality associated with the tuna fishery (Constance and Bonanno 1999).

Sea turtles are one of the species accidentally harvested in the shrimp fishery (Sands 2000). Bycatch of sea turtles in shrimp trawls caused ethical debates between some segments of the American public and foreign shrimp operators who had not installed turtle excluder devices (TEDs). TEDs allow turtles to escape through a trapdoor rather than being drowned in the nets. This ultimately resulted in the modification of shrimp operators' trawl nets to include the TEDs (Sands 2000). As in the dolphin-safe tuna harvesting controversy, the public and political pressure to modify shrimp harvesting gear to accommodate sea turtles underwent a complex negotiation process in the international forum, with the United States initially setting import restrictions that led to discontent among other GATT parties. The issue was finally resolved to the benefit of the sea turtle populations (Sands 2000).

Ethical questions affecting fisheries management have also originated in the use of fishing gear that damages aquatic habitats, such as trawling, which results in sea floor destruction (USGS n.d.) and drift nets, which kill a plethora of non-target species (Hunter et al. 2002: 699–700). Competition between large foreign fishing vessels and local subsistence fishers for a fish stock is also a common ethical debate (see Pollnac, Chapter 9). Ethical conflicts will continue to develop as diverse stakeholders interact with one another, fish populations, and their habitats. Decisions made at the international level, including international law and policy, will determine how ethical dilemmas may be resolved among communities and nations that harvest common fish stocks and can be expected to expand over time.

THE FUTURE OF FISHERIES MANAGEMENT IN A GLOBAL ENVIRONMENT

To be successful in managing our globally influenced fisheries in a sustainable manner, managers and policy-makers need to apply a new way of thinking and a new approach to solving problems. Fisheries managers no longer have the luxury of managing local fisheries in isolation from factors beyond their management jurisdictions that are acting on their fisheries and aquatic ecosystems. Thus fisheries management and policy need to evolve to meet the new challenges and embrace the opportunities presented by an increasingly globalized world. Therefore, we believe that to achieve sustainable fisheries, managers need to be proactive in understanding, utilizing, and expanding global networks that provide information on the social, economic, political, and ecological conditions of world fisheries. The ability to access and understand these networks will enable managers to assess effectively the global influences on local fisheries and their ecosystems, and allow for the design of responsive management plans (Taylor et al. 1995).

A global structure to facilitate fisheries management is imperative for the sustainability of our world's fisheries. This structure requires fisheries professionals to be more inclusive in their identification of stakeholders (Lynch 2001) at all geopolitical scales. They must also become more interdisciplinary in approach and establish communication links among private and public organizations spanning multiple jurisdictions to achieve cooperative management of shared fisheries resources. The creation of these links should result in the development and maintenance of a highly interactive and dynamic global social network among fisheries professionals that allows for discussion and resolution of management challenges and conflicts at all levels of governance. This would allow management institutions to be able to respond rapidly to changes in fisheries ecosystems in a cooperative effort across multiple jurisdictions, thereby facilitating the implementation of joint ventures and enforcement activities among local, regional, national, and international jurisdictions. In the absence of such a structure, the future of sustainable fisheries and healthy ecosystems is unpredictable at best and dismal at worst as the lack of interjurisdictional cooperation will render most local management plans ineffective.

Equally important, globalization will increase stakeholders' and managers' exposure to cultural and ethical differences that affect the management and valuation of fish, their habitats, and their ecosystems. For sustainable fisheries to become a reality, managers will need to improve their understanding of stakeholder differences to incorporate the multifaceted and often divergent values of the public in their management plans. Only by doing so will it be possible to sustain the health of the world's fisheries resources and their ecosystems.

Acknowledgments

The authors thank Dr. Michael Schechter for his guidance and enthusiasm for exploring the effects of globalization on fisheries and aquatic ecosystems. The development of this chapter and our growing appreciation for the direct and subtle influences that global processes have on fish, their habitats, and our fisheries were greatly clarified through our interactions with Dr. Schechter. Additionally, we thank Jack Laurie, past president of the Michigan Farm Bureau, for sharing his perceptions on the impact that globalization has had on agricultural communities and industries. There are many parallels between globalization of the fisheries resource and agribusiness management. Jack's experiences at the state, national, and international levels were extremely helpful to our understanding of events occurring in fisheries that are related to globalization. We received outstanding editorial comments by Krista Ecklin, Sara Hughes, Tracy Kolb, and Katrina Mueller which significantly improved the chapter's clarity and readability. Comments from several anonymous reviewers also provided significant improvement to the chapter. Lastly, we would like to thank the Great Lakes Fishery Commission, which provided us with a venue to interact closely with colleagues dedicated to interjurisdictional management and who manage a large, diverse, and highly valuable fisheries ecosystem that has been and continues to be affected by globalization.

References

Abramovitz, J. N. 1996. *Imperiled Waters, Impoverished Future: The Decline of Freshwater Ecosystems*, Worldwatch Paper 128. Washington, DC: Worldwatch Institute.

Alaska Department of Fish and Game. 2002. *Atlantic Salmon: A White Paper*. Available online at www.state.ak.us/adfg/geninfo/special/AS/docs/AS_white2002.pdf

Anonymous. 2003a. The blue revolution: a new way to feed the world. *The Economist* **368**(8336): 9.

Anonymous. 2003b. Fish farming: the promise of a blue revolution. *The Economist* **368**(8336): 19–22.

Barlow, M. 2001. *Blue Gold: The Global Water Crisis and the Commodification of the World's Water Supply*. Ottawa, Ontario: Council for Canadians. Available online at www.canadians.org/documents/blue_gold-e.pdf

BBC-h2g2. 2004. *The Cuyahoga River, Ohio, USA*, h2g2, edited guide entry ID A2966772, BBC. Available online at www.bbc.co.uk/dna/h2g2/A2966772.

Beeton, A. M., Sellinger, C. E., and Reid, D. E. 1999. An introduction to the Laurentian Great Lakes ecosystem. In *Great Lakes Policy and Management: A Binational Perspective*, eds. W. W. Taylor and C. P. Ferreri. East Lansing, MI: Michigan State University Press, pp. 3–54.

Centers for Disease Control and Prevention. 2006. *Home Page*. Available online at www.cdc.gov

Cohen, A. N. 1999. Invasions status and policy on the U.S. West coast. *Proceedings of the 1st National Conference on Marine Bioinvasions*, pp. 40–45.

Cole, H. 2003. Contemporary challenges: globalization, global interconnectedness and that "there are not plenty more fish in the sea." Fisheries, governance and globalization: is there a relationship? *Ocean and Coastal Management* **46**: 77–102.

Constance, D. H. and Bonanno, A. 1999. Contested terrain of the global fisheries: "dolphin-safe" tuna, the Panama Declaration, and the Marine Stewardship Council. *Rural Sociology* **64**: 597–623.

Cosgrove, W. J. and Rijsberman, F. R. 2000. *World Water Vision*. London: Earthscan Publications. Available online at www.worldwatercouncil.org/Vision/cce1f838f03d073dc125688c0063870f.shtm

Dawson, C. P., Lichtkoppler, F. R., and Pistis, C. 1989. The charter fishing industry in the Great Lakes. *North American Journal of Fisheries Management* **9**: 493–499.

Dochoda, M. R. 1999. Authorities, responsibilities, and arrangements for managing fish and fisheries in the Great Lakes ecosystem. In *Great Lakes Policy and Management: A Binational Perspective*, eds. W. W. Taylor and C. P. Ferreri. East Lansing, MI: Michigan State University Press, pp. 93–110.

EC-Fisheries (European Commission's European Union Policies on Fisheries European Commission). 1998. *Fact Sheet 2.2: The Common Fisheries Policy*.

Brussels: European Commission. Available online at http://europa.eu.int/comm/fisheries/doc_et_publ/factsheets/facts/en/pcp2_2.htm

Environmental Protection Agency. 2006. *Fish Advisory*. Washington, DC: EPA. Available online at http://epa.gov/waterscience/fish/states.htm

FAO Fisheries Department. 2002. *The State of World Fisheries and Aquaculture 2002*. Rome: Food and Agriculture Organization of the United Nations. Available online at www.fao.org/docrep/005/y7300e/y7300e00.htm

Florida Fish and Wildlife Conservation Commission. 2001. New exotic fish now present in Florida's freshwater system, News Release March 22, 2001. Available online at http://199.250.30.114/fishing/news-rel/snakehead.html

Friedman, T. L. 2000. *The Lexus and the Olive Tree*. New York: Anchor Books.

Gleick, P. H., Wolff, G., Chalecki, E. L., and Reyes, R. 2002. Globalization and international trade of water. In *The World's Water 2002-2003*, ed. P. H. Gleick. Washington, DC: Island Press. pp. 33-56.

Hansen, M. J. 1999. Lake trout in the Great Lakes: basinwide stock collapse and binational restoration. In *Great Lakes Fisheries Policy and Management: A Binational Perspective*, eds. W. W. Taylor and C. P. Ferreri. East Lansing, MI: Michigan State University Press, pp. 417-454.

Hardin, G. 1968. The tragedy of the commons. *Science* **162**: 1243-1248.

Harris, M. 1998. *The Lament for an Ocean: The Collapse of the Atlantic Cod Fishery – A True Crime Story*. Toronto, Ontario: McClelland & Stewart.

Hatcher, A. and Robinson, C. 1999. Overcapacity, overcapitalization and subsidies in European fisheries. *Proceedings of the 1st Concerted Action Workshop on Economics and the Common Fisheries Policy*, Portsmouth, UK, October 28-30, 1998.

Heidinger, R. C. 1999. Stocking for sport fishing enhancement. In *Inland Fisheries Management in North America*, 2nd edn, eds. C. C. Kohler and W. A. Hubert. Bethesda, MD: American Fisheries Society, pp. 375-402.

Held, D., McGrew, A., Goldblatt, D., and Perraton, J. 1999. *Global Transformations: Politics, Economics and Culture*. Cambridge, UK: Polity Press.

Hummer, J. 2001. Cuyahoga River area of concern. In *Great Lakes Areas of Concern*. Environmental Protection Agency. Available online at www.epa.gov/glnpo/aoc/cuyahoga.html

Hunter, D., Salzmann, J., and Zaelke, D. 2002. *International Environmental Law and Policy*, 2nd edn. New York: Foundation Press.

Ihonvbere, J. O. 1996. Africa and the new globalization: challenges and options for the future. In *Futurevision: Ideas, Insights, and Strategies*, ed. H. F. Didsbury, Jr. Bethesda, MD: World Future Society, pp. 345-366.

IJC (International Joint Commission). 1989. *Revised Great Lakes Water Quality Agreement of 1978*, as amended by protocol signed November 18, 1987. Ottawa, Ontario: International Joint Commission.

Jones, W. W. 1996. Balancing recreational user demands and conflicts on multiple use public waters. In *Proceedings of the 3rd Reservoir Symposium: Multidimensional Approaches to Reservoir Fisheries Management*, eds. L. E. Miranda and D. DeVries. Bethesda, MD: American Fisheries Society, pp. 179-185.

Kennelly, S. J. and Broadhurst, M. K. 2002. By-catch begone: changes in the philosophy of fishing technology. *Fish and Fisheries* **3**: 340-355.

Keus, B. 1997. *Co-management in Dutch Shellfish Fisheries*. Rijswijk, Netherlands: Dutch Fish Board – Produktschap Vis. Available online at www.waddensea-secretariat.org/news/publications/Wsnl/WSNL97-2/97-2-04Keus.html

Kirshman, H. and Leonard, N. 2003. Operation kingfisher. In *FORUM: Special Law Enforcement Edition*, eds. J. Finster and M. Gaden. Ann Arbor, MI: Great Lakes Fishery Commission, pp. 4-5.

Koel, T. M., Irons, K. S., and Ratcliff, E. 2000. *Asian Carp Invasions of the Upper Mississippi River System*, Project Report PSR 2000-05. La Crosse, WI: USGS Upper Midwest Environmental Sciences Center. Available online at www.umesc.usgs.gov/reports_publications/psrs/psr_2000_05.html

Kulp, M. A. and Moore, S. E. 2000. Multiple electrofishing removals for eliminating rainbow trout in a small Southern Appalachian stream. *North American Journal of Fisheries Management* **20**: 259-266.

Lake Michigan Federation. 2002. *An Advocate's Field Guide to Protecting Lake Michigan*, ch. 1, part II. Available online at www.lakemichigan.org/field_guide/habitat_urban.asp

Laurec, A. and Armstrong, D. 1997. The European common fisheries policy and its evolution. In *Global Trends: Fisheries Management*, eds. E. K. Pikitch, D. D. Huppert, and M. P. Sissenwine. Bethesda, MD: American Fisheries Society, pp. 61-72.

Le Sann, A. 1998. *A Livelihood from Fishing*. London: Intermediate Technology Publications.

Lentner, H. H. 2000. Globalization and power. In *Rethinking Globalization(s)*, eds P. S. Aulakh and M. G. Schechter. New York: St. Martin's Press, pp. 56-72.

Lichtkoppler, F. R. and Hushak, L. J. 1993. Ohio's 1990 Lake Erie Charter Fishing Industry. *Fisheries* **18**(1): 14-21.

Lynch, K. D. 2001. Formation and implications of interorganizational networks among fisheries stakeholder organizations in Michigan's Pere Marquette River Watershed. Ph.D. thesis. Michigan State University, East Lansing, MI.

Manchester, S. J. and Bullock, J. M. 2000. The impacts of non-native species on UK biodiversity and the effectiveness of control. *Journal of Applied Ecology* **37**: 845-864.

Marsden, J. E. and Jude, D. J. 1995. *Round Gobies Invade North America*, OHSU-FS-065. Columbus, OH: Illinois-Indiana Sea Grant Program and the Ohio Sea Grant College Program. Available online at www.ohioseagrant.osu.edu/_documents/publications/FS/FS-065%20Round%20gobies%20invade%20North%20America.pdf

McClanahan, T. R. 2002. The near future of coral reefs. *Environmental Conservation* **29**: 460-483.

Mills, E. L., Leach, J. H., Carlton, J. T., and Secor, C. L. 1993. Exotic species in the Great Lakes: a history of biotic crises and anthropogenic introductions. *Journal of Great Lakes Research* **19**: 1-54.

Miranda, L. E. 2001. A review of guidance and criteria for managing reservoirs and associated riverine environment to benefit fish and fisheries, Section 3, Guidance and criteria for managing fish stocks and fisheries. In *Dams, Fish and Fisheries Opportunities, Challenges and Conflict Resolution*, FAO Fisheries Technical Paper 419, ed. G. Marmulla. Rome: Food and Agriculture Organization of the United Nations. Available online at www.fao.org/docrep/004/y2785e/y2785e04.htm

Nielson, L. A. 1999. History of inland fisheries management in North America. In *Inland Fisheries Management in North America*, eds. C. C. Kohler and W. A. Hubert. Bethesda, MD: American Fisheries Society, pp. 3-30.

NMFS (National Marine Fisheries Service). 2001. *Fisheries of the United States, World Aquaculture and Commercial Catches, 1991-2000*. Silver Springs, MD: National Marine Fisheries Service, National Oceanic and Atmospheric Administration, U.S. Department of Commerce, Available online at www.st.nmfs.gov/st1/fus/current/index.html

NMFS. 2002. *Fisheries of the United States, World Aquaculture and Commercial Catches, 1992-2001*. Silver Springs, MD: National Marine Fisheries Service, National

Oceanic and Atmospheric Administration, U.S. Department of Commerce. Available online at www.st.nmfs.gov/st1/fus/fus01/index.html

NMFS. 2004. *Fisheries of the United States, World Aquaculture and Commercial Catches, 1994–2003*. Silver springs, MD: National Marine Fisheries Service, National Oceanic and Atmospheric Administration, U.S. Department of Commerce. Available online at www.st.nmfs.gov/st1/fus/fus04/index.html

Noble, R. L. and Jones, T. W. 1993. Managing fisheries with regulations. In *Inland Fisheries Management in North America*, eds. C. C. Kohler and W. A. Hubert. Bethesda, MD: American Fisheries Society, pp. 383–402.

Novinger, D. C. and Rahel, F. J. 2003. Isolation management with artificial barriers as a conservation strategy for cutthroat trout in headwater streams. *Conservation Biology* **17**: 772–781.

Parsons, L. S. and Beckett, J. S. 1997. Fisheries management in Canada: the case of Atlantic groundfish. In *Global Trends: Fisheries Management*, eds. E. K. Pikitch, D. D. Huppert, and M. P. Sissenwine. Bethesda, MD: American Fisheries Society, pp. 73–79.

Pollock, K. H., Jones, C. M., and Brown, T. L. 1994. *Angler Survey Methods and Their Applications in Fisheries Management*. Bethesda, MD: American Fisheries Society.

Radonski, G. C. 1991. Politics trumps science. *Fisheries* **16**(5): 4.

Rahel, F. J., Muth, R. T., and Carlson, C. A. 1999. Endangered species management. In *Inland Fisheries Management in North America*, 2nd edn, eds. C. C. Kohler and W. A. Hubert. Bethesda, MD: American Fisheries Society, pp. 375–402.

Responsive Management National Office. 1999. *The future of fishing in the United States: Assessment of needs to increase sport fishing participation, Phase V: Final report recommendations and strategies*, Grant Agreement 1448-98210-98-G048. Harrisonburg, VA: International Associations of Fish and Wildlife Agencies Federal Aid in Sport Fish Restoration. Available online at www.rbff.org/research/fof.pdf

Robbins, J. 2001. *The Food Revolution: How Your Diet Can Help Save Your Life and Our World*. Berkeley, CA: Conari Press.

Roy, N. 1997. The Newfoundland fishery a descriptive analysis. *Symposium on the Efficiency of North Atlantic Fisheries*, Reykjavik, Iceland, September 12–13, 1997. Available online at www.ucs.mun.ca/~noelroy/nffishery.text.html

Salz, R. J. and Loomis, D. K. 2001. *Awareness by New Hampshire Anglers and Hunters of Resource Management Agencies, Programs and Funding Sources: Survey of 1999 Hunting and Fishing License Holders*. Concord, NH: New Hampshire and Game Department.

Sands, P. 2000. "Unilateralism" values and international law. *European Journal of International Law* **11**: 291–302.

Scott, R. J., Noakes, D. L. G., Beamish, F. W. H., and Carl, L. M. 2003. Chinook salmon impede Atlantic salmon conservation in Lake Ontario. *Ecology of Freshwater Fish* **12**: 66–73.

Scott, W. B. and Crossman, E. J. 1998. *Freshwater Fishes of Canada*. Oakville, Ontario: Galt House Publications.

SCRS (Standing Committee on Research and Statistics). 2005. *Report of the Standing Committee on Research and Statistics (SCRS)*, Madrid, October 4–8, 2004. Madrid: International Commission for the Conservation of Atlantic Tunas. Available online at www.iccat.es/SCRS.htm

Serageldin, I. 1999. Looking ahead: water, life and the environment in the twenty-first century. *Water Resources Development* **15**: 17–28.

Simmons, P. J. and Oudraat, C. D. J. 2001. Managing global issues: an introduction. In *Managing Global Issues: Lessons Learned*, eds. P. J. Simmons and C. D. J. Oudraat. Washington, DC: Brookings Institution Press, pp. 3–22.

Song, Y. H. 1998. The common fisheries policy of the European Union: restructuring of the fishing fleet and the financial instrument for fisheries guidance. *International Journal of Marine and Coastal Law* **13**: 537–577.

Southwick Associates. 2002. U.S. – 2001 Economic contributions of sportfishing. Available online at www.southwickassociates.com/freereports/default.aspx

Symes, D. 1997. The European community's common fisheries policy. *Ocean and Coastal Management* **35**: 137–155.

Tanner, H. A. and Tody, W. H. 2002. History of the Great Lakes salmon fishery: a Michigan perspective. In *Sustaining North American Salmon: Perspectives across Regions and Disciplines*, eds. K. D. Lynch, M. L. Jones, and W. W. Taylor. East Lansing, MI: Michigan State University Press, pp. 139–153.

Taylor, W. W., Ferreri, C. P., Poston, F. L., and Robertson, J. M. 1995. Educating fisheries professionals using a watershed approach to emphasize the ecosystem paradigm. *Fisheries* **20**(9): 6–9.

Thiele, W. 2001. *Global Trends in Fishing Technology and their Effects on Fishing Power and Capacity*, Report of the Regional Workshop on the Effects of Globalization and Deregulation on Fisheries in the Caribbean 640. Rome: Food and Agriculture Organization of the United Nations, pp. 69–80.

United Nations. 1982. *United Nations Convention on the Law of the Sea*. Available online at www.oceanlawenet/texts/unfsa.htm

USGS (U.S. Geological Survey). 2000. Geospatial technology, geospatial technology expertise, applications, and highlights: Western Fisheries Research Center Information. Available online at http://biology.usgs.gov/geotech/documents/applications_and_infrastructure/highlights/wfrc.html

USGS. 2002. Zebra mussels fact sheet. Available online at http://nas.er.usgs.gov/zebra.mussel/docs/sp_account.html

USGS. 2003a. *Zebra Mussels*. La Crosse, WI: USGS Upper Midwest Environmental Sciences Center. Available online at www.umesc.gov/invasive_species/zebra_mussels.html

USGS. 2003b. *Round Goby*. La Crosse, WI: USGS Upper Midwest Environmental Sciences Center. Available online at www.umesc.gov/invasive_species/round_goby.html

USGS. 2003c. *Channa argus* (northern snake head) fact sheet. Available online at http://nas.er.usgs.gov/queries/SpFactSheet.asp?speciesID=2265

USGS n.d. *Coral Reefs of the United States Virgin Islands*. Washington, DC: U.S. Geological Survey. Available online at http://biology.usgs.gov/s+t/SNT/noframe/cr134.htm

Warner, W. W. 1983. *Distant Waters: The Fate of the North Atlantic Fisherman*. Boston, MA: Little, Brown.

World Health Organization. 2006. *Regional Office of Europe Home Page*. Available online at www.who.int/en

Wright, M. 2001. *A Fishery for Modern Times: The State and the Industrialization of the Newfoundland Fishery, 1934–1968*. Don Mills, Ontario: Oxford University Press.

JACKIE ALDER AND REG WATSON

2

Fisheries globalization: fair trade or piracy?

INTRODUCTION

The international trade in marine fishery products is big business despite the global decline in marine fish landings. Nearly 40 percent of world fish production is traded globally – much more than for other food staples such as wheat (20 percent) and rice (5 percent) (FAO 2001). The implementation of United Nations Convention on Law of the Sea (UNCLOS) and the trend toward globalization of business, banking, and telecommunications, trade liberalization, and the expansion of fishing fleets over the past 50 years have increased the commerce of fish products with a net flow of fish from "the more needy to the less needy [countries]" (Kent 1983) or from the developing to the developed world (FAO 2002a). Traded fish products are important commodities for developing countries, including those facing food security issues. The trading of fish has not just extended spatially to cover most regions of the world; it has also expanded in volume and value. The earliest global estimate of fish trading in 1963 was 5.3 million tonnes (OECD 1989). Over the past 25 years, the total volume and value have increased steadily since 1976, when detailed fish trade statistics were first recorded (Fig. 2.1). In 1976, nearly 8 million tonnes of fish worth $8 billion were traded. This increased to more than 49 million tonnes (live weight equivalent) of exported fish products worth $56 billion traded globally in 2001 (FAO 2000).

Fish trade at the international scale has been undertaken ever since preservation techniques such as drying and salting were developed so that explorers and colonial settlers could survive, prosper, and trade with their home countries. Since the 1950s, however, with engineering advances in ship design and growing independence of many colonial states, fish trading has expanded rapidly through distinct

Globalization: Effects on Fisheries Resources, ed. William W. Taylor, Michael G. Schechter, and Lois G. Wolfson. Published by Cambridge University Press. © Cambridge University Press 2007.

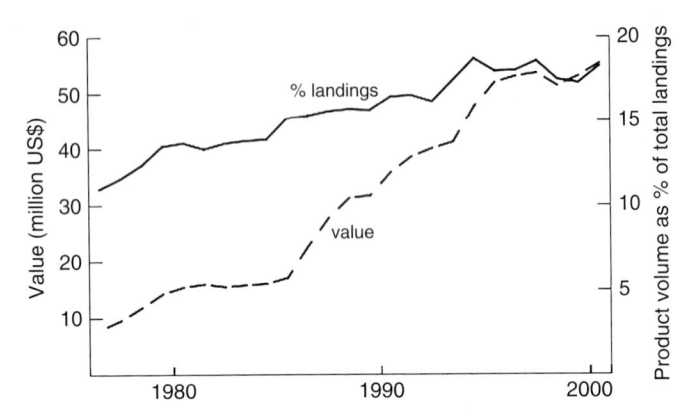

Figure 2.1 Trends in volume and value (US$) of exported marine capture fish products.

stages. The first was the expansion of distant water fleets after World War II through the mid 1970s; the second was the introduction of UNCLOS and the declaration of economic exclusion zones (EEZ) throughout the 1970s and early 1980s; and the current stage is the development of neo-liberal economic polices and the strengthening of the globalization process (Arbo and Hersoug 1997; Kurien 1998; Thorpe and Bennett 2001).

The economic, political, and social dimensions of globalization today (Nierop 1994; Waters 1995; Dunning 1997), make it difficult to provide readers with a short and crisp definition to work with. Many definitions focus on the economic dimension and describe globalization in qualitative and quantitative terms, as exemplified by Oman (1999:37), "the growth or more precisely, the accelerated growth, of economic activity that spans politically defined national and international boundaries." This clearly encompasses the trend in the exporting and importing of fishery products over the past 40 years. Globalization of the fishing sector is not new – as early as the sixteenth century, fish trade was undertaken at an international scale on the Grand Banks of Canada (McCay and Finlayson 1995). The impact of globalization over the past 40 years, however, has increased rapidly and been much more widespread than during the past four centuries. Between 1963 and 2001, exports volume (live weight equivalent) increased from 5.3 million tonnes (OECD 1989) to 49 million tonnes (FAO 2000).

Much of the discussion on traded fish products in the published literature focuses on the monetary value of the trade (Deere 1999), and products are often examined without the context of the landings or stocks from which they originate – that is, the impact that removing the fish has on the ecosystem and the human communities that rely on those ecosystems. Other aspects such as the environmental and social impacts of trade in fish products are also poorly understood (Dommen 1999). Though the monetary value provides some insight into the trends of the products, this information provides few insights into environmental, social, or economic impacts of the trade of fish products.

It is generally agreed that much of the exports in fish products are from developing to developed countries (Kent 1987; FAO 2002a), and that developing countries are increasing their share of the value of products. In 1976, developing countries traded 2.5 million tonnes of marine fish and fish products worth $2.9 billion. In 2000, 13.2 million tonnes worth $28 billion were traded (FAO 2000). In 2002, 50 percent of the traded value of marine products originated in developing countries (FAO 2002a), up from 34 percent in the early 1980s (Kent 1983). However, few details on what fish are being traded between what countries and how the benefits of trade flow between developing and developed countries are reported.

The analysis of the trade of fish products is complex because total production within the fishing sector is divided into marine, brackish, and freshwater, and within these sectors it is divided into capture fisheries and aquaculture. Marine capture landings of fish and invertebrates, however, made up more than 79 percent of total fish production in 2001 (FAO 2000). In addition, marine capture fisheries account for approximately 60 percent of the employment of fishers globally, the largest source of employment within the fishing sector. Given this economic and social reliance on marine capture fisheries, the value of traded fish products is sensitive to changes in landings, and social and economic ramifications ripple from international markets to local food supplies and employment in coastal communities.

Using a rule-based analytical framework, we examined the spatial and temporal changes in the increasing trade of four major marine and brackish capture fisheries between 1976 and 2000; the potential environmental, social, and economic impacts of this increasing international trade of fishery products on global fish stocks; and the consequential implications for the food security of developing countries that rely on many of these fish stocks.

ANALYTICAL FRAMEWORK

The analysis is based on the FAO's Fisheries Commodities Production and Trade database (FAO 2000) and the marine capture fish landings, adjusted for many sources of error including China's misreporting, from the Sea Around Us Project database (Watson et al. 2004). Marine and brackish fisheries were included in the analysis; freshwater and anadromous (e.g., salmon) species were excluded. In the fisheries sector with China included, marine capture fisheries represented 64 percent of the global fish production, and marine aquaculture, inland capture fisheries, and inland aquaculture provided the remaining production (36 percent). The FAO groups of species: demersal, pelagic, crustaceans, and cephalopods represent most of the marine products that are traded globally and are analyzed here. Crustaceans and cephalopods (squids, octopus, etc.) were grouped into invertebrates; other mollusk groups such as bivalves were not included in this analysis because their trade was very limited.

The Sea Around Us Project[1] has disaggregated landing data from the FAO and other sources to a finer spatial scale so that it is possible to estimate annual landings extracted from individual EEZs using a number of criteria, including the distribution of commercially targeted species and the access agreement observed fishing patterns of distant water fleets (Watson et al. 2004). These disaggregated data are necessary to differentiate between domestic and foreign catches within a particular EEZ. The data presented here are based on the annual values averaged over 25 years, unless stated otherwise.

The FAO Aquaculture Production database was also accessed to assist in determining the total landings of crustaceans because the Commodities Production and Trade database does not differentiate between wild and cultured crustaceans (FAO 2000).

The live weight of processed products in the Commodities Production and Trade database is not recorded, only the processed weight (e.g., loins, fillets, meal, oils, etc.). This makes comparisons between total landings and exported landings difficult. All marine products were, therefore, converted to equivalent live weight using conversion factors from various sources.[2] Double-counting of by-products of commodities, such as oils from meals, was avoided by using a conversion factor of zero for the by-product commodity. In general, trade flows of fish products are described in terms of value because more than 90 percent of the fish traded are processed into products with a higher value and, therefore, live weight equivalent is not a meaningful measure for some quantitative comparisons (Ruckes 2000). Live weight equivalent, however, is

---

[1] See www.seaaroundus.org
[2] Available at www.fao.org/fi/statist/fisoft/conv.asp

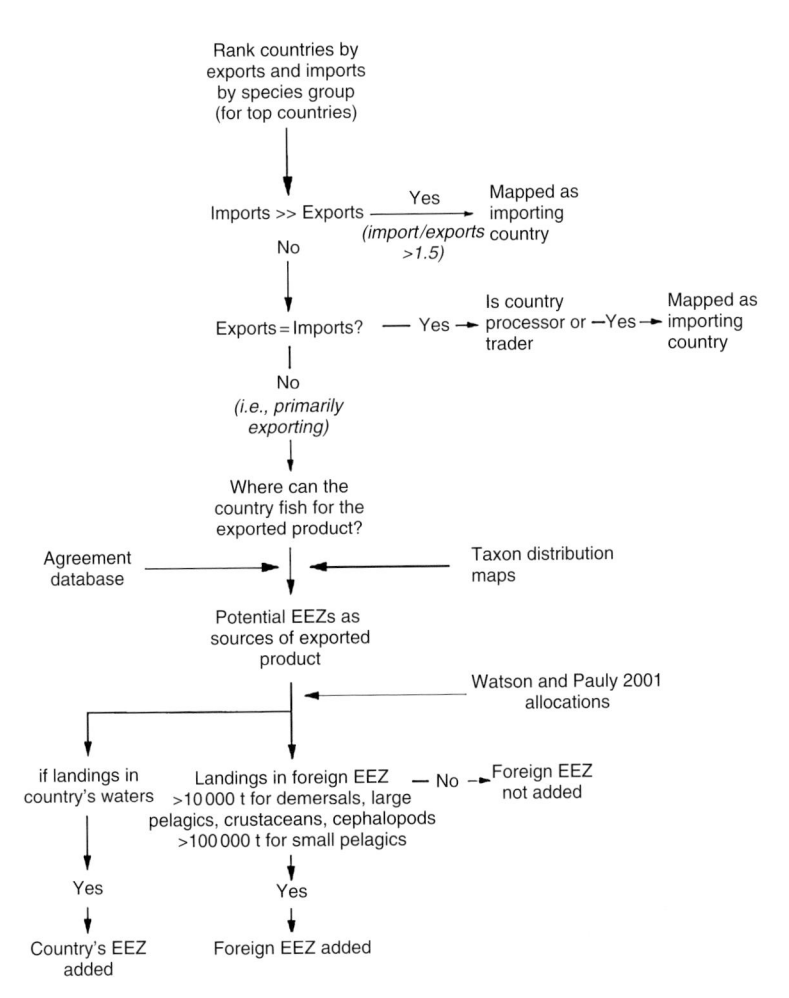

Fisheries globalization: fair trade or piracy? 51

Figure 2.2 Allocation procedure for classifying countries as importers or exporters.

necessary in this study to assess the impact of trade on the ecology of marine ecosystems in addition to monetary value.

A set of rules was developed to allocate the exported products to their likely country or area of origin for the top exporting and importing countries (by volume) (Fig. 2.2). Some countries are major importers and exporters of marine products, such as the United States, which exports high volumes of crustaceans such as lobster and imports high volumes of shrimp. Other countries, such as the Philippines, which is a major processor of canned tuna, are primarily traders or processors and consequently import high volumes of unprocessed fish and export high volumes of processed fish. The import-to-export ratio in the allocation algorithm identifies these anomalies that were excluded from the analysis. The disaggregated landings from the Sea Around Us Project database are used to determine if the exported fish are potentially sourced within a country's internal waters or in another country's EEZ.

TRADE TRENDS: MARINE CAPTURE FISH

Demersal

Demersal fish exports have increased by more than 250 percent by volume and by more than 500 percent by value over the past 25 years (Figs. 2.3 and 2.4). Total value increased from $1.4 million in 1976 to $7.3 million in 2000. Over this time period, live weight equivalent exports increased from 2.6 million tonnes in 1976 to 6.3 million tonnes in 2000. The proportion of demersal fish landings that were exported increased from 19 percent to 36 percent over the same time period.

The exported demersal fish products, much of it "whitefish," which includes cod, haddock, hake, and pollock, are caught in the EEZs of countries in the Northern Hemisphere and exported to countries within the Northern Hemisphere, with any shortfall in supply met by countries from the Southern Hemisphere such as Chile, Argentina, Namibia, South Africa, and New Zealand (Fig. 2.5). Processing much of that fish

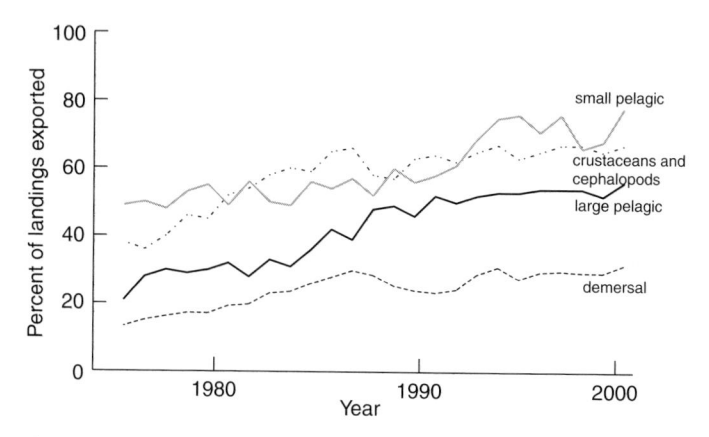

Figure 2.3 Trends in volume (% of landings exported) for major FAO commodity groups 1976 to 2000.

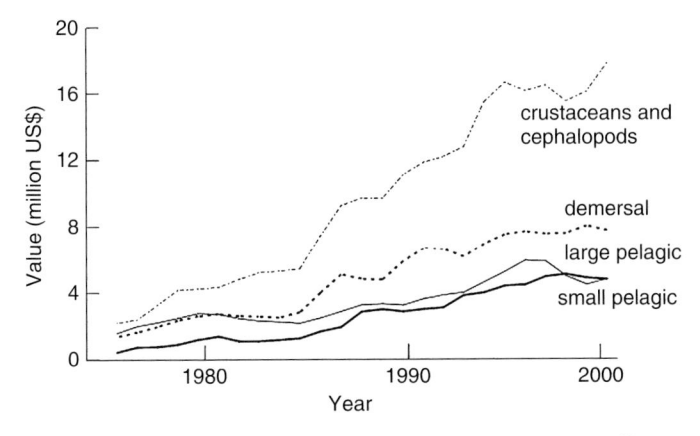

Fisheries globalization: fair trade or piracy? 53

Figure 2.4 Trends in value (million US$) for major FAO commodity groups 1976 to 2000.

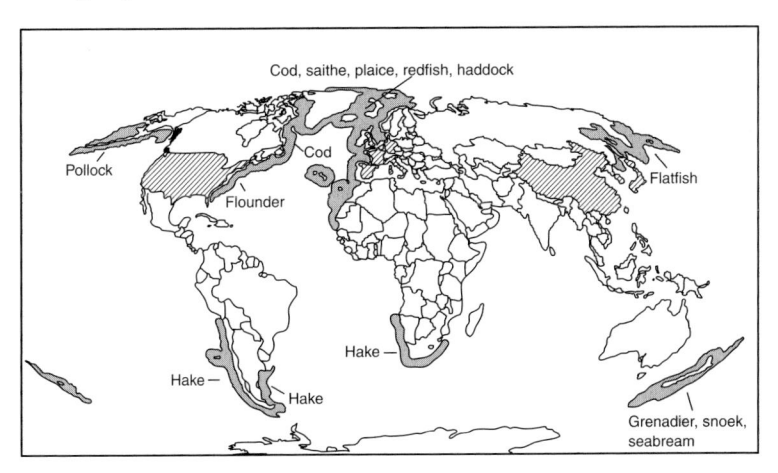

Figure 2.5 Major importing (hatched) countries and the EEZ (gray) where the majority of exports were sourced for demersal fish 1976–2000. For image in colour please see plate section.

into fillets, fish sticks, or surimi increases the landed value several times and prices it beyond the budgets of consumers in developing countries.

Norway, Iceland, Denmark, Canada, and the United States have been the major exporting countries of demersal fish over the past 25 years (Table 2.1), while the United States, Japan, Germany, China, and the United Kingdom have been major importers. The United States is both an importer and an exporter because lower-valued pollock is exported globally for processing into surimi, fish fingers, and fish burger products, and higher-valued products such as fresh or frozen fillets from higher-valued species are imported.

Table 2.1 *Major exporting countries by decade (1976–2000) for FAO fish commodities*

Country[a]	Demersal	Small pelagic	Large pelagic	Crustacea	Cephalopods
	1970s 1980s 1990s	1970s 1980s 1990s	1970s 1980s 1990s	1970s 1980s 1990s	1970s 1980s 1990s
Norway					
Iceland					
Denmark					
Germany					
France					
Spain					
Faeroe Islands					
Ireland					
Netherlands					
UK					
Russia					
Poland					
Canada					
USA					
New Zealand					
Taiwan					
Korea (Rep.)					
Japan					
Peru					
Argentina					
Mexico					
Chile					
Columbia					
Ecuador					
Solomon Islands					
Papua New Guinea					
Philippines					
Thailand					
Indonesia					
India					
China					
Vietnam					
Mauritania					
Morocco					
Namibia					
Ghana					
Côte d'Ivoire					

[a] Countries in italics are classed as developing.

Small pelagics

Small pelagic fisheries are sensitive to changes in environmental conditions such as El Niño which result in highly variable landings, and this is reflected in fluctuating export volumes and values. Despite this uncertainty, fishery exports of small pelagic fish more than doubled over the past 25 years (Fig. 2.3), from 12 million tonnes in 1976 to 26 million tonnes in live weight equivalent in 2000. The proportion of small pelagic landings exported also increased substantially over the same time, from 50 percent of the landed volume in 1976 to 77 percent in 2000. The value of exports increased by 300 percent (Fig. 2.4), from $1.6 million in 1976 to $4.8 million in 2000.

The major exporters of small pelagic marine products have remained the same over the past 25 years, with eight countries consistently in the top ten exporters (Table 2.1). Chilean and Peruvian export volumes are a magnitude higher than those of the remaining countries, and they export more than 90 percent of what is landed in their EEZs. Denmark, Norway, Iceland, and the United States are also major exporters of small pelagic fish products (Fig. 2.6). Germany, Japan, China, Taiwan, and the United Kingdom are the top major importers by volume and value; most of their imports are fishmeal.

The small pelagic products that are exported are primarily fishmeal, which is used for aquaculture feed, poultry or pig feed, or fertilizer. Sixty percent of fishmeal production is exported. A smaller

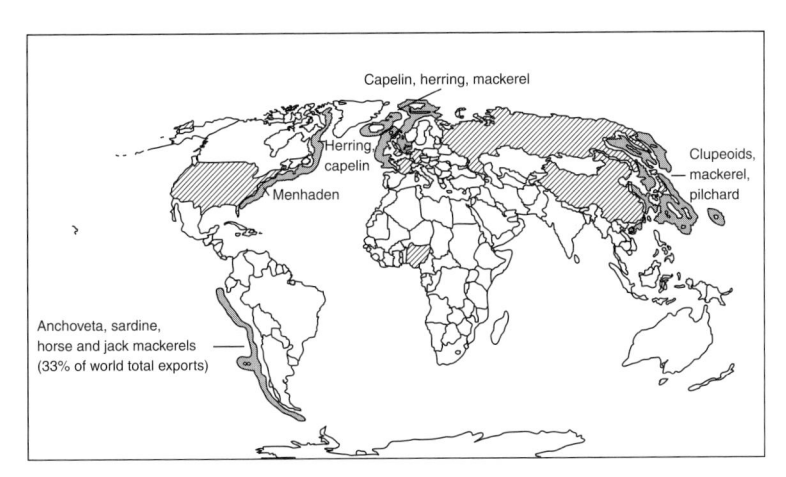

Figure 2.6 Major importing (hatched) countries and the EEZ (gray) where the majority of exports were sourced for small pelagic fish 1976–2000. For image in colour please see plate section.

proportion of small pelagic export products are low-value fish for human consumption as tinned fish or frozen fish blocks. A third of the marine products traded globally are sourced from small to medium pelagic fisheries in the southeast Pacific waters offshore of Chile, Peru, and Ecuador. The fishmeal industry is a major source of export earnings and employment in those countries. Small pelagic fish are also a source of cheap protein for residents of many coastal communities and are closely linked to their food security. Consequently, changes in the demand for fishmeal have the potential to affect their food security.

Countries such as Ecuador (a major exporter) and Ghana (a major importer) are lesser developed and suffer from food deficit (Kurien 2004). There is a strong link between the growth of the aquaculture sector and the demand for fishmeal. As high-valued aquaculture ventures such as salmon and other carnivorous fish species expand, the demand for fishmeal will also expand and the price will increase. Fish that would have been used as cheap sources of human food are diverted to the fishmeal sector, and the food security of locals is reduced. In Peru, the government has restricted industrial fisheries (most of whose harvest is destined for fishmeal) from fishing for some small pelagic species and in certain areas to ensure there is an affordable supply of fish for human consumption (FIS 2003).

The small pelagic fish, such as sardines and herrings, caught in the northern waters of the Atlantic and the northwest Pacific are processed and sold for human consumption in the developed world. Some production of poorer-quality and lower-valued fish is also exported to developing countries such as Nigeria and Ghana. Small pelagic fish not destined for food consumption are processed for fishmeal and sold internationally. There is concern about fishmeal from northeast Atlantic, especially the Baltic, because it has higher levels of pollutants such as dioxins than meal from South America (New and Wijkström 2002).

Large pelagics

Large pelagic fish exports increased substantially in volume and value between 1976 and 2000. Export volumes grew from 0.5 million tonnes in 1976 to 3.3 million tonnes in 2000. The proportion of large pelagic landings also increased, from 21 percent in 1976 to 56 percent in 2000 (Fig. 2.3). The value increased by an order of magnitude, from $0.4 million in 1976 to $4.7 million in 2000 (Fig. 2.4).

The large pelagic fishery differs from the other fisheries in that it is focused on tunas and billfishes, with fishing grounds offshore in the EEZs of countries as well as in the high seas, and that the top exporters

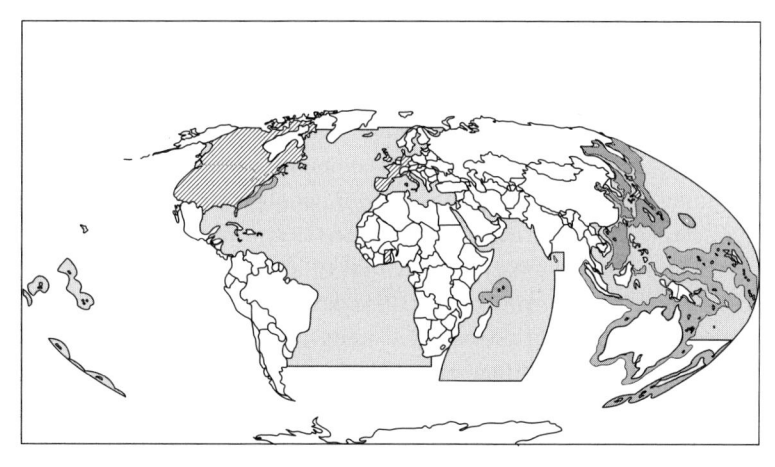

Figure 2.7 Major importing (hatched) countries and the EEZ (gray) where the majority of exports were sourced for large pelagic fish 1976–2000. For image in colour please see plate section.

are also countries with major distant water fleets that fish throughout the world (Fig. 2.7), with the products ultimately exported to developed countries in the Northern Hemisphere.

The major exporters of large pelagic marine products have remained the same over the past 25 years, with six countries consistently in the top ten exporters (Table 2.1). Taiwan, Korea, and Japan have large tuna fleets and fish in the three major oceans for tuna. France also has a large fleet that primarily fishes in proximity of its colonies in the Pacific and Indian oceans. Spain fishes primarily in the Atlantic and the Mediterranean, and, more recently, in the Pacific. The Philippines, Thailand, Senegal, and Côte d'Ivoire are major processing centers for tuna. They import large quantities of frozen tuna, process it for canning, and then export substantial volumes of finished product. Thailand, Senegal, and Côte d'Ivoire import significantly more fish than they land through fishing, so in this study they do not have a significant offshore fishing fleet. The United States, Japan, Italy, France, and the United Kingdom are the top importers by volume and value.

In the Pacific, 90 percent of the tuna catch is harvested by distant water fleets that pay access fees of only 4 percent of their gross revenue of the catch (Gillett *et al.* 2001). The countries of the Pacific benefit from the access fees and the economic activity of distant water fleet vessels, which use local port facilities (Petersen 2003). These values, however, pale in comparison with the benefits they could have derived from having major processing facilities in country through joint ventures with foreign countries, as in Ghana, where a well-developed tuna canning industry is diversifying into fishmeal (Worldfish Report 2004). Similarly, Thailand and the Philippines have well-developed onshore processing facilities and have developed joint ventures. These ventures have helped to develop a range of supporting industries as well as providing employment opportunities for many people.

Increasing most developing nations' share of the trade in large pelagics such as tuna has the potential to improve the economic and social conditions of many fishers and coastal dwellers. Two important factors in sustaining these benefits are ensuring food security and sustaining stocks so that trade is sustainable. According to recent South Pacific Commission reports, the decline of bigeye tuna stocks due to fishing is emerging as a problem in the Western and Central Pacific Ocean (WCPO) (Hampton and Williams 2003). This has the potential to diminish the benefits that countries in the South Pacific derive from selling access rights to distant water fleets. In some cases, conflicts occur between the tuna fleet and small-scale inshore fishers over baitfish resources (e.g., in the Solomon Islands) (Johannes *et al.* 2000). The baitfish is important to local fishers as a source of food and income.

Invertebrates

Total invertebrate (crustacean and cephalopod) exports from marine capture and aquaculture increased in volume by 400 percent between 1976 and 2000. Exports increased from 1.3 million tonnes – representing 38 percent of total production in 1976 – to 6.6 million tonnes in 2000, 62 percent of total production (Fig. 2.3). The value of exports also increased from $2.2 million in 1976 to $17.8 million (Fig. 2.4). Global shrimp production is centered in developing countries, with exports to Northern developed countries in the form of frozen whole (or "headed" shrimp). The exports of invertebrates, particularly high-valued shrimp, include products sourced from the aquaculture sector, so countries with high aquaculture production often dominate the export sector.

The major exporting countries are Thailand, China, Argentina, the United States, and the Democratic Republic of Korea (Fig. 2.8). Some of the main exporting countries changed over time as countries such as Colombia and Vietnam developed their shrimp aquaculture industry (Table 2.1). Shrimp account for a major portion (60 percent) of extracted invertebrates, and most exports are sourced in developing countries with well-developed aquaculture sectors, such as Thailand, China, and Ecuador. Nearly 30 percent of total shrimp production is cultured, and

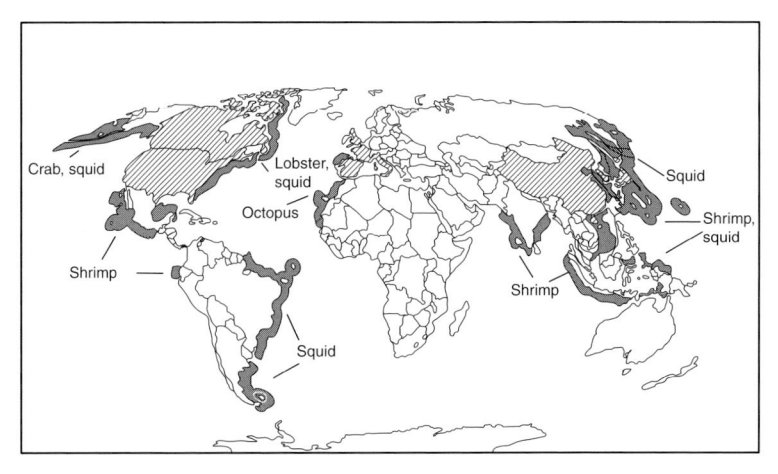

Figure 2.8 Major importing (hatched) countries and the EEZ (gray) where the majority of exports were sourced for crustaceans and cephalopods 1976–2000. For image in colour please see plate section.

80 to 95 percent of cultured shrimp is exported (Rosenberry 1991). The contribution of cultured shrimp to overall production and exports may have leveled off as disease and available land constrain current expansion, though potential exists for expansion in Latin America and the Middle East (Yap 2000). The major importing countries are Japan, Spain, Italy, Korea, and the United States.

DRIVERS OF FISHERIES GLOBALIZATION

Globalization is a complex process that is difficult to define (Oman 1999). Fisheries globalization can be defined as the accelerated growth in the trade of fish products resulting in economic activity that spans politically defined national and international boundaries. Though it is complex, the major drivers and consequential impacts can be described (Fig. 2.9). In the marine fisheries sector, high-valued export-driven fisheries, the need for foreign exchange to service national debts, and vertical integration of a number of fisheries-oriented companies are major factors in the globalization of fisheries.

Export-oriented fisheries

The growth in exported fish products over the past 25 years has been in the small and large pelagic, crustacean and cephalopod fisheries (Table 2.1) in developing countries. The situation for demersal fisheries

1900

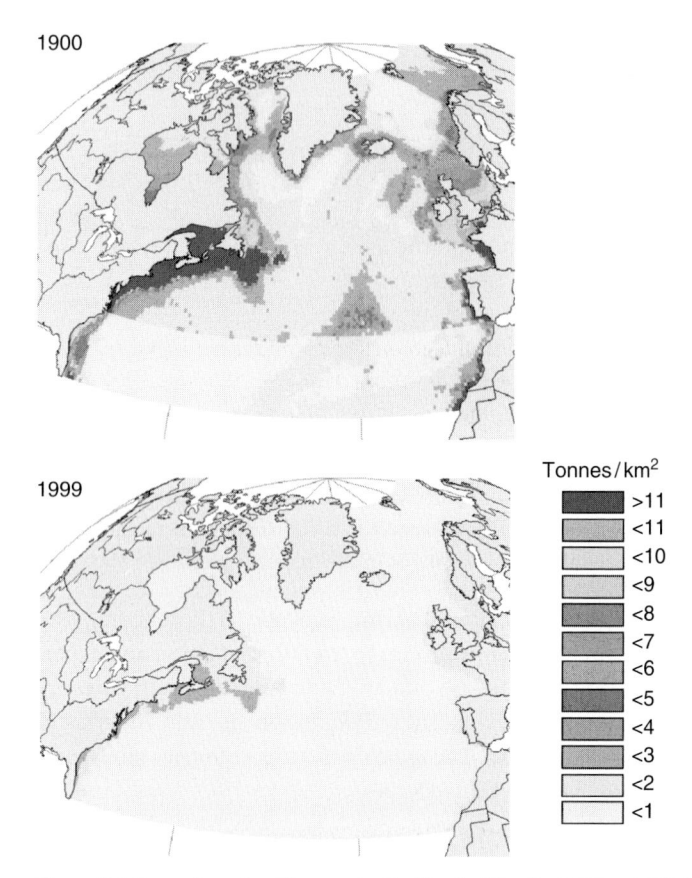

1999

Tonnes/km²
>11
<11
<10
<9
<8
<7
<6
<5
<4
<3
<2
<1

Figure 2.9 Loss of marine biomass in the North Atlantic 1900 to 1999. For image in colour please see plate section. (*Source:* Christensen *et al.* 2003.)

has not changed significantly over the past 25 years; the northern areas of Europe, North America (Canada and United States), and Japan still dominate the sector. Much of the trade in demersal products is among these countries, which are sometimes referred to as the "Triad" (Rugman and Moore 2001): products are imported from the South to meet the shortfall in demand from the North.

Over the last 25 years the growth of export-oriented fisheries can be attributed to meeting the increasing demand for fish in the developed world and in countries where there is significant income growth such as China (Delgado *et al.* 2003), the growth of aquaculture, in particular shrimp and salmon, and the development of Sanitary and Phytosanitary agreements (SPS) using the Hazard Analysis and Critical Control Points system (HACCP). The development of SPS agreements has enabled developing countries to access markets that were closed because of concerns over the quality and safety of the fish products. The HACCP system ensures consumers in the importing countries of a product that is safe for human consumption.

Exports are sourced directly through a country's domestic fleet fishing within its own EEZ or sourced indirectly by selling fishing access rights to foreign fishing fleets. Some of the increase in exports has been possible through the expansion of fishing fleets in shallow coastal shelf waters along with perverse subsidies contributing to the expansion (Dommen 1999). In many countries the expansion of fishing fleets has been through the growth of domestic fishing vessels, many of them small fishing craft (FAO 1998). This growth has resulted in domestic fleets targeting high-valued fisheries for export leading to overcapitalization, overcapacity, and overexploitation of targeted stocks as well as the less-valued stocks since they are heavily exploited to meet domestic demands for fish. In British Columbia, salmon fishery was export-driven until the stocks were severely overexploited (Rees 2000). This is the case for Ghana where high-valued fish are exported and low-valued fish are consumed locally with most coastal fish stocks overexploited (Atta-Mills *et al.* 2004). In many developed countries perverse subsidies also contribute to overcapitalization, overcapacity, and consequently overexploitation since the government underwrites much of the cost of fishing, which distorts the true cost of fishing (Porter 2001).

Globally there was also an expansion of large distant water fishing vessels from the North fishing in the waters of less-developed countries. The expansion of the global fishing fleet peaked in the early 1980s as EEZs came into effect (OECD 1985). Developed countries with established distant water fleets could not dismantle their fleets without significant economic and social consequences; moreover their domestic waters were often overfished and had no capacity to absorb the distant water fleets (Kurien 1998). Rather than working with developing countries to establish joint ventures and develop fishing industries within the host country, access agreements became the preferred option.

Technological advances in vessel design, engines, winches, fishing gear, and navigation equipment made it possible for fishing vessels to fish anywhere in the world with considerable accuracy in terms of location, depth, and gear (Martinez 1995). Rapid advances in communications, banking, capital flows, and transportation made it much easier to conduct financial transactions in real time, which facilitated trading fish and fish products globally (Arbo and Hersoug 1997). These changes combined with a policy of "fleets to the South and fish to the North" in response to the introduction of EEZs and subsidies have maintained many distant water fleets and in some cases this has led to the problem of overcapacity and overexploitation (Kurien 1998). In Spain, most of the Galician fleet sat idle in 2002 as the European Union (EU) negotiated with Morocco to allow EU fishing vessels, in particular the Spanish, to fish in Moroccan waters (FIS 2002). Moreover, many EU distant water fleets are heavily dependent on subsidies. Porter (1997) noted that in the mid 1990s license fees for European vessels in West Africa ranged from 6 percent to 32 percent of the cost of access. Dommen (1999) found that some countries oversubsidized their fisheries and did little to regulate distant water fishing fleets.

The growth of aquaculture has been due to the expansion of shrimp facilities in the tropics and salmon farming in cool temperate waters (FAO 2002a). These products are in demand and high-valued on the international market. Moreover, the demand cannot be met by the current supply of wild capture fisheries. In fact some fisheries such as salmon in the North Atlantic are considered overexploited and after several years of reduced quotas stocks have still not recovered (Alder et al. 2001). Approximately 80 to 95 percent of farmed shrimp are exported (Yap 2000), while the proportion of exported farmed salmon is much more variable. In Europe salmon is one of the few finfish species that is farmed and much of it is traded within Europe (Smit and Taal 2001). In Chile much of the farmed salmon is exported to Japan and the United States (Phyne and Mansilla 2003).

The introduction of Sanitary and Phytosanitary Standards using HACCP systems in the exporting of fish products can constrain and facilitate export-oriented fisheries. Exports from countries that do not apply or meet the standards can be refused, while countries that use the standards may be able to enter new markets. There is a cost to meeting these standards; often infrastructure needs to be established, staff trained, and inspectors paid. These costs can only be recovered through increasing the value of the product or by increasing the volume of production and contributing to overexploitation. Developed countries have the capital and human resources to establish these standards within their fishing export industries; however, developing countries are severely disadvantaged and meeting such standards can be considered a barrier to trade (World Fish Center 2002).

Foreign exchange

The need to generate foreign exchange to supplement the national budget or to service the national debt is common in many developing countries. The fishing sector can help to reduce this need through selling fishing access rights to foreign fleets or by increasing production and exports of fish and fish products. The value of fishing access rights is high in many developing countries. In Tuvalu it generated 50 percent of the country's government revenue in 2001 (Hunt 2003); similarly in Mauritania it makes up 27 percent of the national budget (UNEP 2002a). However, in some cases the amount paid does not necessarily reflect the real resource rent of the fish that are caught by the foreign fleet. In the WCPO it is estimated that resources rents range between 1 percent and 31 percent of gross revenue depending on the vessel technology (Petersen 2003). In Guinea Bissau resource rents paid by the EU were 0.4 percent of the market value of tuna in 1997 and 7.5 percent of the processed value of all fish in 1996 (Kaczynski and Fluharty 2002).

Developing countries have increased their foreign exchange from fisheries exports. In 2002 net foreign exchange from fish and fish products was approximately $16 billion (FAO 2002b), up from $5.1 billion in 1985 (Delgado and Courbois 1999). Not all of this increase can be attributed to increased prices since they have stagnated for developing countries (Delagado and Courbois 1999); however, volumes have increased (Fig. 2.2). Export volumes increased either by increasing exploitation as discussed above in countries such as Ghana or by reducing the domestic supply in countries such as Senegal (UNEP 2002a).

Increasing foreign exchange is also a consequence of trade liberalization, which can result in changes in employment structures and wealth distribution among other things. In the fisheries sector it has mixed impacts since it can be beneficial by allowing domestic fishers access to lower prices for vessels and fishing technology, improving employment opportunities, and in some cases to new fishing grounds, but it can also be detrimental since it can increase the level of exploitation, reduce prices, and disrupt employment. Norway's shift from a net exporter of whitefish to a net importer in the 1990s resulted in a reduction in employment in some coastal areas of the country as fish processing was concentrated in the few areas of the country receiving the imported fish (Arbo and Hersoug 1997). In Namibia, the fisheries sector, which is export-oriented, has developed as a source of employment since independence in 1991 (Armstrong *et al.* 2005). The expansion of shrimp exports in India and Latin America has been at the expense of locals; the developers and owners of the shrimp farms are from outside of the area and do not reinvest the profits in the local community and clear mangroves which support a number of fisheries (Tobey *et al.* 1998).

However, the development of specialized or "boutique" fisheries such as the live fish trade and aquarium fish trade have often benefited local small-scale fishers. In Indonesia fishers receive from 200 to 2500 percent more for live reef fish compared to dead fish (Erdmann and Pet-Soede 1996). The benefits of the live reef fish trade often only benefits fishers who use destructive fishing practices such as explosives that enable fishers to take more fish per unit time but has significant impacts on reef ecosystems. Fishers who choose not to use destructive fishing techniques suffer economically since the long-term sustainability of their coral reef ecosystem is threatened (Soede-Pet et al. 1999). The global trade in aquarium fish was worth $963 million in 1996 and a significant source of commerce for fishers in Southeast Asia (Tomey 1997; Burke et al. 2002).

Vertical integration

Over the last 20 years there has been considerable consolidation in the fishing industry. Many fishing companies have merged resulting in global companies with various subsidiaries responsible for different segments of the operations. These vertically integrated companies manage the catching, processing, and distributing of fish, or in the case of aquaculture feed and fish production, processing, and marketing within the same parent company. When companies move towards vertically integrating their operations the nature of businesses changes since these are often large conglomerates which are managed away from where the resource is exploited. This change affects employment since the company may shift operations to areas where labor costs are cheaper, or access rights are dismissed by exercising more political influence on governments and local economies by economies of scale, or by taking advantage of scale. In some cases the company may catch the fish in one country, process it another, and distribute it to several other countries.

Several Korean companies catch pollock in Russian waters, process it in China where labor costs are less, and then distribute the product (e.g., surimi) to the United States and Europe (Won 2003). Integrated companies also control a substantial proportion of production. Pescanova, a Spanish conglomerate, controls 20 percent of the world's hake production with a fleet of 140 industrial fishing vessels, seven factories, and 25 000 retail outlets around the world (Decoster 2001). Nutreco is the largest aquaculture producer of salmon and it is one of four multinational companies controlling 80 percent of the world's salmon feed market (Charron 1999).

The economic efficiencies of large vertically integrated companies result in cheaper fish for consumers assuming there is fair competition in the marketplace. In New Zealand, individual transferable quotas along with the concentration of fishing companies enables them to distribute peak harvests over several processing plants throughout the country providing stable employment for locals over the fishing season (McClintock and Taylor 2002). There are others that do not benefit. In Norway the consolidation of the whitefish fisheries into a few companies, centrally located, ultimately resulted in closures of fish processing plants in many coastal areas and fishers having to go further out to sea to fish and to travel further to land their catches (Arbo and Hersoug 1997). Prior to the arrival of the factory trawler in Alaska in the 1970s, small operators took 80 percent of the pollock catch, but by the late 1990s their share of the catch decreased to 30 percent resulting in significant job losses in fishing and processing (St. Clair 1997). In Namibia most revenues from fishing go principally to a few large fishing conglomerates that use Namibian companies to accumulate quotas to obtain a large enough share of the resources so they can operate profitably (FIS 2001). This concentration limits the ability of smaller national companies to enter the fishing sector.

Property rights of local fishers are often lost when large vertically integrated companies enter the fishery especially where transferable quotas are used to control access to the resource. A single owner–operator does not have access to the capital required to compete with larger companies for quotas that are available in the market place. In Bangladesh larger producers based outside of the local area who control access to water resources have marginalized producers and broker exclusive deals with large landholders for access to coastal areas (Alauddin and Hamid 1999). The Bangladeshi situation is exacerbated by large outside producers who employ laborers from outside of the local area.

IMPACTS

Globalization of the fishing sector consumes marine resources, changes the nature of business with flow-on effects to employment and property rights, and also affects how resources are managed. Collectively these changes impact marine ecosystems, societies, and economies in various ways and with different consequences. These impacts are discussed below.

Ecosystems

Marine fish catches are declining globally and globalization has contributed to this decline as countries expand their fishing fleets and fishing grounds to meet the global demand for seafood. Declines in specific fish stocks can be attributed to export-driven fisheries such as coral reef fish associated with the live fish trade (Bryant et al. 1998; Hunt 2003). In western Africa several stocks targeted by distant water fleets are also overexploited (Alder and Sumaila 2004). Foreign fishing vessels targeting bigeye tuna in the WCPO are considered responsible for the current overexploited state of the stocks (Hunt 2003). There has been a significant decline in biomass over the last century in the North Atlantic (Fig. 2.9), where once whitefish caught off the Grand Banks of Canada were traded globally (Christensen et al. 2003). Similar declines in the trophic level of commercial catches have been identified there and in other areas of the world (Pauly and Watson 2003) (Fig. 2.10).

The growing demand for fish has seen a rise in the use of destructive fishing practices as well as an increase in the spatial extent of these practices. Destructive fishing practices such as trawling, blasting, and long-lining impact on marine ecosystems directly through disturbing benthic habitats and trapping bycatch, and indirectly by incidental hooking of seabirds (Brothers et al. 1999). Many demersal fish and shrimp are captured using bottom trawl gear which also has considerable impact on the environment. Tuna and

Figure 2.10 Global trophic level change, 1950–2000. Trophic level refers to the feeding level of an organism. Phytoplankton are given a unitless value of 1 and organisms at higher levels are usually dependent on organisms in preceding trophic levels as a source of food (energy). For image in colour please see plate section. (*Source:* Watson and Pauly 2003.)

Patagonian toothfish are export-oriented fisheries, which use a variety of gear including long-lines, gear which has been implicated in the decline of large seabirds such as petrels and albatrosses (Tasker et al. 2000).

The concentration of fishing effort and fish marketing in a few companies can also result in impacts to fish stocks. In New Zealand individual transferable quotas were introduced into the fishing industry. From 1992 to 2002 larger companies bought quotas from smaller companies and individual fishers so that eight companies control 73 percent of the fish quotas. The snapper and orange roughy fisheries, two of the most valuable fisheries nationally, and where only a few companies control fishing operations, are overfished (Bernier 2002). The Patagonian toothfish fishery is one of the most threatened globally and most valuable fisheries globally and it is also controlled by two companies that landed more than 50 percent of the estimated total catch for toothfish in 2000–01 (Wright 2003).

Coastal ecosystems have been destroyed in countries with well-developed shrimp aquaculture industries. The loss of mangroves in Indonesia, Ecuador, and Thailand is due to the expansion of shrimp ponds into critical mangrove habitats. The ponds usually are productive for a limited time (7–15 years) and then converted to other uses such as industrial or housing estates (Stevenson 1997). Other potential uses such as crab grow-out, fish fingerlings, and polychaetes culturing need further investigation. The conversion of mangroves to other uses can impact on survival of young fish as seen in the Caribbean where mangroves influence the community structure of fish on nearby coral reefs (Mumby et al. 2004).

Economic effects

Governments, industry, and individuals are economically impacted by the globalization of the fishing sector. Countries that export fish directly can benefit from governments increasing their foreign exchange and economic development through the establishment of support industries such as fish processing, vessel maintenance, marketing, and transportation. Not all developing countries can capitalize on export markets, they can be constrained by Sanitary and Phytosanitary agreements if they do not have the facilities and expertise to implement the agreement (Dommen 1999). However, in some countries export-oriented production diverts labor, capital, and other resources away from production for local consumption – increasing unemployment and reducing potential food supply especially in developing countries (Kent 1987).

Exports that are generated by selling fishing access provide governments with cash; however, licenses are less likely to generate the same level of economic development as direct expansion of export activities. In western Africa, the sale of fishing access accounts for 27 percent of the Mauritanian government's national budget (UNEP 2002b). In the western central Pacific, over 50 percent of government revenue in Tuvalu is from licenses granted to foreign fishing vessels (Petersen 2003). The revenues generated from selling fishing access rights make a significant contribution to the budgets of some countries. However, the fees do not directly generate any other significant economic activities.

The trend of concentrating fishing in a few multinational companies that are vertically integrated has seen many of these companies increase their profit margins considerably. In turn the shareholders have increased their economic wealth. In some countries, especially where labor and other operating costs are low, these companies can generate subsequent benefits such as employment and the establishment of support services. Improvements in company profits are often at the cost of smaller companies and owner–operator fishers who cannot compete with large operators.

At the individual level, the fishers that take advantage of the high-priced–low-volume fisheries such as the aquarium fish trade and the live fish trade do well economically. However, access to these economic opportunities is limited for most fishers. Processors who also take advantage of the changes in the fisheries can improve their profits. Many of the fish processors in eastern Canada extended their operations to brokering of imported fish and improved their profits despite the moratorium on cod fishing (McCay and Finlayson 1995). When large corporations dominate the fishery including the processing and marketing of the products, small fishers and support workers are economically disadvantaged as they lose access to the resource and support industries are transferred to countries with lower operating costs. In the 1970s small independent operators and several processing factories landed nearly 80 percent of the Alaskan pollock catch. Within a few years of the arrival of factory trawlers owned by multinational companies that processed the fish in China, less than 30 percent of the catch was landed by small operators and most processing factories had closed by the 1980s (St. Clair 1997). The globalization of the whitefish sector has impacted the economies of many Norwegian communities (Arbo and Hersoug 1997).

Social impacts

Globalization of fisheries can lead to a number of social impacts including unemployment, the displacement of local communities, threats to small-scale fishers, price increases, and risks to food security. Globally, the number of people employed in the fishing sector is declining in developed countries and in some developing countries, while in developing countries where unemployment is high it is increasing or stable (FAO 2002a). A number of drivers including economic and social as described above contribute to the decline. Fisheries that expand to meet increasing demands for exported fish can generate employment opportunities directly through fishing activities and indirectly through the industries that support fishing (e.g., processing, repairs, etc.). In some countries such as Namibia and Senegal expansion of the export fisheries has been positive in terms of employment (Alder and Sumaila 2004). If it leads to declining stocks then employment drops, as seen with the collapse of the cod fishery in Newfoundland. In Ghana, overfishing of inshore stocks by the semi-industrial fleet resulted in many fishers shifting to the artisanal fishery, placing further pressure in already depleted stocks (Atta-Mills *et al.* 2004). Weber (1994) estimated that 100 000 fishers lost their jobs in the early 1990s due to overfishing. However, in Norway the centralizing of whitefish processing in one area of Norway resulted in significant unemployment in another area (Arbo and Hersoug 1997).

The expansion of shrimp aquaculture in several developing countries has displaced local residents, especially where land and sea tenure is unclear or absent. In Bangladesh the number of landless people in the coast increased with the expansion of the aquaculture industry as outsiders used force and coercion to acquire land (Alauddin and Hamid 1999). Similarly small-scale fishers are often displaced in inshore areas as larger vessels make it difficult for smaller vessels or deplete the inshore stocks. In West Africa foreign vessels and large domestic vessels often disregard fishing regulations and either fish in areas restricted to small-scale fishers or use illegal gear (Alder and Sumaila 2004; Atta-Mills *et al.* 2004). Similar problems exist in Asia, including Indonesia, India, Thailand, and Malaysia, where the growth of trawling and expanding industrial fleets compete with small-scale and artisanal fishers for resources, often in the same fishing grounds (Sharma 2003).

Over 950 million people, many of them are in developing countries, depend on fish as their primary source of protein (Porter 2001). Globalization puts this supply at risk through lack of fish to catch, as discussed above, or prices increasing beyond the financial reach of people. In Senegal there are protein deficits in the countryside because the supply of cheap fish for local consumption was reduced as fishers shifted to catching priority export species, and the domestic market price more than tripled from 1993 to 2002 (UNEP 2002a). While some argue that the loss of fish for domestic consumption can be offset with imports, this is rarely the case, as seen in Asia and West Africa (Kurien 1998; Alder and Sumaila 2004). The shift from milkfish that were consumed on the local market to cultured shrimp for export reduced the supply of cheap protein for many people in the Philippines (Primavera 1991). Converting rice fields to shrimp ponds reduced the quantity of rice on the domestic market in Thailand (Flaherty et al. 1999).

Domestic supplies have declined in some exporting countries. Many countries that export fish or sell fishing access rights are also Low-Income Food Deficit Countries (LIFDC). For example, in West Africa, Morocco and Senegal are classed as LIFDC and yet they host large distant water fleets which catch quantities of fish much of which is exported or shipped directly to Europe. Ecuador, China, India, Indonesia, and the Philippines do not host large distant water fleets but they are LIFDC and major exporters of high-valued fish products such as shrimp and demersal fish (FAO 2002a).

WHO BENEFITS?

Large multinational corporations, some governments, communities where operating costs are low, and consumers benefit from globalizing the fisheries sector. Marine ecosystems, many communities, and artisanal fishers, especially in developing countries are not benefiting from globalization of fisheries. Globalization of fisheries should have widespread benefits. However, the key to success is that the trade in goods and services is fair and equitable, and that the benefits from globalization are widely distributed throughout the nations involved in the trade of fish products.

How to better distribute the benefits in fisheries is the challenge for industry, economists, policy-makers, and politicians. The list of ways to better distribute the benefits of fisheries globalization is far too long to be discussed here. However, the following are considered the priority actions with considerable potential to affect change in the fisheries sector.

Clearly, as stated by many researchers and managers, the management of fish stocks based on ecosystem management approach and encompassing the precautionary principle is needed if stocks are to recover and be sustainably fished in the future. Perverse subsidies need to be eliminated so that trade is placed in a more equitable context allowing developing countries to compete on the global market. The environmental costs of fishing need to be internalized within the industry and consumers so that the financial resources to manage fisheries, including surveillance and enforcement, are available to managers. Members of the World Trade Organization need to ensure that they incorporate environmental protection mechanism so that trade measures do not negatively impact on marine ecosystems.

Acknowledgment

This work was initiated and funded by the Pew Charitable Trusts of Philadelphia.

References

Alauddin, M. and Hamid, M. A. 1999. Shrimp culture in Bangladesh with emphasis on social and economic aspects. In *Towards Sustainable Shrimp Culture in Thailand and the Region*, ed. P. Smith. Canberra, ACT: Australian Centre for International Agricultural Research, pp. 53–62.

Alder, J. and Sumaila, U. R. 2004. Western Africa: a fish basket of Europe past and present. *Journal of Environment and Development* **13**: 156–178.

Alder, J., Lugten, G., and Ferris, B. 2001. Compliance with international fisheries instruments in the North Atlantic. In *Fisheries Impacts on North Atlantic Ecosystems: Evaluations and Policy Exploration*, Research Report No. 9(5), eds. T. Pitcher, U. R. Sumaila, and D. Pauly. Vancouver, BC: Fisheries Centre, pp. 55–80.

Arbo, P. and Hersoug, B. 1997. The globalization of the fishing industry and the case of Finnmark. *Marine Policy* **21**: 121–142.

Armstrong, C. W., Sumaila, U. R., Erastus, A., and Msiska, O. 2005. Benefits and costs of the Namibianisation policy. In *Namibia's Fisheries: Ecological, Economic and Social Aspects*, eds. U. R. Sumaila, S. I Steinshamn, M. D. Skogen, and D. Boyer. Amsterdam, Netherlands: Eburon, pp. 203–214.

Atta-Mills, J., Alder, J., and Sumaila, U. R. 2004. The unmaking of a regional fishing nation: the case of Ghana and West Africa. *Natural Resources Forum* **28**: 13–21.

Bernier, D. 2002. Canada's fisheries policy: beware fisheries privatization. *Coastal Community News Magazine* **2(3)**. Available online at www.coastalcommunities.ns.ca/v2_i3.html

Brothers, N. P., Cooper, J., and Lokkeborg, S. 1999. *The Incidental Catch of Seabirds by Longline Fisheries: Worldwide Review and Technical Guidelines for Mitigation*, FAO Fisheries Circular No. 937. Rome: Food and Agriculture Organization of the United Nations.

Bryant, D., Burke, L., McManus, J., and Spalding, M. 1998. *Reefs at Risk*. Washington, DC: World Resources Institute.

Burke, L., Selig, E., and Spalding, M. 2002. *Reefs at Risk in Southeast Asia*. Washington, DC: World Resources Institute.

Charron, B. 1999. The salmon feed industry: hard ball game or oligopoly? *Intrafish Report*. Available online at www.intrafish.com/intrafish-analysis/feed_1999_43_eng/

Christensen, V., Guenette, S., Heymans, J. J., et al. 2003. Hundred year decline of North Atlantic predatory fishes. *Fish and Fisheries* **4**: 1-24.

Decoster, J. (ed.) 2001. Challenges facing artisanal fishery in the 21st century. Available online at www.alliance21.org/en/proposals/finals/final_pche_en.pdf

Deere, C. 1999. *Net Gains: Linking Fisheries Management, International Trade and Sustainable Development*. Washington, DC: IUCN.

Delgado, C. and Courbois, C. 1999. Changing fish trade and demand patterns in developing countries and their significance for policy research. In *Fisheries Policy Research in Developing Countries: Issues, Priorities and Needs*, eds. M. Ahmed, C. Delgado, S. Sverdrup-Jensen, and R. A. V. Santos. Washington, DC: International Food Policy Research Institute, pp. 21-33.

Delgado, C., Wada, M., Rosegrant, M. W., Jeijeer, S., and Ahmed, M. 2003. *First to 2020: Supply and Demand in Changing Global Markets*. Washington, DC: International Food Policy Research Institute.

Dommen, C. 1999. *Fish for Thought: Fisheries, International Trade and Sustainable Development*. Geneva, Switzerland: International Centre for Trade and Sustainable Development.

Dunning, J. H. 1997. The advent of alliance capitalism. In *The New Globalism and Developing Countries*, eds. J. H. Dunning and K. A. Hamdani. Tokyo: United Nations University Press. pp. 12-49.

Erdmann, M. V. E. and Pet-Soede C. 1996. How fresh is too fresh? The live reef food trade in eastern Indonesia. *Naga, The ICLARM Quarterly* **19**: 4-8.

FAO (Food and Agriculture Organization). 1998. *State of Fisheries and Aquaculture*. Rome: Food and Agriculture Organization of the United Nations.

FAO. 2000. *FISHSTAT Plus: Universal Software for Fishery Statistical Time Series*, Version 2.3. Fisheries Department, Fishery Information, Data and Statistics Unit. Available online at www.fao.org/fi/statist/FISOFT/FISHPLUS.asp

FAO. 2001. *Multilateral Trade Negotiations on Agriculture: A Resource*, Manual I, Module 1. Rome: Food and Agriculture Organization of the United Nations.

FAO. 2002a. *State of Fisheries and Aquaculture*. Rome: Food and Agriculture Organization of the United Nations.

FAO. 2002b. *Food Outlook*. Rome: Food and Agriculture Organization of the United Nations.

FIS (Fish Information Service). 2001. Namibia: Research Unit recommends changes in fishing policy. *Fish Information and Services*. Available online at www.fis.com/fis/worldnews

FIS. 2002. Fischler questions third country agreement. *Fish Information and Services*. Available online at www.fis.com/fis/worldnews

FIS. 2003. Satellite control will lead to increase in fishery for human consumption. *Fish Information and Services*. Available online at www.fis.com/fis/worldnews

Flaherty, M., Vandergeest, P., and Miller, P. 1999. Rice paddy or shrimp pond: tough decisions in rural Thailand. *World Development* **27**: 2045-2060.

Gillett, R., McCoy, M., Rodwell, L., and Tamate, J. 2001. *Tuna: A Key Economic Resource in the Pacific*, a report prepared for the Asian Development Bank and the Forum Fisheries Agency. Manila, Philippines: Asian Development Bank.

Hampton, J. and Williams, P. 2003. *The Western and Central Pacific Tuna Fishery: 2001 Overview and Status of Stocks*. Noumea, New Caledonia: South Pacific Commission.

Hunt, C. 2003. Economic globalization impacts on Pacific marine resources. *Marine Policy* **27**: 79-85.

Johannes, R. E., Freeman, M. M. R., and Hamilton, R. J. 2000. Ignore fishers' knowledge and miss the boat. *Fish and Fisheries* **1**: 257-271.

Kaczynski, V. M. and Fluharty, D. L. 2002. European policies in West Africa: who benefits from fisheries agreements? *Marine Policy* **26**: 75-93.

Kent, G. 1983. The pattern of fish trade. *ICLARM Newsletter* **6**(2): 12-13.

Kent, G. 1987. *Fish, Food and Hunger: The Potential of Fisheries for Alleviating Malnutrition*. London: Westview Press.

Kurien, J. 1998. *Small-Scale Fisheries in the Context of Globalization*, CDS Working Paper No. 289. Trivandrum, Kerala, India: Centre for Development Studies.

Kurien, J. 2004. *Responsible Fish Trade and Food Security*. Rome: Food and Agriculture Organization of the United Nations.

Martinez, A. 1995. Fishing out aquatic diversity. *Seedling* (July 1995). Available online at www.grain.org/publications/jul951-en.cfm

McCay, B. J. and Finlayson, A. C. 1995. The political ecology of crisis and institutional change: the case of the northern cod. *Annual Meeting of the American Anthropological Association*, Washington DC, November 15-19, 1995. *Arctic Circle*. Available online at www.lib.uconn.edu/ArcticCircle

McClintock, W. L. and Taylor, C. 2002. *Business Ownership in Natural Resource Dependent Industries*, Working Paper No. 31. Christchurch, New Zealand: Taylor Baines.

Mumby, P. J., Edwards, A. J., Arias-González, J. E., et al. 2004. Mangroves enhance the biomass of coral reef fish communities in the Caribbean. *Nature* **427**: 533-536.

New, M. and Wijkström, U. 2002. *Use of Fishmeal and Fish Oil in Aquafeed: Further Thoughts on the Fishmeal Trap*. Rome: Food and Agriculture Organization of the United Nations.

Nierop, T. 1994. *Systems and Regions in Global Politics: An Empirical Study of Diplomacy, International Organization and Trade, 1950-1991*. Chichester, UK: John Wiley.

OECD (Organisation for Economic Co-operation and Development). 1985. *Problems of Trade in Fishery Products*. Paris: OECD.

OECD. 1989. *Fisheries Issues: Trade and Access to Resources*. Paris: OECD.

Oman, C. 1999. Globalization, regionalization and inequality. In *Inequality, Globalization, and World Politics*, eds. A. Hurrell and H. Woods. Oxford, UK: Oxford University Press, pp. 36-65.

Pauly, D. and Watson, R. 2003. Running out of fish. *Scientific American* **289**: 35-39.

Petersen, E. 2003. The catch in trading fishing access for foreign aid. *Marine Policy* **27**: 219-228.

Pet-Soede, C., Cesar, H. J. S., and Pet, J. S. 1999. An economic analysis of blast fishing on Indonesian coral reefs. *Environmental Conservation* **26**: 83-93.

Phyne, J. and Mansilla, J. 2003. Foraging linkages in the commodity chain: the case of the Chilean salmon farming industry, 1987-2001. *Sociologia Ruralis* **43**: 108-127.

Porter, G. 1997. Euro-African fishing agreements: subsidizing overfishing in African waters. In *Subsidies and Depletion of World Fisheries: Case Studies*, ed. S. Burns. Washington, DC: World Wildlife Fund, pp. 7-33.

Porter, G. 2001. *Fisheries Subsidies and Overfishing: Towards a Structured Discussion*. Geneva, Switzerland: United Nations Environment Program.

Primavera, J. H. 1991. Intensive prawn farming in the Philippines: ecological, social and economic implications. *Ambio* **20**: 28-33.

Rees, W. E. 2000. The dark side of the force (of globalization). Available online at www.wwdemocracy.nildram.co.uk/democracy_today/rees.htm

Rosenberry, B. (ed.) 1991. *World Shrimp Farming 1991*. San Diego, CA: Aquaculture Digest.

Ruckes, E. 2000. Evolution of the international regulatory framework governing international trade in fishery products. Paper presented at the International Institute of Fisheries Economics and Trade 2000. Available online at htt://oregonstate.edu/dept/IIFET/2000/abstracts/ruckes.html

Rugman, A. and Moore, K. 2001. The myths of globalization. *Ivey Business Journal* Sep/Oct, 64–69.

Sharma, C. 2003. The impact of fisheries development and globalization processes on women of fishing communities in the Asian region. *Asian Pacific Research Network* **8**(2): 1–5.

Smit, M. and Taal, C. 2001. *EU Market Survey 2001: Fishery Products*. Rotterdam, Netherlands: Centre for the Promotion of Imports from Developing Countries.

Stevenson, N. 1997. Disused shrimp ponds: options for redevelopment of mangrove. *Coastal Management* **25**: 423–435.

St. Clair, J. 1997. Fish business. *In These Times* **21**(14): 14.

Tasker, M. L., Camphuysen, C. J., Cooper, J., et al. 2000. The impacts of fishing on marine birds. *Journal of Marine Science* **57**: 531–547.

Thorpe, A. and Bennett, E. 2001. Globalization and the sustainability of world fisheries: a view from Latin America. *Marine Resource Economics* **16**: 143–164.

Tobey, J., Clay, J., and Vergne, P. 1998. *The Economic, Environmental and Social Impacts of Shrimp Farming in Latin America*. Narragansett Bay, RI: Coastal Resources Center, University of Rhode Island.

Tomey, W. A. 1997. Review of developments in the world ornamental fish trade: update, trends and future prospects. In *Sustainable Aquaculture: Proceedings of the INFOFISH-AQUATECH '96 International Conference on Aquaculture*, eds. K. P. P. Namibar and T. Singh.

UNEP (United Nations Environment Program). 2002a. *Integrated Assessment of Trade Liberalization and Trade-Related Policies: A Country Study on the Fisheries Sector in Senegal*. Geneva, Switzerland: United Nations Environment Program.

UNEP. 2002b. *Environmental Impact of Trade Liberalization and Trade-Linked Measures in the Fisheries Sector*. Nouadhibou, Mauritania: National Oceanographic and Fisheries Research Centre.

Waters, M. 1995. *Globalization*. London: Routledge.

Watson, R. A. and Pauly, D. 2001. Systematic distortions in world fisheries catch trends. *Nature* **414**: 534–536.

Watson, R. A. and Pauly, D. 2003. Counting the last fish. *Scientific American* **289**: 42–47.

Watson, R. A., Kitchingman, A., Gelchu, A., and Pauly, D. 2004. Mapping global fisheries: sharpening our focus. *Fish and Fisheries* **5**: 168–177.

Weber, P. 1994. *Net Loss: Fish, Jobs and the Marine Environment*. Washington, DC: Worldwatch.

Won, K. S. 2003. *Republic of Korea, Fishery Products Annual 2003*, GAIN Report KS3056. Washington, DC: U.S. Department of Agriculture Foreign Agricultural Service.

World Fish Center. 2002. *Fish: An Issue for Everyone*. Penang, Malaysia: World Fish Center.

Worldfish Report. 2004. Ghana cannery diversifies into fishmeal. *Worldfish Report* August 25, 2004. Available online at www.agra-net.com

Wright, L. 2003. Illegal fishing gangs threat. *The Advertiser* (Adelaide, SA) November 30, 2003.

Yap, W. G. 2000. Shrimp culture: a global overview. *Bay of Bengal News* **2**(16): 5–12.

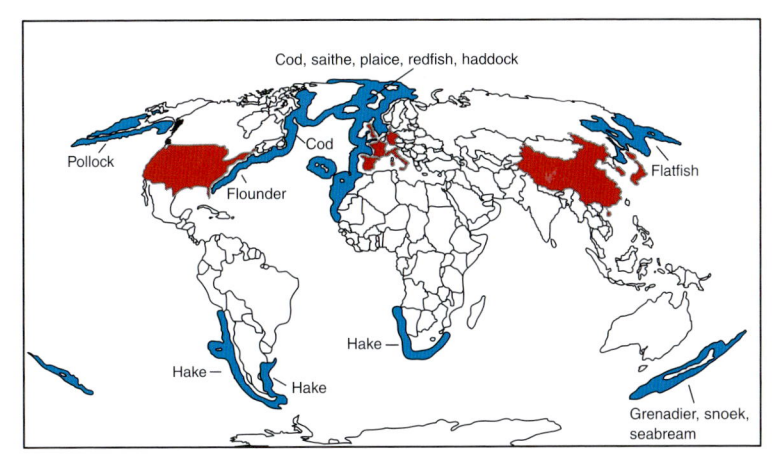

Figure 2.5 Major importing (red) countries and the EEZ (blue) where the majority of exports were sourced for demersal fish 1976–2000.

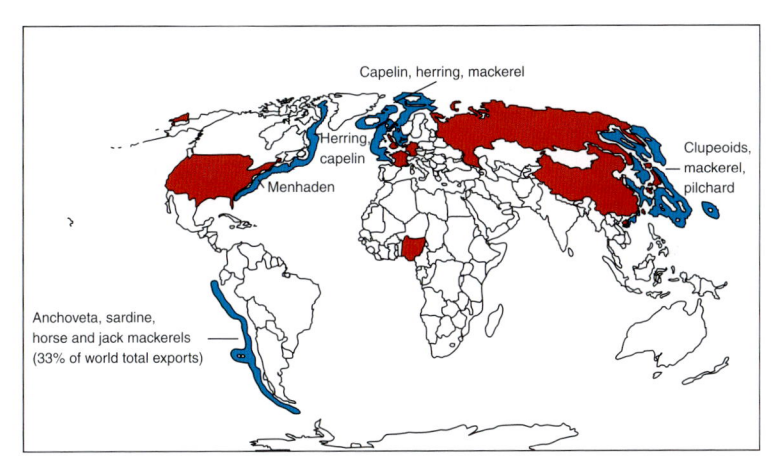

Figure 2.6 Major importing (red) countries and the EEZ (blue) where the majority of exports were sourced for small pelagic fish 1976–2000.

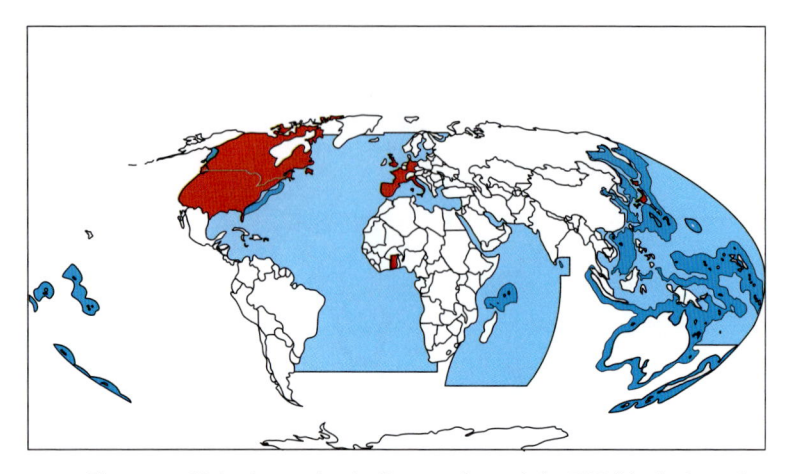

Figure 2.7 Major importing (red) countries and the EEZ (blue) where the majority of exports were sourced for large pelagic fish 1976–2000.

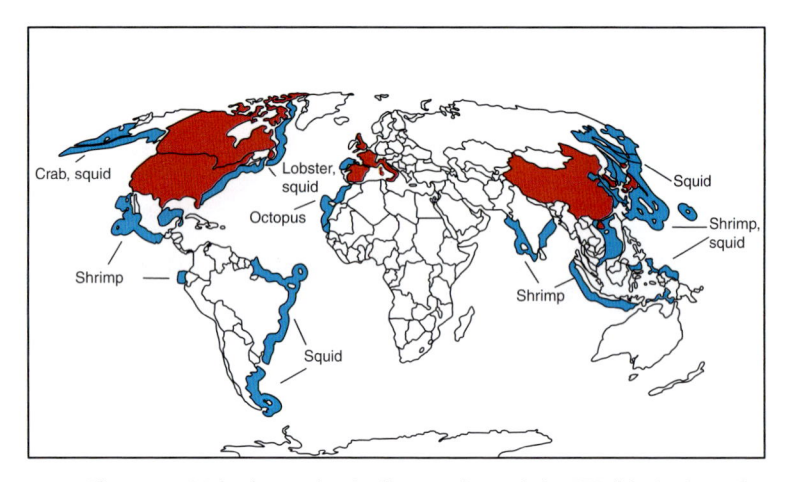

Figure 2.8 Major importing (red) countries and the EEZ (blue) where the majority of exports were sourced for crustaceans and cephalopods 1976–2000.

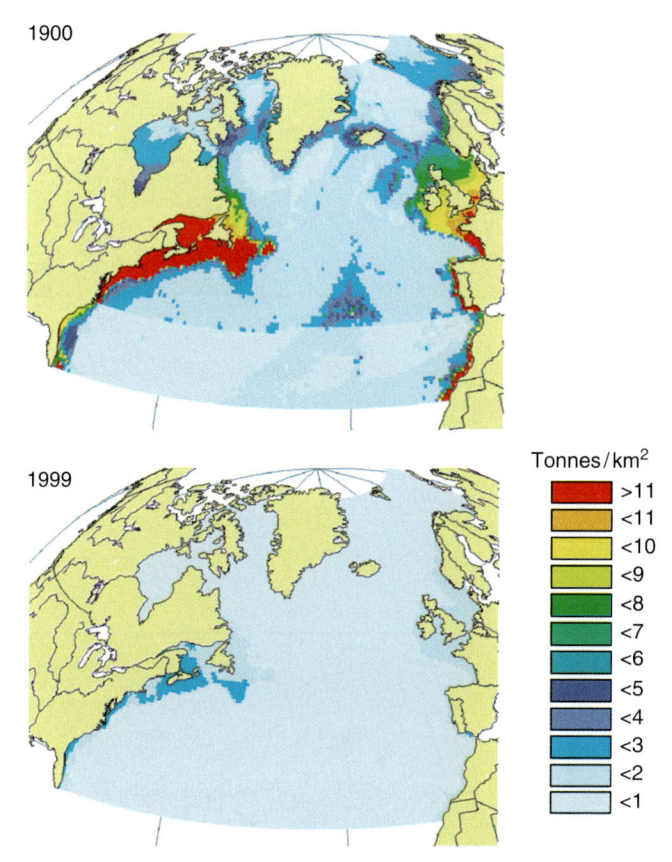

Figure 2.9 Loss of marine biomass in the North Atlantic 1900 to 1999. (*Source:* Christensen *et al.* 2003.)

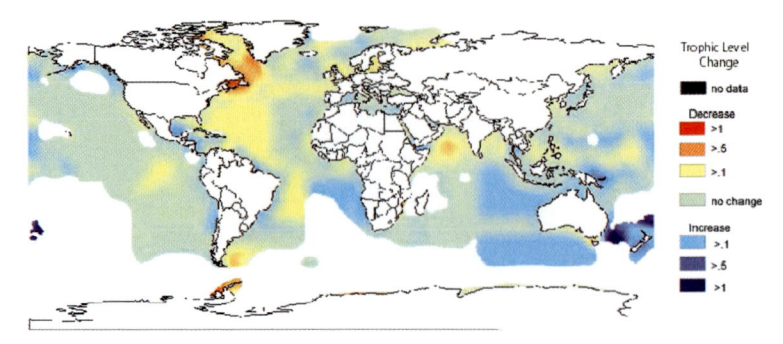

Figure 2.10 Global trophic level change, 1950–2000. Trophic level refers to the feeding level of an organism. Phytoplankton are given a unitless value of 1 and organisms at higher levels are usually dependent on organisms in preceding trophic levels as a source of food (energy). (*Source:* Watson and Pauly 2003.)

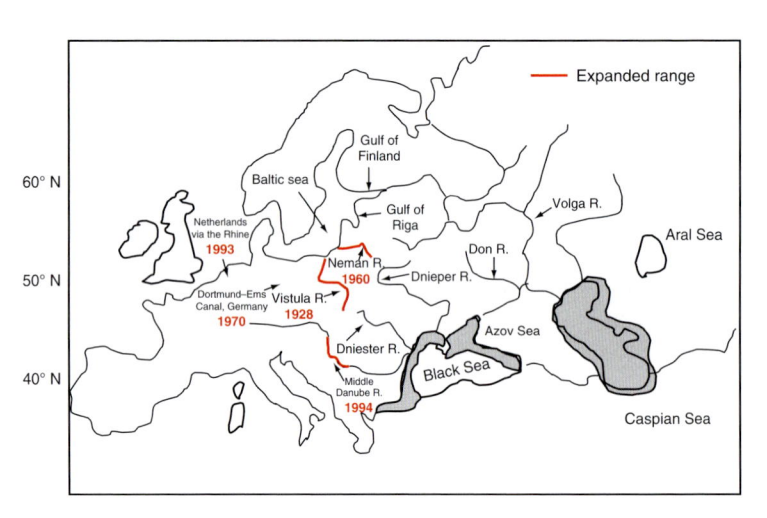

Figure 6.8A Generalized distribution of *Echinogammarus ischnus* in Eurasia.

JESSICA SEARES, KATHERINE SMITH, ELIZABETH ANDERSON,
AND CATHERINE PRINGLE

3

Effects of globalization on freshwater systems and strategies for conservation

INTRODUCTION

Though the effects of globalization on marine systems (i.e., commercial fisheries) have received relatively more attention, freshwater ecosystems have been and continue to be profoundly affected by increasing human populations and changing global processes. Human activities now utilize more than half of the available surface fresh water (Postel 1999), and water consumption is doubling every 20 years (WWI 2003). Currently, four out of every ten persons live in river basins that experience water scarcity and it is estimated that by 2025, at least 50 percent of the world's population will face water scarcity (WRI 2000). The ecological effects of this demand on freshwater systems are already apparent. Biodiversity of freshwater ecosystems has declined by 50 percent globally over the 30-year period from 1970 to 1999 and current extinction rates of many freshwater taxa are more than 1000 times the normal "background" rate (Master *et al.* 1998). The Organisation for Economic Co-operation and Development (OECD) has recognized that globalization, specifically the liberalization of trade and the promotion of markets and investment structures, is likely to have a large but uncertain impact on water resources (OECD 1998). These conditions illustrate the need for continued analysis of how globalization and associated policies drive environmental change in freshwater ecosystems.

"Globalization" is a nebulous concept, with different meanings for different people. Here, we limit our definition to the post-World War II era of transnational legal frameworks designed with the intention of reducing trade barriers between nations (International Forum on Globalization 2002) and the era of multinational corporate expansion (Korten 2001). This definition also includes the era of the large-scale development projects financed by the World Bank and the

Globalization: Effects on Fisheries Resources, ed. William W. Taylor, Michael G. Schechter, and Lois G. Wolfson. Published by Cambridge University Press. © Cambridge University Press 2007.

International Monetary Fund (IMF) (Escobar 1995; Parker 2002). We include in our definition the spread of the ideology of trade liberalization in the latter half of the twentieth century, as well as the push for privatization of resource management and market liberalization (Parker 2002).

THE ENVIRONMENTALIST RESPONSE TO GLOBALIZATION

In 1944, the leading Western economists met at Bretton Woods to plan a global economy that, through its openness, would be more likely to prevent an outbreak of international warfare (Karliner 1997; Sassen 1998; Korten 2001; International Forum on Globalization 2002). With the goal of reducing individual nations' abilities to impose trade barriers on one another, the Bretton Woods attendees laid the foundation for the World Bank and the IMF (International Forum on Globalization 2002). The General Agreements on Tariffs and Trade (GATT) was created subsequently (Korten 2001) to promote non-discrimination in trade among member nations. In its first inception, the signatory nations of the GATT treaty numbered only 23. By 1995, after several rounds of trade negotiations, most countries of the world were members. In 1995, the 135 member nations of the GATT created an institutional, juridical body called the World Trade Organization (WTO) to adjudicate trade disputes. Unlike its predecessor, the WTO had legally binding authority to resolve trade disputes among member nations. Notably, many scholars consider this a crucial turning point in the era of globalization because of the WTO's potential ability to trump national sovereignty (Kiely and Marfleet 1998; Sassen 1998; Joseph 2001).

Despite environmentalist fears of the WTO's potential to undermine national environmental policy and decision-making capacity, by 2001 and 2002, the relevance of the WTO as a global legislative body had been called into question because of the failure of the Seattle round of trade talks in 1999, and, later, the Cancún rounds of trade negotiations in 2003 (Thornton 2003). Conflicts over agricultural subsidies were largely at the heart of the failure of the WTO delegates to achieve consensus at the Cancún meetings (Stokes 2003). However, later cases, in which the WTO ruled against the U.S. subsidies of the steel and cotton industries (Becker 2004), as well as the ability of members to reach a consensus on the Agreement on Agriculture in Geneva in 2004, highlight the continued relevance of this international body, as well as its potential to be responsive to the economic concerns of developing countries.

Since their inception, the North American Free Trade Agreement (NAFTA), the WTO, the IMF, and the World Bank have met with opposition on various fronts. This resistance has been largely grounded in fear of non-democratic, supranational organizations with the potential to push through development projects at great cost to the environment (the World Bank and the IMF) and undo hard-won national and international environmental protections (NAFTA and WTO) such as the U.S. Clean Air Act, the Endangered Species Act, and the Convention on International Trade in Endangered Species (CITES) (Gould et al. 2004). For example, Chapter 11 of NAFTA allows companies to sue for lost profits if they are prevented from locating in a nation because of prohibitive environmental regulations. A California-based company, Sun Belt, has sued the government of Canada under NAFTA, stating that British Columbia's ban on water exports forced the company to forgo profits of $10 billion (Barlow 2002). In response to these and other worst-case scenarios, the environmentalist movement has found a strong foothold in responding to globalization (Gould et al. 2004).

CASE STUDIES

In this chapter, we examine the effects or potential effects of globalization on freshwater resources via three case studies: global expansion of export agriculture, hydropower development in tropical countries, and the privatization and corporate ownership of freshwater resources. To illustrate potential pathways to freshwater resource protection, we include a general history of the environmentalist response to globalization as well as the local environmentalist response to each case study. Other effects of globalization on freshwater systems not covered via these case studies include exotic species transfer, water security, and water scarcity issues (Postel 1999; WWI 2003). An awareness of the global influences and drivers of these environmental changes is essential to the conservation of freshwater systems, and although we cannot cover all topics related to globalization in this chapter, we encourage future scholarship on the overall effects of globalization on freshwater systems.

Case study I: Globalized agriculture – the potential effects of Brazil's case against U.S. cotton subsidies on freshwater ecosystems in developing countries

The WTO continues to provide a stage for conflicts between colonial and post-colonial countries. In April 2004, the WTO ruled against subsidies paid by the U.S. government to its cotton growers. This complaint, based on the $19 billion per year in subsidies paid by the U.S. government to cotton farmers, was brought to the WTO by Brazil, which was joined in the case by third-party countries Argentina, Australia, Benin, Canada, Chad, China, the European Community, India, New Zealand, Pakistan, Paraguay, Taiwan, and Venezuela (Becker 2004). Brazil's ability to create solidarity with developing countries with agriculturally based economies was integral to its successful WTO case against subsidized cotton. Brazil argued that subsidies paid to American cotton farmers have led to overproduction on the global market, undermining the livelihoods of cotton farmers in developing countries. This issue typifies the complaint that developing countries are being unfairly forced into competition with either wealthy multinational corporations or subsidized farms in the North.

Brazil and other developing countries have stated that by preventing cotton dumping in poor nations and allowing them to compete with American exports globally, the removal of subsidies paid to U.S. farmers would help the economies of poor countries with many subsistence farmers. If the United States concedes to the WTO's ruling, reduced impoverishment is expected for small-scale farmers in export-dependent nations. From an environmental perspective, however, the ramifications of eliminating these types of subsidies are unclear. Most export crops are typically highly dependent on pesticides and fertilizers for production (Pesticide Action Network 2001; Stokes 2003). Thus, the removal of subsidies, which distort prices and result in the types of advantages that the WTO was established to prevent, may shift negative environmental consequences of cotton farming from the United States to countries that this decision will favor. Many of the countries that brought this complaint to the WTO have less stringent legal frameworks for environmental protections, and as a result, the environmental impacts of increasing agricultural pressure on freshwater systems may be greater in these countries than in those with stronger regulatory frameworks for protection of natural resources such as the United States.

The negative effects of commercial agricultural expansion on temperate freshwater systems have been well documented. These effects include bank destabilization when riparian zones are cleared for cultivation, eutrophication due to excessive runoff, ecosystem and trophic disruption due to toxin exposure (Barrett 1968; Bellaire and Dubois 1997; Mortensen et al. 1998; McDonald et al. 1999; Lemly et al. 2000), and destabilization of food chains when non-native species are cultivated or introduced into river systems (Tilman 1999; Aparicio *et al.* 2000; Dudgeon 2002). World agriculture consumes approximately 70 percent of the fresh water withdrawn per year (UNESCO 2001) and related water diversions further threaten aquatic ecosystems by increasing inter-basin water transfers that facilitate exotic species spread, stream dewatering, dam construction, and habitat fragmentation (Pringle 2000, 2001). In addition, contamination of fresh waters by irrigation drainage waters negatively affects the biointegrity of freshwater ecosystems (see summary provided by Lemly *et al.* 2000 and Pringle 2003). The effects of nitrogen and phosphorus runoff on freshwater systems can be severe (Gleick 1993). This type of production may threaten the long-term sustainability of both natural and agricultural systems, especially in tropical regions (Vandermeer *et al.* 1995; Tilman 1999; Altieri 2002; Seares 2004).

Cotton production provides no exception to the environmental costs of monocultural production. The genetic uniformity of cotton has left it extremely vulnerable to pests, so growers must apply insecticides, herbicides, miticides, and defoliants prior to harvest, in some extreme cases as many as 30 times per season. Worldwide, 25 percent of all insecticides and 10 percent of all pesticides are applied to cotton fields (Kimbrell 2002). In addition to pesticides, industrial monoculture techniques for cotton typically involve heavy applications of chemical fertilizers. These synthetic fertilizers contaminate drinking wells and pose long-term threats to both farmland productivity and wildlife. Fertilizers are also linked to amphibian declines (Rouse *et al.* 1999), eutrophication and hypoxia (Alexander *et al.* 2000), changes in aquatic food webs, and declines in coastal benthic communities and fisheries (Diaz and Rosenberg 1995).

Despite relatively strong environmental regulation in the United States, farmers in the main cotton-growing states used more than 75 million pounds of pesticides in 2000 (Kimbrell 2002). In California, the only state where full pesticide use reporting is required, approximately 14.5 pounds of pesticides were applied per acre. Furthermore, over 200 chemicals are used on California cotton, including the highly toxic cholinesterase inhibitor aldicarb, paraquat, and chlorpyrifos. Aldicarb, which can cause respiratory failure and death in birds and mammals, is a known groundwater contaminant (Pesticide Action Network 2003). Despite these dangers, California cotton growers applied a total of 250 000 pounds of aldicarb annually between 1999 and 2003 (Pesticide Action Network 2003).

Pesticides may present even larger environmental problems in developing countries because of their often weaker environmental controls and regulations. For example, toxaphene, once a heavily used pesticide in southeastern U.S. cotton farming, was banned in the United States after the World Health Organization (WHO) deemed it a major hazard for many aquatic and terrestrial species (WHO 1984:66). Despite this finding, it was commonly used to protect cotton crops in Nicaragua until 1993. At that time, it was found in Nicaraguan lakes (Calero et al. 1993), well water (Appel 1991:143), and human milk (Boer and Wester 1993). Despite these concerns and its eventual ban, in a 1996 study, toxaphene was found in high concentrations in agricultural soils in Nicaragua (Carvalho et al. 2003). Leaching from soils has contaminated local rivers, and subsequent water-mediated transport has resulted in toxaphene contamination of coastal lagoons and fisheries (Carvalho et al. 2003). Protection of local wetlands and mangroves could reduce the effects of toxic agricultural discharge reaching aquatic systems, but aquacultural and development pressures makes such conservation actions difficult.

Models of economic development based on increasing foreign trade in agricultural goods, particularly cash crops, have in many cases brought tremendous financial benefits to small-scale farmers in developing countries. However, because of the resource concentration of monocultural systems, cash crops tend to be dependent upon vast quantities of pesticides and fertilizers, which pose great risks to ecosystems, farmers, and consumers. Hence, environmental costs are often associated with these types of economic gains (Seares 2004). Thus, the cotton case won by Brazil and other nations at the WTO represents an economic boon to small-scale farmers seeking a market for their goods. However, the environmental costs associated with export production may, along with increasing profits, be shifted to this region as well.

Ultimately, the consumer choices that drive the orientation of regional, national, and international economies can play a positive role in affecting freshwater systems. For example, the recent growth in markets for both organically produced and fair trade goods such as cotton, coffee, and bananas (Moskin 2004) offer hope that globalization of trade may be able to positively affect agricultural production decisions. Selection for these types of commodities may be one way in which global trade and consumer demand may positively affect local environments. Efforts that reorient market production to local and regional rather than global outlets would also theoretically make it easier to achieve local protection of freshwater environmental resources.

Case study II: Dams and development – the expanding role of private, foreign companies in the transformation of tropical rivers

The past two decades have seen a global trend toward the transfer of electricity generation responsibilities from government institutions to private companies (Gleick et al. 2002). New legislation and economic incentives now encourage private companies to generate electricity in more than 70 countries worldwide (Izaguirre 2000; Raphals 2001). In developing countries, these electricity sector reforms have been introduced in response to a lack of domestic capital and rising demands for electricity (Rogers 1991; Dunkerley 1995), and with trade liberalization, international electricity companies have realized the potential of the developing world as an energy market. Thus, the rise in demand for private financing of electricity generation in developing countries has coincided with an increase in the supply of technologies and capital from abroad. Although some economists suggest that electricity sector reforms will increase the economic welfare of developing countries (Ingco 1996; Coes 1998), restructuring may also have substantial environmental impacts.

In tropical regions rich in freshwater resources, electricity sector reform has influenced hydropower development. For example, for countries along the Central American isthmus (Belize, Guatemala, El Salvador, Honduras, Nicaragua, Costa Rica, and Panama), high relief and wet climate translate to considerable hydropower potential; this potential is rapidly being tapped by local and international companies following the passage of legislation during the 1990s that partially or totally privatized electricity generation in all Central American countries (ECLAC 1996; CEPAL 2005). In Costa Rica, nearly 30 private hydropower plants began operation between 1990 and 2000, with most development concentrated on watersheds draining the country's northern Caribbean slope (Anderson 2002; Alvarez 2005; Duran 2005). In 2004, private generators provided approximately 10 percent of Costa Rica's electricity. Additional examples of new private hydropower developments in Central America include the Pasabien and Rio Hondo II hydropower projects, recently constructed by Hydrowest International in the Zacapa region of Guatemala (Stone and Manrique 2002), and the Esti hydroelectric facility in Panama, developed by AES Corporation. Furthermore, as of 2005, numerous dams are either being studied or under construction throughout the region, many by private companies. A compilation of proposed hydropower projects in Central

America prepared by the Conservation Strategy Fund, which documents projects in various stages from "investment opportunity" to "feasibility," counts approximately 400 potential new hydropower projects for the Mesoamerican region (Burgues Arrea 2005).

Small hydropower is likely to be a beneficiary of electricity sector reform in tropical developing countries (Majot 1997). Here we define small hydropower as dams less than 15 m high, with reservoir capacities less than $3 \times 10^3 \, m^3$ (WCD 2000). Small dams may be easier for foreign, private companies to manage than large hydropower projects that combine many stakeholders across many levels of society. In addition, investment in a small dam presents a much lower financial risk for a private company than investment in a large dam. Several examples from developing countries worldwide illustrate the link between small hydropower and electricity privatization (Majot 1997). In Costa Rica, nearly all of the recently constructed private hydropower plants are small dams (Anderson 2002). In Sri Lanka, approximately 30 small hydropower projects are either in operation or planned for the country's rivers (Ljung 2001). Growth of the private electricity industry in Zimbabwe is expected to result in a significant increase in the number of dams as the country looks toward developing chains of small hydropower plants as a means to provide rural electrification (Mungwena 2002). The governments of Honduras and Nicaragua are encouraging foreign, private investment in renewable electricity sources, especially small hydropower plants. In Nicaragua, recently proposed legislation that provides incentives for private companies could result in construction of small hydropower dams throughout the country.

The environmental effects of private hydropower dam developments to tropical freshwater ecosystems must be as considered part of the globalization and development debate. The type of dam, its mode of operation, and its location are just a few of the factors that determine environmental impacts, and make generalizations difficult. With respect to small hydropower, considerable uncertainty exists about the environmental impacts of single small dams or the cumulative impacts of many small dams on river systems (Gleick 1998; Anderson et al. 2006). In some cases, small dams actually dewater or fragment more kilometers of stream, flood more land area, or lose more water to evaporation per unit of energy produced than larger dams (Gleick 1992). Despite their size, small hydropower dams almost always result in significant hydrologic alterations and losses in hydrologic connectivity. Small dam developments in tropical regions have been shown to affect movement patterns and persistence of migratory animals such as shrimps and fishes and to substantially alter physical habitat conditions (Benstead *et al.* 1999; Pringle *et al.* 2000; Anderson *et al.* 2006). Many small private dams in tropical countries are rapidly being constructed on previously unaltered systems draining rural areas. Consequently, the magnitude of ecological changes to freshwater resources resulting from new hydropower development may be much greater in developing nations than in developed countries such as the United States, where relatively few free-flowing rivers remain (Benke 1990).

Potential economic advantages of electricity sector reform are perhaps greatest in tropical developing countries, where many residents lack access to electricity. Environmental risks, however, are also high in developing countries, where governments are often unable to provide adequate regulatory control to protect public interests and the environment (Gleick *et al.* 2002). In light of the global push toward privatization, the issue of private hydropower development is of increasing concern to policy-makers, non-governmental organizations (NGOs), and local communities. Thus, as recommended by the World Commission on Dams, multiple stakeholders should be evaluating the environmental costs and benefits of privatization schemes, particularly those that involve the construction of hydropower dams (WCD 2000). Efforts to coordinate the input of citizens and communities affected by development projects with multilateral development banks, such as those undertaken by the Bank Information Center (2004), offer hope that environmental and social concerns will be voiced to development planners.

Case study III: Globalization and the commodification of water – the Great Lakes/Nova group case

Increasing human demands for water are threatening the biodiversity and stability of many freshwater ecosystems throughout the world. In river systems of the western United States, the private appropriation of fresh water for irrigation and other human activities (Postel and Richter 2004) consumes 75 percent or more of instream flow, resulting in negative environmental impacts to both riverine and riparian ecosystems (e.g., Pringle 2000, 2001). In some cases, water from a river system or lake is piped to locations hundreds of miles away from the source. The potentially devastating ecological effects of large-scale water appropriations are well illustrated by the collapse of Asia's Aral

Sea ecosystem and North America's Colorado delta ecosystem. In these cases, large-scale water diversion led to the loss of economically important fisheries and eventually to the displacement of human populations dependent on them. Controversial projects to divert large amounts of water from freshwater ecosystems are being proposed throughout the world, in river basins both large and small – from southern Africa's Okavango River to the Paraguay River in South America to the Apalachicola–Chattahahoochee–Flint river basin in the southeastern United States.

Economists have forecasted that privatization and sale of fresh water may become one of the most pressing issues of the twenty-first century as this resource comes under increasing pressure to be internationally traded in order to alleviate water scarcity and degradation. Our last case study illustrates the response and successful influence of local communities to protect the world's largest freshwater ecosystem, the Laurentian Great Lakes, from the threat of potentially unsustainable privatization and global trade. In this particular case, the threat of privatization ultimately resulted in stronger environmental protection measures coming into force.

On March 31, 1999, the Ontario Ministry of the Environment granted a permit to the Nova Group, a private Ontario-based corporation, to export up to 156 million gallons (700 million liters) of water from Lake Superior to Asia for crop irrigation annually for up to 5 years. This decision was met with strong protests from both grassroots environmentalists as well as both U.S. and Canadian government officials. Objections to the mandating of international water markets were also voiced by the Canadian trade minister at the 1999 Seattle meetings of the WTO. Concerns over ownership of the water (it is still unclear who actually owns the water in the Lakes), rights to licensing, and sustainable use of the water resources were posed by U.S. representatives and environmental NGOs (Cangelosi 2001). In response to immense public pressure, the Ministry eventually rescinded the permit (Cooper and Miller 2001). In its defense, Nova claimed that the amount of water it planned to export was negligible and did not pose a threat to Lake Superior. Environmentalists voiced concern that granting one permit for export would ultimately pose a threat to the Great Lakes by allowing for hundreds more to follow.

Environmental protection of the Great Lakes has a long tradition and remains a priority among diverse interests groups because the Lakes support tourism, recreation, fisheries, and a rich aquatic biodiversity. Not only is the Great Lakes basin a global hotspot of aquatic biodiversity, it is also one of the largest sources of fresh water in the world. The Lakes cover 94 250 square miles (244 000 km^2) and contain approximately 21% of the world's fresh water. In addition, the Great Lakes basin contains thousands of tributary streams and supports extensive wetlands (U.S. Environmental Protection Agency and Government of Canada 1995). These lakes, rives, and wetlands support more than 300 species of fish, hundreds of species of invertebrates, and one of if not the richest freshwater mussel faunas in the world (Jude and Pappas 1992).

Since the original squall over this issue, several agencies have deemed the measures in place to protect Great Lakes water inadequate and have called for more stringent laws (Council of Great Lakes Governors 2001; Canadian Environmental Law Association 2004). The Institute for Agriculture and Trade Policy (IATP) proposed comprehensive conservation planning by both Canada and the United States that would include limitations on water withdrawals. Because these limitations on withdrawal would be just one aspect of the overall conservation measures, IATP believed they would be more resistant to regulation by international trade bodies such as the WTO (Ritchie 1999). Indeed, many state-based and U.S.–Canada policy measures were introduced that limited water withdrawal from the Great Lakes in response to the Nova permitting issue (Cangelosi 2001). Though Great Lakes water ultimately was not licensed, this case became a major flashpoint in the anti-globalization, pro-environmentalist movement. The manner in which civil society reacted to this request for a license to export fresh water and the extensive conservation planning that emerged as a result of the threat of exporting Great Lakes water illustrate the potential of conservation initiatives to successfully counter threats posed by privatization and commodification of freshwater resources.

CONCLUSIONS: PROTECTING WATER RESOURCES
IN THE AGE OF GLOBALIZATION

Because the story of the destruction of the Aral Sea is linked to the era of development, dam-building, and commercial agricultural expansion, the story of its demise serves as a precautionary tale for the potential negative environmental consequences of globalization. Although not caused necessarily by global factors, the Aral Sea disaster serves as a cautionary reminder of the long-lasting negative effects that may accompany the simplification of ecosystems for export agriculture. Now considered one of the greatest environmental catastrophes in history, the devastation of the Aral Sea is traced to the late 1930s, when the Soviet government began to operate large-scale commercial cotton farms in the region (Calder and Lee 1995). In order to do this, the rivers emptying into the Aral Sea were dammed, and irrigation diversions to surrounding farms were created. After 1960, diversions of water out of the Aral Sea outpaced returns, causing the sea to shrink, setting off a positive feedback loop that increased salinity and evaporation. By the 1980s, the once-abundant fisheries had collapsed, and the climate of the region began to change as the Aral Sea lost its capacity to moderate the Siberian winters. Sensitive river delta habitat was also destroyed when the lake began to recede, and contaminated sediment from the exposed shores, totaling more than 28 000 km^2, was spread as far north as the Arctic via windstorms. Similar destruction of wetlands, albeit on a lesser scale, as a result of diversions and irrigation runoff from agricultural lands has occurred in the Kesterson Wildlife Refuge, the Salton Sea, and the Stillwater National Wildlife Refuge in western Nevada (Pringle 2003).

In response to these environmental tragedies, many environmentalists have recommended that environmental controls and regulations be built into international laws and trade agreements. However, there is now evidence that many of the laws outlined in international treaties on the environment are disobeyed (French 2000). Reasons for flouting the law may range from economic hardship to straightforward resentment of "top–down" conservation efforts by local people. In the case of the Aral Sea, local communities did not have the power to prevent policies driving large-scale environmental alterations. By contrast, the relative strength of civil society, local governments, and regional coalitions was an important component in environmentalists' ability to prevent the licensing and trading of Great Lakes water by a private corporation. These two contrasting examples illustrate the importance of sociopolitical and economic contexts in successful community-led protection of freshwater resources.

Much hope has been placed in the potential of local communities (Goldsmith and Mander 1996) as well as coalitions of diverse groups to positively affect environmental health (French 2000; Gould et al. 2004). For coalitions across borders and ideologies to be successful in protecting resources such as freshwater biodiversity, however, an understanding of and respect for the divisions in world-views, local histories and conflicts, and on the ground realities for people involved in use and protection of water resources is needed. Promising examples of such coalitions do exist, such as the international movements that blocked dam development in Brazil (Rothman and Oliver 1999) and the labor/environmental coalition that emerged at the WTO Seattle Round in 1999 (Gould et al. 2004). As the era of globalization unfolds, protection of freshwater resources will depend not only on the ability of local communities to form effective coalitions that magnify attention to their cause but also on the willingness of governments and international trade and development agencies to enact and maintain legal frameworks that protect freshwater resources.

References

Alexander, R. B., Smith, R. A., and Schwarz, G. E. 2000. Effect of stream channel size on the delivery of nitrogen to the Gulf of México. *Nature* **403**: 758–761.

Altieri, M. 2002. Agroecology: the science of natural resource management for poor farmers in marginal environments. *Agriculture, Ecosystems and Environment* **93**: 1–24.

Alvarez, M. 2005. Privatización de la generación eléctrica: el asalto del siglo. *Ambientico* **137**: 8–10.

Anderson, E. 2002. Electricity sector reform means more dams for Costa Rica. *World Rivers Review* **17**: 3.

Anderson, E. P., Freeman, M. C., and Pringle, C. M. 2006. Ecological consequences of hydropower development in Central America: impacts of small dams and water diversion on neotropical stream fish assemblages. *River Research and Applications* **22**: 397–411.

Aparicio, E., Vargas, M. J., Olmo, J. M., and de Sosota, A. 2000. Decline of native freshwater fishes in a Mediterranean watershed on the Iberian Peninsula: a quantitative assessment. *Environmental Biology of Fishes* **59**: 11–19.

Appel, J. 1991. *Uso, Manejo y Riesgos Asociados a Plaguicidas en Nicaragua*. Managua, Nicaragua: Proyecto Regional de Plaguicidas, Confederación Universitaria Centroamericana.

Bank Information Center. 2004. About the Bank Information Center. Available online at www.bicusa.org/bicusa/issues/about_the_bank_information_center/index.php

Barlow, M. 2002. Water incorporated: the commodification of the world's water. *Earth Island Journal*. Available online at www.globalpolicy.org/globaliz/special/2002/0305water.htm

Barrett, G. W. 1968. The effects of an acute insecticide stress on a semi-enclosed grassland ecosystem. *Ecology* **49**: 1019–1035.

Becker, E. 2004. W. T. O. Rules Against U.S. on cotton subsidies. *New York Times* April 27, 2004.

Bellaire, L. and Dubois, C. 1997. Distribution of thiabendazole-resistant *Colletotrichum musae* isolates from Guadeloupe banana plantations. *Plant Disease* **81**: 1378–1383.

Benke, A. C. 1990. A perspective on America's vanishing streams. *Journal of the North American Benthological Society* **9**: 77–88.

Benstead, J. P., March, J. G., Pringle, C. M., and Scatena, F. N. 1999. Effects of a low-head dam and water abstraction on migratory tropical stream biota. *Ecological Applications* **9**: 656–668.

Boer, J. and Wester, P. G. 1993. Determination of toxaphene in human milk from Nicaragua and in fish and marine mammals from the Northeastern Atlantic and the North Sea. *Chemosphere* **27**: 1879–1890.

Burgues Arrea, I. 2005. *Inventario de Proyectos de Infraestructura en Mesoamérica. Conservation Strategy Fund, Proyecto: Integración de la Infraestructura y la Conservación de la Biodiversidad en Mesoamérica*. Curridabat, Costa Rica: Fondo de Alianzas para los Ecosistemas Críticos, The Nature Conservancy, Conservation Strategy Fund.

Calder, J. and Lee, J. 1995: ICE Case studies: Aral Sea and defense issues. *Inventory on Conflict and the Environment*. Available online at www.american.edu/TED/ice/aralsea.htm

Calero, S., Fomsgaard, I., Lacayo, M. L., Martinex, V., and Rugama, R. 1993. Toxaphene and other organochlorine pesticides in fish and sediment from Lake Xolotlan, Nicaragua. *International Journal of Environmental Analytical Chemistry* **53**: 297–305.

Canadian Environmental Law Association. 2004. *Great Lakes Water Management Initiative*. Available online at www.cglg.org

Cangelosi, A. 2001. Sustainable use of Great Lakes water: the diversion threat's silver-lining? Available online at www.nemw.org/ERGLwaterdivert.htm

Carvalho, F. P., Montenegro-Guillen, S., Villeneuve, J. P., et al. 2003. Toxaphene residues from cotton fields in soils and in the coastal environment of Nicaragua. *Chemosphere* **53**: 627–636.

CEPAL (Comisión Económica para América Latina y el Caribe). 2005. *Istmo Centroamericano: Estadísticas del Subsector Eléctrico*. Available online at www.cepal.org/publications

Coes, D. V. 1998. Beyond privatization: getting the rules right in Latin America's regulatory environment. *Quarterly Review of Economics and Finance* **38**: 525–532.

Cooper, K. and Miller, S. 2001. Selling our water: water taking in Lake Superior. *Intervenor* **23**(2): 12.

Council of Great Lakes Governors. 2001. *Great Lakes Water Management Initiative Overview*. Available online at www.cglg.org/1projects/water/overview.asp

Diaz, R. J. and Rosenberg, R. 1995. Marine benthic hypoxia: a review of its ecological effects and the behavioral responses of benthic macrofauna. *Oceanography and Marine Biology* **33**: 245–303.

Dudgeon, D. 2002. An inventory of riverine biodiversity in monsoonal Asia: present status and conservation challenges. *Water Science and Technology* **45**(11): 11–19.

Dunkerley, J. 1995. Financing the energy sector in developing countries. *Energy Policy* **23**: 929–939.

Duran, O. 2005. La estafa legal de la energía privada. *Ambientico* **137**: 11–14.

ECLAC (Economic Commission for Latin America and the Caribbean). 1996. *Progress in the Privatization of Water-Related Public Services: A Country-by-Country Review for Mexico, Central America, and the Caribbean*. Santiago, Chile: Environment and Development Division, ECLAC.

Escobar, A. 1995. *Encountering Development: The Making and Unmaking of the Third World*. Princeton, NJ: Princeton University Press.

French, H. 2000. *Saving the Planet in the Age of Globalization*. New York: W. W. Norton.

Gleick, P. 1992. Environmental consequences of hydroelectric development: the role of facility size and type. *Energy* **17**: 735–747.

Gleick, P. 1993. *Water in Crisis: A Guide to the World's Fresh Water Resources*. New York: Oxford University Press.

Gleick, P. 1998. *The World's Water: The Biennial Report on Freshwater Resources 1998-1999*. Washington, DC: Island Press.

Gleick, P., Wolff, G., Chalecki, E. L., and Reyes, R. 2002. *The New Economy of Water*. Hayward, CA: Pacific Institute for Studies in Development, Environment, and Security.

Goldsmith, E. and Mander, J. 1996. *The Case against Globalization and for a Turn to Localization*. San Francisco, CA: Sierra Club.

Gould, K. A., Lewis, T. L., and Roberts, J. T. 2004. Blue-Green coalitions: constraints and possibilities in the post 9-11 political environment. *Journal of World-Systems Research* **10**: 91-116.

Ingco, S. P. 1996. Structural changes in the power sector in Asia. *Energy Policy* **24**: 723-733.

International Forum on Globalization. 2002. *Alternatives to Economic Globalization: A Better World is Possible*. San Francisco, CA: Berret-Koehler.

Izaguirre, A. K. 2000. *Private Participation in Energy*, Note No. 208. Washington, DC: Private Sector and Infrastructure Network, The World Bank Group.

Joseph, T. 2001. Decolonization in the era of globalization: the independence experience of St. Lucia. D.Phil. thesis, University of Oxford, UK.

Jude, D. J. and Pappas, J. 1992. Fish utilization of Great Lakes coastal wetlands. *Journal of Great Lakes Research* **18**: 651-672.

Karliner, J. 1997. *The Corporate Planet: Ecology and Politics in the Age of Globalization*. San Francisco, CA: Sierra Club.

Kiely, R. and Marfleet, P. 1998. *Globalization and the Third World*. London: Routledge.

Kimbrell, A. 2002. *Fatal Harvest: The Tragedy of Industrial Agriculture*. Washington, DC: Island Press.

Korten, D. C. 2001. *When Corporations Rule the World*. Bloomfield, CT: Kumarian Press.

Lemly, A. D., Kingsford, R. T., and Thompson, J. R. 2000. Irrigated agriculture and wildlife conservation: conflict on a global scale. *Environmental Management* **25**: 485-512.

Ljung, P. 2001. *Trends in the Financing of Water and Energy Resource Projects*, Thematic Review Paper III.2. Cape Town, South Africa: World Commission on Dams.

Majot, J. 1997. *Beyond Big Dams: A New Approach to Energy Sector and Watershed Planning*. Berkeley, CA: International Rivers Network.

Master, L. L., Flack, S. R., and Stein, B. A. 1998. *Rivers of Life: Critical Watersheds for Protecting Freshwater Biodiversity*, special publication of the Nature Conservancy. Arlington, VA: NatureServe.

McDonald, L., Jebellie, S. J., Madramootoo, C. A., and Todds, G. T. 1999. Pesticide mobility on a hillside soil in St. Lucia. *Agriculture, Ecosystems and Environment* **72**: 181-188.

Mortensen, S. R., Johnson, K. A., Weisskopf, C. R., *et al*. 1998. Avian exposure to pesticides in Costa Rican banana plantations. *Environmental Contamination and Toxicology* **60**: 562-568.

Moskin, J. 2004. Helping third world one banana at a time. *New York Times* May 5, 2004.

Mungwena, W. 2002. Hydropower development on Zimbabwe's major dams. *Renewable Energy* **25**: 455-462.

OECD (Organisation for Economic Co-operation and Development). 1998. *Water Consumption and Sustainable Water Resources Management*. Paris: OECD Publications.

Parker, D. 2002. Privatization and neoliberalism: the European experience and the lessons for Latin America. In *Globalization and Sustainable Development in*

Latin America, eds. S. K. Saha and D. Parker. Cheltenham, UK: Edward Elgar, pp. 51-77.

Pesticide Action Network. 2001. Cotton pesticides cause more deaths in Benin. *Pesticide News* June 2001: 12-14.

Pesticide Action Network. 2003. PAN pesticides database: California pesticide use. Available online at www.pesticideinfo.org

Postel, S. L. 1999. *Pillar of Sand: Can the Irrigation Miracle Last?* New York: W. W. Norton.

Postel, S. L. and Richter, B. 2004. *Rivers for Life: Managing Water for People and Nature*. Washington, DC: Island Press.

Pringle, C. M. 2000. Threats to U.S. public lands from cumulative hydrologic alterations outside of their boundaries. *Ecological Applications* **10**: 971-989.

Pringle, C. M. 2001. Hydrologic connectivity and the management of biological reserves: a global perspective. *Ecological Applications* **11**: 981-998.

Pringle, C. M. 2003. Interacting effects of altered hydrology and contaminant transport: emerging ecological patterns of global concern. In *Achieving Sustainable Freshwater Systems: A Web of Connections*, eds. M. Holland, E. Blood, and L. Shaffer. Washington, DC: Island Press pp. 85-107.

Pringle, C. M., Freeman, M. C., and Freeman, B. J. 2000. Regional effects of hydrologic alterations on riverine macrobiota in the New World: tropical-temperate comparisons. *BioScience* **50**: 807-823.

Raphals, P. 2001. *Restructured Rivers: Hydropower in The Era of Competitive Markets*. Berkeley, CA: International Rivers Network.

Ritchie, M. 1999. Great Lakes water in danger from international trade laws. *Detroit Free Press* June 29, 1999. Available online at www.iatp.org

Rogers, P. 1991. Energy use in the developing world: a crisis of rising expectations. *Environmental Science and Technology* **25**: 580-583.

Rothman, F. D. and Oliver, P. E. 1999. "From local to global" anti-dam movement in Brazil. *Mobilization* **4**: 41-57.

Rouse, J. D., Bishop, C. A., and Struger, J. 1999. Nitrogen pollution: an assessment of its threat to amphibian survival. *Environmental Health Perspectives* **107**: 799-803.

Sassen, S. 1998. *Globalization and Its Discontents: Essays on the New Mobility of People and Money*. New York: New York Press.

Seares, J. 2004. From the local to the global: the ecology of agricultural development in St. Lucia, West Indies. Ph.D. thesis, University of Georgia, Athens, GA.

Stokes, B. 2003. WTO: the cost of failing to change. *National Journal* **35**(11): 839.

Stone, D. and Manrique, J. 2002. U.S. collaboration promotes integrated social development in Guatemala. *Hydropower and Dams* **2002**(4): 102-104.

Thornton, P. 2003. Farm reform vital after WTO failure, says Beckett. *The Independent (London)* March 31, 2003.

Tilman, D. 1999. Global environmental impacts of agricultural expansion: the need for sustainable and efficient practices. *Proceedings of the National Academy of Sciences of the United States of America* **96**: 5995-6000.

UNESCO (United Nations Educational, Scientific, and Cultural Organization). 2001. *Securing the Food Supply*. Paris: UNESCO.

U.S. Environmental Protection Agency and Government of Canada. 1995. *The Great Lakes: An Environmental Atlas and Resource Book*. Toronto, Ontario: Government of Canada and Chicago, IL: EPA.

Vandermeer, J. H., Perfecto, I., and Jody, M. Z. 1995. *Breakfast of Biodiversity: The Truth about Rainforest Destruction*. Oakland, CA: Institute for Food and Development Policy.

WCD (World Commission on Dams). 2000. *Dams and Development: A Framework for Decision-Making*. Available online at www.damsreport.org

WHO (World Health Organization). 1984. *Camphechlor*, Environmental Health Criteria 45. Geneva, Switzerland: WHO.

WRI (World Resources Institute). 2000. *People and Ecosystems: The Fraying Web of Life*. Amsterdam, Netherlands: Elsevier.

WWI (World Watch Institute). 2003. *Vital Signs*. Washington, DC: Worldwatch Press.

JOAN B. ROSE AND STEPHANIE L. MOLLOY

4

Globalization effects on water quality: monitoring the impact on and control of waterborne disease

GLOBALIZATION AND THE NATURE OF THE WATER ENVIRONMENT

Historically, the development of communities near and on coastlines and river banks meant that populations were mobile, utilizing the waterways for travel and trade. The nearby fisheries served as an important component of the food supply, and rivers and lakes provided an abundance of fresh water. This interconnectedness between water and development of civilizations remains as important today as it was throughout history. Population growth correlates directly with increase in water use (Fig. 4.1) (much of this use is non-consumptive), and the water is generally returned to the environment in a poorer quality than it was when it was taken. Studies investigating the relationship between a country's economic status (gross national product, GNP) and water use have shown that, though there is little correlation between these variables within countries grouped by income ranges (low income, middle–lower, middle–higher, and high income), there is a clear relationship between economic development in the four broad categories and per capita water use (World Bank 2000) (Fig. 4.2).

Water resources are part of a global interconnectedness and interdependency and in many cases are now being viewed and will require management under the doctrine of scarcity. It is estimated that water shortages will affect 2.7 billion people in about 20 years, threatening the global food supply as well as the economies of the more than 50 countries where international boundaries in water basins have created a hydrogeopolitical setting (Gleick 1993; Shmueli 1999; World Water Council 2003). There are 215 international river basins, and waters are shared by more than one country on every continent (Gleick 1993). There are 13 international water basins worldwide shared by five or

Globalization: Effects on Fisheries Resources, ed. William W. Taylor, Michael G. Schechter, and Lois G. Wolfson. Published by Cambridge University Press. © Cambridge University Press 2007.

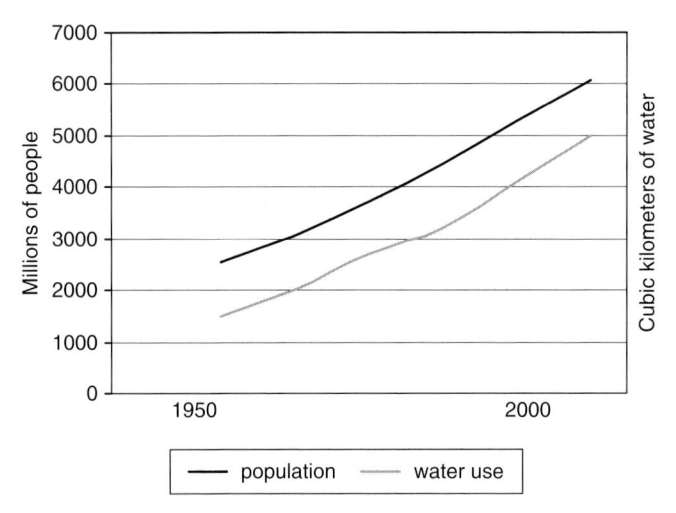

Figure 4.1 Water use and population growth, 1950–2000. (*Source*: World Bank 2000.)

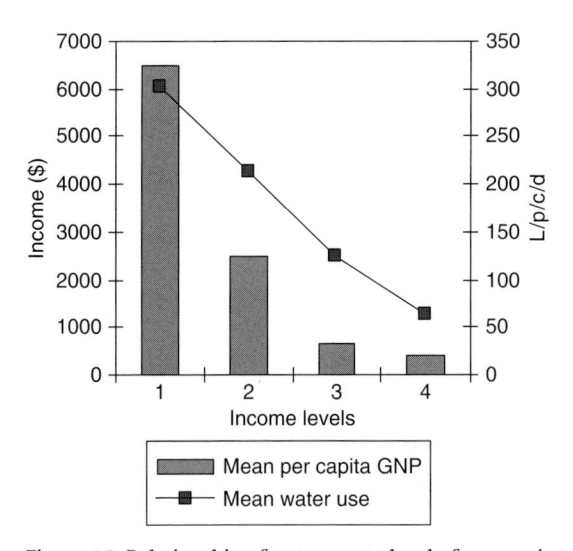

Figure 4.2 Relationship of water use to level of economic prosperity (World Bank 2000). GNP, gross national product; L/p/c/d, Liters per person consumption per day; annual income levels: 4 = low income (< $650), 3 = low-middle income ($650–2500), 2 = upper-middle income (> $2500–6500), 1 = high income (> $6500).

more nations, including the Danube, Niger, Nile, Zaire, Rhine, Ambezi, and Amazon. The Great Lakes water basin is one of the largest volumes of fresh water storage and is shared by Canada and the United States. In addition to maintaining the beneficial uses of water for agriculture, industry, and communities, there is an increasing need to protect the natural features of water systems including the animals and plants that inhabit these ecosystems. Water pollution remains a critical issue that has directly influenced the scarcity of water. As the quality of water has deteriorated in many parts of the world, this means that restoring the waters for their intended purposes will require significant financial investment. Achieving access to both sufficient quantity and quality of water will require local, regional, national and international cooperation.

Pollution of the water environment is tied directly to human activities, types of water use, and pollution prevention strategies employed. The relationship between water quality and human health has been well established. According to the World Health Organization (WHO), 40 percent of the world's population (6 billion people) lack sanitation services and 1 billion people lack access to safe water; up to 2.2 million will die from diarrhea (Brown 2002). The situation is most dire in developing countries. United Nations Secretary-General Kofi Annan has stressed that water-related diseases are responsible for 80 percent of the illnesses and deaths in the developing world and that innovations are needed to address this growing global challenge. Zambia's Central Board of Health estimated that 54 percent of the population (4 million people) were suffering illnesses and dying because of drinking contaminated water from both surface and shallow wells. As many as 250 million cases of water-related diseases in sub-Saharan Africa have been reported each year (United Nations 2003). To address this global calamity, the United Nations Millennium Development Goals stated the following objectives for water: "Halve by 2015 the proportion of people without sustainable access to safe drinking water" and "halve the proportion of people who do not have access to basic sanitation" (United Nations 2005).

Water contamination in affluent countries has also resulted in dramatic outbreaks of waterborne disease, and looms as an increasing and constant reminder of the vulnerability of the water system to significant water quality degradation resulting in disease and death if engineering controls fail (Hrudey and Hrudey 2004). These outbreaks and water quality degradation, particularly along coastlines and beaches, have influenced both local and national policies for protection of the water resource.

Though it is acknowledged that a water crisis exists, much effort has been placed on measuring the quantity of water, when it is the quality of water that is directly linked to health and about which there is a paucity of information. Currently, there is no global database on water quality, and water quality is not addressed in many international water treaties (Shmueli 1999). To meet the water resource challenges of the twenty-first century and the Millennium Development Goals in an era of globalization will require a comprehensive examination of the causes of water pollution and degradation of water quality, along with development of a comprehensive spatial and temporal database on water quality conditions and trends.

Globalization and impacts on water quality

The definition of globalization includes the interconnectedness of our populations, both geographically and socially, technologically, and economically. Many water basins and watersheds traverse local, national, and international jurisdictions. Thus there is a direct connection between water quantity and water quality and the dynamics of the peoples sharing the same water basin. Some population always lives downstream of another, so the water quality of one community is affected by unchecked economic development and associated pollution discharges of another. Population growth, urbanization, and industrialization have resulted in major impacts directly on water quality via increased volumes and spatial distribution of sewage and manufacturers' discharges. Globalization also has many subtle indirect effects on water quality.

The globalization of goods, services, labor, capital, and technology has a strong influence on water quality at all scales from the very small to the largest. The growing population and economics associated with a global food market have led to intensification of agriculture, and an increased demand for protein has meant that more animals and their wastes are produced. More fruits and vegetables may enter the world market from areas where irrigation waters are laden with sewage and animal wastes.

The growth in world trade has correlated with an increase in the shipping industry and has resulted in the transport of water from one area of the globe to another through the collection and release of ballast waters. Travel in general has increased, and with the ability to travel has come the enhancement of the tourism industry with significant impacts on local and national economies. This increase in tourism has particularly affected coastlines, which have experienced rapid development and increased pollution compromising the aquatic resources that attracted visitors in the first place.

Finally, the increase in population and energy needs has influenced the climate via release of greenhouse gases. This climate change is realized in changes in precipitation and temperatures as well as the timing and duration of droughts, storms, and other extreme events that directly and indirectly influence water quality.

Table 4.1 describes some of these indicators of globalization and their potential impacts on water quality. The processes and factors involved in indicators of globalization such as population growth, sewage discharges, and the relationship of the resulting water quality influencing human health are complex. To begin to understand these processes, we must first define waterborne disease.

Waterborne disease

Waterborne disease is a global problem. According to the latest estimate by the WHO, some 250 million cases of water-related diseases occur annually in addition to 4.37 billion cases of diarrhea associated with contaminated water (UNDP 2003; WHO 2003a). *Waterborne* disease is defined as an illness in humans by which the etiological agent (the organism responsible for the disease) is transmitted through ingestion of contaminated water via a variety of activities (drinking and recreational). *Water-washed* disease is caused by poor personal hygiene and skin or eye contact with contaminated water, and includes scabies, trachoma, and flea-, lice-, and tick-borne diseases. *Water-based* disease is caused by parasites found in intermediate organisms living in water, and includes dracunculiasis, schistosomiasis, and conditions caused by other helminths. Finally, *water-related* diseases are caused by insect vectors that breed in water, and include dengue, filariasis, malaria, onchocerciasis, trypanosomiasis, and yellow fever.

It is the first two categories that are primarily associated with water contamination and water quality. The most common sources of the microbial agents associated with waterborne disease are animal and human waste. The infectious agents that cause these diseases are enteric bacteria, parasites, and viruses. For many human illnesses, it is the transmission of the agent from feces of infected humans or animals to water and then back to humans that is the primary risk. Fecal contamination can enter waterways via sewage discharges, animal waste discharges, septic tanks, and stormwater. Fecal contamination of water and exposure to sewage have long been associated with

Table 4.1 *Globalization indicators and some impacts on the water environment*

Indicator	Result	Impact	Spatial scale
Increases in trade	Increases in shipping	Transport of exotic species and disease agents from ballast waters	Spreads quickly to the larger basin scale
Tourism	Increases in infrastructure, sewage production, and water use	Chemical and pathogen discharge to the water environment; overutilization of the resource	Local impact at the beach and coastline
Global food supply	Intensification of agriculture and aquaculture; application of antibiotics and fertilizers	Non-point source discharge of pathogens from associated animal wastes (manure) and chemicals; eutrophication; antibiotic resistance	Watershed level
Climate change	Changes in temperature, precipitation, storms	Enhanced transport of contaminants; changes in pathogen survival; greater conflict over resources more recycling; extreme events increasing loading	Local and large basin-scale impact
Rapid transport	Movement of people	Disease and infections transported globally	Local community impacts; global epidemics
Population change	Growth and diversity of populations; urbanization; large metropolises	Sensitive subpopulations; increased populations and wastes and wastewater discharges; increases in point and non-point pollution; loss of natural buffers for contaminant attenuation	Localized areas of impact and concern; downstream impacts

```
EXPOSURE                    TRANSPORT                   SOURCES

┌─────────────────┐      ┌─────────────────┐      ┌─────────────────┐
│ Consumption and │      │ Microbial fate  │      │ Sources of      │
│ use: community  │◄─────│ and climate     │◄─────│ microbial       │
│ growth; tourism;│      │ change as the   │      │ agents and their│
│ global food     │      │ driver of change│      │ contributions   │
│ supply          │      │                 │      │ to water        │
└─────────────────┘      └─────────────────┘      └─────────────────┘

┌─────────────────┐      ┌─────────────────┐      ┌─────────────────┐
│ Beaches,        │      │ Transport via   │      │ Land use:       │
│ swimming pools, │      │ rain and runoff │      │ population and  │
│ recreational    │      │                 │      │ animal numbers  │
│ activities      │      │                 │      │ and distributions│
└─────────────────┘      └─────────────────┘      └─────────────────┘

┌─────────────────┐      ┌─────────────────┐      ┌─────────────────┐
│ Seafood and     │      │ Survival and/or │      │ Return flows    │
│ shellfish;      │      │ regrowth via    │      │ from communities│
│ vegetables      │      │ temperature     │      │ (Sewage)        │
│ and fruits      │      │                 │      │                 │
└─────────────────┘      └─────────────────┘      └─────────────────┘

┌─────────────────┐      ┌─────────────────┐      ┌─────────────────┐
│ Drinking water  │      │ More or less    │      │ Urbanization:   │
│ supplies: need  │      │ dilution and    │      │ increase storm  │
│ for new supplies│      │ attenuation     │      │ drains; loss of │
│ and greater     │      │                 │      │ natural buffers │
│ quantity        │      └─────────────────┘      └─────────────────┘
└─────────────────┘      ┌─────────────────┐
                         │ Extreme events  │      ┌─────────────────┐
                         │ and increased   │      │ Agricultural    │
                         │ loading         │      │ intensification:│
                         └─────────────────┘      │ animal manure;  │
                                                  │ greater         │
                                                  │ irrigation      │
                                                  └─────────────────┘

                                                  ┌─────────────────┐
                                                  │ Discharge of    │
                                                  │ contaminated    │
                                                  │ ballast waters  │
                                                  │ or wastewaters  │
                                                  │ from ships      │
                                                  └─────────────────┘
```

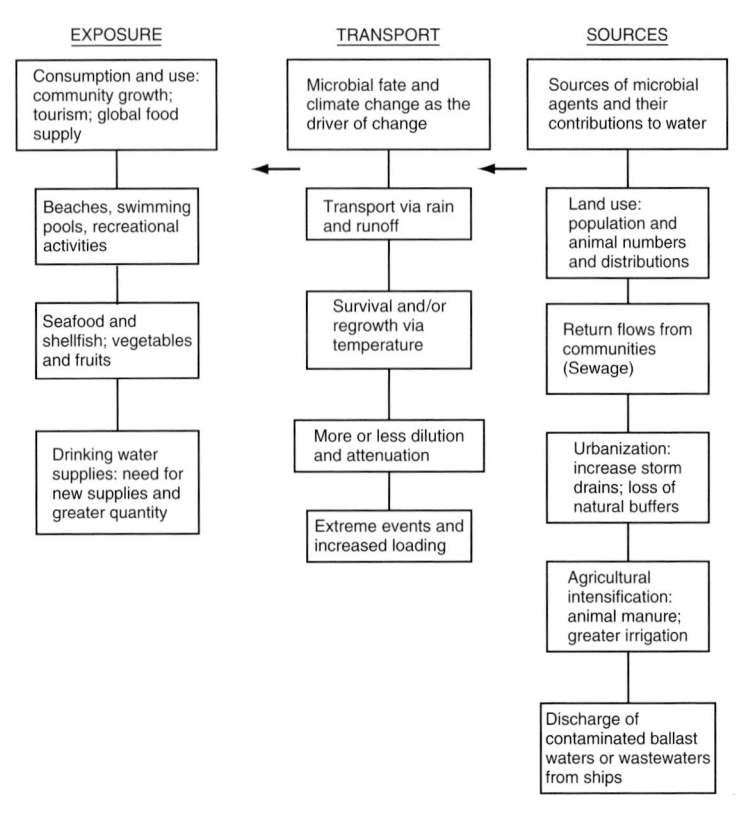

Figure 4.3 Waterborne disease potential: globalization factors influencing the source to exposure.

disease. Thus the presence of feces in and of themselves has been a water quality indicator for defining the potential for waterborne disease.

The probability of an illness occurring from any waterborne pathogen will depend on the type of contact made, that is, the exposure, the duration of the exposure, and the concentration of the organisms in the contaminated water, the survival and transport of the pathogen from the source to the contact point, and the level of individual or population susceptibility to the pathogen (Meyland and Machlin 2001).

To better understand the risks to the water system and human health associated with the globalization factors referred to in Table 4.1, it is helpful to address the interconnectedness of these factors, from the sources of the pathogens to the exposure (Fig. 4.3).

The major direct transmission routes of waterborne disease are associated with exposure to contaminated drinking water and recreational water. In this context, recreational exposure will be assumed to

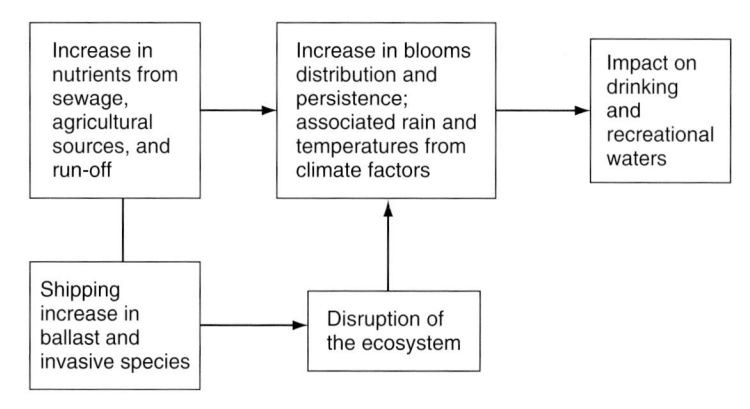

Figure 4.4 Harmful algal blooms and factors influencing their appearance and persistence.

be through ambient waters. In freshwater systems, rivers and lakes may serve multiple purposes as drinking water supplies and recreational waters.

Drinking water and treatment had been the major focus for understanding and preventing waterborne disease, but the worldwide increase in tourism as an economically important industry, particularly for coastal communities, has also increased the interest in coastal pollution and recreational waterborne disease. The inclusion of access to sanitation and watershed protection as an expansion to the paradigm for providing "safe water" has meant that greater emphasis is needed on the sources of the pathogens and the factors that influence the loading, transport, and fate of pathogens in the water environment.

The exposure via water to toxins produced by microorganisms has not been generally included or readily described by any of the previous waterborne disease definitions. It is clear that water quality degradation in many countries has been measured by increases in nutrients from both sewage and agricultural sources, which have been suggested to be related to increases in harmful algal blooms (HABs) (Chorus and Bartram, 1999). These harmful algal blooms may affect recreational uses of water ("red tide") and drinking water (cyanobacteria, blue-greens), and their frequency and distribution may be related to some of the same globalization factors as those of the traditional waterborne disease agents of fecal origin (Fig. 4.4).

The greater recognition and awareness of waterborne disease has led to much public concern, particularly for the members of our society who are very susceptible to severe outcomes, including death (Haas *et al.* 1999). Infants, young children, elderly persons, the immunocompromised, and those with underlying disease conditions are at greatest risk when exposed to water pollution hazards. This has led to a movement to better understand the causes so that potential solutions for mitigating the risk, through both practical engineering methods and societal and political approaches, can be implemented.

UNDERSTANDING THE SOURCES, FATE, AND IMPACT OF WATERBORNE AGENTS

Sewage: an important source of waterborne pathogens

One of the most important indicators of the global change is the unprecedented and rapid increase in population and the subsequent increase in sewage discharges contributing to widespread water quality degradation. Waterborne pathogens are found in varying concentrations in wastewater, whether untreated, primary treated, or secondary treated. Only advanced treatment and/or adequate disinfection will eliminate or dramatically decrease the levels of pathogens in sewage (NRC 1999). It has been suggested that sewage discharges represent one of the most significant sources of health risk associated with waterborne disease because of the concentrations and the variety of pathogens that can be present. Table 4.2 describes some of the pathogens that may be found in untreated sewage and the diseases they can cause (Rose and Grimes 2001). Hundreds of types of bacteria, parasites, and viruses can be found in untreated sewage, and worldwide there is no program to monitor for these pathogens in sewage discharges. Typically, these pathogens are highly infectious; even with great dilution in the environment the large concentrations found in sewage make the level in polluted waters capable of causing illness. This is particularly true of the viruses and parasites. In addition, viruses and parasites are more robust than bacteria, surviving for extensive times in water, and they are more resistant to water disinfection used in wastewater or in drinking water.

Globally, about 40 percent of the population has no access to any type of adequate sanitation system (sewers, flush latrines, septic tanks, and communal toilets in urban areas, and pit privies, pour-flush latrines, septic tanks, and communal toilets in rural areas) (Gleick 1998), so untreated wastes are either directly or indirectly washed into waterways. In the United States, 98 percent of the wastewater undergoes secondary treatment, and in the past decade, the European Union has mandated that secondary treatment be used. However, in most countries with sewer coverage, less than 1 percent of the flow is

Table 4.2 Examples of pathogens found in sewage

Pathogen	Infectivity[a]	Disease(s)	Sources, fate, and transport issues
Bacteria			
Campylobacter	High	Diarrhea, chronic outcomes, Gullain–Barré syndrome	Susceptible to water disinfection; ambient recreational water exposure a risk. Can come from animals and humans
E. coli (including O157: H7)	Moderate (to high)	Bloody diarrhea, uremic hemolytic syndrome	Cattle a source of O157 in addition to sewage
Salmonella	Moderate	Gastroenteritis and septicemia	Also animal fecal sources
Shigella	High	Bacillary dysentery	Important cause of recreational outbreaks in fresh ambient waters
Vibrio cholerae	Low	Asiatic cholera	Drinking water due to sewage. Only source is human fecal wastes
Enteric viruses			
Adenoviruses	High	Acute respiratory infection, pharyngitis, acute hemorrhagic cystitis, diarrhea	Very stable in water
Coxsackie viruses, enteroviruses, poliovirus	High	Myocarditis, pleurodynia, rash, meningitis, paralysis, diarrhea	Most often identified in water
Norovirus	High	Diarrhea	Recent increase in cruise ship outbreaks and recreational outbreaks. Also found in animal fecal sources
Protozoa			
Cryptosporidium	High	Diarrhea, high risk of death in immunocompromised	Drinking and recreational risks; highly resistant to wastewater and drinking water disinfection
Giardia	High	Diarrhea, chronic infections, failure to thrive in the young	Drinking and recreational risks; moderately resistant to water disinfection; found in the highest concentrations in sewage

[a] Probability of infection: high infectivity, low numbers are likely to initiate infection; low infectivity, higher concentrations are needed to initiate infection.

treated prior to discharge. China, for example discharges 40.1 billion cubic meters of wastewater and sewage every year, of which 51 percent is from the cities. China has approximately 400 city sewage treatment plants in operation, with a total daily treatment capacity of 25.34 million cubic meters (Zhang n.d.).

Separate sanitary sewers intended to collect municipal and industrial wastewater and carry it to a wastewater treatment plant are not meant to carry a superfluous amount of water, such as storm water. But in many places, including the United States, sanitary sewer overflows (SSO) occur as discharges of sewage from a sanitary sewer collection system before reaching a wastewater treatment plant, both intentionally and unintentionally (Meyland and Machlin 2001).

Discharges that enter waterways from diffuse sources (rather than the end of a pipe) are considered non-point and are currently of significant concern as a major cause of impaired water quality. A specific water pollution source directly related to globalization is the increased use of septic tanks along coastlines where tourism and population growth are rising.

Globalization is resulting in increased travel and tourism. This industry and population shift to coastal communities puts pressure on coastal environments. Often wastewater from these communities is disposed directly or indirectly into coastal waters. Approximately 9.3×10^7 people (37 percent of the total U.S. population) reside in U.S. coastal areas and discharge about 4.5×10^{10} liters of treated wastewater per day (NRC 1993). In the year 2000, the United States had more than 11 000 beach closings or advisories (freshwater and marine beaches), almost twice as many in the previous year, and a majority of these closings were due to wastewater pollution (NRDC 2001). On a global scale, coastal development is twice that of inland sites, and about 90 percent of the generated wastewater is released untreated into marine waters (Crosette 1996; Henrickson et al. 2001).

There is no adequate estimate of the numbers of septic tanks worldwide, but in the United States, approximately 25 percent of all households and one of every three new houses are served by these on-site waste disposal systems. Contamination of nearby coastal waters and groundwater has been demonstrated. In particular, pathogenic viruses, because of their colloidal nature, have been shown to contaminate both coastal waters and drinking water (Borchardt et al. 2003; Griffin et al. 2003). The use of viral tracers was particularly successful in the study of septic tanks in Florida. Bacteriophages were flushed down toilets and in some cases appeared in adjacent surface waters in 8 hours and in nearshore waters in 23 hours. Estimated migration rates were as high as 24.2 meters per hour, a rate 500-fold greater than that seen using subsurface flow meters (Paul et al. 1995). This illustrates the significant contribution that septic tank discharges can make to the deterioration of regional coastal surface water quality.

Animal waste contributions

Over the past 25 years, per capita meat consumption in developing countries grew at three times the rate in developed countries, and it is estimated that global livestock production will have to double by 2020 to supply needs (Bradford et al. 1999; Delgado et al. 1999).

In traditional systems manure is an important fertilizer, but manure produced in industrial systems is usually in excess of agronomic requirements and is a major disposal issue in intensive animal production facilities. Today, the world contains approximately 1.2 billion cattle, 800 million pigs, and 10 billion chickens. Livestock numbers are increasing at the same time as agricultural land has been decreasing at a rate of 7 percent per decade because of urbanization, commercial forces, and land degradation (Oldeman et al. 1991). Livestock densities in both developed and developing countries range from 5 kg/km^2 to more than 6000 kg/km^2 (AGA 2006) with the greatest concentrations in India, China, and Europe.

Manure disposal via land application has been used since early human history, but it has caused problems such as nitrate leaching to groundwater and to surface waters with excess rain and sandy soils. Runoff of manure into waters has caused increased microbial growth and, along with high biochemical oxygen demand, has led to anoxic conditions or altered aquatic environments (Hooda et al. 2000). Most commonly, feces on beef and dairy farms are usually collected as solids and later applied to land. There is an anaerobic slurry storage period that could decrease microorganisms, but it has been shown that certain pathogenic organisms can survive beyond the designated period (Pell 1997; Hooda et al. 2000).

Industrialized systems of animal agriculture currently produce 8 billion tonnes of waste per year. These intensive systems may continue to grow by 4 percent per year and it is estimated that in 2020 they will have 20 billion tonnes of animal waste for disposal (De Haan et al. 1997).

Numerous bacterial and protozoan pathogens can be found in animal wastes that affect human health and can affect the health of other animals. The application of animal manures to agricultural land is one route by which pathogens may be introduced into the environment and eventually end up in waters through poor irrigation practices and/or rainfall. Some of the bacteria of concern are *Salmonella* spp., *Campylobacter* spp., *Listeria monocytogenes*, *Mycobacterium paratuberculosis*, and *Escherichia coli* O157; the protozoa include *Cryptosporidium parvum* and *Giardia intestinalis* (Mawdsley et al. 1995; Pell 1997; Nicholson et al. 2000; Hill 2003).

The following pathogens have been detected in 10 to 50 percent of the animals during various surveys:

Cattle: *Salmonella*, *Listeria*, *E. coli* O157, *Campylobacter*, *Cryptosporidium*, and *Giardia* have all been found in cattle manures.

Pigs: *Salmonella*, *Listeria*, *E. coli* O157, *Campylobacter*, *Cryptosporidium*, and *Giardia* have all been isolated from pig manures.

Poultry: The most commonly found pathogens in poultry manure are *Salmonella* and *Campylobacter*.

Sheep: *Salmonella*, *E. coli* O157, *Campylobacter*, and *Cryptosporidium* have all been isolated from sheep manure.

Great strides have been made in the mapping of food consumption, human dimensions, and livestock-oriented production systems as well as water use (Gerbens-Leenes et al. 2002; Kruska et al. 2003). The tonnage of animal waste has been estimated on the basis of nitrogen in animal excreta (Bouwman and van der Hoek 1997); little effort has been made to address the microbial and zoonotic infectious disease potential associated with loading of manure onto land and water.

Impacts of shipping on transport of waterborne pathogens and indirect effects on aquatic ecosystems

The increase in the transportation of goods by water has increased the numbers of ships dramatically since World War II, from 80 292 large commercial vessels to 605 218 by 2003 (MBS 2005). This increase also means an increase in ballast water. Estimates suggest that 95 billion liters of ballast water enter U.S. waters per year. Ballast is defined as any solid or liquid that is brought on board a vessel to replace cargo, thereby stabilizing the ship's center of gravity. In most cases, coastal water picked up as ballast in one area carries with it a multitude of organisms that can then be disseminated globally and released as non-native and, in many cases, harmful species.

The 1990s gave rise to one of the largest cholera epidemics in Latin America, much of it waterborne. A total of 1 076 372 cases and 10 098 deaths due to cholera in the Americas were reported by June 1995, according to the Pan American Health Organization (PAHO). In 1994, cholera was detected for the first time from the Bug River (fresh water) in Poland, and soon afterwards Hong Kong reported two outbreaks of cholera (Lee *et al.* 1996). Although the cause of the Hong Kong outbreaks was not clearly identified, increasing pollution of coastal waters has been implicated. In 1991 and 1992, toxigenic *Vibrio cholerae* 01, serotype Inaba, biotype El Tor, was recovered from ballast water from five cargo ships docked in ports of the U.S. Gulf of Mexico (McCarthy and Khambaty 1994). Four of the ships had taken on ballast water in cholera-infected countries; the fifth took on ballast in a non-infected country. The isolates of *V. cholerae* detected in the ballast water were indistinguishable from the Latin America epidemic strain, but they were significantly different from an endemic strain found along the Gulf Coast.

The introduction of foreign aquatic life in national waters – indeed, the issue of non-indigenous species generally – is of growing concern. Government authorities have begun to regulate the discharge of ballast water from cargo ships entering all ports in the United States in an attempt to reduce the adverse environmental problems caused by the inadvertent introduction of foreign species (U.S. Department of Homeland Security 2003). The problem of non-indigenous species gained prominence in California where in the San Francisco Bay and Los Angeles–Long Beach ports more than 200 and 46 non-native species, respectively, are known to have been introduced (Cohen and Carlton 1998).

Although non-native species arrive by a variety of means, ballast water is a common way for exotic animals, as well as potentially pathogenic microorganisms, to reach domestic ports and waterways (Ruiz *et al.* 2000). Species that are unintentionally introduced into a new coastal environment come in many shapes and sizes and include algae, shellfish, developing larvae, eggs, and other microscopic organisms. Some well-documented "stowaway" species are zebra mussels, Eurasian ruffes, and Chinese mitten crabs (Mills *et al.* 1993; Ruiz *et al.* 2000).

Most of the aquatic invasive species would not be considered to have direct human health effects, though there are some indications that the zebra mussel invasion of the Great Lakes has increased the occurrence of harmful algal blooms associated with cyanobacteria and the presence of toxins. Thus, in addition to the role of nutrients (related to increases in non-point sources and storm waters), some evidence suggests that exotic herbivores interact in promoting the abundance and relative dominance of harmful algal blooms in the Great Lakes (Sarnelle et al. 2005).

Harmful algal blooms: complexities of bloom-associated health risks

The increased stresses on aquatic ecosystems associated with increased eutrophication from anthropogenic inputs along with climate change have been suggested as major causes for increasing reports of harmful algal blooms (Epstein 1998). Filamentous and colonial cyanobacteria are the most important taxa causing harmful phytoplankton blooms (harmful algal blooms, HABs) in lakes, rivers, and low-salinity estuaries (Fogg et al. 1973; Reynolds 1984; Paerl 1988). Cyanobacterial blooms reduce water transparency and recreational value, cause odor and taste problems, and can be toxic to both terrestrial and aquatic organisms (Carmichael and Falconer 1993; Chorus and Bartram 1999).

Blooms have disastrous short-term (toxicity and taste/odor problems, hypoxia/anoxia) and long-term (loss of fisheries and recreational resources, biogeochemical and trophic alterations) consequences for water quality and resource utilization of affected waters. These blooms also have an effect on human health (Carmichael 2001).

Human health concerns have been related to cyanobacteria blooms and the production of toxins. Microcystins and cyclic heptapeptides are the most common cyanobacterial toxins in fresh water and are often produced by *Microcystis* (Sivonen and Jones 1999; Codd 2000). Several cases of blue-green algae toxicosis in domestic animals have been recorded. A number of cattle died in a herd of 175 Hereford and Angus cattle in Burlington, Colorado, after ingesting water containing an algal bloom (Puschner et al. 1998). In February 1996, at a dialysis center in Caruaru, Brazil, 52 patients died from a syndrome now known as Caruaru syndrome, in which high concentrations of microcystin toxins were detected in the water used for the drinking supply. Individuals experienced visual disturbances, nausea, and vomiting after dialysis treatments, and 100 patients developed acute liver failure. Three main types of toxins have been identified: neurotoxins, hepatotoxins, and contact irritants (Repavich et al. 1990). Types of neurotoxins include anatoxin-a, anatoxin-a(s), and saxitoxin. Hepatotoxic poisoning results in anorexia, diarrhea, vomiting, and weakness (Carmichael *et al.* 2001). Hepatotoxins inhibit protein phosphatase enzymes, resulting in liver damage that can be severe enough to lead to fulminant hepatic failure. The most common hepatotoxin is microcystin, a cyclic heptapeptide with at least 65 known variants (Codd 2000). There is some evidence that microcystin may act as a tumor promoter (Falconer 1996).

The WHO has set guidelines based on acute animal toxicity studies of 1 μg/l for drinking water (Chorus and Bartram 1999). Management and prevention in the future must address the increase in nutrients from increased sewage, agricultural sources, and stormwater as well as the role of invasive species.

Climate impacts on water quality and health

"Climate" refers to average meteorological conditions over a specified time period (usually at least a month), including the frequency and intensity of extreme events and other statistical characteristics of the weather (NRC 2001). Climate change may have several latent effects, such as extreme heat, extreme storms, air pollution, and ecological shifts. As a result, global climate change could affect the geographic distribution of harmful pathogens, the pathogens' virulence, and their incidence, and cause increased occurrence and persistence of enteric pathogens in surface waters (Rose *et al.* 2001). Extreme storms usually cause an overburdening of wastewater treatment facilities and increased stormwater runoff, which in turn increases enteric pathogen levels nearby surface waters and possibly in groundwater. Intense rainfall events have already increased over the past century (Twilley *et al.* 2001). A statistical association between extreme precipitation and disease outbreaks via drinking water has been demonstrated in the United States (Curriero *et al.* 2001). In addition, winds and rain were shown to be key factors in the degradation of water quality in the Great Lakes and the Gulf of Mexico (Whitman *et al.* 1999; Lipp *et al.* 2001; Olyphant and Whitman 2004).

MONITORING THE WATER ENVIRONMENT AND IMPACTS ON HUMAN HEALTH

Water quality, in international basins in particular, and the ability or inability to monitor and influence protection have been shown to be associated with the severity of the water quality problem, the physical

> **Box 4.1** Disease detection
>
> Health surveillance: cases of disease diagnosed by the physician and/or laboratory and reported to state or national system.
> *Problem: most cases do not go to a physician or report the illness.*
> *No association with exposure identified.*
> Outbreak investigation: cases reporting more than one person affected at the same time and from the same water, usually a community outbreak.
> *Problem: outbreaks are extreme events and are not always recognized until a large proportion of the population is affected; underestimates amount of endemic disease.*
> Epidemiological studies that statistically relate the transmission of the disease to the health of the population exposed.
> *Problem: studies are too limited in time and space to relate water quality changes to disease; often water quality and exposure are not adequately addressed.*

and human setting, level of competitive uses, economic development, location/political setting, and agreements in place between the transboundary powers (Shmueli 1999). The protection of human health and imminent danger to health have always been drivers of action. There are a number of approaches for monitoring water quality impacts on the health of human populations. Measuring from the disease endpoint – that is, the number of people who are ill as a result of exposure to contaminated waters – has a number of problems (Box 4.1 illustrates some of the systems in place). These systems are passive and limited in recording the "true" amount of disease – that is, most illnesses are not reported. In most cases, the exposures and the type of microorganisms associated with the illnesses are not identified. These systems are after-the-fact and are not preventative or useful in examining the factors that led to the water quality degradation.

Direct water quality monitoring for fecal contamination and waterborne pathogens has been suggested and used for protection of watersheds, drinking water supplies, and recreational waters (NRC 2004). No global assessment of water quality has ever been undertaken, however, and data are fragmented both spatially and temporally. A consistent monitoring method has not even been advocated. And though reports of pathogen levels and indicator levels in waters throughout the world continue, no central clearing house for reporting or compiling these data exists.

The lack of data has been suggested as the main deterrent to developing a vested interest in water systems and the implementation of policies and programs to improve water quality at the local and national levels (Barker 2004).

To meet the Millennium Development Goals, access both to safe drinking water and to sanitation will be critical. Comprehensive watershed assessment, planning, and integrated programs to address both water quality and quantity are likely to be effective at the community level and may be much more cost-effective with long-term sustainability than plans that address temporary access only at the household level through household treatment (Souter et al. 2003). This approach underscores the need to enhance and refine the detection of contaminants in water, monitor changes in observable water quality, and assess exposure and the potential public health impacts and improvements made with investment in public works.

The indicator concept

Water quality assessment has been used for identifying designated uses of water bodies and impairment. The assessment of water quality for safe recreation and fishing uses very different approaches than assessment for use as a water supply, even though the same body of water may be used for all three activities. For recreation and drinking water, indicator bacteria as indicators of fecal contamination and potential risks to human health are used to determine safety. Chemical bioaccumulation is used to determine water safety for fishing.

For more than 100 years, the safety or the level of "purity" of waters has been measured via indicators. An indicator microorganism is defined as an organism normally found in the intestines of animals and humans, routinely shed in the feces in large numbers, and easily measured in water. An indicator is usually not a pathogen, but its presence indicates the potential for the presence of pathogenic microorganisms and hence the potential for disease. Historically, the indicator used was a group of enteric bacteria (defined by the ability to ferment lactose) known collectively as "total coliform bacteria." The total coliform bacteria are currently used in the United States and many other countries as an indicator of drinking water purity and safety via protection and treatment. The fecal coliform bacteria are a subgroup of the total coliform bacteria (defined by the ability to ferment lactose at an increased temperature of 44.5 °C) and were thought to be more indicative of fecal contamination of the water.

The fecal coliform bacteria have been used to monitor recreational waters and some wastewater discharges that may affect these types of waters. *Escherichia coli* is a specific bacterium that belongs to the fecal coliform group and is present in the gastrointestinal tracts of warm-blooded mammals as a non-pathogenic organism, though pathogenic strains are now recognized.

Another alternative fecal indicator, the enterococci bacteria, has been proposed specifically for recreational water. *Enterococcus* bacteria are a subgroup of the gram-positive fecal *Streptococcus*. They are characterized by their ability to grow at low and elevated temperatures (10 °C and 45 °C), at elevated pH (9.5) and in 6.5 percent sodium chloride (Schleifer and Kilpper-Batz 1987).

One of the largest single studies to date using indicators to examine water quality was carried out in 1914 between Canada and the United States and focused on total coliform bacterial concentrations in water. The goals were to assess the extent and causes and identify the localities in the boundary waters between the United States and Canada that had become so polluted as to be injurious to public health and unfit for domestic and other uses. At the time, large numbers of cases of typhoid fever were found throughout the basin, which supported a thriving 7 million people. Waterborne disease was associated with sewage discharges, but more evidence was needed to begin to ban untreated sewage discharges to key areas in the Great Lakes. In 1909, the United States and Great Britain (Canada) signed the Boundary Waters Treaty aimed at preventing and resolving disputes between the United States and Canada over waters forming the boundary between the two countries and also established a formal binational body, the International Joint Commission (IJC). The treaty was originally intended to protect lake levels and navigability, but it also formed the basis for the IJC to get involved in pollution problems in the boundary waters (Bilder 1972). Article IV of the treaty states that the "boundary waters and waters flowing across the boundary shall not be polluted on either side to the injury of health or property on the other." The interest in water quality and impacts on human waterborne disease instigated one of the most comprehensive bacteriological studies ever conducted in the Great Lakes region in 1913 by the IJC at the request of the two governments to address Article IV of the 1909 treaty. The study examined more than 2000 miles (3200 km) of sampling transects and collected more than 19 000 water samples. This study is possibly the most extensive characterization of bacterial pollution ever conducted, yet few scientists know of it (Durfee and Bagley 1997).

Microbial source tracking is now an emerging field of investigation. It uses the phenotypic or genetic specificity of the indicator bacteria and attempts to determine the source of the fecal pollution in waterways, whether it be human or animal (Sinton et al. 1998; Scott et al. 2002, 2005; Simpson et al. 2002; Griffith et al. 2003). These types of data can now be used within the global context to assign responsibility to upstream polluters.

Indicators for the protection of recreational waters

To meet the needs of the growing globalized tourist industry, a consistent and comparable water quality goal would seem imperative for protecting the health of visitors from all over the world who may frequent local beaches. A number of epidemiological studies performed since the 1980s on recreational waters found that diarrhea and respiratory illness in people using polluted water were best correlated with levels of the indicator organisms *E. coli* and enterococci. A meta-analysis of the studies over the past few decades supports the relationship of these two bacterial indicators with illness (Wade et al. 2003). Though there is a correlation (the higher the level of these bacteria in the water, the greater the risk of becoming ill while swimming), the specific relationships were shown to vary greatly by water type, indicator, and location (Pruss 1998). This was true in both marine and freshwater systems, though fewer studies have been undertaken in fresh waters. Therefore, developing a common threshold number below which the water quality is deemed "safe" has been challenging.

Clear deficiencies in the indicator system were recognized early on. Indicator bacteria did not always correlate with risk of disease or predict the potential for an outbreak. The indicators do not correlate with the presence of parasites and viruses in ambient waters – these pathogens can be detected in waters in the absence of the indicator – partly because of differences in survival and transport. The indicators as currently used can not indicate or differentiate the source of the fecal contamination (wildlife, farms, or urban sewage) (NRC 2004).

Despite the limitations, standards and guidelines for water quality for recreational waters have been developed using indicators. The U.S. Environmental Protection Agency (EPA 1986) developed new criteria and suggested guidelines for U.S. recreational waters based on geometric mean values of several (generally five or more) equally spaced samples over a 30-day period:

> Fresh water: *E. coli* not to exceed 126/100 ml
> Enterococci not to exceed 33/100 ml
> Marine water: Enterococci not to exceed 35/100 ml.

These standards were based on the log-linear relationships between mean *Enterococcus* or *E. coli* density/100 ml and swimming associated rates of gastrointestinal symptoms per 1000 persons (EPA 1986). Single sample maximum allowable densities were also developed. These values were based on risk levels of 8 and 19 gastrointestinal illnesses per 1000 swimmers at freshwater and marine beaches, respectively, and they were estimated to be equivalent to the risk levels for an older standard used by many (200/100 ml fecal coliform). Only about 30 percent of the states have adopted these as state recreational standards to date.

The WHO has taken a slightly different approach to regulation of recreational waters (WHO 2003b) (Table 4.3). The guidance was based on the epidemiological studies that supported the use of enterococci, but rather than a single number and mean, the 95th percentile value was used. This again was for mainly marine systems; and freshwater recreational waters were not addressed.

The WHO has also recognized the importance of sewage as a source of pathogens and has suggested that, in the case of "microbial presence of human sewage" (e.g., due to a pipeline breakage), the 95th percentile value of intestinal enterococci/100 ml greater than 500 (or greater than 200 if the source is mainly from human fecal pollution) in consecutive samples is a source of high risk. Beyond fecal contamination, the WHO guidelines also consider "algal and cyanobacterial, presence of scums or detection of 100,000 cells/ml" a health hazard.[1]

---

[1] In the United States, for drinking water, only the total coliform is used as a standard test after treatment and in the distribution system and should be zero in 100-ml samples. Yet the language in the rule is such that it is the measurement of *E. coli* which would lead to boil orders and more significant ramifications. The WHO has recommended that *E. coli* must not be detectable in any 100-ml sample. For treated water entering or in the distribution system the same recommendation is also given for total coliform bacteria, with a provision that allows up to 5 percent positive samples within the distribution system (WHO 2001). There have been no specific standards or guidelines set for ambient water indicator levels for fresh waters used for potable supplies.

Table 4.3 *WHO guidelines for marine recreational waters*

95th percentile enterococci/100 ml value	Estimated risk per exposure	
	Gastrointestinal illness	Acute febrile respiratory illness
≤40	<1%	<0.3%
41–200	1–5%	0.3–1.9%
201–500	5–10%	1.9–3.9%
>500	>10%	>3.9%

Direct pathogen monitoring

An understanding of the ecology of the infectious agents themselves is often necessary particularly given that indicators are not correlated with viruses and parasites. These data are necessary to address risk and prevention strategies and have been used to examine prioritization of large expenditures of funds for the upgrading of sewage treatment and disposal methods. Pathogen monitoring will be a part of the future for characterizing biological hazards in water (NRC 2004). Any pathogen of concern can now be monitored for in water. Filtration techniques, followed by clarification methods (e.g., immunomagnetic separation), cultivation, antibody-based detection, or molecular detection, have been used to detect bacteria, viruses, and parasites. Molecular methods can now be developed for any microorganism of interest. Most recently the National Academy of Sciences, the Organisation for Economic Co-operation and Development, and the American Academy of Microbiology have all summarized the methods available for monitoring any pathogen of interest in water (OECD 1998; Rose and Grimes 2001; NRC 2004).

Cryptosporidium and *Giardia* sampling have been used to address ambient waters and have been the focus of the adequacy of watershed protection schemes. The monitoring was used directly in decisions in New York City about whether to invest in a large water filtration facility for the water supply or in watershed protection measures (Okun *et al.* 1997). In addition, viruses have been the major driver of efforts to restore water quality in coastal systems worldwide and in recreational waters, such as the reef environment and the Florida Keys (Griffin *et al.* 2001, 2003).

FUTURE DIRECTIONS FOR THE PROTECTION OF HUMAN HEALTH, WATER QUALITY, AND DESIGNATED USE

It has been suggested that water is key to improving the economic status of nations, as well as people's health and protection of ecosystems (Falkenmark 2002). As globalization makes its mark on the waters of the world, it is important to understand the stresses put on aquatic systems to facilitate better strategies for understanding:

Indicators of aquatic and water-related human health impacts
Pollution prevention at the watershed level
Sustainability of the water resource

Waterborne disease has not abated. Microbial contamination and human health risks have increased for both recreational and drinking waters. In the past, there was considerable discussion about the quantity of water, which was often addressed as the critical issue, regardless of water quality, for improving the health status of populations. Independent of sanitation issues, it was thought that waterborne infectious disease could be addressed through drinking water treatment, vaccination, and improved medical care (antibiotics). However, the world is much different several decades later than anticipated. Global changes have contributed not only to emerging and re-emerging infectious diseases but also to increased risk of waterborne disease due to increases in populations, increases in wastewater loading to the water environment, increases in non-point source discharges (e.g., ballast), and lack of infrastructure. This situation has been exacerbated by climate variations and ecosystem disruptions.

There is no doubt that water quality degradation – in particular, fecal contamination from humans and/or animals – has contributed to the global burden of waterborne disease. Improved waterborne disease surveillance and reporting are needed on a global level. This has been encouraged by the WHO, but investment in public health infrastructure is needed. The AIDS crisis has mobilized efforts to improve diagnosis and treatment of infectious diseases. Along with these efforts, a focus on environmental health should be encouraged because this will improve the plight of millions of people.

Improved water quality testing and reporting are also necessary. Information and access to information lead to empowerment and change. It is acknowledged that a water crisis exists, but the emphasis has been on measuring water quantity, when it is water quality that directly affects human health. The most significant cause of deaths and illnesses in the developing world is the contamination of water. Conventional methods (APHA 2005) as well as the development and application of new tools for the assessment of water quality (Rose and Grimes 2001) can be used at the watershed level to identify sources of contamination. Significant advances in molecular technology allow for characterization of water quality and related public health risks (OECD 1998). The impact of these new tools on decisions made to protect and treat water would be immense, and could protect public health at a grand scale. Water projects are implemented at the local level, often with local solutions, but harmonization of approaches is needed. Knowledge is power. Local decisions can be integrated at the watershed level, and local data can feed into a large system database. These actions will mobilize the global community and the political will toward the choices that will need to be made as we face the impacts of further globalization on water resources.

References

AGA. 2006. *Livestock Atlas*, series 1, *Livestock Geography: New Perspectives on World Resources*. Available online at http://ergodd.zoo.ox.ac.uk/livatl2/Livestock.htm

APHA (American Public Health Association, American Water Works Association, and Water-Environment Federation) 2005. *Standard Methods for the Examination of Water and Wastewater*, 21st edn. Denver, CO American Water Works Association.

Barker, S. 2004. Staying on target. *Environmental Protection* February: 23-27.

Bilder, R. B. 1972. Controlling Great Lakes pollution: a study in the United States Canadian environmental cooperation. *Michigan Law Review* **70**: 469-556.

Borchardt, M. A., Bertz, P. D., Spencer, S. K., and Battigelli, D. A. 2003. Incidence of enteric viruses in groundwater from household wells in Wisconsin. *Applied Environmental Microbiology* **69**: 1172-1180.

Bouwman, A. F. and van der Hoek, K. W. 1997. Scenarios of animal waste production and fertilizer use and associated ammonia emission for the developing countries. *Atmospheric Environment* **31**: 409-412.

Bradford, E., Baldwin, R. L., Blackburn, H., *et al.* 1999. *Animal Agriculture and the Global Food Supply*. Ames, IA: Council on Agriculture, Science and Technology.

Brown, K. 2002. Water security: forecasting the future with spotty data. *Science* **297**: 926-927.

Carmichael, W. W. 2001. Health effects of toxin-producing Cyanobacteria: "The CyanoHABs". *Human and Ecological Risk Assessment* **7**: 1393-1407.

Carmichael, W. W., and Falconer, I. R. 1993. Diseases related to freshwater blue-green algal toxins and control measures. In *Algal Toxins in Seafood and Drinking Water*, ed. I. R. Falconer. London: Academic Press, pp. 187-209.

Carmichael, W. W., Azevedo, S. M. F. O., An, J. S., *et al.* 2001. Human fatalities from Cyanobacteria: chemical and biological evidence for cyanotoxins. *Environmental Health Perspectives* **109**: 663-668.

Chorus, I. and Bartram, J. 1999. *Toxic Cyanobacteria in Water: A Guide to Their Public Health Consequences, Monitoring and Management*. Geneva, Switzerland: World Health Organization.

Codd, G. A. 2000. Cyanobacterial toxins, the perception of water quality, and the prioritization of eutrophication control. *Ecological Engineering* **16**: 51-60.

Cohen, A. N. and Carlton, J. T. 1998. Accelerating invasion rate in a highly invaded estuary *Science* **279**: 555-558.

Crosette, B. 1996. Hope, and pragmatism, for U.N. cities conferences, *New York Times*, June 3, 1996, p. A3.

Curriero, F. C., Patz, J. A., Rose, J. B., and Lele, S. 2001. The association between extreme precipitation and waterborne disease outbreaks in the United States, 1948-1994. *American Journal of Public Health* **91**: 1194-1199.

De Haan, C., Steinfeld, H., and Blackburn, H. 1997. *Livestock and the Environment: Finding a Balance – Issues and Options*. Rome: Food and Agriculture Organization of the United Nations.

Delgado, C., Rosegrant, M., Steinfeld, H., Ehui, S., and Courbois, C. 1999. *Livestock to 2020: The Next Food Revolution*. Food, Agriculture, and the Environment Discussion Paper No. 28. Rome: Food and Agriculture Organization of the United Nations. Available online at www.fao.org/waicent/faoinfo/agricult/AGA/agal/lvst2020/20201.pdf

Durfee, M. and Bagley, S. T. 1997. Bacteriology and diplomacy in the Great Lakes 1912-1920. Paper prepared for the 1997 meeting of the American Society for Environmental History, Baltimore, MD, March 6-9, 1997.

EPA (Environmental Protection Agency). 1986. *Ambient Water Quality Criteria for Bacteria – 1986*, EPA440/5-84-002. Washington, DC: Environmental Protection Agency.

Epstein, P. R. 1998. *Marine Ecosystems: Emerging Diseases as Indicators of Change*. Boston, MA: Center for Health and the Global Environment, Harvard Medical School. Available online at www.med.harvard.edu/chge

Falconer, I. R. 1996. Potential impact on human health of toxic cyanobacteria. *Phycologia* **35**(6): 6-11.

Falkenmark, M. 2002. Summary and conclusion of the 2001 Stockholm Water Symposium. *Water Science Technology* **45**(8): 1-4.

Fogg, G. E., Stewart, D. P., Fay, P., and Walsby, A. E. 1973. *The Blue-Green Algae*. London: Academic Press.

Gerbens-Leenes, P. W., Nonhebel, S., and Ivens, W. P. M. F. 2002. A method to determine land requirements relating to food consumption patterns. *Agriculture, Ecosystems and Environment* **90**: 47-58.

Gleick, P. H. 1993. *Water in Crisis: A Guide to the World's Fresh Water Resources*. New York: Oxford University Press.

Gleick, P. H. 1998. *The World's Water 1998-1999*. Washington, DC: Island Press.

Griffin, D. W., Lipp, E. K., McLaughlin, M. R., and Rose, J. B. 2001. Marine recreation and public health microbiology: quest for the ideal indicator. *BioScience* **51**: 817-825.

Griffin, D. W., Donaldson, K. A., Paul, J. P. and Rose, J. B. 2003. Pathogenic human viruses in coastal waters. *Clinical Microbiology Review* **16**: 129-143.

Griffith, J. F., Weisberg, S. B., and McGee, C. D. 2003. Evaluation of microbial source tracking methods using mixed fecal sources in aqueous test samples. *Journal of Water and Health* **01**: 141-151.

Haas, C. N., J. B. Rose, and C. P. Gerba. 1999. *Quantitative Microbial Risk Assessment*. New York: John Wiley.

Henrickson, S. E., Wong, T., Allen, P., Ford, T., and Epstein, P. R. 2001. Marine swimming-related illness: implications for monitoring and environmental policy. *Environmental Health Perspectives* **109**: 645-650.

Hill, V. R. 2003. Prospects for pathogen reductions in livestock wastewaters: a review. *Critical Reviews in Environmental Science and Technology* **33**: 187-235.

Hooda, P. S., Edwards, A. C., Anderson, H. A., and Miller. 2000. A review of water quality concerns in livestock farming areas. *Science of the Total Environment* **250**: 143-167.

Hrudey, S. E. and Hrudey, E. J. 2004. *Safe Drinking Water: Lessons from Recent Outbreaks in Affluent Nations*. London: International Water Association.

Kruska, R. L., Reid, R. S., Thornton, P. K., Henninger, N., and Kristjanson, P. M. 2003. Mapping livestock-oriented agricultural production systems for the developing world. *Agricultural Systems* **77**: 39-63.

Lee, S. H., Lai, S. T., Lai, J. Y., and Leung, N. K. 1996. Resurgence of cholera in Hong Kong. *Epidemiology and Infection* **117**: 43-49.

Lipp, E. K., Schmidt, N., Luther, M. E., and Rose, J. B. 2001. Determining the effects of El Niño-Southern Oscillation events on coastal water quality. *Estuaries* **24**: 491-497.

Mawdsley, J. L., Bardgett, R. D., and Merry, R. J. 1995. Pathogens in livestock waste, their potential for movement through soil and environmental pollution. *Applied Soil Ecology* **2**: 1-15.

MBS (Maritime Business Strategies). 2005. *Growth of the the World Fleet since WWII*. Available online at www.coltoncompany.com/shipping/statistics/wldfltgrowth.htm

McCarthy, S. A. and Khambaty, F. M. 1994. International dissemination of epidemic *Vibrio cholerae* by cargo ship ballast and other nonpotable waters. *Applied Environmental Microbiology* **60**: 2597-2601.

Meyland, S. J. and Machlin, A. 2001. *The Plain Language Guide to Sanitary Sewer Overflows: A Brief Introduction to the Basics of SSOs and What You Can Do to Eliminate them*. Farmingdale, NY: Citizens' Environmental Research Institute.

Mills, E. L., Leach, J. H., Carlton, J. T., and Secor, C. L. 1993. Exotic species in the Great Lakes – a history of biotic crises and anthropogenic introductions. *Journal of Great Lakes Research* **19**: 1-54.

NRC (National Research Council). 1993. *Managing Wastewater in Coastal Urban Areas*. Washington, DC: National Academy Press.

NRC. 1999. *Viability of Augmenting Potable Water Supplies with Reclaimed Water*. Washington, DC: National Academy Press.

NRC. 2001. *Under the Weather*. Washington, DC: National Academy Press.

NRC. 2004. *Indicators of Waterborne Pathogens*. Washington, DC: National Academy Press.

NRDC (National Resources Defense Council). 2001. *Oceans: In Brief – Increased Monitoring and Reporting Reveals Widespread Pollution at U.S. Beaches*. Washington, DC: National Resources Defense Council.

Nicholson, F. A., Hutchison, M. L., Smith, K. A., et al. 2000. *A Study on Farm Manure Applications to Agricultural Land and an Assessment of the Risks of Pathogen Transfer into the Food Chain*. London: Ministry of Agriculture Fisheries and Food.

OECD (Organisation for Economic Co-operation and Development). 1998. *Molecular Technologies for Safe Drinking Water: Results from the Interlaken Workshop*, Switzerland, 5-8 July 1998. Available online at www.oecd.org/dataoecd/34/8/2097510.pdf

Okun, D. A., Craun, G. F., Edzwald, J. K., Gilbert, J. B., and Rose, J. B. 1997. New York City: to filter or not to filter? *Journal of the American Water Works Association* **89**(3): 62-74.

Oldeman, L. R., Hakkeling, R. T. A., and Sombroek, W. G. 1991. *World Map of the Status of Human-Induced Soil Degradation: Global Assessment of Soil Degradation*. Wageningen, Netherlands: International Soil Reference and Information Centre.

Olyphant, G. A. and Whitman, R. L., 2004. Elements of a predictive model for determining beach closures on a real time basis: the case of 63rd Street Beach Chicago. *Environmental Monitoring and Assessment* **98**: 175-190.

Paerl, H. W. 1988. Nuisance phytoplankton blooms in coastal, estuarine, and inland waters. *Limnology and Oceanography* **33**: 823-847.

Paul, J. H., Rose, J. B., Brown, J., et al. 1995. Viral tracer studies indicate contamination of marine waters by sewage disposal practices in Key Largo, Florida. *Applied Environmental Microbiology* **61**: 2230-2234.

Pell, A. N. 1997. Manure and microbes: public and animal health problems? *Journal of Dairy Science* **80**: 2673-2681.

Pruss, A. 1998. Review of epidemiological studies on health effects from exposure to recreational water. *International Journal of Epidemiology.* **27**: 1-9.

Puschner, B., Galey, F. D., Johnson, B., et at. 1998. Blue-green algae toxicosis in cattle. *Journal of the American Veterinary Medicine Association* **213**: 1605-1607.

Repavich, W. M., Sonzogni, W. C., Standridge, J. H., Wedepohl, R. E., and Meisner, L. F., 1990. Cyanobacteria (blue-green algae) in Wisconsin waters: acute and chronic toxicity. *Water Research* **24**: 225-231.

Reynolds, C. S. 1984. *The Ecology of Freshwater Phytoplankton*. Cambridge, UK: Cambridge University Press.

Rose, J. B. and Grimes, D. J. 2001. *Reevaluation of Microbial Water Quality: Powerful New Tools for Detection and Risk Assessment*. Washington, DC: American Academy of Microbiology.

Rose, J. B., Epstein, P. R., Lipp, E. K., et al. 2001. Climate variability and change in the United States: potential impacts on water- and foodborne diseases caused by microbiologic agents. *Environmental Health Perspectives* **109** (S2): 211-221.

Ruiz, G. M., Fofonoff, P. W., Carlton, J. T., Wonham, M. J., and Hines, A. H. 2000. Invasion of coastal marine communities in North America: apparent patterns, processes, and biases. *Annual Review of Ecology and Systematics* **31**: 481-531.

Sarnelle, O., Wilson, A. E., Hamilton, S. K., Knoll, L. B., and Raikow, D. E. 2005. Complex interactions between exotic zebra mussels and the noxious phytoplankter, *Microcystis aeruginosa*. *Limnology and Oceanography* **50**: 896-904.

Schleifer, K. H., and Kilpper-Batz, R. 1987. Molecular and chemotaxonomic approaches to the classification of streptococci, enterococci, and lactococci: a review. *Systematic and Applied Microbiology* **10**: 1-19.

Scott, T. M., Rose, J. B., Jenkins, T. M., Farrah, S. R., and Lukasik, J. 2002. Microbial source tracking: current methodology and future directions. *Applied Environmental Microbiology* **68**: 5796-5803.

Scott, T. M., Jenkins, T. M., Lukasik, J., and Rose, J. B. 2005. Potential use of a host associated molecular marker in *Enterococcus faecium* as an index of human fecal pollution; *Environmental Science and Technology* **39**: 283-287.

Shmueli, D. F. 1999. Water quality in international river basins. *Political Geography* **18**: 437-476.

Simpson, J. M., Santo Domingo, J. W., and Reasoner, D. J. 2002. Microbial source tracking: state of the science. *Environmental Science and Technology* **36**: 5279-5288.

Sinton, L. W., Finlay, R. K., and Hannah, D. J. 1998. Distinguishing human from animal faecal contamination in water: a review. *New Zealand Journal of Marine Freshwater Research* **32**: 323-348.

Sivonen, K. and Jones, G. 1999. Cyanobacterial toxins. In *Toxic Cyanobacteria in Water: A Guide to Their Public Health Consequences, Monitoring and Management*, eds. I. Chorus and J. Bartram. London: E. & F. N. Spon, pp. 41-111.

Souter, P. F., Cruickshank, G. D., Tankerville, M. Z., et al. 2003. Evaluation of a new water treatment for point-of-use household applications to remove microorganisms and arsenic from drinking water. *Journal of Water and Health* **1**(2): 73–84.

Twilley, R. R., Barron, E. J., Gholz, H. L., et al. 2001. *Confronting Climate Change in the Gulf Coast Region: Prospects for Sustaining Our Ecological Heritage*. Cambridge, MA: The Union of Concerned Scientists, and Washington, DC: Ecological Society of America.

UNDP (United Nations Development Programme). 2003. *Human Development Report 2003*. New York: United Nations.

United Nations. 2003. *World Environment Day 2003, June 5th*. Available online at www.un.org; www.iwahq.org.uk

United Nations. 2005. *Millennium Development Goals*. Available online at http://unstats.un.org/unsd/mi/pdf/MDG%20Book.pdf

U.S. Department of Homeland Security. 2003. Coast Guard Mandatory Ballast Water Management Program for U.S. Waters. *Federal Register* July 30, **68**(146): 44691–44696.

Wade, T. J., Pai, N., Eisenberg, J. N., Colford, J. M., Jr. 2003. Do U.S. Environmental Protection Agency water quality guidelines for recreational waters prevent gastrointestinal illness? A systematic review and meta-analysis. *Environmental Health Perspective* **111**: 1102–1109.

Whitman, R. L., Nevers, M. B., and Gerovac, P. J. 1999. Interaction of ambient conditions and fecal coliform bacteria in southern Lake Michigan beach waters: Monitoring program implications. *Natural Areas Journal* **19**: 166–171.

WHO (World Health Organization). 2001. *Water Quality Guidelines, Standards and Health: Assesment of Risk and Risk Management for Water-Related Infectious Disease*, eds. J. Fewtrell, and J. Bartram. Avaliable online at www.who.int/water_sanitation_health/dwq/whoiwa/en/index1.html

WHO. 2003a. *World Health Report 2003: Shaping the Future*. Available online at www.who.int/whr/en/

WHO. 2003b. *Guidelines for Safe Recreational Waters*, vol. 1, *Coastal and Fresh Waters*. Available online at www.who.int/water_sanitation_health/bathing/srwe1/en/

World Bank. 2000. *World Development Report 1999/2000*. Available online at www.worldbank.org/wdr/2000/fullreport.html

World Water Council. 2003. *The 3rd World Water Forum Opens March 16th: Crucial Water Issues to Be Addressed*. Marseilles, France: World Water Council. Available online at www.worldwatercouncil.org/fileadmin/wwc/News/WWC_News/

Zhang, H. n.d. *Nine Dragons, One River: The Role of Institutions in Developing Water Pricing Policy in Beijing*. Available online at http://chs.ubc.ca/china/PDF%20Files/Zhang/Box%204.1.pdf

5

Health challenges to aquatic animals in the globalization era

INTRODUCTION

As the globalization era progresses, societies and world economies have been transformed dramatically, mainly through increased international trade and cultural exchange. Advances in technology, major biotechnological innovations, and greater international movements of people and commodities have created a system that metaphorically unites the world into one global village; goods and services produced in one part of the world can easily be transported and made available in all parts of the globe. As much as the globalization era has created new challenges, it has also inherited many unresolved challenges of the past. How will a world, divided into over 200 sovereign countries and territories, cope with these unthinkably fast globalization processes, while preserving the environmental integrity of our planet? In particular, achieving a balance among globalization processes, the growing economy, preservation of biodiversity, protection against biosecurity threats, and safeguarding the health of the fragile ecosystem seems to be a distant dream that would be difficult to realize.

Concomitant with the emergence of the globalization era, mass mortalities have been observed over a wide range of farmed and wild aquatic animals, including fish, mollusks, crustaceans, harbor seals, manatees, turtles, frogs, coral organisms, and sea urchins (Heide-Jorgensen and Harkonen 1992; Lafferty and Kuris 1993; Moyer et al. 1993; Littler and Littler 1995; Altstatt et al. 1996; Rahimian and Thulin 1996; Jones et al. 1997; Bossart et al. 1998; Ford et al. 1999; Harvell et al. 1999; Rosenberg and Loya 2004). In addition to these mortalities, many aquatic animals are at the brink of extinction, and many more cannot reproduce, thereby causing their fisheries to collapse. Although a direct cause-and-effect relationship between globalization and aquatic animal health

Globalization: Effects on Fisheries Resources, ed. William W. Taylor, Michael G. Schechter, and Lois G. Wolfson. Published by Cambridge University Press. © Cambridge University Press 2007.

deterioration cannot be proven with precision, there are many correlations through which multiple links can be drawn. Needless to say, dramatic increases in demand for the live aquatic animal trade, explosive expansion in aquaculture, increased urbanization and industrialization, invasion by non-indigenous species, catastrophic climatic events, and continuous influxes of toxic chemicals into water bodies top the list of links. In this chapter, the risk of transferring microbes that are pathogenic to aquatic animals through domestic and international trade will be discussed. In addition, efforts by the international community to minimize the risks of pathogen transfer without unduly impeding live aquatic animal trade will be addressed.

The aquatic environment differs substantially from the terrestrial environment, even though many broad types of disease-causing organisms occur both on land and in the water. Some emerging disease problems in aquatic environments are associated with pathogens moving from terrestrial to aquatic systems such as toxoplasmosis in sea otters (Kreuder *et al.* 2003) and aspergillosis in sea fans (Kim and Harvel 2004). Major differences between aquatic and terrestrial environments include the following:

(1) Taxonomic diversity, with its associated anatomical features and physiological functions, is far more evident in aquatic rather than terrestrial animals. This diversity definitely influences disease transmission modes and many other aspects of the disease process.

(2) Aquatic populations are reproductively typically more open than terrestrial ones, with the potential for long-distance dispersal of larvae. As a result, the rates with which epidemics spread in the aquatic environment are more rapid than those observed for terrestrial pathogens. For example, herpes virus ravaged pilchard populations in the Southern Ocean and the morbillivirus epidemic spread rapidly among marine mammals (McCallum *et al.* 2003). The rapid spread of the pilchard epidemic in Australia was not merely a result of directional transport in currents, as it spread against the prevailing currents (McCallum *et al.* 2003).

(3) In the case of non-motile, colonial aquatic animals, such as sponges and coral organisms, rates of infection transmission of pathogens are higher than in other motile aquatic or terrestrial animals due to the ease of pathogen transmission among adjacent susceptible hosts.

(4) Modes of disease transmission are different between aquatic and terrestrial organisms. For example, vertical transmission, which is important for many terrestrial diseases, has been proven in only a few aquatic diseases such as the bacterial kidney disease of salmonids. In the same context, transmission through a vector, which is common in mammalian diseases, appears to be rare in aquatic animals diseases, in spite of well-documented examples of blood parasites of fish which use leeches and gnathiid isopods as vectors (Davies and Smit 2001), fireworms which spread infection of *Vibrio* sp. among corals (Sussman *et al.* 2003), and invertebrate intermediate hosts for fish worms.

Proper understanding of aquatic animal diseases is compounded with a number of constraints. For example, baseline health data for the presence of pathogens and diseases of aquatic animals is lacking and disease databases are either non-existent or relatively primitive. This lack of knowledge impedes accurate disease risk analysis, increases the difficulty of differentiating between exotic and endemic infections, and hinders the selection of disease management options. In the same context, information on host ranges (i.e., all species susceptible to infection) is vastly lacking for most pathogens of aquatic animals. There is an urgent need to develop more sensitive diagnostic tools to detect subclinical carriers, at least for pathogens that have a significant economic impact on production and trade.

WIDESPREAD MOVEMENTS OF AQUATIC ANIMALS AND THEIR PATHOGENS

As long as archives of natural history have been kept, movements of species ranging in size from microscopic organisms to gigantic animals have been well documented. The majority of these movements occurred naturally through migration patterns and often was controlled by climatic events. In the last few decades, however, technological advances coupled with exponential increases in world trade have led to countless intentional and unintentional movement of species to habitats where they are not native. Parallel to species movements, symbionts, parasites, and pathogenic microorganisms have invaded new geographic ranges and encountered native species, often leading to catastrophic consequences.

Movements of aquatic animals exceed those of their terrestrial counterparts. The current wide-scale international movement of aquatic animals species, fish and shellfish in particular, is far higher than terrestrial animals (FAO 2005). In 1996, Bartley and Subasinghe published results of a worldwide survey which demonstrated that national governments were responsible for 40 percent of fish and shellfish species introductions, with the private sector accounting for 18 percent, individuals another 15 percent, and international organizations 7 percent, with the remaining 20 percent being of unknown source. The same survey showed that aquaculture is the major motive for most introductions, followed by international trade in live and frozen fish and shellfish, creation of new sport and commercial fisheries, and the aquarium trade. Further, the discharge of ballast water introduces harmful aquatic organisms, including diseases, bacteria, and viruses, to both marine and freshwater ecosystems, thereby degrading commercially important fisheries and recreational opportunities.

Aquaculture and pathogen introduction

Concomitant with the eruption of the globalization era, the aquaculture segment of animal protein production worldwide has increased at an annual rate of over 10 percent, and this increase is more than double the growth of other terrestrial animal commodities (Moffitt 2005). Currently, aquaculture contributes to over 30 percent of the global food fish production (FAO 2005). Since most international aquaculture practices are based on species diversification, introduction and transfers of live non-indigenous aquatic animals with aquaculture potentials continues to be on the rise to the extent that some species such as the Nile tilapia (*Oreochromis niloticus*) became global in their distributions and is being raised in almost every country in the world (Coward and Little 2001).

From historic and recent records, it is clear that aquaculture has been a channel through which many parasites and pathogens of aquatic animals have spread (Bartley and Subasinghe 1996; Lightner *et al.* 1997; Naylor *et al.* 2000; Murray and Peeler 2005). In addition to pathogen introduction, fish stressed from high densities and other aquaculture practices become more vulnerable to infection by microbes of low virulence to which they would have otherwise been more resistant. These pathogens can then spread to wild fish populations through water or escapees (Baringa 1990; Blanc *et al.* 1997), thereby causing losses in neighboring wild stocks.

Some fish pathogens exist in asymptomatic carrier states for long periods and can be transmitted via vertical modes and some resist topical egg disinfections. One such pathogen is *Renibacterium salmoninarum*, the causative agent of bacterial kidney disease in salmonines. This slowly progressing systemic infection often causes high losses among susceptible cultured and wild populations (Fryer and Sanders 1981; Faisal and Hnath 2005). The vertical transmission nature of *R. salmoninarum* coupled with the ease of egg transport contributed to the global spread of bacterial kidney disease among salmonid populations. *Renibacterium salmoninarum* is believed to have spread from the United States to Japan through import of eggs from Oregon and Washington (Yoshimizu 1996). Other pathogens capable of causing heavy mortalities in salmonines, such as *Flavobacterium psychrophilum*, the causative agent of cold-water disease, infectious pancreatic necrosis virus (IPNV, Birnaviridae), infectious hematopoeitic necrosis virus (IHNV, Rhabdoviridae), and viral erythrocyte inclusion body syndrome are also believed to have been introduced to Japan via importation of eggs from Oregon and Washington (Yoshimizu 1996). Surface disinfection of eggs with iodophors failed to eliminate viruses from contaminated eggs (Sano 1973).

Through importation of fish for aquaculture purposes into the Asia Pacific region from multiple sources some diseases and pathogens, including the copepod *Lernaea cyprinacea*, myxosporeans of the genus *Myxobolus*, and the epizootic ulcerative syndrome, have spread throughout much of South and Southeast Asia (Djajadiredja et al. 1983; Tonguthai 1985; Arthur and Shariff 1991; Lilley et al. 1992). It is estimated that epizootic ulcerative syndrome caused tremendous financial losses in Thailand during 1982–83 alone.

Taura syndrome of penaeid shrimp is another example of an emerging disease that has been distributed transcontinentally through aquaculture. The syndrome is caused by the taura syndrome virus (TSV, Picornaviridae), which causes devastating losses in cultured and wild *Penaeus vannamei*. The disease was first discovered in Ecuador in 1992 and within 2 years the virus had spread rapidly and caused massive production losses in most shrimp-growing countries in the Americas. The disease reached Florida through shrimp shipments from Central America (Hasson et al. 1995; Lightner et al. 1995; Lightner 1996). Another example of shrimp disease introduction involves the infectious hypodermal and hematopoietic necrosis virus (IHHNV, Parvoviridae) which causes serious epizootics in cultured *Penaeus stylirostris* juveniles. Lightner et al. (1992a) and Lightner (1996) documented many case histories of IHHNV spread into new geographical regions with the movement of shrimp stocks for aquaculture. Some of the accidental introductions of IHHNV into Hawaii and Mexico have resulted in serious consequences for the shrimp industry in those locations (Moore 1991; Moore and Brand 1993).

In France, multiple uncontrolled introductions of the Pacific oyster (*Crassostrea gigas*) from Japan, Korea, British Columbia, and California are believed to have spread an iridovirus which devastates the native Portuguese oyster (*Crassostrea angulata*) populations. The Pacific oyster also carried the microsporidian *Marteilia refringens* which has decimated European flat oyster (*Ostrea edulis*) populations and the bivalve-pathogenic protozoan *Bonamia ostreae*. With continuous movements of bivalves for aquaculture and conservation purposes, these three diseases spread from France to many European countries leaving trails of heavy losses to bivalve growers (Renault 1996).

International trade in live and frozen food fish

Through trade involving infected live or frozen fish, many trans- and intercontinental movements of important parasites and pathogens have been documented (Hoffman 1970, 1990). For example, the eel swimbladder nematode (*Anguillicola crassus*) which caused serious losses in European eel (*Anguilla anguilla*) populations throughout Europe in the late 1980s and early 1990s (Molnar *et al.* 1991) has been introduced through imports of live Japanese eels (*A. japonica*) into Germany from Taiwan (Haenen 1995) and into Italy from New Zealand (Ghittino *et al.* 1989).

Gaffkemia, caused by *Aerococcus viridans* var. *homari*, is a bacterial disease that is enzootic in North America causing no or negligible harm to the American lobster (*Homarus americanus*), but when introduced to Europe through lobster wholesale and retail trades, it caused severe mortalities in the European lobster (*H. gammarus*). Alderman (1996) identified one such example where an infection was introduced via a shipment of live American lobsters to a holding site in Ireland. The imported American lobsters were held along with a group of European lobsters that were later shipped to Wales. A portion of these European lobsters was further shipped to a facility in the Netherlands. Within a few days most European lobsters in both Welsh and Dutch facilities suffered severe mortalities due to gaffkemia. It was determined that unlike the American lobster, the European lobster is extremely vulnerable to *A. viridans* infection.

Through multiple trans-Atlantic movements of the North American signal crayfish (*Pacifastacus leniusculus*) and Louisiana swamp crayfish (*Procambarus clarkii*), the fungus *Aphanomyces astaci* has been introduced to Europe and has consequently spread across the continent (Rahe 1987; Alderman et al. 1990). North American crayfish species are resistant carriers to *A. astaci*, while in the European crayfish (*Astacus astacus*), the pathogenic fungus causes an acute infection, known as crayfish plague, associated with devastating mortalities (Alderman 1996). Crayfish plague vastly reduced numbers of native crayfish in Europe.

Live bait (e.g., minnow, frogs, worms, and squid) is routinely used by recreational and commercial fishermen to catch fish but no health certification is required for the transfer of live bait from one region to another. This practice has the potential to spread pathogens into non-infected geographical areas (Hoole et al. 2001; Gaughan 2002). Even dead bait may include live fellow travelers, such as parasites or diseases, whether in the bait itself or in the carrying medium, and many fishermen dump their bait after a fishing trip.

Frozen food can also introduce pathogens to new environments. For example, the importation of frozen shrimp from enzootic regions can transmit devastating viral diseases such as TSV and IHHNV which remain infectious after one or more freeze–thaw cycles (Berry et al. 1994; Lightner 1996). Whirling disease of salmonids, caused by the protozoan *Myxobolus cerebralis*, was first introduced to the United States through a shipment of frozen trout from Denmark (Marnell 1986). The parasite was first detected in Pennsylvania and has since spread to Midwestern and western states through fish stocking, fishing equipment and gear, and recreational boats (Yoder 1972; Faisal and Garling 2004).

The aquarium and water garden trade is another pathway through which aquatic animal pathogens have been introduced to new geographical locations resulting in serious harm to native species. For example, it has been demonstrated that through ornamental fish imports from Asia (China in particular), spring viremia of carp virus (SVCV) has been introduced to the United States. The disease has subsequently spread to koi and goldfish farms in Washington State, North Carolina, and Missouri and to wild common carp populations in Illinois (Goodwin 2002; Dikkeboom et al. 2004; Liu et al. 2004).

Creation of sport and commercial fisheries

Each year many regulatory agencies worldwide release billions of juvenile fish and shellfish into public streams, bays, creeks, rivers, lakes, ponds, and designated marine sites. These fish and shellfish are released for stock enhancement of recreational fisheries and restoration of native species, for creation of new fisheries, and for biological control purposes. Over 700 fish and shellfish species are involved in these stockings, with some being brought from one region to another within the same country, or, more often, being brought from one continent and introduced to another. Salmonid and cyprinid fish species constitute the majority of stocked fish (FAO 2005).

As one would expect, this practice has resulted in the introduction of fish pathogens to new habitats. For example, bacterial kidney disease was introduced in Wyoming, with the stocking of infected non-native brook, brown, and rainbow trout (Mitchum *et al.* 1979). Atlantic salmon smolts imported from Sweden for stock enhancement introduced the monogenean parasitic fluke *Gyrodactylus salaris* into Norwegian rivers. This parasite devastated native Norwegian salmon populations (Johnsen and Jensen 1986; Sattuar 1988) and endangered the existence of the entire Norwegian population of Atlantic salmon to the extent that Norwegian authorities destroyed all resident fish stocks in 30 rivers to eradicate the parasite, or at least to control its spread (Hindar *et al.* 1991). The Norwegian salmon stocks also suffered from a new extremely virulent strain of furunculosis that was introduced with the stocking of salmon smolts imported from Scotland, infecting 72 rivers in 7 years (Johnsen and Jensen 1994).

IMPEDIMENTS FACING THE DEVELOPMENT OF EFFECTIVE AQUATIC ANIMAL HEALTH PLANS

The expansion of international trade in aquatic animals and products along with the history of past pathogen introduction have increased the need for the development of more stringent legislations to control pathogen introductions. Most of the current legislations were issued in response to outbreaks associated with non-aquatic animal introductions, primarily fish and shellfish. For example, the 1937 Diseases of Fish Act of Great Britain was introduced in response to several outbreaks of furunculosis disease, caused by *Aeromonas salmonicida*, in wild salmon and other fish species in the rivers of England, Wales, and Scotland, attributed to the importation of infected live rainbow trout from Germany (reviewed in Hill 1996). The continuous growth in aquatic animal trade, however, stimulated the development of a number of regional, national, and international disease control policies (Bernoth *et al.* 1999; Mitchell and Stoskopf 1999). Unfortunately, such efforts have been impeded by the weakness of the aquatic animal health infrastructure in most nations. There are a limited number of laboratories that can accurately diagnose aquatic animal diseases, and there is a huge shortage in aquatic animal health professionals. This problem is particularly evident in the Asia Pacific region which produces approximately 79 percent in value and 88 percent in volume of aquaculture worldwide, yet it is equipped with very few professional health services. It appears that, while aquaculture has been growing rapidly in many countries, there has been no matching expansion of the supporting aquatic animal health infrastructure (Bondad-Reantaso et al. 2005; Primavera 2005).

The problem is compounded with the current knowledge gap on the life cycle, host range, and ecology of the most serious fish and shellfish pathogens and parasites, a matter that impedes the accuracy of disease surveillance programs and the reliability of risk analysis, and control measures. The absence of baseline data on disease epidemiology urged Hedrick (1996, 1997) to emphasize the need to differentiate between presumed new introductions of disease and first observations of infections already present or even well established in the geographical region of concern. For example, the initial discovery of the intranuclear microsporidian parasite *Enterocytozoon salmonis* in Chile was assumed to be associated with the introduction of salmon eggs from the Pacific Northwest region of the United States and Canada, the only zone in which the parasite was identified at that time (Chilmonczyk et al. 1991). However, further molecular analyses including sequencing of the small subunit ribosomal ribonucleic acid gene demonstrated that Chilean strains are different from those of North America (Barlough et al. 1995). Additional analysis also suggested the presence of a non-salmonid reservoir as the protozoan was demonstrated among halibut in Norway and lumpfish in Newfoundland (Nilsen et al. 1995).

The same pattern was repeated in 1988 when viral hemorrhagic septicemia (VHS) virus was detected for the first time in North America in the Pacific salmon (*Onchorhynchus tshawytscha*) in the state of Washington (Batts et al. 1993). At the time, the virus was believed to have been introduced from Europe. Radical procedures were implemented to control the further spread of the disease including destruction of stocks in two hatcheries and elimination of all resident fish within several miles of the stream that supplied one facility. Subsequent occurrences of infection at other sites such as Alaska and in other fish species (e.g., cod and herring) were believed to be evidence of the further spread of the virus (Eaton et al. 1991; Meyers et al. 1992, 1994).

However, subsequent molecular characterization of the virus established that North American isolates are clearly different from European strains (Bernard et al. 1992; Batts et al. 1993). The sum of these findings suggest that VHS virus is enzootic among cod and herring populations from Washington to Alaska and in British Columbia, Canada, where it was established long before the first observations of this virus in Pacific salmon (reviewed in Hedrick 1996). The cause of the new outbreaks was undetermined but may have been associated with changes in pathogen virulence or host susceptibility.

The number of health surveys on wild aquatic animal stocks is limited and when performed are focused on only a handful of diseases. Such a shortage has led to undesirable consequences that could have been avoided. For example, the introduction of IHHNV into the Gulf of California in Mexico, in 1987, was followed by serious epizootics of IHHNV in farmed *P. stylirostris* stocks in the Mexican states of Sonora and Sinaloa in 1989 and 1990 (Lightner et al. 1992a). It was suggested that cross-contamination among farms was the means by which the virus was transmitted (Lightner et al. 1992b). A few years later, a disease survey conducted on wild shrimp stocks in the commercial fishery of the northern Gulf of California revealed that IHHNV infections were present at high prevalences (Pantoja-Morales and Lightner 1991), probably causing the 50 percent decrease in shrimp landings observed in the years prior to the virus discovery in farms (Moore 1991). The broodstock used by affected farms had been collected from IHHNV-infected wild stocks. The lack of baseline data in these regions made it impossible to determine whether disease was transmitted from the wild to the farms or vice versa (reviewed in Lightner 1996).

One of the most important constraints in pathogen detection is the lack of diagnostic tools to enhance detection capabilities, particularly in cases of subclinical and carrier infections. For example, cell lines routinely used to isolate intracellular pathogens of vertebrates are currently lacking for both marine mollusks and crustaceans. This has been a significant constraint to the detection and understanding of the epidemiology of viral and other intracellular microbial infections affecting these animals. In the same context, there is an urgent need to develop more sensitive serological and molecular assays for detecting and comparing pathogens, a step that will enable tracking the movements of these agents between continents, countries, and states. Indeed, the use of molecular epidemiological tools has already helped us to determine geographical origin of pathogens. However, many of the procedures have not been fully validated (Cunningham 2002) and the interpretation of the results can be problematic. Additionally, the development of more sensitive diagnostic techniques has focused on pathogens that have a significant economic impact on production and trade. New pathogens, or those of more regional significance, rely on more "conventional" but less sensitive diagnostic tests. The detection of other potential pathogens that may be significant to wild/cultured resources will not be possible if detection is based solely on pathogen-specific diagnostic tools.

In brief, the impediments of pathogen detection described above are reflections of the weakness in the aquatic animal health infrastructure prevalent in most countries. Therefore, there is an urgent need to develop national and international pathogen control policies, to develop standardized protocols to diagnose fish and shellfish pathogens with precision, and to improve aquatic animal health infrastructure.

NATIONAL, REGIONAL, AND INTERNATIONAL EFFORTS TO CONTROL PATHOGEN TRANSFER

Globalization in fish trade and its link to disease transmission is one area that national, regional, and international regulatory fishery bodies have not addressed as extensively as other traditional fishery issues. The current situation, however, mandates that fishery organizations, at all levels, join forces to balance unimpeded trade with low risk of pathogen introduction. The consensus is that this goal can be achieved through the following principles:

- developing, harmonizing, and enforcing appropriate and effective national, regional, and interregional policies and regulatory frameworks on introduction and movement of live aquatic animals and products to reduce the risks of introduction, establishment, and spread of aquatic animal pathogens;
- developing and implementing effective national disease reporting systems, databases, and other mechanisms for collecting and analyzing aquatic animal disease information;
- improving technology through research to develop, standardize, and validate accurate and sensitive diagnostic methods, safe therapeutants, and effective disease control methodologies; and
- promoting a holistic system approach to aquatic animal health management, emphasizing preventative measures and maintaining a healthy ecosystem.

Guided by these principles, a number of nations have developed national health plans. The Diseases of Fish Act issued in 1937 in Great Britain is the longest-standing example of national legislation specifically devised to control fish diseases (Hill 1996). The Act prohibited the importation of live salmonids into Great Britain, and made it illegal to import salmonid ova and all live freshwater fish species without a license. Moreover, the Act enabled any disease to be designated as "notifiable," meaning that even the suspicion of its presence in any waters must be reported to the official services. While these principles remain the core of most national policies, other countries amended and expanded their policies as needed to face emerging challenges (Brückner 1996; Campos Larrain and Valenzuela Alfaro 1996; Carey 1996; Doyle et al. 1996; Hill 1996; Schlotfeldt 1996).

In addition to efforts and legislations at the national level, two regional health plans were developed, for the European Union (Daelman 1996) and for states and provinces sharing the Great Lakes basin (Hnath 1993). This regional approach has significantly extended the scope of aquatic animal health legislation to include recommendation directives and decisions to ensure movement of live fish and their products, while guaranteeing a high level of animal health. Both national and regional plans included lists of notifiable diseases, a system of certification, and description of reliable protocols for use in laboratory diagnostic testing.

There are a number of international organizations that are involved in the prevention of pathogen introductions and spread (Box 5.1), the most comprehensive of which is the United Nations World Organization of Animal Health.

The World Organization of Animal Health (OIE)

Established in 1924, the World Organization of Animal Health, formerly known as the Office International des Epizooties (OIE), is an intergovernmental organization that is responsible for promoting animal health worldwide. The OIE implements international agreements among its 167 member countries, thereby enabling nations to work together in order to reduce the risks of aquatic animal pathogen dissemination. The OIE collects and analyzes information on animal diseases (aquatic animals included) and uses these analyses in developing standards, guidelines, and recommendations for member countries. Indeed, the OIE recommendations are considered the uncontested international reference in animal diseases. The World Trade

> **Box 5.1** International organizations and agreements involved in developing standards for prevention of pathogen introductions through live fish and shellfish trade
>
> - United Nations
>   Convention on Biological Diversity
>   Convention on International Trade in Endangered Species (CITES)
> - International Council for the Exploration of the Sea (ICES)
>   Code of Practice for Introduction and Transfer of Marine Organisms
> - World Animal Health Organization (Office International des Epizooties, OIE)
>   International Aquatic Animal Health Code
> - General Agreement on Tariffs and Trade (GATT)
> - World Trade Organization (WTO)
> - Sanitary/Phytosanitary (SPS) Agreement
> - Canada–U.S. Free Trade Agreement (FTA)
> - North American Free Trade Agreement (NAFTA)
> - North American Commission (NAC) (Canada and United States)
> - North Atlantic Conservation Organization (NASCO)
> - International Joint Commission under the Boundary Waters Treaty Act

Organization (WTO) Agreement on the Application of Sanitary and Phytosanitary Measures (SPS Agreement) recognizes the OIE as the authorized international organization responsible for the development and promotion of international animal health standards, guidelines, and recommendations affecting trade in live animals and animal products, whether aquatic or terrestrial in origin.

The OIE relates to member countries the latest science-based recommendations on aquatic animal pathogens through two major documents: the *Aquatic Animal Health Code* (OIE 2005) and the *Manual of Diagnostic Tests for Aquatic Animals* (OIE 2003). Information within these documents is provided as guidelines for the preparation of veterinary regulations for import and export of live animals and products. Within OIE, the Aquatic Animal Health Standards Commission (AAHSC) is the body responsible for developing methods for surveillance, diagnosis, control, and prevention of infectious aquatic animal diseases.

The OIE Aquatic Code (OIE 2005) has listed a number of diseases (Tables 5.1–5.3) that devastate fish, mollusks, and crustaceans. Once one of the OIE-listed diseases or pathogens has been suspected or confirmed, each member country has the obligation to report the disease(s) it detects in its territory. Notification is also required for other significant diseases of aquatic animals of which the international community needs to be aware. Criteria for designating a disease as notifiable to the OIE are summarized in Fig. 5.1.

The OIE has designated 15 fish disease laboratories, four mollusk laboratories, and two crustacean laboratories as OIE Reference Laboratories to pursue all the scientific and technical problems relating to aquatic

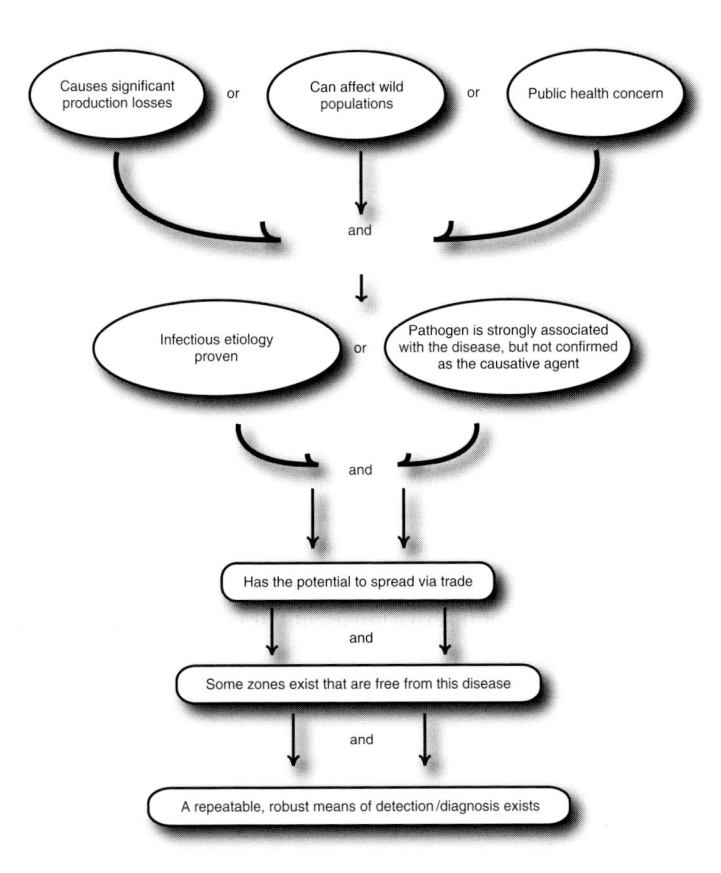

Figure 5.1 Criteria necessary to determine if a disease should be notified to the Office International des Epizooties (OIE).

Table 5.1 Fish diseases listed by the Office International des Epizooties (OIE 2006)

Disease	Host	Reference laboratories and experts
Epizootic hematopoietic necrosis	Redfin perch (*Perca fluviatilis*), rainbow trout (*Oncorhynchus mykiss*), Macquarie perch (*Macquaria australasica*), silver perch (*Bidyanus bidyanus*)	Dr. A. Hyatt: CSIRO, Australian Animal Health Laboratory, Australia
	Mountain galaxies (*Galaxias olidus*), sheatfish (*Silurus glanis*), catfish (*Ictalurus melas*), mosquito fish (*Gambusa affinis*) and other species belonging to the family Poeciliidae	Dr. R. Whittington: University of Sydney, Australia
Infectious hematopoietic necrosis	Rainbow or steelhead trout (*Oncorhynchus mykiss*), Pacific salmon including chinook (*O. tshawytscha*), sockeye (*O. nerka*), chum (*O. keta*), masou (*O. masou*), and coho (*O. kisutch*), and Atlantic salmon (*Salmo salar*)	Dr. J.R. Winton: Western Fisheries Research Center, Washington, USA
Spring viremia of carp	Common carp and koi carp (*Cyprinus carpio*), grass carp (*Ctenopharyngodon idellus*), silver carp (*Hypophthalmichthys molitrix*), bighead carp (*Aristichthys nobilis*), crucian carp (*Carassius carassius*), goldfish (*Carassius auratus*), tench (*Tinca tinca*), sheatfish (*Silurus glanis*)	Dr. B.J. Hill: The Centre for Environment, Fisheries, and Aquaculture Science (CEFAS), Weymouth, United Kingdom
Viral hemorrhagic septicemia	Fish belonging to the family Salmonideae, grayling (*Thymallus thymallus*), white fish (*Coregonus* spp.), pike (*Esox lucius*), turbot (*Scophthalmus maximus*), herring and sprat (*Clupea* spp.), Pacific salmon (*Oncorhynchus* spp.), Atlantic cod (*Gadus morhua*), Pacific cod (*G. macrocephalus*), haddock (*G. aeglefinus*), rockling (*Onos mustelus*)	Dr. N.J. Olesen: Danish Institute for Food and Veterinary Research, Community Reference Laboratory for Fish Diseases, Denmark

Infectious salmon anemia	Atlantic salmon (*Salmo salar*), rainbow trout (*Oncorhynchus mykiss*), brown trout (*Salmo trutta*)	Dr. F. Kibenge: University of Prince Edward Island, Canada Dr. B. Dannevig: National Veterinary Institute, Norway
Epizootic ulcerative syndrome	Genera *Channa, Mastacembelus, Puntius, Trichogaster, Catla, Mugil, Labeo*	Dr. S. Kanchanakhan: Aquatic Animal Health Research Institute, Department of Fisheries, Kasetsart University Campus, Thailand
Gyrodactylosis (*Gyrodactylus salaris*)	Salmonids including Atlantic salmon parr (*Salmo salar*), rainbow trout (*Oncorhynchus mykiss*), Arctic char (*Salvelinus alpinus*), North American brook trout (*S. fontinalis*), grayling (*Thymallus thymallus*), North American lake trout (*Salvelinus namaycush*), brown trout (*Salmo trutta*)	Dr. T. A. Mo: National Veterinary Institute, Fish Health Section, Norway
Red sea bream iridoviral disease	Red sea bream (*Pagrus major*), as well as other cultured marine fish including yellowtail (*Seriola quinqueradiata*), sea bass (*Lateolabrax* sp.), and Japanese parrotfish (*Oplegnathus fasciatus*)	Dr. K. Nakajima: Fisheries Research Agency, Headquarters, Queen's Tower B 15F, Yokohama, Japan
Koiherpesvirus disease	Carp (*Cyprinus carpio*)	

Table 5.2 Diseases of mollusks listed by the Office International des Epizooties (OIE 2006)

Disease	Host	Reference laboratories and experts
Infection with *Bonamia ostreae*	*Ostrea edulis, O. angasi, O. denselammellosa, O. puelchana, Ostreola conchaphila* (= *O. lurida*), and *O. chilensis* (= *Tiostrea lutaria*)	Dr. I. Arzul: IFREMER, Laboratoire de Génétique at Pathologie, France
Infection with *Bonamia exitiosa*	*Ostrea chilensis* and *O. angasi*	Dr. I. Arzul
Infection with *Marteilia refringens*	*Ostrea edulis, O. angasi*, and *Ostrea chilensis*	Dr. I. Arzul
Infection with *Perkinsus marinus*	*Crassostrea virginica* and *C. gigas*	Dr. E. M. Burreson: Virginia Institute of Marine Science, Virginia, USA
Infection with *Perkinsus olseni*	*Haliotis ruber, H. cyclobates, H. scalaris, H. laevigata, Ruditapes philippinarum, R. decussates*, and *Austrovenus stutchburyi*	Dr. E. M. Burreson
Infection with *Xenohaliotis californiensis*	Members of the genus *Haliotis* including black abalone (*H. cracherodii*), red abalone (*H. rufescens*), pink abalone (*H. corrugate*), green abalone (*H. fulgens*), and white abalone (*H. sorenseni*)	Dr. C. Friedman, University of Washington, Washington, USA
Abalone viral mortality	*Haliotis diversicolor supertexta*	

Table 5.3 Diseases of crustaceans listed by the Office International des Epizooties (OIE 2006)

Disease	Host	Reference laboratories and experts
Taura syndrome	Pacific white shrimp (*Penaeus vannamei*), Pacific blue shrimp (*P. stylirostris*), and Gulf white shrimp (*P. setiferus*)	Dr. D. V. Lightner: University of Arizona, USA
White spot disease	Most commercially cultivated penaeid (family Penaeidae) shrimps and prawns	Dr. D. V. Lightner Dr. G. Lo: National Taiwan University, China
Yellowhead disease	Black tiger shrimp (*Penaeus monodon*), Pacific white shrimp (*P. vannamei*), Pacific blue shrimp (*P. stylirostris*), Gulf white shrimp (*P. setiferus*), Gulf brown shrimp (*P. aztecus*), Gulf pink shrimp (*P. duorarum*), and Kuruma prawn (*P. japonicus*)	Dr. P. Walker: CSIRO, Aquaculture and Aquatic Animal Health (AAHL), Australia
Tetrahedral baculovirosis (*Baculovirus penaei*)	Various shrimps from the family Penaeidae	Dr. D. V. Lightner
Spherical baculovirosis (*Penaeus monodon*-type baculovirus)	Various prawns and shrimps from the family Penaeidae including black tiger shrimp (*P. monodon*)	Dr. D. V. Lightner Dr. G. Lo
Infectious hypodermal and hematopoietic necrosis	Various shrimps from the family Penaeidae including *P. stylirostris* and *P. vannamei*	Dr. D. V. Lightner
Crayfish plague (*Aphanomyces astaci*)	Various Decapoda including ones from Astacidae, Cambaridae such as the signal crayfish (*Pacifastacus leniusculus*) and the Louisiana swamp crayfish (*Procambarus clarkii*)	Dr. R. Hoffmann: Institute of Zoology, Fish Biology and Fish Diseases, Germany Dr. D. J. Alderman: The Centre for Environment. Fisheries, and Aquaculture Science (CEFAS), Weymouth, United Kingdom

animal diseases on the OIE lists. The mission of these laboratories is to standardize diagnostic techniques for their designated diseases. The OIE has also designated an "expert" for each of the listed diseases who provides scientific and technical assistance and expert advice on topics linked to surveillance and control of the diseases for which the Reference Laboratory is responsible (Tables 5.1–5.3).

The urgency of dispatching information varies according to the nature of the disease. Countries are required to notify the OIE within 24 hours of the occurrence of an outbreak of a notifiable aquatic animal disease, a disease likely to have serious repercussions on public health, wild populations, or the economy of aquatic animal production. The OIE then dispatches these data directly to member countries so that necessary preventive actions can be taken. Information is sent out immediately or periodically depending upon the seriousness of the disease. Reporting is required not only when an infectious disease breaks out but also when the disease is eradicated (OIE 2005).

COMPONENTS OF NATIONALLY AND INTERNATIONALLY ACCEPTED PROCEDURES TO ELIMINATE RISKS OF AQUATIC ANIMAL PATHOGENS TRANSFER

Over the last three decades national and international agencies presented several proposals to develop a universal policy to minimize the risk of disease transmission without forming barriers against the trade of live aquatic animals and their products. Several of these proposals progressed further to become the basis of national legislations specifically designed for diseases of fish, mollusks, and crustaceans. Regardless of the source, all proposals agreed on the need to establish accurate disease surveillance system, create a system of certification, establish disease-free zones, start assessment of pathogen introduction risks associated with the trade, and quarantine for introduced species before permitting access into the importing country (Box 5.2).

Disease surveillance

Surveillance for diseases and their etiologic agents in wild and cultured aquatic animals is the cornerstone of any health control program at the regional, domestic, and international levels. For this reason, The OIE Aquatic Code (OIE 2005) emphasized the profound importance of sound surveillance practices, which are based on the standards set by the OIE. The code defines *surveillance* as "the continuous investigation of

> **Box 5.2** Important components designed to prevent pathogen introduction through live fish and shellfish trade
>
> - Identification of notifiable diseases
> - Standardization of diagnostic techniques for each of the notifiable diseases
> - Implementation of health certification for each shipment
> - Development of a surveillance program
> - Identification of disease-free zones
> - Implementation of a quarantine system

a given population to detect the occurrence of disease for control purposes, which may involve testing of a part of population." The code defines *monitoring* as "an on-going program directed at the detection of changes in the prevalence of disease in a given population and in its environment." The term *surveillance program* encompasses both surveillance and monitoring tasks.

Most scientists and managers consider the primary purpose of aquatic animal disease surveillance to be to provide scientifically accurate and cost-effective information for assessing and managing risks of disease transfer associated with trade (intra- and international) in aquatic animals, animal production efficiency, and public health (Subasinghe et al. 2004). There are a number of benefits to a country, or group of countries, to design and implement sound surveillance programs for aquatic animals diseases and pathogens. These benefits include:

- providing early warning of serious and emerging disease outbreaks;
- providing evidence of freedom from diseases relevant to movement of aquatic animals and their products;
- revealing the potential source and subsequent spread of diseases and their pathogens; and
- assessing the efficacy of control or eradication measures for a particular disease or diseases.

Disease surveillance as a discipline has been revisited over the years and therefore has a number of definitions, reflecting its multiple uses for various objectives (Cameron 2002; Scudamore 2002). Throughout the literature one finds terms such as passive, active, scanning, general, and targeted surveillance used interchangeably (Scudamore 2002). A comprehensive surveillance program, however, can be made up of a combination of many approaches for gathering surveillance data (Cameron 2002).

General surveillance is considered an ongoing investigation or observation of the endemic disease profile of a population so that unexpected and/or unpredicted changes can be quickly recognized. General surveillance is very useful for early detection of emerging diseases and often provides a general picture of the disease situation in a population. This type of surveillance cannot be used to demonstrate reliably the absence of a particular disease from a given area. Routine gathering of information on disease incidents from reports of farmers or field officers or from specimens submitted to diagnostic or research laboratories is a classical example of what epidemiologists consider passive surveillance because disease information is a "by-product" of more general disease investigations (Scudamore 2002).

Targeted surveillance, on the other hand, collects information about a specific disease or condition so that its presence in a defined population can be measured or its absence reliably substantiated. This kind of surveillance provides the data required to prove that a specified population is free of a specific disease. Sampling techniques are aimed at maximizing the likelihood of pathogen detection, based on available epidemiological information (Gustafson *et al.* 2005; McClure *et al.* 2005; Murray and Peeler 2005; Subasinghe 2005). Both general and targeted surveillance can complement each other. Stephenson *et al.* (2003) provided an excellent example of how a general surveillance can lead to a targeted one. In 2002, the Shellfish Health Unit at the Canadian Gulf Fisheries Center received oysters from an oyster grower in Cape Breton, Nova Scotia, with a history of up to 80 percent mortality. It was determined that the cause of oyster death was the protozoan *Haplosporidium nelsoni*, the causative agent of a disease known as multinucleated sphere X (MSX). That was the initial detection of this OIE-notifiable pathogen in Canada. Such a discovery prompted a targeted surveillance for MSX that involved not only Cape Breton but also other locations throughout Nova Scotia, New Brunswick, Prince Edward Island, and Quebec. The choice of surveillance sites was based on historical data as well as current oyster transfers throughout the Atlantic region. The precise design and structure of surveillance programs vary with their exact purpose but all share some basic common features given in Box 5.3.

> **Box 5.3** Common features of surveillance programs
>
> - Clearly stated objectives
> - Agreed-upon list of diseases of concern
> - Development of specific protocols for collection of the information required
> - Capability to recognize a disease outbreak with the required level of diagnostic certainty
> - A system to record, collate, and report findings.
>
> (Adopted from Subasinghe et al. 2004.)

Certification

The OIE requests that an official health certificate accompany each shipment of live fish or shellfish. Because of likely variations in aquatic animal health situations in the exporting country, in the transit country, and in the importing country, diagnosis and official certification should be based upon OIE standards, guidelines, and recommendations. For example, the OIE *Manual of Diagnostic Tests for Aquatic Animals* (OIE 2003) provides a uniform approach to diagnose OIE-listed diseases which can be followed in laboratories all over the world, thus increasing efficiency and promoting improvements in aquatic animal health worldwide. OIE criteria for aquatic animal health certification are displayed in Box 5.4.

Zoning

Disease zoning is the process of delineating infected and uninfected populations within a country or group of countries. "Infected zone" and "uninfected zone" usually apply to a specific disease (Murray 2002; Cockings and Martin 2005). An uninfected zone can be established for a specific disease within a particular geographic or hydrographic area within a country. The OIE Aquatic Code recommends that zones for diseases of concern to international trade be established to meet internationally accepted standards (OIE 2005).

The concept of zoning is extremely difficult to apply to the aquatic environment, however. In general, catchment areas and rivers may be used to define continental zones. Coastal zonation for specific diseases is often complicated by a host's home range and migratory

> **Box 5.4** Principles used for the preparation of international aquatic animal health certificates
>
> (1) Issued by the Competent Authority and signed by a qualified and authorized certifying official.
> (2) Written in terms that are as simple, unambiguous, and easy to understand as possible, without losing their legal meaning.
> (3) Written in the language of the importing country. In such circumstances, they should also be written in a language understood by the certifying official.
> (4) Require appropriate identification of aquatic animals and aquatic animal products.
> (5) Not require a certifying official to certify matters that are outside his/her knowledge or that he/she cannot ascertain and verify.
> (6) Accompanied, when presented to the certifying official, by notes of guidance indicating the extent of enquiries, tests, or examinations expected to be carried out before the certificate is signed.
> (7) Text should not be amended.
> (8) Only original certificates are acceptable.

pattern as well as the lack of knowledge about infection reservoirs. The simplest freshwater zonation system is a farm that obtains incoming water from an unshared surface or groundwater source. In inland situations, however, most aquaculture facilities are connected to each other through shared waterways through which the disease agent can be transmitted to wild aquatic animal populations. In such case, the entire river system or water catchment area is considered as a zone. Natural barriers determined by geography or climate are more effective in containing diseases of aquatic animals than are political borders (Subasinghe et al. 2004).

The types of zones that are recognized are listed below.

Free zone

A free zone can be established within a country or a region where an OIE-notifiable disease is present. In the free zone, there must be knowledge of the location of all aquaculture facilities and wild susceptible

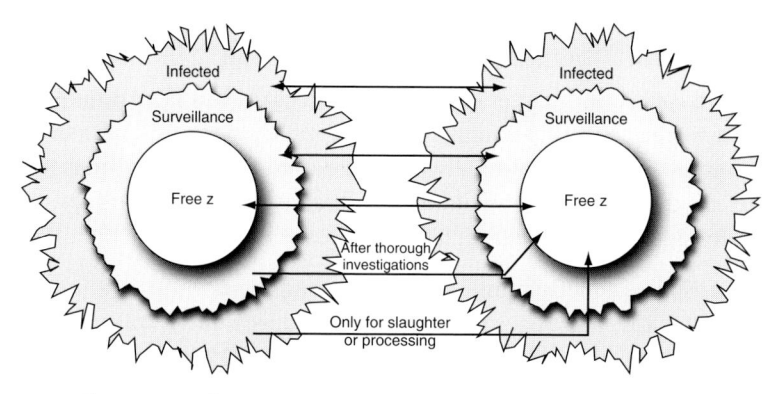

Figure 5.2 A diagram illustrating the disease zoning concept in which a surveillance zone separates a free and an infected zone. Importation of aquatic animals from a surveillance zone into the free zone takes place only after thorough investigation confirming absence of pathogen. No live aquatic animals may leave the infected zone except for transport to slaughtering premises.

species. Suspected outbreaks of a disease must be investigated immediately by the Competent Authority and reported to the OIE. Ideally, a free zone should be separated from an infected neighboring zone by a buffer zone where surveillance is constantly performed. Importation of aquatic animals from other parts of the country, or from countries where the disease is present, into the free zone must take place under strict controls established by the Competent Authority (Fig. 5.2).

Surveillance zone

A surveillance zone acts as a buffer zone between a free and an infected zone. Aquatic animal movements must be controlled within the surveillance zone and rigorous disease prevention and control measures must be practiced. A mechanism for immediate reporting to the Competent Authority must be in place. Suspected outbreaks of a disease must be investigated immediately and, if confirmed, must be eliminated. Adequate surveillance activities must be implemented in order to determine the potential spread of any outbreaks. Accordingly, it may be necessary to modify the boundaries of the zone. Importation of susceptible aquatic animals into the surveillance zone from parts of the country or from other countries where the disease exists can only take place under suitable controls established by the Competent Authority. Freedom from infection should be confirmed by appropriate tests.

Infected zone

An infected zone is a zone where the disease is present in an otherwise disease-free country. Movement of susceptible aquatic animals out of the infected zone into the disease-free parts of the country must be strictly controlled. It is recommended that the following alternatives be considered in managing a disease-infected zone (Fig. 5.2):

(1) No live aquatic animals may leave the zone, or
(2) Aquatic animals can be moved by mechanical transport to special fish slaughtering premises or mollusk and shrimp production facilities located in the surveillance zone for immediate slaughter, or
(3) As an exception, live aquatic animals from the infected zone can enter the surveillance zone under suitable controls established by the Competent Authority. For diseases in which the disease agent constitutes a surface pathogen, appropriately disinfected eggs can enter a surveillance zone. Freedom from infection of these aquatic animals must be confirmed by appropriate tests before being allowed to enter the zone, or
(4) Live aquatic animals can leave the infected zone if the epidemiological conditions are such that disease transmission cannot occur.

It is essential to prevent live aquatic animals from being transported from infected to uninfected zones, including into buffer zones. Moreover, it is necessary to control shipments of known or suspected vectors of the disease agent.

Risk assessment

The OIE, the Food and Agriculture Organization (FAO), the International Council for the Exploration of the Sea (ICES), and the European Inland Fishery Advisory Commission (EIFAC) recommend that in the case of international trade involving aquatic animals or their products including genetic and pathological samples, a risk assessment for the possibility of pathogen introduction into the importing country should be conducted prior to the arrival of shipments. These agencies stipulate a process for conducting risk assessment that starts with the development of a proposal by the entity moving an exotic species that specifies the location of the facility, planned use, and source of the exotic species. Based on the information provided, a group of independent experts review and evaluate the proposal and potential impacts and risk/benefits of the proposed introduction. Pathogen introduction, ecological requirements and interactions, genetic concerns, socioeconomic impacts, and identification of local species most affected are some of the considerations evaluated by the review panel. A regulatory agency may require that a proposal contain an evaluation of the risk/benefits. This evaluation would then be forwarded to an independent review or advisory panel, or the advisory panel could make the first evaluation of the proposal.

Advice and comments would then be communicated among the proposers, evaluators, and decision-makers. The independent review panel advises to accept, refine, or reject the proposal. If approval to introduce a species is granted, then quarantine, containment, monitoring, and reporting systems are implemented. Importation of the (formerly) exotic species becomes subject to review and inspection to check the general condition of the shipments, verify the absence of pathogens, and ensure that the correct species is being shipped. Competent Authorities may require quarantine procedures to be explicitly described in the proposal before approval to move an exotic species is granted.

Risk assessment should be conducted by the importing country in liaison with the exporting country. The OIE Aquatic Code specifies the health measures to be used by the veterinary administrations and competent authorities in the importing and exporting countries in order to avoid the transfer of any agents that are pathogenic to aquatic animals or humans by the exotic species. Concomitantly, measures are taken to prevent the imposition of unjustified sanitary barriers that might delay or impede the process. The analysis should be transparent and the exporting country should be provided with clear justifications for the imposition of an import.

Quarantine

There is a consensus among aquatic animal health experts that an effective quarantine system is the cornerstone of any health plan aimed at preventing pathogen introduction. The term "quarantine" refers to retaining animals in facilities designed specifically to prevent the release of animals, or the pathogens they may carry, into regions where these animals or their pathogens do not exist (Arthur 1996; Doyle *et al.* 1996). ICES and EIFAC have developed codes of practice on the use of introduced species (Bartley *et al.* 1996; ICES 1998; Cowx 2000).

These codes call for linking exotic species introduction to the availability of a sound quarantine system in the importing country. The codes recommend a sequence of events that starts with the importing country allowing import of eggs, not live animals, delivered directly to an approved quarantine facility. These eggs will then be used as seed to develop a broodstock population. Fish produced from the eggs and the broodstock will be examined regularly for the presence of pathogens following established guidelines. If no pathogens become evident then the first-generation progeny (F_1), but not the original import, can be released to culture sites or the natural environment. Routine disease testing should continue on transplanted individuals. The codes further emphasize that no additional animals of the introduced species be imported and that F_1 individuals should be used to establish local broodstocks.

Australia, for example, has a fully operational quarantine system for imported fish species. Australian regulations were established in 1984 to prohibit the entry of live fish into the country unless the species to be imported is listed by the Australian Nature Conservation Agency (NCA) as being among those aquarium fishes that may be freely imported, or that can be imported under special permit for scientific purposes (Lehane 1993; Humphrey 1995; Doyle *et al.* 1996). In the late 1990s, the NCA list was found to be in violation of the WTO's Sanitary and Phytosanitary (SPS) Agreement. As a result, the Australian authorities expanded the NCA list to include fish species that were originally banned from importation.

In the case of freshwater ornamental fish, Australia's quarantine system requires fish shipments to be accompanied by a health certificate issued by the exporting country. The premises of the exporter must be inspected regularly, and the exporter certified as competent, by government fish health inspectors of the exporting country. Fish are held for a minimum of 14 days, during which time they are examined regularly for clinical signs of diseases. Marine species are not quarantined upon arrival, but are inspected to determine their identity, to verify that they are among the species approved for importation, to ensure that they exhibit no clinical signs of disease, and to confirm that no other organisms are present in the shipment (Lehane 1993). The practice of relying upon clinical signs to determine disease status has been heavily criticized because fish carrying bacterial and viral diseases have escaped detection (Langdon *et al.* 1986; Anderson *et al.* 1993; Humphrey 1995). The authors recommended that inspection for clinical disease during quarantine should be extended to specific health accreditation or random sampling of imported batches for laboratory analyses.

LESSONS LEARNED AND RESEARCH NEEDED

There is no doubt that the rapid pace of globalization processes has presented the modern world with mounting challenges. This is particularly true in the case of aquatic animals with the recent worldwide spread of pathogens and the emergence of new devastating diseases. This is, however, not a single event, as many infectious diseases of humans, animals, and plants have lately emerged in an unprecedented magnitude. This phenomenon reflects the complex social, economic, political, environmental, ecological, and microbiological factors that are globally linked. Unlike other health disciplines, the aquatic animal health field was not prepared for this recent, global disease surge. Knowledge gaps are huge, the number of well-trained aquatic animal health professionals is inadequate, laboratories are not well equipped, commercial diagnostic reagents are largely unavailable, and investments in building diagnostic capacity on private and governmental levels are not commensurate with the magnitude of the problem. In brief, there is a severe need to learn from the past in order to be prepared for the future. If the current trend continues, effective control of the international spread of fish and shellfish pathogens will remain a distant and elusive goal to achieve.

Aquaculture, considered by some as the major channel through which many pathogens were introduced to new geographical locations, is filling the gap produced by declining fisheries (freshwater and marine species alike). The expansion in aquaculture will, and should, continue to provide the growing human population with high-quality proteins at affordable cost, particularly in developing countries. Likewise, governmental conservation and restoration programs will continue to intensify, even though some of the efforts may require species introduction, or reintroduction. Given all of the above, it is imperative to focus efforts on finding ways to eliminate the risks associated with the international trade of live fish and shellfish. These efforts have to be designed and implemented at both the national and the global levels.

International organizations such as OIE, FAO, EIFAC, and ICES have done an excellent job in coming up with guidelines and models for prevention of disease spread. Most of their recommendations came from the more structured and well-studied examples in human and

Table 5.4. *Immediate research needs for most emerging aquatic animal pathogens*

Area	Knowledge gaps to be filled
Detection method	Sampling and enrichment techniques to enable pathogen identification even in subclinical and carrier status
	Enhanced diagnostics with high specificity and sensitivity and at affordable costs
Disease ecology	Infection reservoirs
	Route(s) of transmission
	Geographical range and seasonality
	Life cycle
Pathogenicity	Effects at subcellular, cellular, organ, and individual levels
	Infectious dose
	Host range
	Subpopulations/populations at risk
Growth characteristics	Free-living versus obligate parasite
	Growth requirements
	Temperature
	pH
	Oxygen
Survival characteristics	Heat and acid resistance
	Susceptibility to antibiotics
	Sensitivity to disinfectants, desiccation, ionizing radiation, and ultraviolet radiation

veterinary medicine. It is currently premature to judge the effectiveness of national and international Codes of Practice in controlling the introduction of aquatic animal pathogens. Indeed, it is questionable how much of the legislation can in practice be applied. For example, how can a quarantine system be effective when it relies primarily on the appearance of clinical signs within a relatively short time? In the same context, how effective is zoning, considering interconnectedness of surface and groundwater? These two questions, and many others, remain to be addressed not only by aquatic animal health experts, but also by managers, economists, and politicians. To this end, there is an acute need to translate the codes into reasonable, scientifically correct, and economically affordable protocols to implement worldwide and in both developed and developing countries.

Concurrent efforts are needed to narrow knowledge gaps, to find innovative ways to implement health control measures, and to be prepared in case an emerging disease erupts. These efforts include the establishment of comprehensive databases on pathogens and diseases and related issues essential for the understanding of national disease status, for use in risk assessment studies, and to serve as a decision support system for introductions and national quarantine policies, guidelines, and strategies. There is a need to provide financial and technical assistance to developing countries to enhance surveillance and to build diagnostic capabilities. Finally, the events that lead to the emergence of a disease are often complex, with the cause often being obscure and only indirectly related to the new agent. Therefore, global efforts should be focused on preparing a contingency research plan that can immediately be followed should an emerging infection erupt, in order to expedite the identification of effective control measures. One such plan is displayed in Table 5.4.

References

Alderman, D.J. 1996. Geographical spread of bacterial and fungal diseases of crustaceans. *Revue scientifique et technique de l'Office International des Epizooties* **15**: 603–632.

Alderman, D.J., Polglase, J.L., and Reeve, I. 1990. Signal crayfish as vectors in crayfish plague in Britain. *Aquaculture* **86**: 3–6.

Altstatt, J.M., Ambrose, R.F., Engle, J.M., et al. 1996. Recent declines of black abalone *Haliotis cracherodii* on the mainland coast of central California. *Marine Ecology Progress Series* **142**: 185–192.

Anderson, I.G., Prior, H.C., Rodwell, B.G., and Harris, G.O. 1993. Iridovirus-like virions in imported dwarf gourami (*Colisa lalia*) with systemic amoebiasis. *Australian Veterinary Journal* **70**: 66–67.

Arthur, J. R. 1996. Fish and shellfish quarantine: the reality for Asia-Pacific. In *The Health Management in Asian Aquaculture*, FAO Fisheries Technical Paper No. 360, eds. R. P. Subasinghe, J. R. Arthur, and M. Shariff. Rome: Food and Agriculture Organization of the United Nations, pp. 11-28.

Arthur, R. and Shariff, M. 1991. Towards international fish disease control in Southeast Asia. *Proceedings of INFOFISH International*, Kuala Lumpur, vol. **3**, pp. 45-48.

Baringa, M. 1990. Fish, money and science in Puget Sound. *Science* **247**: 631.

Barlough, J. E., Mcdowell, T. D., Milani, A., *et al.* 1995. Nested polymerase chain reaction for detection of *Enterocytozoon salmonis* genomic DNA in chinook salmon *Oncorhynchus tshawytscha*. *Diseases of Aquatic Organisms* **23**: 17-23.

Bartley, D. M. and Subasinghe, R. P. 1996. Historical aspects of international movement of living aquatic species. *Revue scientifique et technique de l'Office International des Epizooties* **15**: 387-400.

Bartley, D. M., Subasinghe, R., and Coates, D. 1996. *Draft Framework for the Responsible Use of Introduced Species*. Dublin: European Inland Fisheries Advisory Commission (EIFAC).

Batts, W. N., Arakawa, C. K., Bernard, J., and Winton, J. R. 1993. Isolates of viral hemorrhagic septicemia virus from North America and Europe can be detected and distinguished by DNA probes. *Diseases of Aquatic Organisms* **17**: 67-71.

Bernard, J., Bremont, M., and Winton, J. R. 1992. Nucleocapsid gene sequence of a North American isolate of viral haemorrhagic septicaemia virus, a fish rhabdovirus. *Journal of General Virology* **73**: 1011-1014.

Bernoth, E. M., Murray, G., Rickard, M. D., and Hurry, G. 1999. Approaches to managing aquatic animal health in Australia. *Revue scientifique et technique de l'Office International des Epizooties* **18**: 228-238.

Berry, T. M., Park, D. L., and Lightner, D. V. 1994. Comparison of the microbial quality of raw shrimp from China, Ecuador, or Mexico at both wholesale and retail levels. *Journal of Food Protection* **57**: 150-153.

Blanc, G., Bergot, F., and Vigneux, E. 1997. L'introduction des agents pathogènes dans les ecosystèmes aquatiques: aspects théoriques et réalités. *Bulletin français de la peche et de la pisciculture* **344/345**: 489-515.

Bondad-Reantaso, M. G., Subasinghe, R. P., Arthur, J. R., *et al.* 2005. Disease and health management in Asian aquaculture. *Veterinary Parasitology* **132**: 249-272.

Bossart, G. D., Baden D. G., Ewing, R., Roberts, B., and Wright, S. 1998. Brevetoxicosis in manatees (*Trichechus manatus latirostris*) from the 1996 epizootic: gross, histologic, and immunohistochemical features. *Toxicologic Pathology* **26**: 276-282.

Brückner, G. K. 1996. Review of disease control in aquaculture in the Republic of South Africa. *Revue scientifique et technique de l'Office International des Epizooties* **15**: 703-710.

Cameron, A. 2002. *Survey Toolbox for Aquatic Animal Diseases: A Practical Manual and Software Package*. Canberra, ACT: Australian Centre for International Agricultural Research.

Carey, T. G. 1996. Finfish health protection regulations in Canada. *Revue scientifique et technique de l'Office International des Epizooties* **15**: 647-658.

Campos Larrain, M. C. and Valenzuela Alfaro, M. E. 1996. Chilean legislation for the control of diseases of aquatic species. *Revue scientifique et technique de l'Office International des Epizooties* **15**: 675-686.

Chilmonczyk, S., Cox, W. T., and Hedrick, R. P. 1991. *Enterocytozoon salmonis* n.s.: an intranuclear microsporidian from salmonid fish. *Journal of Protozoology* **38**: 264–269.

Cockings, S. and Martin, D. 2005. Zone design for environment and health studies using pre-aggregated data. *Social Science and Medicine* **60**: 2729–2742.

Coward, K. and Little, D. C. 2001. Culture of the "aquatic chicken": present concerns and future prospects. *Biologist (London)* **48**: 12–16.

Cowx, I. G. 2000. *Management and Ecology of River Fisheries*. Oxford, UK: Fishing News Books.

Cunningham, C. O. 2002. Molecular diagnosis of fish and shellfish diseases: present status and potential use in disease control. *Aquaculture* **206**: 19–55.

Daelman, W. 1996. Animal health and the trade in aquatic animals within and to the European Union. *Revue scientifique et technique de l'Office International des Epizooties* **15**: 711–722.

Davies, A. J. and Smit, N. J. 2001. The life cycle of *Haemogregarina bigemina* (Adeleina: Haemogregarinidae) in South African hosts. *Folia Parasitologica* **48**: 169–177.

Dikkeboom, A. L., Radi, C., Toohey-Kurth, K., *et al.* 2004. First report of Spring Viremia of Carp Virus (SVCV) in wild common carp in North America. *Journal of Aquatic Animal Health* **16**: 169–178.

Djajadiredja, R., Panjaitan, T. H., Rukyani, A., *et al.* 1983. Country reports: Indonesia. In *Fish Quarantine and Fish Diseases in Southeast Asia*, eds. F. B. Davy and A. Chouinard. Ottawa, Ontario: International Development Research Centre, pp. 19–30.

Doyle, K. A., Beers, P. T., and Wilson, D. W. 1996. Quarantine of aquatic animals in Australia. *Revue scientifique et technique de l'Office International des Epizooties* **15**: 659–674.

Eaton, W. D., Hulett, J. L., Brunson, R., and True, K. 1991. The first isolation in North America of infectious hematopoietic necrosis virus (IHNV) and viral hemorrhagic septicemia virus (VHSV) in coho salmon from the same watershed. *Journal of Aquatic Animal Health* **3**: 114–117.

Faisal, M. and Garling, D. 2004. *What is Whirling Disease?* Fact Sheet Series No. 113. East Lansing, MI: North Central Regional Aquaculture.

Faisal, M. and Hnath, J. G. 2005. Fish health and diseases issues in the Laurentian Great Lakes. In *Health and Diseases of Aquatic Organisms: Bilateral Perspectives*, eds. R. C. Cipriano, I. S. Shchelkunov, and M. Faisal. East Lansing, MI: Michigan State University Press, pp. 331–350.

FAO (Food and Agriculture Organization). (2005). *Yearbook of Fishery Statistics*. Rome: Food and Agriculture Organization of the United Nations.

Ford, S. E., Schotthoefer, A., and Spruck, C. 1999. *In vivo* dynamics of the microparasite *Perkinsus marinus* during progression and regression of infections in eastern oysters. *Journal of Parasitology* **85**: 273–282.

Fryer, J. L. and Sanders, J. E. 1981. Bacterial kidney disease of salmonid fish. *Annual Review of Microbiology* **35**: 273–298.

Gaughan, D. J. 2002. Disease-translocation across geographic boundaries must be recognized as a risk even in the absence of disease identification: the case with Australian *Sardinops*. *Reviews in Fish Biology and Fisheries* **11**: 113–123.

Ghittino, C., Ghittino, P., and Marin de Mateo, M. 1989. Adjournment on anguillicolosis, a common parasitic aerocystitis of eel. *Rivista italiana di acquacoltura (Verona)* **24**: 125–136.

Goodwin, A. E. 2002. First report of spring viremia of carp virus (SVCV) in North America. *Journal of Aquatic Animal Health* **14**: 161–164.

Gustafson, L. L., Ellis, S. K., and Bartlett, C. A. 2005. Using expert opinion to identify risk factors important to infectious salmon-anemia (ISA) outbreaks on salmon farms in Maine, USA and New Brunswick, Canada. *Preventative Veterinary Medicine* **70**: 17–28.

Haenen, O. 1995. *Anguillicola crassus* (Nematoda, Dracunculoidea) infections of European eel (*Anguilla anguilla*) in the Netherlands: epidemiology, pathogenesis and pathobiology. Ph.D. thesis. University of Wageningen, Netherlands.

Harvell, C. D., Kim, K., Burkholder, J. M., et al. 1999. Emerging marine diseases: climate links and anthropogenic factors. *Science* **285**: 1505–1510.

Hasson, K. W., Lightner, D. V., Poulos, B. T., et al.1995. Taura syndrome in *Penaeus vannamei*: demonstration of a viral etiology. *Diseases of Aquatic Organisms* **23**: 115–126.

Hedrick, R. P. 1996. Movements of pathogens with the international trade of live fish: problems and solutions. *Revue scientifique et technique de l'Office International des Epizooties* **15**: 523–532.

Hedrick, P. 1997. How microbial diseases of salmonids impact aquaculture. *Microbiology Australia* **18**: 26–30.

Heide-Jorgensen, M. P. and Harkonen, T. 1992. Epizootiology of the seal disease in the eastern North Sea. *Journal of Applied Ecology* **29**: 99–107.

Hill, B. J. 1996. National legislation in Great Britain for the control of fish diseases. *Revue scientifique et technique de l'Office International des Epizooties* **15**: 633–645.

Hindar, K., Ryman, N., and Utter, F. 1991. Genetic effects of cultured fish on natural fish populations. *Canadian Journal of Fisheries and Aquatic Sciences* **48**: 945–957.

Hnath, J. G. 1993. *Great Lakes Fish Disease Control Policy and Model Program*, Special Publication No. 93-1. Ann Arbor, MI: Great Lakes Fishery Commission.

Hoffman, G. L. 1970. Intercontinental and transcontinental dissemination and transfaunation of fish parasites with emphasis on whirling disease *Myxobolus cerebralis*. In *A Symposium on Diseases of Fish and Shellfish* Special Publication No. 5, ed. S. F. Snieszko. Washington, DC: American Fisheries Society, pp. 69–81.

Hoffman, G. L. 1990. *Myxobolus cerebralis*, a worldwide cause of salmonid whirling disease. *Journal of Aquatic Animal Health* **2**: 30–37.

Hoole, D., Bucke, D., Burgess, P., and Wellby, I. 2001. *Diseases of Carp and other Cyprinid Fish*. Oxford, UK: Fishing News Books.

Humphrey, J. D. 1995. *Australian Quarantine Policies and Practices for Aquatic Animals and Their Products: A Review for the Scientific Working Party on Aquatic Animal Quarantine*. Canberra, ACT: Bureau of Resource Sciences.

ICES (International Council for the Exploration of the Seas). 1998. *Code of Practice on the Introduction and Transfer of Marine Organisms*. Copenhagen: International Council for the Exploration of the Sea.

Johnsen, B. O. and Jensen, A. J. 1986. Infestations of Atlantic salmon (*Salmo salar*), by *Gyrodactylus salaris* in Norwegian rivers. *Journal of Fish Biology* **29**: 233–241.

Johnsen, B. O. and Jensen, A. J. 1994. The spread of furunculosis in salmonids in Norwegian rivers. *Journal of Fish Biology* **45**: 47–55.

Jones, J. B, Hyatt, A. D., Hine, P. M., et al. 1997. Special topic review: Australasian pilchard mortalities. *World Journal of Microbiology and Biotechnology* **3**: 383–392.

Kim, K. and Harvell, C. D. 2004. The rise and fall of a six year coral-fungal epizootic. *American Naturalist* **164**: S52–S63.

Kreuder, C., Miller, M. A., Jessup, D. A., *et al.* 2003. Patterns of mortality in southern sea otters (*Enhydra lutris nereis*) from 1998-2001. *Journal of Wildlife Diseases* **39**: 495-509.

Lafferty, K. D. and Kuris, A. M. 1993. Mass mortality of *Abalone Haliotis cracherodii* on the California Channel Islands: tests of epidemiologic hypotheses. *Marine Ecology Progress Series* **96**: 239-248.

Langdon, J. S., Humphrey, J. D., Copland, J., *et al.* 1986. The disease status of Australian salmonids: viruses and viral diseases. *Journal of Fish Diseases* **9**: 129-135.

Lehane, I. 1993. Risks of fish imports: the 1993 aquatic animal quarantine review. *Australian Veterinary Journal* **70**: 202-204.

Lightner, D. V. 1996. Epizootiology, distribution and the impact on international trade of two penaeid shrimp viruses in the Americas. *Revue scientifique et technique de l'Office International des Epizooties* **15**: 579-601.

Lightner, D. V., Bell, T. A., Redman, R. M., *et al.* 1992a. A review of some major diseases of economic significance in penaeid prawns/shrimps of the Americas and Indo-Pacific. In *Diseases in Asian Aquaculture*, vol. 1, eds. M. Shariff, R. Subasinghe, and J. R. Arthur. Manila: Fish Health Section, Asian Fisheries Society, pp. 57-80.

Lightner, D. V., Poulos, B. T., Bruce, L., *et al.* 1992b. New developments in penaeid virology: application of biotechnology in research and disease diagnosis for shrimp viruses of concern in the Americas. In *Diseases of Cultured Penaeid Shrimp in Asia and the United States*, eds. W. Fulks and K. Main. Makapuu Point, Honolulu, HI: Oceanic Institute, pp. 233-263.

Lightner, D. V., Redman, R. M., Hasson, K. W., and Pantorja, C. R. 1995. Taura syndrome in *Penaeus vannamei* (Crustacea: Decapoda): gross signs, histopathology and ultrastructure. *Diseases of Aquatic Organisms* **21**: 53-59.

Lightner, D. V., Redman, R. M., Poulos, B. T., *et al.* 1997. Risk of spread of penaeid shrimp viruses in the Americas by the international movement of live and frozen shrimp. *Revue scientifique et technique de l'Office International des Epizooties* **16**: 146-160.

Lilley, J. H., Phillips, M. J., and Tonguthai, K. 1992. *A Review of Epizootic Ulcerative Syndrome (EUS) in Asia*. Bangkok: Aquatic Animal Health Research Institute.

Littler, D. S. and Littler, M. M. 1995. Impact of CLOD pathogen on Pacific coral reefs. *Science* **267**: 1356-1360.

Liu, H., Gao, L., Shi, X., *et al.* 2004. Isolation of spring viremia of carp virus (SVCV) from cultured koi (*Cyprinus carpio* koi) in P.R. China. *Bulletin of the European Association of Fish Pathologists* **24**: 194-202.

Marnell, L. F. 1986. Impacts of hatchery stocks on wild fish populations. In *Fish Culture in Fisheries Management*, ed. R. H. Stroud. Bethesda, MD: American Fisheries Society, pp. 339-347.

McCallum, H., Harvell, D., and Dobson, A. 2003. Rates of spread of marine pathogens. *Ecology Letters* **6**: 1062-1067.

McClure, C. A., Hammell, K. L., Stryhn, H., Dohoo, I. R., and Hawkins, L. J. 2005. Application of surveillance data in evaluation of diagnostic tests for infectious salmon anemia. *Diseases of Aquatic Organisms* **63**: 119-127.

Meyers, T. R., Sullivan, J., Emmeneger, E., *et al.* 1992. Identification of viral hemorrhagic septicemia virus isolated from Pacific cod *Gadus macrocephalus* in Prince William Sound, Alaska, USA. *Diseases of Aquatic Organisms* **12**: 167-175.

Meyers, T. R., Short, S., Lipson, K., *et al.* 1994. Association of viral hemorrhagic septicemia virus with epizootic hemorrhages of the skin in Pacific herring *Clupea harengus* from Prince William Sound and Kodiak, Alaska, USA. *Diseases of Aquatic Organisms* **19**: 27-31.

Mitchell, H. and Stoskopf, M. K. 1999. Guidelines for development and application of aquatic animal health regulations and control programs. *Journal of the American Veterinary Medical Association* **14**: 1786–1789.

Mitchum, D. L., Sherman, L. E., and Baxter, G. T. 1979. Bacterial kidney disease in feral populations of brook trout (*Salvelinus fontinalis*), brown trout (*Salmo trutta*), and rainbow trout (*Salmo gairdneri*). *Journal of the Fisheries Research Board of Canada* **36**: 1370–1376.

Moffitt, C. M. 2005. Environmental, economic and social aspects of animal protein production and the opportunities for aquaculture. *Fisheries* **30**: 36–38.

Molnar, K., Szekely, C., and Baska, F. 1991. Mass mortality of eel in Lake Balaton due to *Anguillicola crassus* infections. *Bulletin of the European Association of Fish Pathologists* **11**: 211–212.

Moore, D. W. 1991. A virus attacks shrimp: IHHN shrimp virus – its introduction into Mexico and the Sea of Cortez. *CEDO News* **3**(2): 7–13.

Moore, D. W. and Brand, C. W. 1993. The culture of marine shrimp in controlled environment superintensive systems. In *CRC Handbook of Mariculture, 2nd edn, vol. 1, Crustacean Aquaculture*, ed. J. P. McVey. Boca Raton, FL: CRC Press, pp. 315–348.

Moyer, M., Blake, N. J., and Arnold, W. S. 1993. An ascetosporan disease causing mass mortalities in the Atlantic calico scallop, *Argopecten gibbus* (Linnaeus 1758). *Journal of Shellfish Research* **12**: 305–310.

Murray, A. G. 2002. Making the case for zoning. *Australian Veterinary Journal* **80**: 458.

Murray, A. G. and Peeler, E. J. 2005. A framework for understanding the potential for emerging diseases in aquaculture. *Preventative Veterinary Medicine* **67**: 223–235.

Naylor, R. L., Goldburg, R. J., Primavera, J. H., et al. 2000. Effect of aquaculture on world fish supplies. *Nature* **405**: 1017–1024.

Nilsen, R., Ness, A., and Nylund, A. 1995. Observations on an intranuclear microsporidian in lymphoblasts from farmed Atlantic halibut larvae (*Hyppoglossus hippoglossus* L.). *Journal of Eukaryotic Microbiology* **42**: 131–135.

OIE (Office International des Epizooties). 2003. *Manual of Diagnostic Tests for Aquatic Animals* 4th edn. Paris: Office International des Epizooties.

OIE. 2006. *Aquatic Animal Health Code*. 6th edn. Paris: Office International des Epizooties.

Pantoja-Morales, C. R. and Lightner, D. V. 1991. Status of the presence of IHHN virus in wild penaeid shrimp from the coast of Sonora, Mexico. In *Abstracts of the 24th Annual Meeting of the Society for Invertebrate Pathology*, August 4–9, Flagstaff, AZ.

Primavera, J. H. 2005. Global voices of science: mangroves, fishponds, and the quest for sustainability. *Science* **310**: 57–59.

Rahe, R. 1987. Geschichte und derzeitiger Stand der Krebspest in der Türkei. *Fischerei und Teichwirt* **6**: 174–177.

Rahimian, H. and Thulin, J. 1996. Epizootiology of *Ichthyophonus hoferi* in herring population off the Swedish west coast. *Diseases of Aquatic Organisms* **27**: 187–195.

Renault, T. 1996. Appearance and spread of diseases among bivalve molluscs in the northern hemisphere in relation to international trade. *Revue scientifique et technique de l'Office International des Epizooties* **15**: 551–562.

Rosenberg, E. and Loya, Y. 2004. *Coral Health and Disease*, New York: Springer-Verlag.

Sano, T. 1973. Studies on viral diseases of Japanese fishes. V. Infectious pancreatic necrosis of amago trout. *Bulletin of the Japanese Society of Scientific Fisheries* **39**: 477–480.

Sattuar, O. 1988. Parasites prey on wild salmon in Norway. *New Scientist* **120**: 21.

Schlotfeldt, H. J. 1996. Synopsis of freshwater aquaculture legislation in Germany since national reunification. *Revue scientifique et technique de l'Office International des Epizooties* **15**: 687–702.

Scudamore, J. 2002. *Partnership, Priorities and Professionalism: A Proposed Strategy for Enhancing Veterinary Surveillance in the UK*. London: Veterinary Surveillance Division, Department for Environment Food and Rural Affairs.

Stephenson, M. F., McGladdery, S. E., Maillet, M., Veniot, A., and Meyer, G. 2003. First reported occurrence of MSX in Canada. *Journal of Shellfish Research* **22**: 355.

Subasinghe, R. P. 2005. Epidemiological approach to aquatic animal health management: opportunities and challenges for developing countries to increase aquatic production through aquaculture. *Preventative Veterinary Medicine* **67**: 117–124.

Subasinghe, R. P., McGladdery, S. E., and Hill, B. J. 2004. *Surveillance and Zoning for Aquatic Animal Diseases*, FAO Fisheries Technical Paper No. 451. Rome: Food and Agriculture Organization of the United Nations.

Sussman, M., Loya, Y., Fine, M., and Rosenberg, E. 2003. The marine fireworm *Hermodice carunculata* is a winter reservoir and spring–summer vector for the coral bleaching pathogen *Vibrio shiloi*. *Environmental Microbiology* **5**: 250–255.

Tonguthai, K. A. 1985. *Preliminary account of ulcerative fish diseases in the Indo-Pacific region (a comprehensive study based on Thai experiences)*. Bangkok: Department of Fisheries, Ministry of Agriculture and Cooperation,

Yoder, W. G. 1972. The spread of *Myxosoma cerebralis* into native trout population in Michigan. *Progressive Fish-Culturist* **34**: 103–106.

Yoshimizu, M. 1996. Disease problems of salmonid fish in Japan caused by international trade. *Revue scientifique et technique de l'Office International des Epizooties* **15**: 533–550.

KRISTEN T. HOLECK, EDWARD L. MILLS,
AND HUGH J. MACISAAC

6

Globalization, biological invasions, and ecosystem changes in North America's Great Lakes

INTRODUCTION

Globalization, in the context of biological invasions, is the increased movement of species around the world. In this chapter, non-indigenous species (NIS) are defined as taxa moved from one geographic location of the world to another from which they were historically absent. The largest geographic barriers to species dispersal, the world's oceans, have been circumvented by the development of a global economy. Increased demand for and transport of goods has resulted in the transfer – both intentional and unintentional – of NIS on unprecedented scales. For example, colonization rates of European crustaceans in North America are estimated to be 50 000 times background levels associated with natural dispersal (Hebert and Cristescu 2002). A number of dispersal vectors are responsible for transport of aquatic NIS, though transoceanic shipping has played a particularly important role as the global economy has expanded.

Establishment of NIS represents one of the most significant threats to the world's indigenous biota (Mooney and Drake 1989; Mack et al. 2000), in addition to adverse ecological and economic effects that they impart on lakes throughout the world (e.g., Hall and Mills 2000). For example, establishment of Nile perch (*Lates niloticus*) in Lake Victoria and peacock bass (*Cichla ocellaris*) in Gatun Lake resulted in extirpation or decline of native fish species (Zaret and Paine 1973; Ogutu-Ohwaya 1990; Witte et al. 1992). In Lake Titicaca, the introduction of brown trout (*Salmo trutta*) and rainbow trout (*Oncorhynchus mykiss*) led to a disease outbreak (*Ichthyophthirius multifiliis* or Ich), a short-lived commercial fishery, and competition with native species (Hall and Mills 2000). Pimentel et al. (2005) estimated that NIS in the

Globalization: Effects on Fisheries Resources, ed. William W. Taylor, Michael G. Schechter, and Lois G. Wolfson. Published by Cambridge University Press. © Cambridge University Press 2007.

United States cause approximately $120 billion per year in economic damage.

The worldwide problem of NIS has intensified with the development of a global economy. Because approximately 98 percent of world trade by weight is transported by sea (Reeves 1999), shipping is a strong vector for the movement of NIS worldwide. For example, shipping has accounted for 38 of 60 unintentional introductions to the Baltic Sea during the past 200 years (Leppäkoski et al. 2002). NIS in the Baltic Sea have caused economic damage to fisheries, shipping and industry, and include the hydrozoan *Cordylophora caspia*, the barnacle *Balanus improvisus*, the cladoceran *Cercopagis pengoi*, and the bivalve *Dreissena polymorpha*. In the Ponto-Caspian region, Grigorovich et al. (2002) identified 136 free-living and 27 parasitic invertebrate NIS that had established reproducing populations and emphasized that activities related to global trade – both shipping and the construction of canals and reservoirs – provided dispersal opportunities to and within the region. The recent addition of the North American ctenophore *Mnemiopsis leidyi* to the Caspian Sea portends catastrophic ecological and economic shifts in this basin (Shiganova 1998).

The Laurentian Great Lakes (Fig. 6.1), collectively the world's largest freshwater resource, are among the best studied and are well documented with respect to transport vectors and impacts of NIS (Mills et al. 1993, 1994; Hall and Mills 2000; Ricciardi 2001). Over 176 (183 reported as of 2007) non-indigenous aquatic plants and animals have been recognized as introduced and established in the Great Lakes basin (Mills et al. 1993; Ricciardi 2001, 2006; Nicholls and MacIsaac 2004), and the arrival of most of these species can be linked either directly or indirectly to globalization. Here we discuss how globalization has affected the structure and function of Great Lakes ecosystems, focusing on transoceanic shipping as a primary transport vector of NIS to the Great Lakes. We pay particular attention to five species introduced from the Ponto-Caspian region of Eurasia (Fig. 6.1), an area that has been identified as an important source of NIS to the Great Lakes (Ricciardi and MacIsaac 2000). Also, we discuss the effects these NIS have had – and may have in the future – on Great Lakes fisheries.

HISTORY OF GLOBALIZATION AND BIOLOGICAL INVASIONS IN THE GREAT LAKES

The first human-mediated introductions of NIS to the Great Lakes likely occurred through the activities of native peoples. However, the

158 Kristen T. Holeck et al.

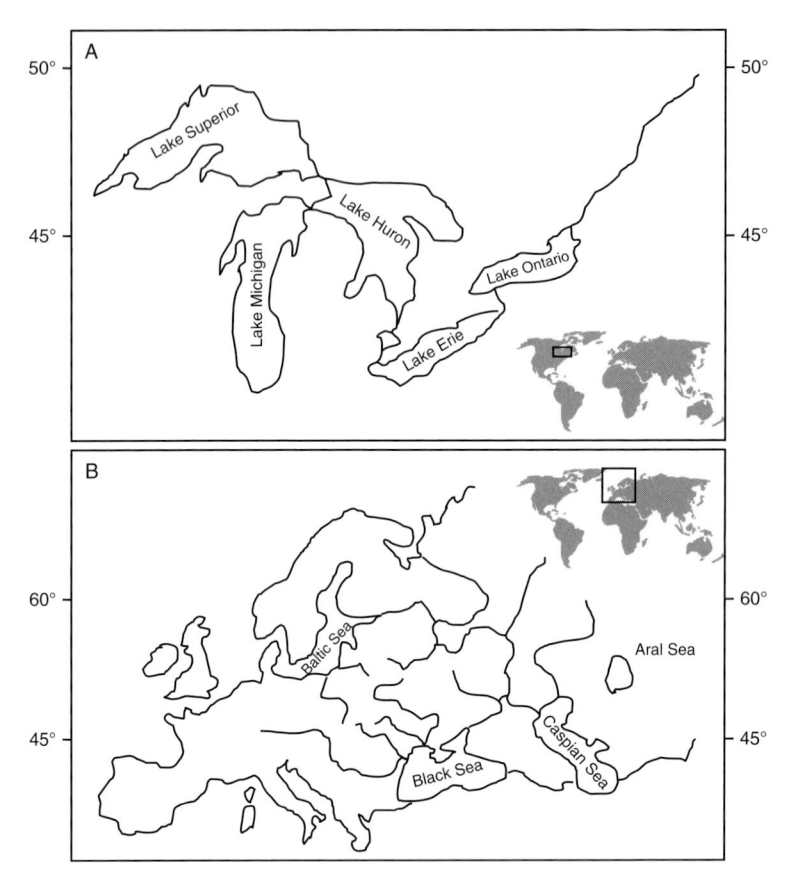

Figure 6.1 The North American Great Lakes (A) and Eurasia (B).

intensification of globalization's effect on the region did not begin until the arrival of European settlers about four centuries ago. The settlers brought animals and plants intentionally for cultivation and unintentionally in ship solid ballast and animal feeds. For example, purple loosestrife (*Lythrum salicaria*) arrived at ports on the Atlantic coast with imported sheep, in solid ballast, or as a cultivated plant (Mills et al. 1993). Three mollusk species (*Valvata piscinalis, Pisidium amnicum,* and *Bithynia tentaculata*) of Eurasian origin that were introduced prior to the turn of the twentieth century are believed to have been transported either in the solid ballast of ships or in straw and marsh grass packaging materials used to protect fragile articles during their overseas journey (Mills et al. 1993). During the late nineteenth century and early twentieth century, human population growth and concomitant development of lands surrounding the Great Lakes resulted in deforestation of the watershed, excessive nutrient loading, and the overharvesting of fishes. The detrimental effects of these large-scale changes on Great Lakes fisheries were amplified by the introduction of NIS via a range of vectors, including deliberate and accidental release, migration through canals, and shipping activities. Shipping vectors transitioned from solid to liquid ballast around the turn of the century, and NIS that established thereafter switched from primarily terrestrial to aquatic-based taxa.

Although some species have entered the Great Lakes from adjacent watersheds or the Atlantic coast, their arrivals are an indirect result of globalization. For example, two fish species that have had profound effects on Great Lakes fisheries, alewife (*Alosa pseudoharengus*) and sea lamprey (*Petromyzon marinus*), gained access to the lakes because of the construction of and improvements to connecting channels (the St. Lawrence Seaway and the Welland Canal), the purpose of which was to provide passage for large ships from the Atlantic Ocean to inland ports. Other species have been introduced intentionally. For example, common carp (*Cyprinus carpio*) was imported as early as 1831 for propagation in a private pond (DeKay 1842), and stocking of these fish in the Great Lakes basin by the U.S. Fish Commission ensued sometime after 1879 (Mills *et al.* 1993). Brown trout were intentionally released into the Pere Marquette River, a tributary of Lake Michigan, in 1883 (Emery 1985) as a potential sportfish; in the same year the fish was released accidentally from a fish hatchery into the Genesee River, a tributary of Lake Ontario. Most species introductions, however, have been unintentional and are linked directly to the development of the global economy. Population growth spurred the need to transport goods to and from the Great Lakes basin and ultimately led to the opening of the St. Lawrence Seaway in 1959. The seaway provided a new pathway for the transport of goods from all parts of the world, but has inadvertently facilitated the introduction of many species that have altered the ecological nature of the Great Lakes.

GLOBALIZATION AND TRANSOCEANIC SHIPPING

Transoceanic shipping has been the most important vector of introduction of NIS to the Great Lakes since the completion of the St. Lawrence Seaway, accounting for about 65 percent of all introductions since that time (data from Mills *et al.* 1993; Ricciardi 2001, 2006) (Fig. 6.2). The discovery of Eurasian ruffe (*Gymnocephalus cernuus*) and zebra mussel (*Dreissena polymorpha*) in the late 1980s prompted the Great Lakes

160 Kristen T. Holeck et al.

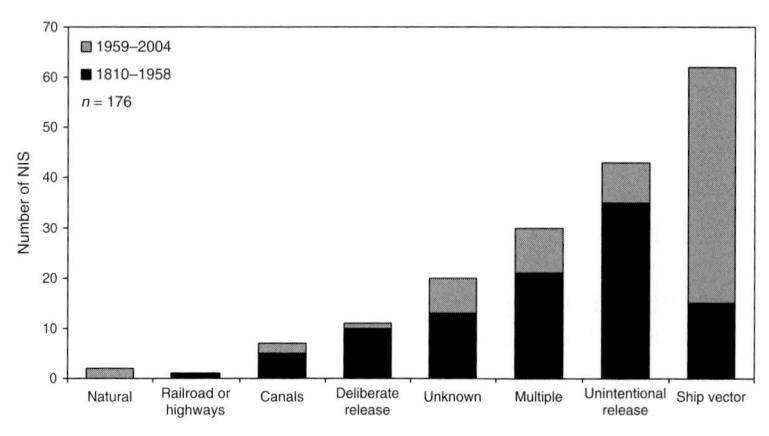

Figure 6.2 Vectors of introduction of non-indigenous species (NIS) to the Great Lakes prior to (black bars) and since (gray bars) the opening of the St. Lawrence Seaway in 1959.

Fishery Commission and the International Joint Commission to respond by calling the governments of the United States and Canada to develop a policy to reduce introductions of NIS from ballast water (Reeves 1999). Voluntary ballast water exchange guidelines were issued by Canada in 1989, and mandatory regulations were issued by the United States in 1993 (U.S. Coast Guard 1993). The legislation, specific to vessels entering the Great Lakes, mandated that vessels arriving from outside the exclusive economic zone (200 nautical miles [370 km] from shore) with declarable ballast water on board (BOB) must conduct open-ocean ballast exchanges if the water was to be subsequently discharged within the Great Lakes system; post-exchange ballast water must possess a salinity of no less than 30 parts per thousand (Locke et al. 1991, 1993; U.S. Coast Guard 1993). The premise behind ballast water exchange was that most freshwater organisms resident in ballast tanks would be purged and remaining organisms would be killed by osmotic stress when saltwater was loaded into the tanks. Ongoing tests to assess the efficacy of ballast water exchange for BOB vessels exchanging freshwater for saline water indicated that the process is likely highly effective and should dramatically reduce risk of invasion of the Great Lakes via this mechanism (Gray et al. 2006).

Contrary to expectations, the discovery rate of ship-vectored NIS increased following implementation of the ballast exchange policy (Holeck et al. 2004). The reasons for this increase are not clearly understood, but one possible explanation is that more than 90 percent of vessels that entered the Great Lakes during the 1990s declared "no ballast on board" (NOBOB) and were not required to exchange ballast, although their tanks contained residual sediments and water that would be discharged in the Great Lakes (Colautti et al. 2003). NOBOB vessels carry cargo, reflecting the transport and globalization of goods, but can carry up to 60 tonnes of sediment and water that may harbor organisms and viable resting stages (Bailey et al. 2003, 2004, 2005). Recent studies demonstrate that NOBOB ships carry diverse assemblages of non-indigenous invertebrates (as free-swimming adults and resting eggs) at low densities in residual water and residual sediments (Bailey et al. 2003, 2004, 2005). It is still not clear whether these sources are large enough to seed new populations in the Great Lakes.

Though the historical trend in reported discovery of new NIS in the Great Lakes has been linear (Fig. 6.3A), the number of NIS introduced via the ship vector has accelerated since the opening of the St. Lawrence Seaway (Fig. 6.3B). It remains unclear whether this pattern reflects actual establishment rates of NIS in the lakes, or is an artifact. Two issues that may cloud the establishment rate are research intensity and time lags. Increased research intensity could partially account for the elevated rate of discovery during the 1990s because the issue of NIS introductions has received considerably more attention in both the scientific sector and popular press in recent decades. However, some of the species (e.g., *Cercopagis pengoi*) discovered in this recent period include taxa that are easily found and identified, thus it is unlikely that they existed in the lakes for years prior to discovery. Similarly, time lags could have an influence on discovery rates if the interval between establishment and discovery has changed through time. This interval may have been reduced in recent years owing to increased scientific investigation of NIS in the lakes. Alternatively, invasion rate might increase if positive interactions involving established NIS or native species facilitated entry of new NIS. Ricciardi (2001) suggested that such a scenario of "invasional meltdown" is occurring in the Great Lakes, although Simberloff (2006) cautioned that most of these cases have not been proven. Of all the Great Lakes (including Lake St. Clair), Lake Superior receives a disproportionate number of discharges by both BOB and NOBOB ships (70 percent), yet it has sustained surprisingly few initial invasions (Colautti et al. 2003). This lake may pose a formidable environment for potential invaders and thus remains poorly invaded despite intense propagule pressure exerted upon it. Conversely, the waters connecting Lakes Huron and Erie are an invasion "hotspot" despite receiving disproportionately few ballast discharges (Grigorovich et al. 2003). If invasional meltdown is occurring

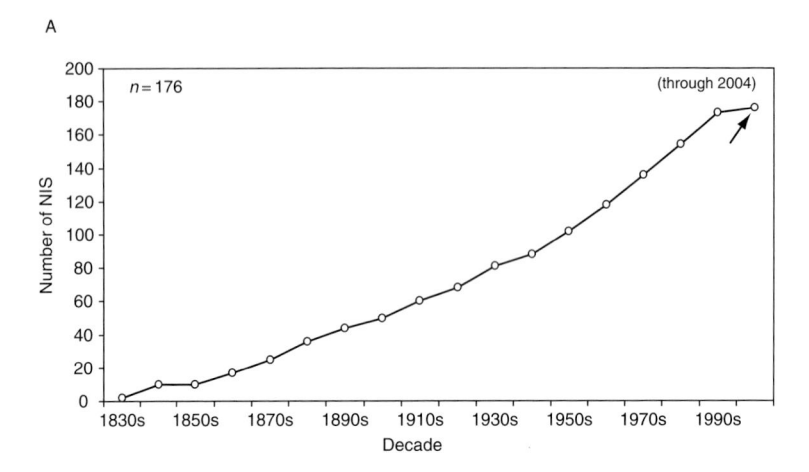

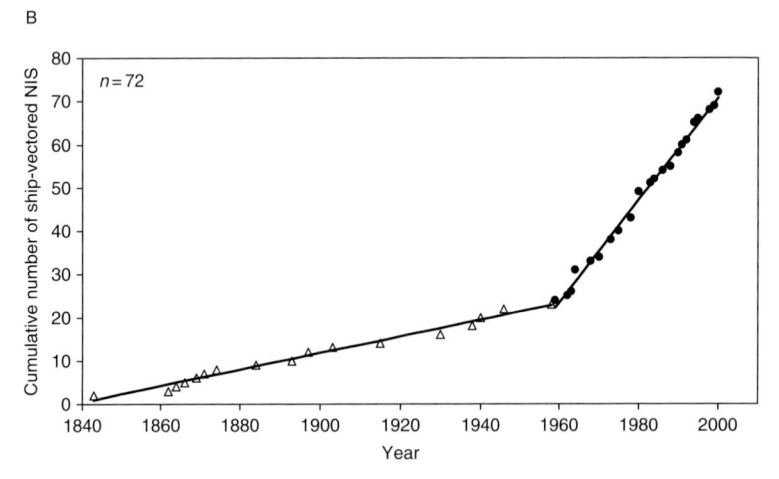

Figure 6.3 Accumulation of NIS in the Great Lakes since 1843 introduced by all vectors (A) and the ship vector only (B). The rate of ship-vectored introductions from 1959 to 2000 is more than five times the rate during the period 1843–1958.

it appears to be limited to the lower lakes. It should be noted, however, that many other hypotheses can account for the same pattern as that developed by invasional meltdown.

The transport of Ponto-Caspian species via the ship vector is consistent with patterns of transoceanic shipping. Between 1986 and 1998, most transoceanic vessels entering the Great Lakes arrived from European ports in the lower River Rhine region, on the North Sea and on the Baltic Sea (Colautti et al. 2003). However, most NIS discovered in

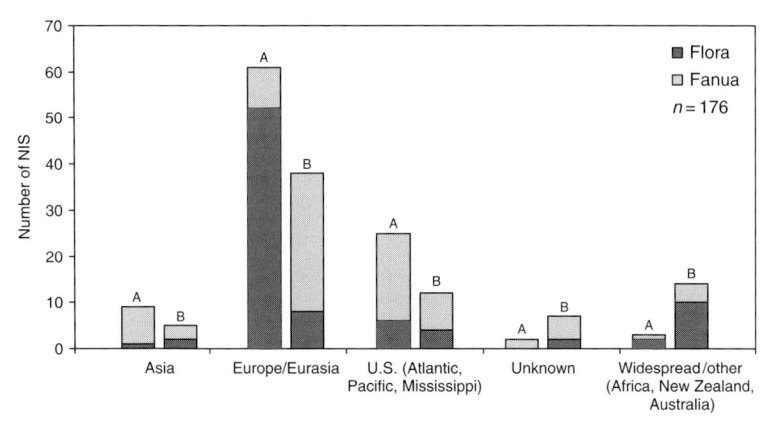

Figure 6.4 Donor regions for floral (black bars) and faunal (gray) NIS introduced to the Great Lakes from 1810 to 1958 (A) and 1959 to 2004 (B). The St. Lawrence Seaway opened in 1959.

the Great Lakes during that time were native to the Ponto-Caspian region (Ricciardi and MacIsaac 2000; MacIsaac et al. 2001; Ricciardi 2006) (Fig. 6.4). Ongoing invasion of key port areas on the North Sea (e.g., Rotterdam, Antwerp) and the southern coast of the Baltic Sea by species native to the Ponto-Caspian region provide opportunities for these taxa to invade the Great Lakes in secondary invasions (Cristescu et al. 2001, 2004; Bij de Vaate et al. 2002; Leppäkoski et al. 2002). Although the last port of call for NOBOB vessels was more likely a ballast water recipient than a ballast water donor, ships operating in the Rhine–North Sea–Baltic Sea region visit several ports there, and this could explain why NOBOB ships arriving from those regions could still represent a strong vector for the entry of Ponto-Caspian NIS to the Great Lakes, as freshwater residuals and accumulated sediment in these vessels represent a mixture from ports recently visited. It should be noted that only a fraction of NOBOBs entering the Great Lakes do so with freshwater residuals, but this sub-vector appears to pose greater risk than that associated with invertebrates living in ballast sediments or which are present as viable, diapausing eggs (Duggan et al. 2005).

INVASION HISTORIES OF PONTO-CASPIAN SPECIES IN NORTH AMERICA AND EURASIA

The dispersal of Ponto-Caspian species to Eurasian and European inland waters has spanned the course of several thousand years, while the invasion of North American waters began only decades ago.

Dispersal rates differ greatly because of differences in the type and degree of human facilitation; despite this fact, patterns of geographic dispersal of Ponto-Caspian invaders in the Great Lakes show similarities to those observed in Europe and Eurasia. Zebra mussel, *Echinogammarus*, round goby, and *Cercopagis* have all exhibited widespread and rapid expansion of their geographic ranges since their introduction to North American waters. Conversely, quagga mussel (*Dreissena rostriformis*) dispersal has been limited. Where quagga mussels have invaded areas occupied by zebra mussels, the former has generally become the dominant dreissenid (Mills *et al*. 1996, 2003). Ricciardi and Rasmussen (1998) argued that invasion history can be used as a predictive criterion for determining whether a species can invade a target region. We suggest that the pattern of dispersal of a species will also be similar in both donor and recipient regions regardless of geography if these organisms utilize human-mediated dispersal mechanisms. Indeed, at the same time that the quagga mussel dispersed in North America, it began to spread by ship-mediated vectors up and down the Volga River in Russia (Orlova *et al*. 2005).

Zebra and quagga mussels

By the early Holocene, the range distribution of the zebra mussel encompassed a small portion of its contemporary distribution in the Black, Azov, and Caspian drainage basins (Starobogatov and Andreeva 1994). The zebra mussel expanded its range during the Holocene in association with human activities. Ancient boat traffic on the Danube, Dnieper, Don, and Volga Rivers likely facilitated its dispersal into these drainages. In 1769, Pallas first described populations of this species from the Caspian Sea and the Ural River. By the late eighteenth century and early nineteenth century, zebra mussel had spread to most major drainages of Europe using the vast network of canals. Zebra mussels appeared in Great Britain in 1824, and then in Leiden (1826), the lower Doru River in Portugal (1829), and many locations throughout the Netherlands and Belgium, France (Nor), Denmark (Copenhagen), Sweden, Finland, Italy, and the rest of Europe (Strayer 1991; Starobogatov and Andreeva 1994). Despite its extensive range throughout Europe and Eurasia (Fig. 6.5A), expansion of *D. polymorpha* northward has been limited by low water temperatures beyond latitudes 58° N (Starobogatov and Andreeva 1994), although Orlova and Panov (2004) reported its establishment in the Neva Estuary in the eastern Gulf of Finland.

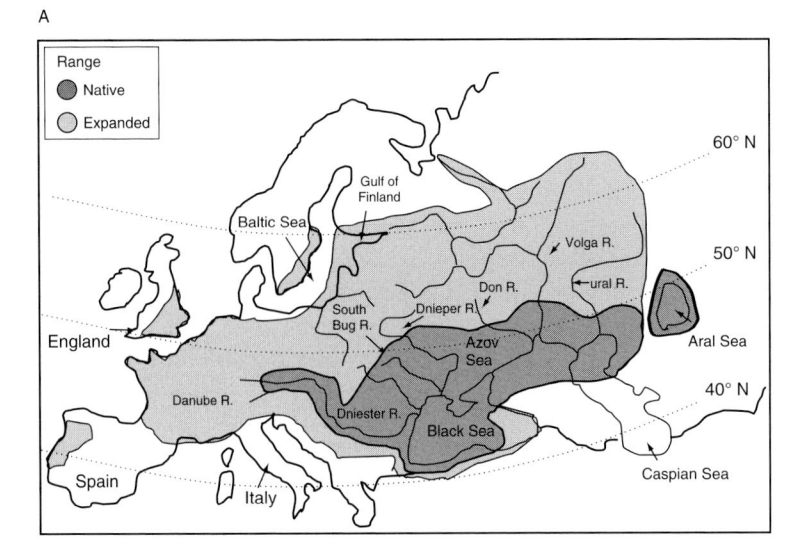

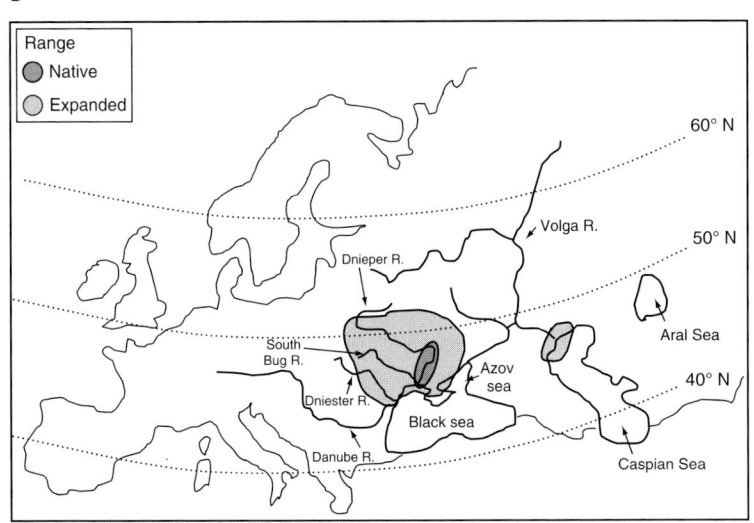

Figure 6.5 Generalized distribution of zebra mussel (A, C) and quagga mussel (B, D) in Eurasia (A, B) and the Great Lakes (C, D).

The quagga mussel, *Dreissena rostriformis bugensis*, was discovered in the Bug portion of the Dnieper–Bug estuary near Nikolaev in the Ukraine by Andrusov (1890). According to Zhuravel (1967), this mussel was first introduced into the Dnieper River attached to ship hulls in 1941. Since the 1940s, the quagga mussel has spread from the South

166 Kristen T. Holeck et al.

C

D

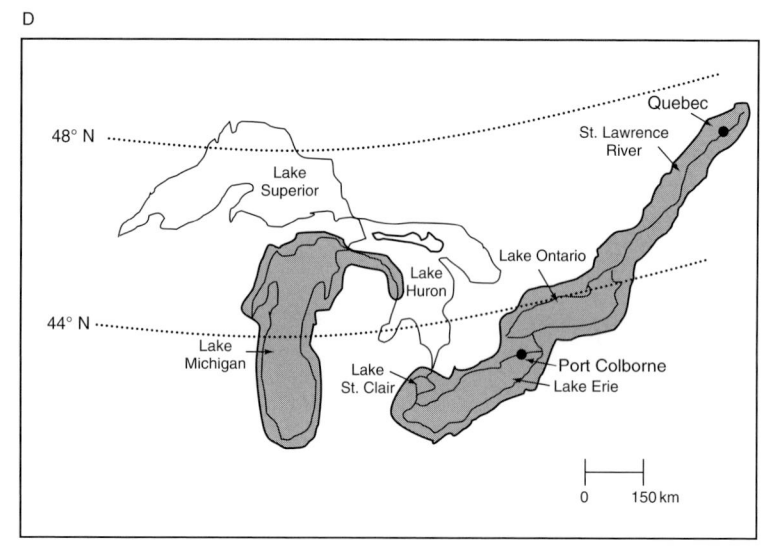

Figure 6.5 (Cont.)

Bug River and the lower Ingulets River into the Dnieper River drainage to regions that earlier had only *D. polymorpha* (Mills et al. 1996). It subsequently expanded its range elsewhere north of the Black Sea and into the northern Caspian Sea and Volga River Delta (1994–97) as well as to the middle (Samara region, 1992) and upper Volga River drainage (Orlova et al. 2005). To date, the geographic distribution of the quagga mussel in Eurasia has been limited to the Ponto-Caspian region, and its range expansion has been relatively slow. Despite its limited geographic distribution (Fig. 6.5B), the quagga mussel can dominate habitats that were once dominated by zebra mussels (Mills *et al.* 1996; but see Zhulidov *et al.* 2006).

Zebra mussels, first collected from the Canadian waters of Lake St. Clair in 1988 (Hebert *et al.* 1989) (Fig. 6.5C), had been found in each of the Great Lakes by 1990. After 1992, populations of zebra mussels rapidly spread throughout the eastern United States. In contrast, the quagga mussel, first collected at Port Colborne in Lake Erie in 1989 (Mills *et al.* 1999), had expanded its range only from Lake St. Clair eastward to Quebec City on the St. Lawrence River (Fig. 6.5D) in the 4 years after its discovery. However, habitats once dominated by zebra mussels are now dominated by quagga mussels, a trend that parallels that in the Dnieper River basin in Ukraine (Mills *et al.* 1996, 1999, 2003).

Round goby

Round gobies (*Apollonia (Neogobius) melanostomus*) are endemic to the Ponto-Caspian region, occurring in the open parts of the Caspian Sea, along the Black Sea and Azov Sea shelf, and in their lagoons and river estuaries (Miller 1986) (Fig. 6.6A). They were first found in the Baltic Sea (Puck Bay, Gulf of Gdansk) in 1990 (Skora and Stolarksi 1993). In the Great Lakes, round gobies were first collected in the St. Clair River on the Michigan–Ontario border in 1990 (Jude *et al.* 1992) (Fig. 6.6B). By 1994, the round goby had spread into northern Lake St. Clair and was abundant in the St. Clair River. By 1997, they had spread to all the Great Lakes, three inland rivers in Michigan (Flint, Shiawassee, and Saginaw), and the Chicago Sanitary and Shipping Canal (Steingraeber *et al.* 1996). In 1998, round gobies were reported from numerous sites along the east shore of Michigan in Lake Huron and in Michigan's Upper Peninsula at Port Inland and in Little Bay de Noc. In 1994, they appeared in southern Lake Michigan in the Calumet–Chicago area of Illinois, and in 1999 near the confluence of the Calumet Sag Channel and the Chicago Sanitary and Shipping Canal. In Indiana, round gobies were found in the Grand Calumet River in 1993. Trawls fished 200 individuals in October 1994 at Fairport, Ohio, in Lake Erie, and by 1995, more than 3000 individuals were collected, indicating that the population had expanded greatly. Round gobies expanded eastward to just west of Erie, Pennsylvania, in 1996, in the eastern basin of Lake Erie at Buffalo, New York, by 1998, and they now have been identified at numerous sites along the north shore of Lake Erie. Round gobies

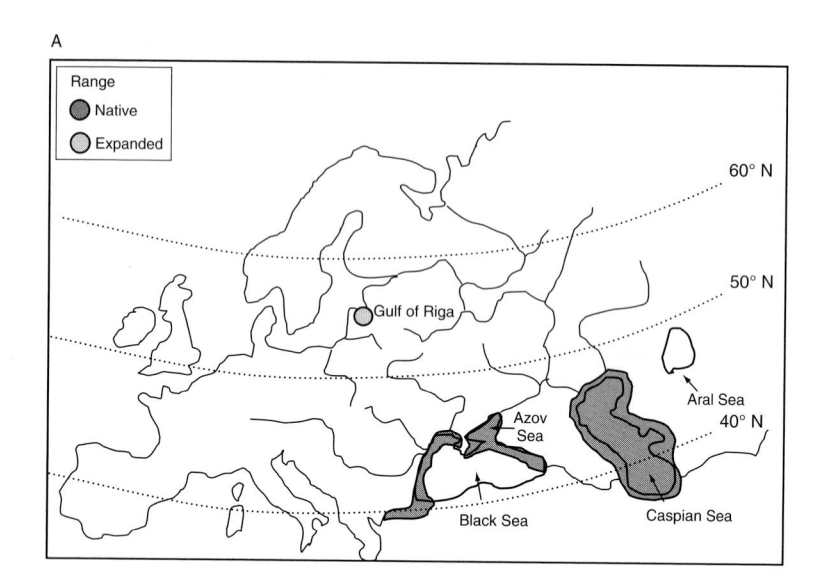

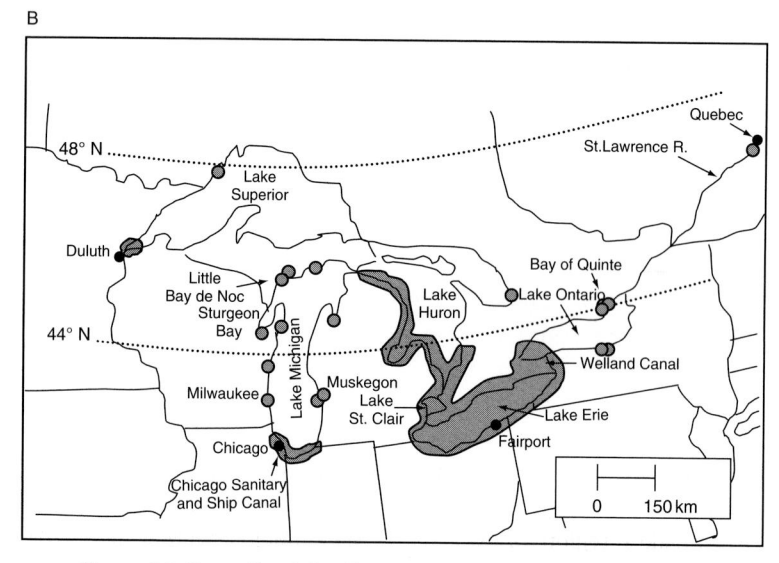

Figure 6.6 Generalized distribution of the round goby in Eurasia (A) and the Great Lakes (B).

have continued to expand their eastward distribution with sightings in the Welland Canal (1998), the St. Lawrence River near Quebec (1997), and northeastern Lake Ontario in the Bay of Quinte (1999). Discontinuity in the geographic spread of the round goby became evident in July 1995 when a single individual was collected in trawls from the Wisconsin waters of St. Louis Bay, Lake Superior. Adult specimens were also found in Duluth Harbor, Minnesota, from 1996 to 1999.

Cercopagis pengoi

Cercopagis pengoi, commonly known as the fishhook flea (MacIsaac *et al.* 1999), is endemic to fresh and brackish waters of the Black, Azov, Caspian, and Aral Sea basins, and to coastal lakes and reservoirs on the Don and Dnieper Rivers (see Mordukhai-Boltovskoi and Rivier 1987) (Fig. 6.7A). Damming, construction of reservoirs and canals, shipping, and the intentional stocking of invertebrates in the Dnieper, Don, and Manych Rivers have facilitated dispersal of this organism in eastern Europe (Grigorovich *et al.* 2000). *Cercopagis pengoi* has established permanent populations in the fresh waters of the Kakhovka, Zaporoozhsk, Kremenshug, Tsimlyansk, and Veselovsk reservoirs on the Don and Dnieper rivers (Krylov *et al.* 1999; Grigorovich *et al.* 2000). This onychopod has invaded the lower reaches of the Danube, Dniester, southern Bug, Dnieper, and Volga Rivers as well as coastal lakes fringing the Black Sea (Mordukhai-Boltovskoi 1968; Mordukhai-Boltovskoi and Rivier 1987; Rivier 1998). In 1992, *C. pengoi* was reported in the Baltic Sea (Parnu Bay and the Gulf of Riga) (Ojaveer and Lumberg 1995). Its range has continued to expand in Europe, as evidenced by its presence in coastal areas off Kotka in the Gulf of Finland (1995) (Uitto *et al.* 1999), the Neva Estuary and the open sea area in the eastern Gulf of Finland (1995 and 1996) (Avinski 1997), and the Baltic Sea proper (1997) (Gorokhova 1998; MacIsaac *et al.* 1999).

Cercopagis pengoi was first detected in the Great Lakes in Lake Ontario in 1998 (MacIsaac *et al.* 1999). It was observed throughout the lake in 1998 and 1999, with density peaks occurring in August and September (Makarewicz *et al.* 2001). In 1999, it appeared in Lake Michigan in two areas: Waukegon Harbor (Illinois) and Grand Traverse Bay (Michigan) (Charlebois *et al.* 2001). It also spread into the Finger Lakes region of New York State (Makarewicz *et al.* 2001). *Cercopagis* was found in western Lake Erie in 2001 (Therriault *et al.* 2002) and central Lake Erie (Fairport, Ohio, and Erie, Pennsylvania) in 2002 (Fig. 6.7B).

Echinogammarus ischnus

The euryhaline gammarid amphipod *Echinogammarus ischnus* is native to the Caspian Sea, fresh and estuarine regions of the Black and Azov Seas,

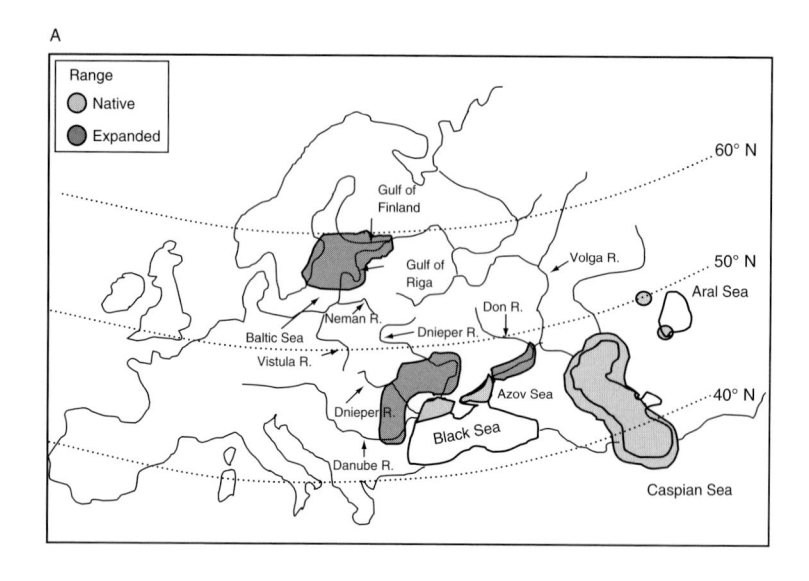

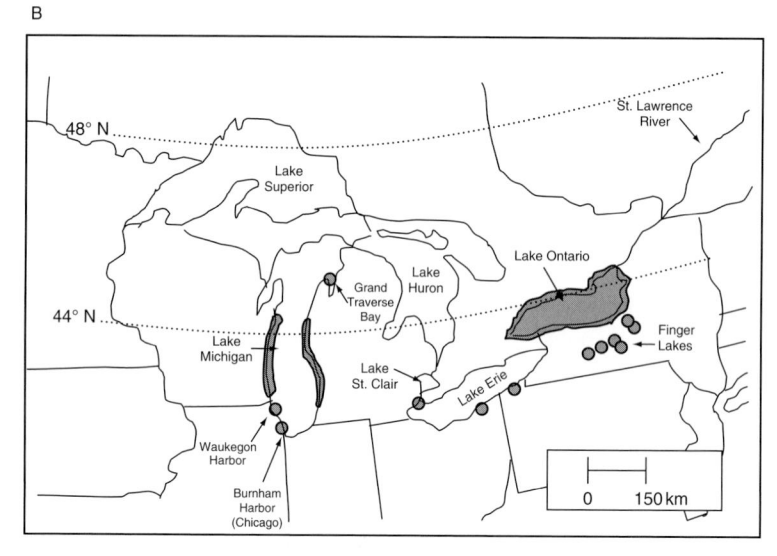

Figure 6.7 Generalized distribution of *Cercopagis pengoi* in Eurasia (A) and the Great Lakes (B).

areas north of the Black Sea (Jazdzewski 1980), and both Russia and the Ukraine (Konopacka and Jesionowska 1995) (Fig. 6.8A). Relict populations occur in some Romanian and Bulgarian lakes. *Echinogammarus* was first observed outside its native range in 1928, when it entered the Pripet-Bug canal system and was discovered in the Vistula River

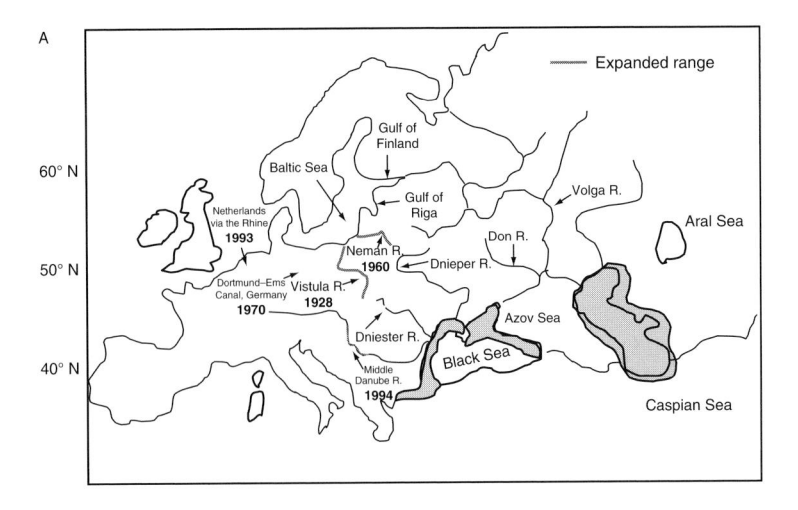

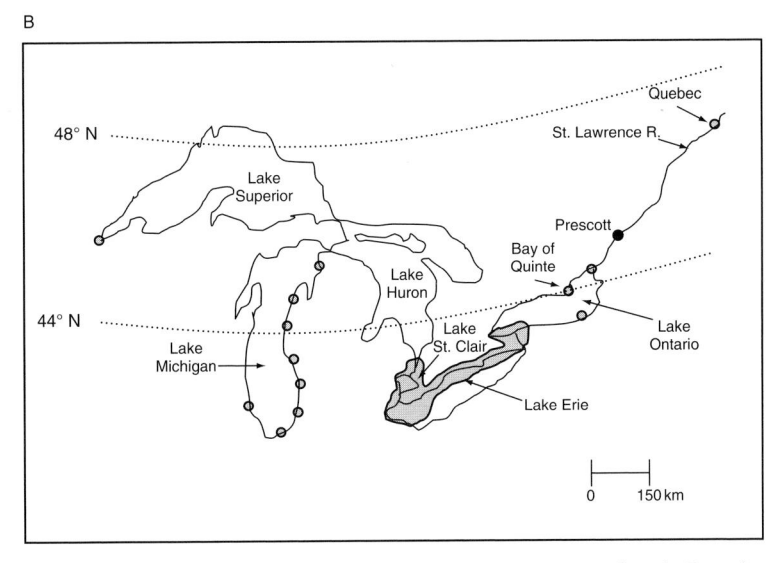

Globalization, invasions, and ecosystem changes 171

Figure 6.8 Generalized distribution of *Echinogammarus ischnus* in Eurasia (A) (for image in colour please see plate section) and the Great Lakes (B).

below Warsaw (Jarocki and Demianowicz 1931). Over 30 years later (1960) *Echinogammarus* had colonized the lower Neman River via the Neman–Pripet Canal. In the 1970s, this amphipod was discovered in the Dortmund-Ems Canal in Germany. By the late 1970s, it had established populations in the Dnieper, Dniester, Southern Bug, Danube, Don, and Volga Rivers (Mordukhai-Boltovskoi 1960, 1979a). *Echinogammarus* has also been widely introduced (deliberately and unintentionally) in Eurasian fresh water by human activities including canal and reservoir construction, shipping, and intentional transplants (Mordukhai-Boltovskoi 1960, 1979a, b). It is now in the middle Danube (Musko 1994) and recently migrated to the Netherlands from Germany via the River Rhine (Van den Brink *et al.* 1993).

The first published report of *Echinogammarus ischnus* in North America was from the Detroit River in 1995 (Witt *et al.* 1997) (Fig. 6.8B). However, analysis of archived samples revealed the presence of the species from western Lake Erie in 1994 and possibly as early as 1993 (van Overdijk *et al.* 2003). Dermott *et al.* (1998) reported this species to be "widespread" from the south end of Lake Huron to Lake Erie and Lake Ontario and the St. Lawrence River at Prescott. *Echinogammarus* was collected from several locations in Lake Michigan, during 1998–99, and established populations were reported in Lake Huron, Lake Ontario, and the upper St. Lawrence River in 2000 (USGS 2005). This amphipod is now common in nearshore areas of the western two-thirds of Lake Ontario, and its eastern range extends to Quebec City on the St. Lawrence River (Vanderploeg *et al.* 2002).

IMPACTS OF PONTO-CASPIAN SPECIES ON GREAT LAKES FISHERIES

The invasion of the Great Lakes by Ponto-Caspian species has altered ecosystem structure and function in several ways. First, colonization by zebra and quagga mussels has increased structural complexity and deposition of organic matter, benefiting benthic invertebrates in eastern Lake Erie, Lake St. Clair, Lake Ontario, and Lake Michigan (Griffiths *et al.* 1991; Stewart and Haynes 1994; Dermott and Kerec 1997; Kuhns and Berg 1999; Bially and MacIsaac 2000). Also, filtering activity by *Dreissena* spp. has increased water clarity, leading to increased vectoring of lower trophic level production to benthic habitats (Nalepa *et al.* 2000). With increased light penetration, dreissenids have indirectly increased benthic algal production (Fahnenstiel *et al.* 1995a; Lowe and Pillsbury 1995), enhanced benthic–pelagic coupling (MacIsaac *et al.* 1999), and increased macrophyte growth (Skubinna *et al.* 1995). In Saginaw Bay, Lake Huron, a reduction in primary production by phytoplankton was nearly compensated by increased benthic algal production (Fahnenstiel *et al.* 1995a, b). Finally, dreissenid mussel establishment may facilitate the success of coevolved Ponto-Caspian invaders (round goby and *Echinogammarus ischnus*) (Ricciardi 2001) by providing food and habitat, respectively. Round gobies prey on native fish eggs (e.g., smallmouth bass) (Steinhart et al. 2004) and have the potential to negatively affect restoration of lake trout (*Salvelinus namaycush*) (Chotkowski and Marsden 1999) and lake sturgeon (*Acipenser fulvescens*) (Nichols et al. 2003). *Echinogammarus* has replaced the native amphipod *Gammarus fasciatus* in the St. Clair, Detroit, and Niagara Rivers and in Lake Ontario at the mouth of the Welland Canal (Dermott et al. 1998), the effects of which are currently not clear. *Cercopagis pengoi* predation on zooplankton has been documented in the Gulf of Riga, where Leppäkoski et al. (2002) correlated its arrival with a decline in *Bosmina coregoni maritima*. Similar declines of small zooplankton, mainly juvenile cyclopoid and calanoid copepods and *Bosmina*, have been reported in Lake Ontario is association with elevated densities of *C. pengoi* (Benoît et al. 2002; Laxson et al. 2003). Fish in both systems feed on *Cercopagis*, although young-of-the-year alewife (less than 66 mm total length) in the Great Lakes have difficulty feeding on *C. pengoi* because of its long caudal spine (Bushnoe et al. 2003). Thus *Cercopagis* may depress zooplankton prey, resulting in decreased alewife growth and lower overwinter survival (O'Gorman et al. 1997).

Increased water clarity, resulting first from phosphorus abatement and enhanced later by dreissenid mussel grazing/filtering, represents one of the most dramatic ecological changes in the recent history of the Great Lakes. Great Lakes fishes are now exposed to much higher light levels than existed prior to dreissenid mussel establishment, and this can have profound effects on predator–prey interactions. For example, the opossum shrimp (*Mysis relicta*) exhibits remarkable diel migration behavior, moving from deep water to feed on epilimnetic and metalimnetic zooplankton during the night (Beeton and Bowers 1982). With increased light penetration, *Mysis* could remain deeper in the water column because of their high sensitivity to light (Gal et al. 1999), thus affecting their availability to prey on fish. Increases in water clarity have also altered fish distribution, with species such as alewife and rainbow smelt shifting to deeper waters in the springtime (O'Gorman et al. 2000). We can only speculate that large-scale changes in light conditions have had significant impact on Great Lakes fish communities by altering prey and predator spatial and temporal distributions and predator–prey interactions.

Ecological surprises can occur in ecosystems from both compounded effects of multiple perturbations (e.g., climate warming and invasion by NIS: Paine et al. 1998) and interactions between established invaders (Ricciardi 2001). For example, dreissenid mussels and round goby may be linked to recent outbreaks of Type E botulism in lakes Erie and Ontario which have resulted in both fish kills and the deaths of tens of thousands of waterfowl (primarily loons, mergansers, and gulls). The diets of some birds have become increasingly dominated by round gobies, which likely transfer the botulin toxin – produced by the bacterium *Clostridium botulinum* – from the quagga mussels upon which they feed. Scavenging gulls are affected through their consumption of dead fish that have already succumbed to the toxin themselves, either directly through the ingestion of dreissenid mussels or indirectly through the ingestion of other mussel-eating fish. Dreissenid mussels not only concentrate the toxin as they filter water proximal to the sediments that contain the *Clostridium* bacterium, but they may also contribute to anoxic conditions that favor the proliferation of the bacterium by generating large amounts of fecal deposits.[1] Finally, the success of round goby and *Echinogammarus* was likely enhanced by the presence of dreissenid mussels (Ricciardi and MacIsaac 2000). The result is an accelerated rate of invasion, or invasional meltdown (Ricciardi 2001), enhanced by facilitative interactions between coevolved NIS, that leads to future uncertainty about how large-scale ecosystems such as the Great Lakes will respond.

CONCLUSIONS AND FUTURE CHALLENGES

The effects of globalization have taken the Great Lakes on an unpredictable ecological path, particularly through the introduction of NIS. Since the early nineteenth century, over 176 NIS have established, and the apparent invasion rate has increased in association with an expanding range of anthropogenic activity. Global trade (i.e., international shipping) has been identified as the primary mechanism for the introduction of NIS over the past four decades. The prevalence of Ponto-Caspian introductions in the past two decades, combined with the fact that most ships arrive from European ports, indicates that secondary transfer routes (where Ponto-Caspian species establish first in ports on the Baltic Sea before being transported to North America) are an important factor.

The biological stressors associated with globalization, including the invasion by dreissenid mussels and other exotic species, have caused profound ecological changes in Great Lakes ecosystems. Responses to these stressors have led to significant changes in Great

---

[1] Helen Domske, New York Sea Grant, Buffalo, New York, personal communication.

Lakes fish communities. For example, dramatic springtime shifts in alewife and rainbow smelt distribution have occurred in response to changes in water clarity, stocks of native lake whitefish populations have dwindled, and the partitioning of energy to fish in the face of newly established exotics has become more complex, especially in Great Lakes' nearshore habitats. Globalization and biological invasions will continue to affect Great Lakes fisheries in the future. At present, the food webs of the Great Lakes are composed of native species and naturalized NIS that share no evolutionary history. With so many Great Lakes fish communities and supporting food web organisms considered non-native, historical food webs are no longer useful for understanding species interrelationships and future outcomes of biological stressors. Consequently, as long as globalization, biological invasions and large-scale ecological changes continue to dominate Great Lakes ecosystems, the need for scientific understanding and the desire to manage fisheries in these large water bodies in the coming decades will be enormous. To this end, the challenge in an era of globalization is clear: scientists and managers must engage in the development of new ecological paradigms and seek new insights into our understanding of functionality if there is any hope of meeting the need to manage fish and other elements of Great Lakes ecosystems in the future.

Acknowledgments

We would like to thank the Great Lakes Fishery Commission and Michigan State University for providing funding for this project, and William Taylor and Lois Wolfson for organizing this volume. This is contribution #225 of the Cornell Biological Field Station.

References

Andrusov, N. I. 1890. *Dreissena rostriformis* Desh. in the Bug River. *Vestnik Estestvoznaniya* **6**: 261–262.

Avinski, V. A. 1997. *Cercopagis pengoi*: a new species in the Eastern Gulf of Finland ecosystem. In *Proceedings of the Final Seminar of the Gulf of Finland Year 1996*, ed. J. Sarakkula, March 17–18, 1997, Helsinki, pp. 247–256.

Bailey, S. A., Duggan, I. C., van Overdijk, C. D. A., Jenkins, P. T., and MacIsaac, H. J. 2003. Viability of invertebrate diapausing eggs collected from residual ballast sediment. *Limnology and Oceanography* **48**: 1701–1710.

Bailey, S. A., Duggan, I. C., van Overdijk, C. D. A., *et al.* 2004. Salinity tolerance of diapausing eggs of freshwater zooplankton. *Freshwater Biology* **49**: 286–295.

Bailey, S. A., Duggan, I. C., Jenkins, P., and MacIsaac, H. J. 2005. Invertebrate resting stages in residual ballast sediment of transoceanic ships. *Canadian Journal of Fisheries and Aquatic Sciences* **62**: 1090–1103.

Beeton, A. M. and Bowers, J. A. 1982. Vertical migration of *Mysis relicta*. *Hydrobiologia* **93**: 53-62.

Benoît, H., Johannsson, O. E., Warner, D. M., Sprules, W. G., and Rudstam, L. G. 2002. Assessing the impact of a recent predatory invader: the population dynamics, vertical distribution and potential prey of *Cercopagis pengoi* in Lake Ontario. *Limnology and Oceanography* **47**: 626-635.

Bially, A. and MacIsaac, H. J. 2000. Fouling mussels (*Dreissena*) colonize soft sediments in Lake Erie and facilitate benthic invertebrates. *Freshwater Biology* **43**: 85-98.

Bij de Vaate, A., Jażdżewski, K., Ketelaars, H. A. M., Gollasch, S., and van der Velde, G. 2002. Geographical patterns in range extension of Ponto-Caspian macroinvertebrate species in Europe. *Canadian Journal of Fisheries and Aquatic Sciences* **59**: 1159-1174.

Bushnoe, T. M., Warner, D. M., Rudstam, L. G., and Mills, E. L. 2003. *Cercopagis pengoi* as a new prey item for alewife (*Alosa pseudoharengus*) and rainbow smelt (*Osmerus mordax*) in Lake Ontario. *Journal of Great Lakes Research* **29**: 205-212.

Charlebois, P. M., Raffenberg, M. J., and Dettmers, J. M. 2001. First occurrence of *Cercopagis pengoi* in Lake Michigan. *Journal of Great Lakes Research* **27**: 258-261.

Chotkowski, M. A. and Marsden, J. E. 1999. Round goby and mottled sculpin predation on lake trout eggs and fry: field predictions from laboratory experiments. *Journal of Great Lakes Research* **25**: 26-35.

Colautti, R. I., Niimi, A. J., van Overdijk, C. D. A., *et al.* 2003. Spatial and temporal analysis of transoceanic shipping vectors to the Great Lakes. In *Invasive Species: Vectors and Management Strategies*, eds. G. M. Ruiz and J. T. Carlton. Washington, DC: Island Press, pp. 227-246.

Cristescu, M. E. A., Hebert, P. D. N., Witt, J. D. S., MacIsaac, H. J., and Grigorovich, I. A. 2001. An invasion history for *Cercopagis pengoi* based on mitochondrial gene sequences. *Limnology and Oceanography* **46**: 224-229.

Cristescu, M. E. A., Witt, J. D. S., Grigorovich, I. A., Hebert, P. D. N., and MacIsaac, H. J. 2004. Dispersal of the Ponto-Caspian amphipod *Echinogammarus ischnus*: invasion waves from the Pleistocene to the present. *Heredity* **92**: 197-203.

DeKay, J. E. 1842. *Natural History of New York*, vol. I, *Zoology: Reptiles and Fishes*, part IV, *Fishes*. Albany, NY: W. A. White and J. Visscher.

Dermott, R. and Kerec, D. 1997. Changes to the deepwater benthos of eastern Lake Erie since the invasion of Dreissena: 1979-1993. *Canadian Journal of Fisheries and Aquatic Sciences* **54**: 922-930.

Dermott, R., Witt, J., Um, Y. M., and Gonzalez, M. 1998. Distribution of the Ponto-Caspian amphipod *Echinogammarus ischnus* in the Great Lakes and replacement of the native *Gammarus fasciatus*. *Journal of Great Lakes Research* **24**: 442-452.

Duggan, I. C., van Overdijk, C. D. A., Bailey, S. A., *et al.* 2005. Invertebrates associated with residual ballast water and sediments of cargo carrying ships entering the Great Lakes. *Canadian Journal of Fisheries and Aquatic Sciences* **62**: 2463-2474.

Emery, L. 1985. *Review of Fish Species Introduced into the Great Lakes, 1819-1974*, Technical Report No. 45. Ann Arbor, MI: Great Lakes Fishery Commission.

Fahnenstiel, G. L., Bridgeman, T. B., Lang, G. A., McCormick, M. J., and Nalepa, T. F. 1995a. Phytoplankton productivity in Saginaw Bay, Lake Huron: effects of zebra mussel (*Dreissena polymorpha*) colonization. *Journal of Great Lakes Research* **21**: 465-475.

Fahnenstiel, G. L., Lang, G. A., Nalepa, T. F., and Johengen, T. H. 1995b. Effects of zebra mussel (*Dreissena polymorpha*) colonization on water quality parameters in Saginaw Bay, Lake Huron. *Journal of Great Lakes Research* **21**: 435–448.

Gal, G., Loew, E. R., Rudstam, L. G., and Mohammadian, A. M. 1999. Light and diel vertical migration: spectral sensitivity and light avoidance by *Mysis relicta*. *Canadian Journal of Fisheries and Aquatic Sciences* **56**: 311–322.

Gorokhova, E. 1998. Zooplankton spatial distribution and potential predation by invertebrate zooplanktivores. *Proceedings of 2nd BASYS Annual Science Conference* September 23–25, 1998, Stockholm, Abstract 7.

Gray, D. K., van Overdijk, C. D. A., Johengen, T., Reid, D. F., and MacIsaac, H. J. 2006. Does open-ocean ballast exchange prevent the transfer of invertebrates between freshwater ports? *Proceedings of 14th International Conference on Aquatic Invasive Species*, May 14–19, Key Biscayne, FL.

Griffiths, R. W., Schloesser, D. W., Leach, J. H., and Kovalak, W. P. 1991. Distribution and dispersal of the zebra mussel *Dreissena polymorpha* in the Great Lakes region. *Canadian Journal of Fisheries and Aquatic Sciences* **48**: 1381–1388.

Grigorovich, I., MacIsaac, H. J., Rivier, I. K., Aladin, N. V., and Panov, V. E. 2000. Comparative biology of the predatory cladoceran *Cercopagis pengoi* from Lake Ontario, Baltic Sea, and Caspian Sea. *Archives of Hydrobiology* **686**: 23–50.

Grigorovich, I. A., MacIsaac, H. J., Shadrin, N. V., and Mills, E. L. 2002. Patterns and mechanisms of aquatic invertebrate introductions in the Ponto-Caspian region. *Canadian Journal of Fisheries and Aquatic Sciences* **59**: 1189–1208.

Grigorovich, I. A., Colautti, R. I., Mills, E. L., et al. 2003. Ballast-mediated animal introductions in the Laurentian Great Lakes: retrospective and prospective analyses. *Canadian Journal of Fisheries and Aquatic Sciences* **60**: 740–756.

Hall, S. R. and Mills, E. L. 2000. Exotic species in large lakes of the world. *Aquatic Ecosystem Health and Management* **3**: 105–135.

Hebert, P. D. N. and Cristescu, M. E. A. 2002. Genetic perspectives on invasions: the case of the Cladocera. *Canadian Journal of Fisheries and Aquatic Sciences* **59**: 1229–1234.

Hebert, P. D. N., Muncaster, B. W., and Mackie, G. L. 1989. Ecological and genetic studies on *Dreissena polymorpha* (Pallas): a new mollusc in the Great Lakes. *Canadian Journal of Fisheries and Aquatic Sciences* **46**: 1587–1591.

Holeck, K. T., Mills, E. L., MacIsaac, H. J., et al. 2004. Bridging troubled waters: biological invasions, transoceanic shipping, and the Laurentian Great Lakes. *BioScience* **54**: 919–929.

Jarocki, J. and Demianowicz, A. 1931. Über das Vorkommen des ponto-kapischen Amphipoden *Chaetogammarus tellenus* (G. O. Sars) in der Wisla (Weischel). *Bulletin de l'Académie polonaise des sciences* **BII**: 513–530.

Jazdzewski, K. 1980. Range extensions of some gammaridean species in European inland waters caused by human activity. *Crustaceana* **6**(Suppl.): 84–107.

Jude, D. J., Reider, R. H., and Smith, G. R. 1992. Establishment of Gobiidae in the Great Lakes basin. *Canadian Journal of Fisheries and Aquatic Sciences* **49**: 416–421.

Konopacka, A. and Jesionowska, K. 1995. Life history of *Echinogammarus ischnus* (Stebbing, 1898) (Amphipoda) from artificially heated Lichenskie Lake (Poland). *Crustaceana* **68**: 341–349.

Krylov, P. I., Bychemkov, D. E., Panov, V. E., Rodionova, N. V., and Telesh, I. V. 1999. Distribution and seasonal dynamics of the Ponto-Caspian invader *Cercopagis pengoi* (Crustacea, Cladocera) in the Neva Estuary (Gulf of Finland). *Hydrobiologia* **393**: 227–232.

Kuhns, L. A. and Berg, M. B. 1999. Benthic invertebrate community responses to round goby (*Neogobius melanostomus*) and zebra mussel (*Dreissena polymorpha*) invasion in Lake Michigan. *Journal of Great Lakes Research* **25**: 910-917.

Laxson, C. L., McPhedran, K. N., Makarewicz, J. C., Telesh, I. V., and MacIsaac, H. J. 2003. Effects of the invasive cladoceran *Cercopagis* on the lower food web of Lake Ontario. *Freshwater Biology* **48**: 2094-2106.

Leppäkoski, E., Gollasch, S., Gruszka, P., et al. 2002. The Baltic: a sea of invaders. *Canadian Journal of Fisheries and Aquatic Sciences* **59**: 1175-1188.

Locke, A., Reid, D. M., Sprules, W. G., Carlton, J. T., and van Leeuwen, H. C. 1991. Effectiveness of mid-ocean exchange in controlling freshwater coastal zooplankton in ballast water. *Canadian Technical Report for Fisheries and Aquatic Sciences* 1822.

Locke, A., Reid, D. M., van Leeuwen, H. C., Sprules, W. G., and Carlton, J. T. 1993. Ballast water exchange as a means of controlling dispersal of freshwater organisms by ships. *Canadian Journal of Fisheries and Aquatic Sciences* **50**: 2086-2093.

Lowe, R. L. and Pillsbury, R. W. 1995. Shifts in benthic algal community structure and function following the appearance of zebra mussels (*Dreissena polymorpha*) in Saginaw Bay, Lake Huron. *Journal of Great Lakes Research* **21**: 558-566.

MacIsaac, H. J., Grigorovich, I. A., Hoyle, J. A., Yan, N. D., and Panov, V. E. 1999. Invasion of Lake Ontario by the Ponto-Caspian predatory cladoceran *Cercopagis pengoi*. *Canadian Journal of Fisheries and Aquatic Sciences* **56**: 1-5.

MacIsaac, H. J., Grigorovich, I. A., and Ricciardi, A. 2001. Reassessment of species invasions concepts: the Great Lakes basin as a model. *Biological Invasions* **3**: 405-416.

Mack, R. N., Simberloff, D., Lonsdale, W. M., et al. 2000. Biotic invasions: causes, epidemiology, global consequences, and control. *Ecological Applications* **10**: 689-710.

Makarewicz, J. C., Grigorovich, I. A., Mills, E. L., et al. 2001. Distribution, fecundity, and genetics of *Cercopagis pengoi* (Ostrumov) (Crustacea, Cladocera) in Lake Ontario. *Journal of Great Lakes Research* **27**: 19-32.

Miller, P. J. 1986. Gobiidae. In *Fishes of the Northeast Atlantic and Mediterranean*, eds. P. J. P. Whitehead, M. L. Bauchot, J. C. Hureau, J. Nielsen, and E. Tortonese. Paris: UNESCO, pp. 1019-1095.

Mills, E. L., Leach, J. H., Carlton, J. T., and Secor, C. L. 1993. Exotic species in the Great Lakes: a history of biotic crises and anthropogenic introductions. *Journal of Great Lakes Research* **19**: 1-54.

Mills, E. L., Leach, J. H., Carlton, J. T., and Secor, C. L. 1994. Exotic species and the integrity of the Great Lakes: lessons from the past. *BioScience* **44**: 666-676.

Mills, E. L., Rosenberg, G., Spidle, A. P., et al. 1996. A review of the biology and ecology of the quagga mussel (*Dreissena rostriformis*), a second species of freshwater dreissenid introduced to North America. *American Zoologist* **36**: 271-286.

Mills, E. L., Chrisman, J. R., Baldwin, B., et al. 1999. Changes in the dreissenid community in the lower Great Lakes with emphasis on southern Lake Ontario. *Journal of Great Lakes Research* **25**: 187-197.

Mills, E. L., Casselman, J. M., Dermott, R., et al. 2003. Lake Ontario: food web dynamics in a changing ecosystem (1970-2000). *Canadian Journal of Fisheries and Aquatic Sciences* **60**: 471-490.

Mooney, H. A. and Drake, J. A. 1989. Biological invasions: A SCOPE program overview. In *Biological Invasions: A Global Perspective*, eds. J. A. Drake, H. A. Mooney, F. diCastri, R. H. Groves, F. J. Kruger, M. Rejmanek, and M. Williamson. New York: John Wiley, pp. 491-508.

Mordukhai-Boltovskoi, F. D. 1960. *Caspian Fauna in the Azov and Black Sea Basins.* Leningrad, Russia: Nauka. (In Russian.)

Mordukhai-Boltovskoi, F. D. 1968. *Kaspijskaya Fauna v. Azovo-Chernomorskom Bssejne.* Leningrad, Russia: Nauka. (In Russian.)

Mordukhai-Boltovskoi, F. D. 1979a. *The River Volga and Its Life.* The Hague, Netherlands: Dr. W. Junk.

Mordukhai-Boltovskoi, F. D. (ed.) 1979b. Composition and distribution of Caspian fauna in the light of modern data. *Internationale Revue der gesamte Hydrobiologie* **64**: 1–38.

Mordukhai-Boltovskoi, F. D. and Rivier, I. K. 1987. *Predatory Cladocerans of the World Fauna.* Leningrad, Russia: Nauka. (In Russian.)

Musko, I. B. 1994. Occurrence of amphipods in Hungary since 1853. *Crustaceana* **66**: 145–151.

Nalepa, T. F., Fahnenstiel, G. L., and Johengen, T. L. 2000. Impacts of the zebra mussel (*Dreissena polymorpha*) on water quality: a case study in Saginaw Bay, Lake Huron. In *Nonindigenous Freshwater Organisms: Vectors, Biology, and Impacts*, eds. R. Claudi and J. H. Leach. New York: Lewis Publishers, pp. 255–271.

Nicholls, K. H. and MacIsaac, H. J. 2004. Euryhaline, sand-dwelling, testate rhizopods in the Great Lakes. *Journal of Great Lakes Research* **30**: 123–132.

Nichols, S. J., Kennedy, G., Crawford, E., et al. 2003. Assessment of lake sturgeon (*Acipenser fulvescens*) spawning efforts in the lower St. Clair River, Michigan. *Journal of Great Lakes Research* **29**: 383–391.

O'Gorman, R., Johannsson, O. E., and Schneider, C. P. 1997. Age and growth of alewives in the changing pelagia of Lake Ontario, 1977–1992. *Transactions of the American Fisheries Society* **126**: 112–126.

O'Gorman, R., Elrod, J. H., Owens, R. W., et al. 2000. Shifts in depth distributions of alewives, rainbow smelt, and age-2 lake trout in southern Lake Ontario following establishment of dreissenids. *Transactions of the American Fisheries Society* **129**: 1096–1106.

Ogutu-Ohwaya, R. 1990. The decline of native fishes of Lake Victoria and Kyoga (East Africa) and the impact of introduced species, especially Nile perch, *Lates niloticus* and the Nile tilapia, *Oreochromis niloticus*. *Environmental Biology of Fishes* **27**: 81–96.

Ojaveer, H. and Lumberg, A. 1995. On the role of *Cercopagis (Cercopagis) pengoi* (Ostroumov) in Parnu Bay and the part of the Gulf of Riga ecosystem. *Proceedings of the Estonian Academy of Science: Ecology* **5**: 20–25.

Orlova, M. I. and Panov, V. E. 2004. Establishment of the zebra mussel, *Dreissena polymorpha* (Pallas), in the Neva Estuary (Gulf of Finland, Baltic Sea): distribution, population structure and possible impact on local unionid bivalves. *Hydrobiologia* **514**: 207–217.

Orlova, M. I., Muirhead, J. R, Antonov, P. I., et al. 2005. Range expansion of quagga mussels *Dreissena rostriformis bugensis* in the Volga River and Caspian Sea basin. *Aquatic Ecology* **38**: 561–573.

Paine, R. T., Tegner, M. J., and Johnson, E. A. 1998. Compounded perturbations yield ecological surprises. *Ecosystems* **1**: 535–545.

Pimentel, D., Zuniga, R., and Morrison, D. 2005. Update on the environmental and economic costs associated with alien-invasive species in the United States. *Ecological Economics* **52**: 273–288.

Reeves, E. 1999. *Exotic Policy: An IJC White Paper on Policies for the Prevention of Invasion of the Great Lakes by Exotic Organisms.* Available online at www.ijc.org/rel/milwaukee/wrkshps/exoticpolicy.htm

Ricciardi, A. 2001. Facilitative interactions among aquatic invaders: is an "invasional meltdown" occurring in the Great Lakes? *Canadian Journal of Fisheries and Aquatic Sciences* **58**: 2513–2525.

Ricciardi, A. 2006. Patterns of invasion of the Laurentian Great Lakes in relation to changes in vector activity. *Diversity and Distributions* **12**: 425–433.

Ricciardi, A. and MacIsaac, H.J. 2000. Recent mass invasion of the North American Great Lakes by Ponto-Caspian species. *Trends in Ecology and Evolution* **15**: 62–65.

Ricciardi, A. and Rasmussen, J.B. 1998. Predicting the identity and impact of future biological invaders: a priority for aquatic resource management. *Canadian Journal of Fisheries and Aquatic Sciences* **55**: 1759–1765.

Rivier, I.K. 1998. The predatory Cladocera (Onychopoda: Podonidae, Polyphemidae, Cercopagidae and Leptodoridaa) of the world. In *Guides to the Identification of the Micro-Invertebrates of the Continental Waters of the World*, vol. 13, ed. H. Dumont. Leiden, Netherlands: Backhuys.

Shiganova, T.A. 1998. Invasion of the Black Sea by the ctenophore *Mnemiopsis leidyi* and recent changes in pelagic community structure. *Fisheries Oceanography* **7**: 305–310.

Simberloff, D. 2006. Invasional meltdown 6 years later: important phenomenon, unfortunate metaphor, or both? *Ecology Letters* **9**: 912–919.

Skora, K.E. and Stolarski, J. 1993. New fish species in the Gulf of Gdansk, *Neogobius* sp. [cf. *Neogobius melanostomus* (Pallas 1811)]. *Notes Bulletin of the Sea Fisheries Institute* **1**: 83.

Skubinna, J.P., Coon, T.G., and Batterson, T.R. 1995. Increased abundance and depth of submersed macrophytes in response to decreased turbidity in Saginaw Bay, Lake Huron. *Journal of Great Lakes Research* **21**: 476–488.

Starobogatov, Y.I. and Andreeva, S.I. 1994. Range and its history. In *Zebra Mussel* Dreissena polymorpha *(Pall.) (Bivalvia, Dreissenidae): Systematics, Ecology, Practical Importance*, ed. Y.I. Starobogatov. Moscow, Russia: Nauka, pp. 47–55.

Steingraeber, M., Runstrom, A., and Theil, P. 1996. *Round goby* (Neogobius melanostomus) *Distribution in the Illinois Waterway System of Metropolitan Chicago*. U.S. Fish and Wildlife Service report.

Steinhart, G.B., Marschall, E.A., and Stein, R.A. 2004. Round goby predation on smallmouth bass offspring in nests during simulated catch-and-release angling. *Transactions of the American Fisheries Society* **133**: 121–131.

Stewart, T.W. and Haynes, J.M. 1994. Benthic macroinvertebrate communities of southwestern Lake Ontario following invasion of Dreissena. *Journal of Great Lakes Research* **20**: 479–493.

Strayer, D.L. 1991. Projected distribution of the zebra mussel, *Dreissena polymorpha*, in North America. *Canadian Journal of Fisheries and Aquatic Sciences* **48**: 1389–1395.

Therriault, T.W., Grigorovich, I.A., Kane, D.D., et al. 2002. Range expansion of the exotic zooplankter *Cercopagis pengoi* (Ostroumov) into western Lake Erie and Muskegon Lake. *Journal of Great Lakes Research* **28**: 698–701.

Uitto, A., Gorokhova, E., and Valipakka, P. 1999. Distribution of the non-indigenous *Cercopagis pengoi* in coastal waters of the eastern Gulf of Finland. *ICES Journal of Marine Science* **56**(Suppl. 1): 49–57.

U.S. Coast Guard. 1993. *Ballast Water Management for Vessels Entering the Great Lakes*, Code of Federal Regulations 33-CFR Part 151.1510. College Park, MD: U.S. Coast Guard.

USGS (U.S. Geological Survey). 2005. *Echinogammarus ischnus collection information*. Available online at http://nas.er.usgs.gov/queries/CollectionInfo.asp?SpeciesID=23

Van den Brink, F.W., Paffer, B.G., Oesterbroek, F.M., and Van der Velde, G. 1993. Immigration of *Echinogammarus ischnus* (Stebbing, 1906) (Crustacea,

Amphipoda) into the Netherlands via the lower Rhine. *Bulletin Zoölogisch Museum Universiteit van Amsterdam* **13**: 167–170.

Vanderploeg, H. A., Nalepa, T. F., Jude, D. J., *et al.* 2002. Dispersal and emerging ecological impacts of Ponto-Caspian species in the Laurentian Great Lakes. *Canadian Journal of Fisheries and Aquatic Sciences* **59**: 1209–1228.

van Overdijk, C. D. A., Grigorovich, I. A., Mabee, T., *et al.* 2003. Microhabitat selection by the amphipods *Echinogammarus ischnus* and *Gammarus fasciatus* in laboratory and field experiments: displacement patterns and mechanisms. *Freshwater Biology* **48**: 567–578.

Witt, J. D. S., Hebert, P. D. N., and Morton, W. B. 1997. *Echinogammarus ischnus*: another crustacean invader in the Laurentian Great Lakes basin. *Canadian Journal of Fisheries and Aquatic Sciences* **54**: 264–268.

Witte, F., Goldschmidt, T., Wanink, J., *et al.* 1992. The destruction an endemic species flock: quantitative data on the decline of the haplochromine cichlids of Lake Victoria. *Environmental Biology of Fishes* **34**: 1–28.

Zaret, T. M. and Paine, R. T. 1973. Species introductions in a tropical lake. *Science* **182**: 449–455.

Zhulidov, A. V., Nalepa, T. F., Kozhara, A. V., Zhulidov, D. A., and Gurtovaya, T. Y. 2006. Recent trends in relative abundance of two dreissenid species, *Dreissena polymorpha* and *Dreissena bugensis*, in the Lower Don River system, Russia. *Archiv für Hydrobiologie* **165**: 209–220.

Zhuravel, P. A. 1967. On the colonization of artificial water bodies by *Dreissena bugensis*. *Gidrobiologicheskii Zhurnal* **3**: 87–90. (In Russian.)

Part II Case studies of globalization and fisheries resources

A. C. J. VINCENT, A. D. MARSDEN, AND U. R. SUMAILA

7

Possible contributions of globalization in creating and addressing sea horse conservation problems

GLOBAL CONTEXT

For economic, ecological, and social reasons, it is important to explore how globalization might be changing conservation and management of fisheries. In international arenas, little attention has been given to the exploitation of species for non-food purposes ranging from medicines to bioremediation to souvenirs. Such fisheries range from very small-scale catches for personal use (e.g., bait: McPhee and Skilleter 2002) to the cumulatively large and valuable (e.g., traditional medicine: Vincent 1996), with catches commonly traded internationally. These fisheries often extract species that are little studied. Unmonitored extraction makes it difficult to deduce the economic, social, or cultural consequences of fishing.

In general, fishery resources are being exploited at a rate that clearly endangers sustained access to market and non-market values that people attribute to them. Overfishing has reduced the biomass of many of the world's major marine living resources to only a small fraction of their former levels (e.g., Jackson *et al.* 2001; Pauly *et al.* 2002; Myers and Worm 2003). As a result, many marine fish species are now considered threatened (IUCN 2003), with most such analyses dating from 1996. In 2002, the Convention on International Trade in Endangered Species of Wild Fauna and Flora (CITES) decided that exports of some marine fishes of commercial importance should be brought under management for the first time (CITES 2004). In addition to numerical declines, the compositions of fish communities (Pauly *et al.* 1998) and whole ecosystems (Jackson *et al.* 2001) have changed dramatically through overexploitation. Such overexploitation is associated with an unprecedented increase in global fishing effort, which

Globalization: Effects on Fisheries Resources, ed. William W. Taylor, Michael G. Schechter, and Lois G. Wolfson. Published by Cambridge University Press. © Cambridge University Press 2007.

continues unabated even with the depletion of major world fish stocks (Garcia and Newton 1997; Mace 1997; Pauly et al. 2002).

Among the many factors driving the decline of fisheries, the coincidence of timing with increasing globalization, especially since the 1980s, merits further consideration. Key elements of globalization are the increasing flows of money, materials, information, and people across national boundaries. Such flows have grown in both volume and diversity. At the same time, the diversity of and distance between sources and destinations have expanded (Crafts 2000). All of these flows are intertwined in complex feedback loops, with the growth of one flow leading to the establishment and growth of other flows, which then further encourage the growth of the original flow. Each flow is facilitated by technological developments in transportation and communication, and these technologies are in turn spurred to higher levels of development by the flows themselves. We have chosen to focus on these flows but acknowledge that many other aspects of globalization may also affect resource management.

The impacts of globalization on resource management are widely debated, with mixed evidence emerging. Arguments in favor of globalization rely heavily on the notion that all nations can become wealthier if each specializes in what it does best (David Ricardo's theory of comparative advantage: see Sraffa and Dobb 1951–73). Increasing wealth should then allow people to spend more on environmental and social programs (OECD 1997; Dollar and Kraay 2002) and to adopt more environmentally friendly production techniques (Antweiler et al. 2001). In addition, some argue that free trade agreements and organizations offer the opportunity to force removal of subsidies that promote overexploitation (Stone 1997; Yu et al. 2002), and should allow environmentally friendly technologies to diffuse more rapidly (Antweiler et al. 2001). In theory, the World Trade Organization (WTO) is working to discipline fisheries subsidies, persuading member countries to sign an agreement on subsidies and countervailing measures (WTO 1994).

The most common concerns about globalization tend to focus on the impacts of free trade. First, critics argue that free trade, by allowing greater production and consumption, will increase depletion of natural resources, an assertion for which there is substantial empirical evidence (Neumayer 2000). This depletion and degradation will be exacerbated if management is ineffective, if damage is transboundary in nature or if property rights are poorly defined (Brander and Taylor 1997, 1998; Neumayer 2000). A second concern is that countries may lower environmental standards to attract investment, leading to a "race to the bottom" where all countries drop their standards in competition with one another (Daly 1993; Neumayer 2000). This hypothesis, however, lacks empirical support (Nordstrom and Vaughan 1999; Neumayer 2000; Antweiler et al. 2001). A third concern is that free trade rules, such as those implemented under the North American Free Trade Agreement (NAFTA), the General Agreement on Tariffs and Trade (GATT), and the WTO, will undermine environmental legislation (Neumayer 2000). Such rules do make allowances for national environmental legislation and regulations and for international environmental agreements (Neumayer 2000; Yu et al. 2002), but the effectiveness of these provisions is greatly diminished by the heavy burden of proof, by the lengthy process required, and by the often weak outcomes (Kibel 2001; Nogales 2002; Valley 2004). Finally, it is argued that producers and traders very seldom absorb the full environmental costs (externalities) of their actions; if these costs were considered more comprehensively, much international trade would probably be uneconomical (Daly 1993; Goldsmith 1996).

In light of arguments about the environmental impacts of globalization, it is worth addressing the means by which globalization might affect the health and sustainability of fisheries. Given the enormous international trade in fisheries products – about 40 percent of total landings (FAO 2002) – might the expansion of trade lead to changing patterns of exploitation? Might expansion of demand to a global scale encourage fishers to take better care of their resources or increase exploitation and devastate fish stocks? Might globalization affect industrial and subsistence fishers similarly or differently? What role might flows of information and technology play in changing fisheries practices? Might human migration and cultural integration lead to the spread of more or less sustainable exploitation methods and consumption patterns?

Most analytical effort has focused on species in industrialized fisheries, but globalization might also affect the vast number of species landed as bycatch or caught in small-scale or subsistence fisheries. Sea horses are a group of such species, with an international trade of many millions of animals. We will use this trade in sea horses as a case study to explore how globalization might both hinder and help the conservation of marine populations, and then try to extract lessons that can be applied on a wider basis.

Given the dearth of formal records and historic trade analyses, we are forced to speculate on how globalization might have influenced both demand for and supply of sea horses. We are well aware that our inferences would require considerable assessment and evaluation before a causal link could be asserted. Moreover, we realize that habitat loss and degradation (among other pressures) may well be adding to the effects of fishing to promote population declines, and that these pressures, too, may be affected by globalization. Our hope, however, is that by examining the topical question of globalization from a less familiar angle – that of exploitation of sea horses – we may provoke new consideration of its potential impacts.

SEA HORSE BIOLOGY, TRADE, AND CONSERVATION

Sea horses (*Hippocampus* spp.) are globally distributed, occurring in the marine waters of at least 100 countries (Lourie et al. 1999). The more than 34 species of these fishes (Lourie et al. 1999; Kuiter 2000, 2001; Lourie and Randall 2003) have a life history that apparently makes them ill-suited to heavy exploitation. Strong mate and site fidelity, combined with essential parental care and limited mobility, are expected to make sea horse populations vulnerable to disruption and then slow to recover (Foster and Vincent 2004).

Our understanding of the trade in sea horses comes from a combination of extensive trade surveys, qualitative evidence, and very sparse official data (available only from Taiwan and mainland China until the late 1990s). The first surveys of sea horse trade occurred in 1993 and 1995 (Vincent 1996); a new, extensive global review is now in preparation. Collection of trade data involves visits to fisheries landing sites, surveys of stocks at merchants' premises, and many interviews with a wide array of stakeholders and other interested parties. Rigor is maintained in data collection through semi-structured interviews by using three forms of cross-validation for consistency: comparing answers to the same key questions, recast at least three times during an interview; comparing answers from different respondents at the same trade level in any given region; and comparing answers across trade levels in any given region, allowing for magnification of trade volumes (Vincent 1996). Although imprecise, deductions from previous surveys of sea horse trade have usually been supported by subsequent data, including those from new formal trade records.

Sea horses are dried for use in traditional medicines (TM) and as curiosities, and/or sold live for ornamental display. A great majority of sea horses are consumed in traditional Chinese medicine (TCM) for treatments of many ailments (Vincent 1996). Trades for the three markets function independently of one another, usually diverging as

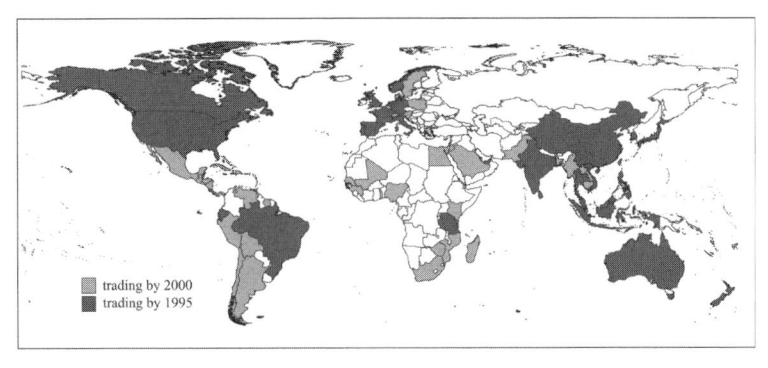

Figure 7.1 Countries known to trade sea horses and/or pipefish (many data do not distinguish) by 1995 (dark gray) and by 2000 (light gray). (Source: Vincent [1996] for 1995 data; A. Vincent and A. Perry, unpublished data, for 2000 data.)

soon as the fisher sells the sea horses to the primary buyer (although live sea horses that die may enter the other trades) (Vincent 1996). The majority of the sea horses sold dried are landed as incidental bycatch, primarily by shrimp and prawn trawlers (Baum and Vincent 2005; Giles *et al.* 2005; A. Vincent and A. Perry, unpublished data). Sea horses for both dried and live consumption are also target-caught, primarily by subsistence fishers (Vincent 1996).

At least 32 countries had traded sea horses and/or their immediate pipefish relatives (also family Syngnathidae) by 1995 (Vincent 1996), but this total had risen to nearly 80 countries by 2000 (A. Vincent and A. Perry, unpublished data) (Fig. 7.1). Much of the expansion occurred in Africa (McPherson and Vincent 2004) or Latin America (Baum and Vincent 2005) and represented genuinely new trades. More countries were involved in dried than live trades. The largest exporters of dried syngnathids were Thailand, India, Mexico, the Philippines and Vietnam (Vincent 1996; Baum and Vincent 2005; Giles *et al.* 2005; A. Vincent and A. Perry, unpublished data). The largest importers were China, Hong Kong, Taiwan, and Singapore (Vincent 1996; Baum and Vincent 2005; Giles *et al.* 2005; A. Vincent and A. Perry, unpublished data). For the live trade, Indonesia, the Philippines, and Brazil were the major exporters, primarily to the United States and the European Union (Vincent 1996; Baum and Vincent 2005; A. Vincent and A. Perry, unpublished data). Trade routes are complex and erratic. For example, sea horses seized by Customs in Belgium had arrived from Mali and were destined for China (E. Fleming in litt. to A. Vincent 1997) as Mali is landlocked, so they probably originated from coastal west African countries such as Guinea

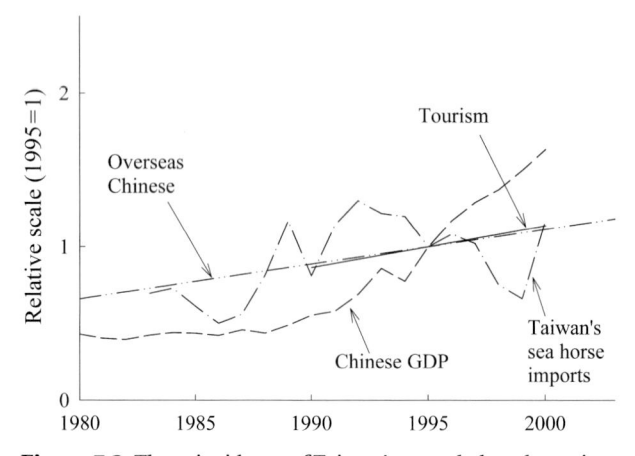

Figure 7.2 The coincidence of Taiwan's recorded sea horse imports with globalization-related variables that may have affected sea horses. All data are on a per capita basis and are presented in standardized units so that the value in 1995 equals 1 for each indicator. The data sources and unstandardized values in 1995 are: (1) international tourist arrivals in all countries – World Tourism Organization (2003), 550 million in 1995; (2) overseas Chinese – Cox (2003) and Poston *et al.* (1994), 34.5 million people in 1995 (interpolated from linear regression through 1980, 1990, 2003); (3) Chinese GDP – IMF (2000b), $700 billion in 1995; (4) Taiwan's sea horse imports – Republic of China (Taiwan) Statistics, 9506 kg in 1995.

or Senegal, both of which were reported as countries of origin by the Census and Statistics Department, Hong Kong.

Volumes of sea horses in trade increased greatly from the mid 1980s into the 1990s (Vincent 1996). The international trade in dried sea horses through China, Hong Kong, Singapore, and Taiwan alone exceeded 45 tonnes (about 16 million animals) in 1995 (Vincent 1996). Trade through these areas was at least as great in 2000 but from a wider range of countries of origin (A. Vincent and A. Perry, unpublished data). Indeed, trade volumes in 2000 probably exceeded 50 tonnes even without (1) comprehensive data for China, a known major consumer; (2) documentation of the apparently large and growing prepackaged patent medicines market; (3) consideration of trade among other regions; and/or (4) accounting for domestic consumption (A. Vincent and A. Perry, unpublished data). Unfortunately, given China's apparent role as a dominant consumer of sea horses (Vincent 1996), Chinese official data on imports were unreliable (A. Vincent and A. Perry, unpublished data). By far the longest time series of data on sea horse trade came from Taiwan (Fig. 7.2). This reveals significant increases from 1983 until the mid

1990s, when volumes leveled off before dipping around the time of the Asian economic crisis in the late 1990s.

Although smaller than the dried trade, the aquarium trade was the primary purchaser of sea horses in some regions (Vincent 1996). Information from source countries indicated that many hundreds of thousands of sea horses were exported for aquarium display in 1995, with trade unabated in 2000 (Vincent 1996; A. Vincent and A. Perry, unpublished data). However, import records and industry data suggest much lower levels. Focused research is needed to reconcile these figures. Dried sea horses command high prices up to $1200 per kilogram at retail outlets in Hong Kong, while individual live sea horses can fetch up to $50 at retail in the United States (Vincent 1996).

Fishers and other informants reported considerable numeric declines in their sea horse catches and trade, even while fishing effort continued to increase (Vincent 1996; Baum and Vincent 2005; Giles *et al.* 2005; A. Vincent and A. Perry, unpublished data). Trade analyses indicated that the main pressures on sea horse populations were overexploitation by incidental and target fisheries, and degradation of sea horse habitats. Many sea horses are listed as threatened under international or national criteria. For example, the *2003 IUCN Red List of Threatened Species* listed nine species of sea horses as vulnerable, one species as endangered, and all others as data deficient (IUCN 2003). In these evaluations, trade was considered the primary pressure for most exploited species, although habitat damage was also important for some. At least ten jurisdictions and the European Union have local legislation that specifically monitors and/or controls sea horse exploitation, exports, or imports, while others have general measures that include sea horses (Lourie *et al.* 1999). For example, Australian sea horse populations were moved under the Australian Wildlife Protection Act in 1998, then placed under the Environmental Protection and Biodiversity Conservation Act in 2001; export now requires permits granted only for sustainable extraction or cultured specimens (Martin-Smith and Vincent 2006).

Despite growing conservation concern, increasing globalization of sea horse trade proceeded largely without regulation until 2004. Until 2000, when discussions began about regulating international trade in sea horses, many governments were unaware that sea horses were exploited or traded in their countries. Moreover, the official trade data that were gathered, by very few countries, exhibited serious gaps and discrepancies (Vincent 1996). Only now, with advent of export controls for sea horses, will the international trade in sea horses be scrutinized on a global scale.

HOW GLOBALIZATION MAY INCREASE PRESSURES
ON SEA HORSES

In this section, we speculate on aspects of globalization that might have contributed to depletion of sea horse populations. We link these issues to the types of flows (money, materials, information, and people) that change with globalization (see Table 7.1). Our focus is on the period since 1980, when both globalization and sea horse trade increased greatly.

Role of globalization in promoting increased demand for sea horses

A first hypothesis is that globalization may have helped prompt greater demand for dried sea horses in TM, largely mediated through China's economic growth. During the mid 1980s, China's economic approach changed from essentially "command and control" to partly market-based (Bell et al. 1993; Tseng 1994). Such a transition led to a greater flow of both money and information in and out of the country (IMF 2000a). China's economy has grown at an average annual rate of greater than 9 percent since 1982 (Fig. 7.2); growth rates for the United States over the same period have been about 3 percent. This high rate of growth in the Chinese economy since the mid 1980s led to greater disposable income (IMF 2000b), which, informants told us during trade interviews, promoted an increase in the quantity of sea horses sought for medicinal purposes (Vincent 1996; A. Vincent and A. Perry, unpublished data). However, official import data for Taiwan also showed reported increases from the mid 1980s to the mid 1990s, suggesting either that Taiwan was considerably affected by mainland China's economic changes or that other changes were occurring (Vincent 1996; A. Vincent and A. Perry, unpublished data). On a more subtle level, TCM importers and retailers noted that, with greater economic activity, higher incomes, and a faster-paced life for many mainland Chinese, TM consumers began to find that time was more of a constraint than money (Vincent 1996), as predicted by standard labor economic theory (Borjas 1999). Such a shift in attitude might explain why many consumers turned from the traditional approach of selecting medicinal ingredients individually to a greater reliance on prepackaged forms of TM, in which ground sea horses are included before the product reaches the consumer (Vincent 1996). Prepackaged medicines allow for use of sea horses that consumers might previously have

Table 7.1 *Potential linkages between issues affecting sea horse conservation and the main four types of flows associated with globalization. Conservation issues are listed in the first column; types of flows are column headings. Each cell shows the potential link between the issue and the flow. Some linkages might apply to multiple flows, as indicated with arrows. We hypothesize that global flows might affect the conservation issue that we cite*

Conservation Issue	Flows			
	Information	Material	Money	People
Conservation challenges related to increased demand for sea horses				
Greater affordability of TCM in China	←——————— Economic and income growth from all four flows ———————→			
Increased awareness of TCM worldwide	Communication of TCM knowledge	←——————— Trade in TCM products ———————→		Chinese migration
Increased awareness of exotic fishes	Dive skills, media			Tourism
Improved technology and skills for keeping marine aquaria	Husbandry skills	Aquarium-supplies trade		
Increased disposable income of aquarium hobbyists	←——————— Economic and income growth from all four flows ———————→			
Conservation challenges related to increased supply of sea horses				
Increased trawling and trawl bycatch	Trawl technology	Trawl equipment, trawl-targeted products	Trade in trawl-targeted products, e.g., shrimp	
Depletion of other marine resources	←——————— All of the above issues and flows might lead to increased pressure on and depletion of other marine resources, leaving sea horse fishing as an alternative livelihood ———————→			
Greater interactions among people for business, labor, and tourism	Easier arrangement of trade; greater awareness of market opportunities	More transport for other purposes on which sea horses can piggyback		Increased travel

Table 7.1 (cont.)

Conservation Issue	Flows			
	Information	Material	Money	People
More efficient sea horse targeted fishing	Fishing gear technology	Fishing equipment, e.g., hookah rigs; other products, e.g., sea cucumbers	Trade in other products, e.g., sea cucumbers	
Improved transport of live animals	Fish-keeping technology	Fish-keeping equipment		Ethnic linkages that facilitate trade in animals
Piggybacking of sea horses on other trades	Communication regarding new trade opportunities	Trade in other products, e.g., sea cucumbers	Trade in other products, e.g., sea cucumbers	Movement of traders in search of new opportunities
Proposed contributions of globalization to sea horse conservation				
Improved collaboration over long distance	Improved communication; media; internet databases	Sharing and transfer of scientific equipment		Easier travel
Greater public awareness of conservation issues abroad	Media	Eco-labeling; alternative livelihoods	Donor support for conservation work	Volunteers on conservation projects; eco-tourism
Greater efforts to coordinate management efforts	IUCN ←――――――――――――― CITES ――――――――――――→			
Increasing incomes of fishers	←――――――――― Economic and income growth from all four flows ―――――――――→			

rejected as too small, too spiny, or too dark, thus adding to the pressure on wild populations.

A second possibility is that the diaspora of Chinese people around the world may well have contributed to considerable interest and training in TCM (including the use of sea horses) on a global scale. Chinese emigration gathered speed in the late nineteenth century as manual laborers moved overseas (Poston et al. 1994), and numbers of overseas Chinese continue to grow (Fig. 7.2). The twentieth century was punctuated by Chinese migrations in response to political upheaval and as a result of China's decision to relax emigration regulations in 1979 to qualify for "most-favored-nation" status with the United States (Lintner 1999). At least 44.5 million overseas Chinese currently live in about 136 countries around the world (Poston et al. 1994). Migrants took the practice of TCM with them, and TCM has grown in popularity in these countries (Cameron et al. 2001). In the United Kingdom, for example, "at least 300 registered TCM practitioners were not of Asian descent and about 50 percent of patients receiving treatment from TM practitioners were not of Asian origin" (Cameron et al. 2001). As well, nearly 70 percent of Canadians and 42 percent of Americans had tried complementary alternative medicine by the 1990s, much of it derived from TM and TCM (WHO 2002). In 1993, TCM products, with an export value of $400 million, were exported to at least 120 countries. The growth in prepared forms of TCM (which are much easier to ship) may have allowed TCM to access new pools of consumers in areas such as North America with relatively little TCM knowledge and/or a greater familiarity and comfort with medicines in packaged form. Prepackaged TCM containing sea horses (in 11 formulations) and/or dried sea horses as raw medicinal materials were available in 90 of 101 U.S. shops surveyed in 1999 (Sheetz and Seligmann 2000).

A third proposal is that greater flows of people and products associated with globalization may have prompted the growing demand for live sea horses and other ornamental marine fishes, which are now traded all over the world (Wood 2001; Sadovy and Vincent 2002). Aquarium hobbyists appear to have gained an appreciation of exotic fish species, sea horses among them, partly as a result of travel to exotic locations, media coverage of these animals, and the growing number of public aquariums; more than 100 significant free-standing aquariums have opened globally since the early 1990s (H. Koldewey, in litt. to A. Vincent 2004), each with exhibits of exotic fish. As just one measure of greater exposure to distant environments, international tourist arrivals increased from 457 million in 1990 to 699 million in 2000, an annual average increase of 4.3 percent (World Tourism Organization 2003) (Fig. 7.2). Concurrent with the rise in interest in exotic fishes, technology and information dissemination developed sufficiently to allow consumers to establish functional marine aquariums at home (Baquero 1999; Larkin and Degner 2001). As this technology emerged, greater disposable income (sometimes argued to be one result of globalization) could have helped make it possible for more hobbyists to afford the expense of keeping marine fishes at home. Marine fish owners in the United States report a higher household income than freshwater owners (American Pet Products Manufacturer's Association 2002). Certainly, whatever its sources, the greater demand for live sea horses encouraged an increase in the quantity supplied to the market, either from increased target fishing or from increased retention of sea horses that survived in live-bait trawls (see below). It should also be noted that similar appreciation of the exotic might explain why so many people acquired sea horses as souvenirs and curiosities, particularly during travel.

The effect of increasing demand on wild populations will vary by source region. Where bycatch was the primary supplier of sea horses, usually for the dried trade (Vincent 1996; A. Vincent and A. Perry, unpublished data), greater demand might lead to increased retention of non-target species. Greater retention is most likely to affect populations where sea horses are brought up relatively undamaged and might otherwise survive if released, as in the live-bait trawls in Florida (Baum et al. 2003). If the gear brings up sea horses dead, then retention for trade might not significantly increase the mortality, injury, or disruption the gear already inflicts on wild populations (see below). Where sea horses are primarily target-caught, usually for the live trade, greater demand may affect the number and characteristics of sea horses landed. Higher prices for sea horses or a loss of other income earning opportunities as catches of other species decline (decreased opportunity costs) might both drive greater exploitation of sea horses.

Role of globalization in promoting increased supply of sea horses

Our first proposition related to supply is that globalization has played an important role in increasing the availability of sea horses from trawl bycatch. Total world shrimp and prawn landings, most commonly obtained by trawling, doubled from 1.6 million tonnes in 1980 to 3 million tonnes in 2001 (FAO 2001a). Much of this expansion was

Sea horse conservation problems 197

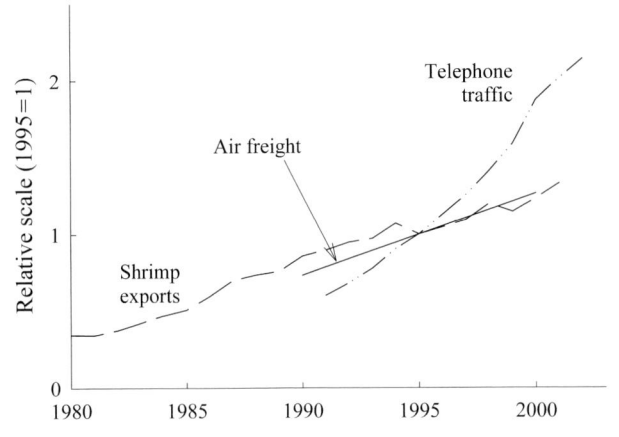

Figure 7.3 Globalization-related variables that may have affected the supply of sea horses. All data are on a per capita basis and are presented in standardized units so that the value in 1995 equals 1 for each indicator. The data sources and unstandardized values in 1995 are: (1) air freight traffic – UNESCAP (2003), 85 billion tonne–kilometers in 1995; (2) shrimp exports – FAO (2001b), 1.2 million tonnes in 1995; (3) international telephone traffic – International Telecommunications Union (2003), 63 billion minutes in 1995.

spurred by the lucrative international trade in shrimp and prawns: exports of shrimp and prawns quadrupled from 400 000 tonnes in 1980 to 1.6 million tonnes in 2001 (FAO 2001b) (Fig. 7.3). Other global flows, including the transfer of trawling technology to developing countries and international investment in trawler fleets, also contributed to the expansion of trawling. Each trawler caught relatively few sea horses per day, but the cumulative totals were large. For example, trawlers in Kien Giang province, Vietnam, caught an average of only about one to two sea horses per day, but the annual total landings in the province would have exceeded 6.5 tonnes, or perhaps 2.28 million sea horses (Vincent 1996; Giles et al. 2005). In Florida, too, the 31 live-bait trawlers operating out of the small port of Hernando Beach each caught about 9.6 sea horses on each of 240 nights of fishing, for a total of approximately 72 000 sea horses per annum (Baum et al. 2003). Trawl-caught sea horses usually enter the dried market, for TCM or curios (Vincent 1996), but trawls for live shrimp (to be used as fishing bait in Florida, for example) bring up their nets often enough to provide sea horses for the live trade (Baum et al. 2003). The global increase in trawling not only catches sea horses, but also exacerbates significant damage to their habitats (Jennings and Kaiser 1998; Kaiser et al. 2002).

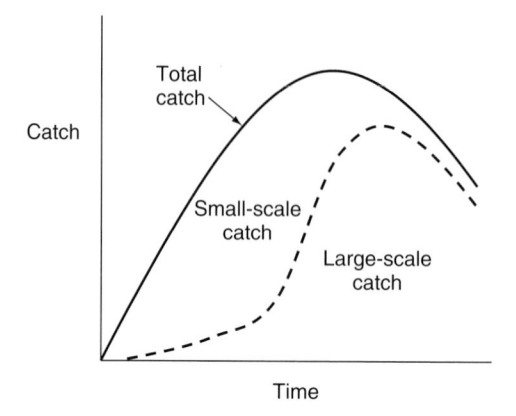

Figure 7.4 Change in distribution of catch between small- and large-scale fisheries over time. Early in the development of a fishery, small-scale fishers take most of the catch. Over time the fishery becomes more industrialized, and large-scale fishers take a larger proportion of the catch. At some point, the fishery is overexploited and the stock begins to decline, reducing the catch of both groups. (Modified from Pauly 1997.)

Second, we would suggest that the increasing global tendency to utilize bycatch from non-selective fisheries (Clucas 1997) might have enhanced expansion of international trade in sea horses. Explanations for greater utilization of bycatch include the general decline in fisheries resources (forcing exploitation of any marketable or consumable resource), increasing demand for feed in the growing aquaculture industry, and some government pressure to reduce discards (Clucas 1997). Increasing demand for sea horses, as one more marketable commodity, might have helped promote additional sorting of bycatch to extract these fishes and foster continued global expansion of trawling. Where the value of retained bycatch diminishes the incentive to reduce such bycatch, it might eventually even influence the locations where trawling occurs, promoting trawl expansion even into areas where target fishing would, on its own, be unprofitable.

Third, we suggest that internationalization of fisheries might have prompted small-scale fishers to catch more sea horses. As industrial-scale fishing (mainly trawling but also purse-seining and long-lining) is introduced, it enters into competition with small-scale or artisanal fisheries for resources (Fig. 7.4), especially in tropical shallow waters (10 to 100 meters); indeed, shrimp trawlers often discard the very fish that are important for local food (Clucas 1997; Pauly 1997). The global expansion of such fishing, coupled with declines in food fish stocks, may contribute to pressures on sea horses. Small-scale fishers turn to previously unexploited species (targeting them or retaining them as bycatch) as other stocks disappear through overexploitation (Pauly et al. 1998) or through sales of fishing rights to foreign fleets (Iheduru 1996; Kaczynski and Fluharty 2002). Many subsistence fishers end up exploiting any marketable marine species that are available, including sea horses. These fishers would have low opportunity costs because other species are in decline, sea horses can be taken in conjunction with other fishing activities at almost no cost, and other income-earning opportunities are rare. They will, therefore, probably continue to catch and supply sea horses to the market even if they are poorly paid for their efforts.

Fourth, we consider it probable that technological improvements might have played an integral role in promoting sea horse trade. For example, international telephone (and fax) traffic grew globally at an annual rate of 13 percent between 1991 and 2001, to 13 billion minutes (Fig. 7.3), and the number of Internet users grew at an astonishing 61 percent annually over the same time period, to 502 million (International Telecommunications Union 2003). Such communication technology allowed rapid, easy negotiations between buyers and sellers in both the live and the dead sea horse trades. Live traders, for example, used telephone, fax, and the Internet to coordinate fish shipments. Dried seafood traders in Hong Kong also depended heavily on telephone and fax communications, although they used computers and the Internet only infrequently (Clarke 2002). In addition, Web-based e-marketplaces advertised sea horses for sale to both merchants and the general public (e.g., Ecrobot 2004).

Fifth, it seems likely that globalization of technology, in the form of new fishing gear, helped fishers to access new resources to meet growing demand. For example, in 2001, fishers in one Philippines village began adopting hookah rigs (surface air supply) to access a deep-water sea cucumber with commercial value in East Asia. This technology allowed fishers to stay underwater longer (increasing effort) and to obtain other species that were previously inaccessible, including a species of sea horse not previously sold in the region (S. Morgan in litt. to A. Vincent 2004). The result was "gold rush" exploitation, with gross depletion of the sea cucumbers and marked overexploitation in the associated new sea horse fishery (H. Panes in litt. to A. Vincent 2004).

Sixth, we are reasonably confident that increased global transportation would have increased sea horse supply. Such links appear to be key determinants of the sea horse trade, particularly for live sea horses (Vincent 1996). These fish must be supported with air supply, and sometimes food, until they arrive at their destinations if they are to survive until sale. Given the distance between the (primarily) developing country sources and the (primarily) developed country destinations (Vincent 1996), international air transport is essential. Indeed, one of the foremost contributors to the increase in trade of marine ornamental species, which is primarily dependent on wild collection, is the improved availability of air transport from remote locations (Larkin and Degner 2001). Air freight traffic on commercial air carriers doubled from 62 billion to 121 billion tonne-kilometers from 1990 to 2000, an average annual growth rate of 7.0 percent (UNESCAP 2003) (Fig. 7.3). Air links are particularly important for the aquarium trade. For example, Indonesia exported marine fish from West Papua (Irian Jaya) until flights to the United States stopped touching down at its Biak airport in the early 1990s; thereafter, it exported most live sea horses through Jakarta and Bali, its two large-volume international airports (Vincent 1996; A. Vincent and A. Perry, unpublished data).

Although they were more easily exported than live fish, dried sea horses were apparently also increasingly traded as global transport links were enhanced. As one example, sea horses from Tamil Nadu in southern India were commonly exported to Singapore, often for transshipment to Hong Kong or Taiwan (Vincent 1996). The large Tamil population in Singapore fostered trade links and also facilitated transport, with apparently reliable stories of travelers between Tamil Nadu and Singapore being offered free flights if they carried cases of sea horses as their checked baggage (Vincent 1996). Thus the flow of people increased the flow of goods.

Seventh, we conjecture that global sea horse supply (from both bycatch and target sources) might have increased as a result of what we call "piggybacking," where flow of one product opens channels for trade in other products. Sea horses may be particularly vulnerable to piggyback trade, in which they accompany other, more valuable commodities. These fishes were destined for the same markets that received exports of the larger and more lucrative sea cucumber and shark fin trades. For example, Hong Kong, the world's largest entrepôt for sea horses (Vincent 1996; A. Vincent and A. Perry, unpublished data), also received about 80 percent of the world's sea cucumbers and 50 to 85 percent of the world's shark fin imports (Clarke 2002). Between 1980 and 2000, the global production of sea cucumber and shark fin increased 18-fold and three-fold, respectively. Shark fin imports to Hong Kong increased by 5.3 percent annually during the late 1980s and 1990s, after allowing for re-exports to China, although sea cucumber imports appear to have declined in the same time period (Clarke 2002). Trading such vast quantities of similar goods would make it easy to foster the trade in sea horses. Our proposal is more than speculation: it appears, for example, that the arrival of ethnic Chinese traders seeking a wide variety of goods was the catalyst that led to sea horse exports from Latin American countries from Mexico to Peru (J. Baum, unpublished data). In the central Philippines, too, a new sea cucumber fishery led quickly to a new sea horse fishery (see above). By adding to the profitability of new trade routes, the piggybacked sea horses might even indirectly foster continued exploitation of their populations through a positive feedback effect.

It seems plausible that the increase in sea horse supply from far-flung sources (largely through retention of shrimp trawl bycatch, married to piggyback trade) has helped markets cope with demand for sea horses that far exceeds what local populations in the same region can supply. As domestic sea horse populations became depleted (apparently largely through overexploitation), some traders reported turning to supplies of sea horses from elsewhere (Vincent 1996; A. Vincent and A. Perry, unpublished data). Indeed, some traders reported certainty that they would achieve greater profit by importing sea horses, perhaps because their supply now became more reliably available but perhaps also because the imported species was more valuable (A. Vincent and A. Perry, unpublished data). As one example of possible long-term depletion, China has used sea horses medicinally for perhaps 12 centuries but now lists *Hippocampus kelloggi* under wildlife protection laws, allowing exploitation and trade only with permits (Law of Wild Animal Protection of the People's Republic of China, 1988). As its domestic sources became depleted, China apparently turned early to Southeast Asian nations for supplies. Then, as sea horse catches (and presumably populations) declined in these nearby regions (Vincent 1996), East Asia began to import sea horses from yet farther afield: from India, then Africa and South America (McPherson and Vincent 2004; A. Vincent and A. Perry, unpublished data). As with other fisheries (e.g., reef fishes traded live for food: Sadovy and Vincent 2002), such global geographic expansion of sea horse fisheries and trades could well be masking population declines in particular regions by securing continued supply until stocks from many sources have all been depleted.

HOW GLOBALIZATION MAY HELP REDUCE PRESSURES ON SEA HORSES

Some aspects of globalization might be linked to improvements in the status of sea horse populations. Our hypothesis is that, as well as fostering demand, greater international flows of information, people, money, and materials have improved opportunities for conservation and management of many species, including sea horses (Table 7.1). Certainly, ever since alarm bells about sea horse exploitation were first sounded (Vincent 1995, 1996), these fishes have benefited from considerable international dialog and collaboration among direct stakeholders, government departments, conservation agencies, and the general public.

First, we suggest that advanced international communication is playing a significant role in alerting stakeholders and larger communities to problems posed by overexploitation and then engaging them in finding conservation responses. In the case of sea horses, extensive international use of media, Web sites, shared databases, exhibits (especially in public aquariums), and academic outlets have enhanced flows of information to and from a wide array of interested parties. As one example of such internationalization, sea horse conservation efforts by one team (Project Seahorse 2004) attracted coverage in newspapers, magazines, and broadcast media in at least 13 languages and 30 countries during the decade since the mid 1990s. As another example, sea horses are displayed in almost all public aquariums around the world, with at least ten institutions in North America and Europe mounting significant focal exhibits on sea horse biology and conservation reaching at least 10 million people (H. Koldewey in litt. to A. Vincent 2004).

Second, we hypothesize that the same international flow of travelers that might have led people to buy sea horses as aquarium fish or as curiosities may also have promoted a sense of identity with exotic fish and locations that could encourage public engagement in conservation. In some cases, this engagement might be by the same people who create conservation concerns through their demand for exotic products; the challenge is to prompt recognition of the connection between demand and conservation consequences. When conservation concern does mount, international travel apparently allows professional expertise, volunteer support, and the media to reach problem sites. Income from tourism has been argued to provide an incentive for conservation among those who benefit from the visitors, albeit with the recognition that tourism can also put new pressures on marine ecosystems (Dixon *et al.* 1993).

Third, we know that global connectedness, through travel and communication, has already offered opportunities to formulate management plans and institutionalize responses to sea horse population declines in a coordinated manner, especially when encouraged by public opinion. Some examples include:

(1) An aquarium professionals' network for conservation (comprising nearly 200 people in 75 institutions in 17 countries) has collaborated to produce husbandry manuals, guidelines on acquiring wild sea horses responsibly, educational materials, public presentations, and primary research (Hall and Warmolts 2003).

(2) A syngnathid researchers' network (comprising 130 people in 95 institutions in 32 countries) has shared knowledge and specimens (e.g., Foster and Vincent 2004), engaged in supporting conservation legislation, promoted further conservation efforts, and assisted with validating conservation status assessments.

(3) Consultation among the researchers' network and other far-flung volunteer specialists has allowed evaluation of species' conservation status for the 2003 *IUCN Red List of Threatened Species* (IUCN 2003), which serves as a warning device. All sea horses and some of their pipefish relatives have now been assessed with the best available composite global information, thus prompting more research and action.

(4) Consultation among disparate stakeholder groups (fisheries scientists, Philippines fishers, Hong Kong traditional medicine traders, North American aquarium professionals, a national policy group in the Philippines, and an international policy group) led to consensus on management measures for wild sea horse populations in the Philippines (Martin-Smith *et al.* 2004); all groups favored no-take marine protected areas and minimum size limits as acceptable means of adaptive management.

Fourth, we argue that good communication helps promote global flows of money and materials. Donor support for sea horse conservation efforts comes from many sectors, including government, private enterprise, non-governmental organizations, private foundations, and individuals (Project Seahorse 2004). This funding generally originates in the West and flows to developing countries, where it may be used to build community capacity and to set up institutional frameworks for conservation. For example, small donations from community groups and individuals in the West are used to support high school marine conservation apprentices, children of sea horse fishers, in impoverished Philippines fishing villages (Project Seahorse 2004). Larger funds have been used to foster regional fishers' organizations and national technical working groups (Project Seahorse 2004).

Finally, there is much speculation that conventional market transactions in an increasingly international arena might help to alleviate extractive pressure on resources if they allow for development of alternative or supplementary livelihoods (John 1994; Allison and Ellis 2001). Many income-generating schemes are arguably possible only because of global markets, transport and communication. For example, when global production of seaweed increased 21-fold from 1950 to 2000, reaching 11.9 million tonnes (FAO 2001a, b), many sea horse fishers in the Philippines began macroalgae culture and reduced the time they spent fishing. Other villagers became engaged in export of handicrafts made with local skills and local materials in ways that were environmentally and socially sustainable (Project Seahorse 2004). The idea was that the supplemental income they received in exchange would help to reduce fishing pressure. The reality was more complicated because of constraints that included villagers' business capacity, as a result externally generated livelihood ventures need to be rethought.

A GLOBAL RESPONSE TO PRESSURES ON SEA HORSES

Our explorations have led us to propose that many elements of increasing globalization might have influenced the trade and apparent overexploitation of sea horses. Greater globalization, however, may well also have provided a positive impetus, framework, and support for conservation. For sea horses, trade volumes appear largely to be maintained by geographic expansion of sources, allied to greater retention and sale of incidental landings from non-selective fishing gear (A. Vincent and A. Perry, unpublished data). There are, of course, limits to such expansion, and measures will need to be found to adjust trade to sustainable levels without simply deflecting the pressure onto other resources. Moreover, we cannot afford to deplete wild fish populations to the point where they reach supposed economic extinction (Clark 1990), given a demonstrable lack of recovery in many such species (Hutchings 2000). Indeed, the common argument that depletion of resources should lead to rising prices and, therefore, a reduction in consumption has not been supported in at least some other luxury marine commodities (Sadovy and Vincent 2002). We now turn to broader international approaches and actions that might contribute to conservation of sea horses and many other marine fishes.

Ecological certification on a global scale may be one way to proceed with improving the prospects for populations of sea horses, as for other exploited marine life. Such schemes should allow consumers to express their regard for non-market values, such as the value (to them) of knowing that their decisions in the marketplace contribute to the maintenance of healthy ecosystems. The Marine Stewardship Council (2004) and the Marine Aquarium Council (2004), two international certification bodies, advance accreditation that should theoretically lead to improved trade management. For example, Marine Aquarium Council decisions are based on standards in ecosystem and fishery management; collection, fishing, and holding; and handling, husbandry, and transport. The few certifications currently in place are, however, too new to allow their costs and benefits to be evaluated thoroughly.

The Convention on International Trade in Endangered Species of Wild Fauna and Flora (CITES) is engaged in what is essentially a long-enduring form of certification, now involving sea horses. This group of more than 165 signatory nations bans or regulates trade in species that are or may become threatened by international export by placing them on lists called Appendix I and II, respectively. Proposals to bring marine fish of commercial importance onto either list have historically been highly disputed and eventually defeated. In November 2002, however, the 12th meeting of the Conference of the Parties to CITES voted to add all sea horses and two species of sharks (basking shark *Cetorhinus maximus* and whale shark *Rhincodon typus*) to Appendix II (CITES 2004). For the first time, then, fully marine species of economic importance became subject to international trade controls (since May 2004 for sea horses and February 2003 in the case of sharks).

The challenge now is to ensure that the CITES Appendix II listings are implemented in such a way as to secure sustainable trade. Under Appendix II, each nation is now charged with making "non-detriment findings," which are declarations that its exports of these fishes do not damage their wild populations. Regulating sea horse trade will be complicated because more than 20 million fish of perhaps 24 species are traded annually among nearly 80 countries. In response to a directive from parties (Decision 12.55, CITES 2004), the CITES technical committee on animals recommended a possible universal minimum size limit (10 cm) for all sea horse exports (Foster and Vincent 2005), which parties may choose to adopt as a move toward making international trade sustainable. Such an approach represents a new form of global collaborative action for fisheries management.

Most certification schemes, including CITES listings, are likely to find it challenging to address management problems posed by trawls and other non-selective fishing gear (which are significant sources of sea horses). Despite considerable evidence about its deleterious impacts on marine populations and habitats (e.g., Alverson et al. 1994; Watling and Norse 1998), shrimp trawling continues to be a dominant form of extraction. Given the limitations of excluder devices, reduction of bycatch for many species will require spatial or temporal restrictions and/or closures (e.g., Broadhurst 2000; Hall et al. 2000). At present, plans and calls to action for the "use of selective, environmentally safe and cost-effective fishing gear and techniques" (FAO 1995) exceed any real action. Even where trawling is banned or restricted, compliance remains problematic (e.g., Butcher 2002). For this form of extraction, then, globalization of fishing technology and trade is running well ahead of globalization of effective management, to the detriment of populations and ecosystems.

GLOBALIZATION IN THE BALANCE: MOVING ON FROM SEA HORSES

If globalization plays the roles in influencing sea horse fisheries that we conjecture – and this certainly remains an open question – then declining populations would seem to suggest that the costs of globalization currently exceed its benefits, at least for sea horses. One reason for such imbalances – and one area open to global influence and change – might be the number of perverse incentives that remain in place (Milazzo 1998). Removals of subsidies would help improve the health of fisheries worldwide, whether these currently promote the use of destructive gear such as trawling, reduce the effective cost of fishing, or increase the price received by fishers (Milazzo 1998; Munro and Sumaila 2002). Disciplining subsidies is one area where global organizations such as the WTO could have positive impacts on the conservation of fishery resources. Political considerations, however, make global wholesale change in perverse subsidies unlikely (Stone 1997).

One problem perhaps relevant to sea horse exploitation but certainly applicable to most fisheries is that much of the economic theory supporting free trade is undermined to some extent by the violation of basic assumptions essential to its success. In particular, the problems of ineffective management, transboundary impacts (Munro 1979), and poorly defined property rights (Gordon 1954) continue to plague most fisheries (Neumayer 2000). In theory, enhanced international trade should allow small-scale fishers to manage resources more sustainably if, as projected, national income growth and diversification of employment opportunities (Nordstrom and Vaughan 1999) benefit fishers and their communities. In practice, however, globalization has not improved the lot of all groups and sectors in a society (IMF 2000b). Because small-scale fishers often end up losing out to their large-scale counterparts (Iheduru 1996; Atta-Mills et al. 2004) (Fig. 7.4), they are pushed into more intense and less selective extraction to meet their needs, with worrying consequences for species, ecosystems, and, ultimately, the fishers themselves.

Despite our uncertainties about how sea horse declines are linked to globalization, management of all fisheries – for sea horses and other species – must be improved at the national and international levels to ensure that fish stocks are exploited sustainably (avoiding the paradigm of shifting baselines: Pauly 1995) and ecosystems remain fully functional. Such management would need to incorporate into decision-making all forms of market as well as non-market values of fish stocks and their ecosystems (Goulder and Kennedy 1997; Sumaila et al. 2002). The transboundary nature of some fisheries resources and issues should be addressed through multilateral agreements such as the United Nations Agreement on Straddling Fish Stocks and Highly Migratory Fish Stocks. Ecological certification and CITES regulations are other tools of benefit to transboundary stocks. As outlined above, CITES can help promote fisheries management for sustainability where other national and multilateral instruments are inadequate.

For some fisheries, management could be improved by the judicious assignment of fishing rights to various entities (e.g., coastal communities, associations, firms, and individuals). For globalization to be of net benefit to small-scale fishers, they may need to have extraction rights and management responsibility, and thus a stake in preserving the resource. Most sea horse fisheries are completely unmanaged in an open-access system with little understanding of international connections (Vincent 1996). One consequence is that fishers in the central Philippines, while willing to engage in selective takes of sea horses by sex and size, become concerned that their neighbors might take the fish they leave behind (A. Vincent, personal observation). In this kind of situation, assignment of property rights is likely to improve conservation, provided they are implemented in a manner that mitigates most of their negative effects. It is important to note that individual transferable quotas do well in economic performance, improve but do not eliminate conservation concerns, and perform poorly socially (see Copes 1986; Hannesson 1996; Sumaila and Watson 2002). Ultimately, the challenge is to obtain good local management in a world where boundaries are being blurred through ever greater globalization, to the detriment of the fish and the fishers alike.

Improved management, better international cooperation on transboundary stocks and tenurial rights will be insufficient to restore fisheries unless overall fishing effort is also reduced. At present, domestic and international consumption combine with large family sizes in fishing communities to push small-scale fishers into overexploitation and the use of destructive fishing methods (Pauly 1990). Consumption and population growth will both need to be adjusted if they are not to overwhelm any gains made by better management. In addition, fishing communities should be exposed to new opportunities for alternative and supplementary forms of employment and income generation (particularly those independent of marine resources), so that they may reduce fishing effort or fish more selectively.

In closing, it is worth emphasizing the profound implications of some international policies for local resource management. One of the lead sea horse exporting nations, the Philippines, has domestic legislation that interprets CITES regulations for sustainable use (Appendix II) as a trade ban (equivalent to CITES Appendix I) and consequently bars all exploitation of the affected species. Such action will rebound heavily on subsistence fishers who depend on sea horse exploitation without reducing sea horse bycatch. In November 2002, CITES directed parties (such as the Philippines) that ban fishing and trade in species listed in the Appendices to allow sustainable trade under the provisions of CITES (Decision 12.53, CITES 2004), but this may be difficult to achieve. Without the purported economic benefits of globalization – indeed, with incidental damage from globalization – our best hope is for fishers to improve their stewardship of marine resources. Using legislation to push fishers from species to species in a piecemeal fashion is unlikely to make them identify with a global community, nor is it likely to benefit the marine environment.

In finding ways to manage exploitation of sea horses, we will be forced to address global issues that increasingly link distant and diverse communities. Given the international nature of sea horse trade, we decided in this chapter to seek out global changes that might conceivably affect sea horse exploitation. From these possible interrelationships we have created hypotheses on how globalization might influence sea horse conservation and management. It would now be appropriate to test these predictions by examining the dynamics of specific sea horse fisheries and trades at a local scale, and how they change over time. In so doing, further consideration can also be given to the possible role of globalization in degrading the sea horses' inshore habitats.

Acknowledgments

This is a contribution from Project Seahorse, the University of British Columbia Fisheries Economics Research Unit, and the Sea Around Us Project. We sincerely thank Project Seahorse's trade research team: Julia Baum, Brian Giles, Boris Kwan Sai-Ping, Patrick Lafrance, Keith Martin-Smith, Jana McPherson, Marivic Pajaro, and Allison Perry. We are also very grateful to Sarah Foster, Brian Giles, Melissa Grey, Bob Hunt, and Ierece Lucena Rosa for their input on this manuscript. We thank William Taylor and Lois Wolfson for organizing the *Globalization: Effects on Fisheries Resources* symposium at the American Fisheries Society Annual Meeting in August 2003, at which this work was first presented. We also acknowledge generous input from three anonymous referees. Project Seahorse research associated with this paper was supported by the John G. Shedd Aquarium, Guylian Chocolates Belgium, United Kingdom Department of Environment, Food and Rural Affairs, GlaxoSmithKline, the People's Trust for Endangered Species, and New England Biolabs Foundation. The Sea Around Us Project is supported by the Pew Charitable Trusts, Philadelphia.

References

Allison, E. H. and Ellis, F. 2001. The livelihoods approach and management of small-scale fisheries. *Marine Policy* **25**: 377–388.

Alverson, D. L., Freeberg, M. K., Murawski, S. A., and Pope, J. G. 1994. *A Global Assessment of Fisheries Bycatch and Discards*, Fisheries Technical Paper No. 339. Rome: Food and Agriculture Organization of the United Nations.

American Pet Products Manufacturer's Association. 2002. *APPMA National Pet Owners Survey*. Greenwich, CT: American Pet Products Manufacturer's Association.

Antweiler, W., Copeland, B. R., and Taylor, M. S. 2001. Is free trade good for the environment? *American Economic Review* **91**: 877–908.

Atta-Mills, J., Alder, J., and Sumaila, U. R. 2004. The decline of a fishing nation: the case of Ghana and West Africa. *Natural Resources Forum* **28**: 13–21.

Baquero, J. 1999. *Marine Ornamentals Trade: Quality and Sustainability for the Pacific Region.* Suva, Fiji: South Pacific Forum Secretariat, Trade and Investment Division.

Baum, J. K. and A. C. J. Vincent. 2005. Magnitude and inferred impacts of the seahorse trade in Latin America. *Environmental Conservation* **32**: 305–319.

Baum, J. K., Meeuwig, J. J., and Vincent, A. C. J. 2003. Bycatch of lined seahorses (*Hippocampus erectus*) in a Gulf of Mexico shrimp trawl fishery. *Fisheries Bulletin* **101**: 721–731.

Bell, M., Khor, H. E., and Kochlar, K. 1993. *China at the Threshold of Market Economy, Occasional Paper No. 107.* Washington, DC: International Monetary Fund.

Borjas, G. J. 1999. *Labour Economics*, 2nd edn. London: McGraw-Hill.

Brander, J. A. and Taylor, M. S. 1997. International trade and open-access renewable resources: the small open economy case. *Canadian Journal of Economics* **30**: 526–552.

Brander, J. A. and Taylor, M. S. 1998. Open access renewable resources: trade and trade policy in a two-country model. *Journal of International Economics* **44**: 181–209.

Broadhurst, M. K. 2000. Modifications to reduce bycatch in prawn trawls: a review and framework for development. *Reviews in Fish Biology and Fisheries* **10**: 27–60.

Butcher, J. G. 2002. Getting into trouble: the diaspora of Thai trawlers, 1965–2002. *International Journal of Maritime History* **14**: 85–121.

Cameron, G., Pendry, S., and Allan, C. 2001. *Traditional Asian Medicine Identification Guide for Law Enforcers.* London: Her Majesty's Customs and Excise; and Cambridge, UK: TRAFFIC International.

CITES. 2004. Convention on International Trade in Endangered Species of Wild Fauna and Flora Website. Available online at www.cites.org

Clark, C. W. 1990. *Mathematical Bioeconomics: The Optimal Management of Renewable Resources.* New York: Wiley-Interscience.

Clarke, S. 2002. *Trade in Asian Dried Seafood: Characterization, Estimation and Implications for Conservation.* London: Wildlife Conservation Society.

Clucas, I. 1997. *A Study of the Options for Utilization of Bycatch and Discards from Marine Capture Fisheries*, Fisheries Circular No. 928. Rome: Food and Agriculture Organization of the United Nations.

Copes, P. 1986. A critical review of the individual quota as a device in fisheries management. *Land Economics* **62**: 278–291.

Cox, W. 2003. *Demographia: Hong Kong Population History.* Belleville, IL: Wendell Cox Consulting. Available online at www.demographia.com/db-hkhist.htm

Crafts, N. 2000. *Globalization and Growth in the Twentieth Century*, Working Paper No. WP/00/44. Washington, DC: International Monetary Fund.

Daly, H. 1993. The perils of free trade. *Scientific American* **269**: 50–57.

Dixon, J., Scura, L. F., and van't Hof, T. 1993. Meeting ecological and economic goals: marine parks of the Caribbean. *Ambio* **22**: 117–125.

Dollar, D. and Kraay, A. 2002. Spreading the wealth. *Foreign Affairs* **81**: 120–33.

Ecrobot. 2004. Ecrobot worldwide trading service Website. Available online at www.ecrobot.com

FAO (Food and Agriculture Organization) 1995. *Declaration of the Kyoto Conference on the Sustainable Contribution of Fisheries to Food Security.* Rome: Food and Agriculture Organization of the United Nations.

FAO. 2001a. *FAO Yearbook, Fishery Statistics, Capture Production 2001*, vol. 92/1. Rome: Food and Agriculture Organization of the United Nations.

FAO. 2001b. *FAO Yearbook, Fishery Statistics, Commodities 2001*, vol. 93. Rome: Food and Agriculture Organization of the United Nations.

FAO. 2002. *The State of World Fisheries and Aquaculture.* Rome: Food and Agriculture Organization of the United Nations.
Foster, S. J. and Vincent, A. C. J. 2004. The life history and ecology of seahorses: implications for conservation and management. *Journal of Fish Biology* **65**: 1-61.
Foster, S. J. and Vincent, A. C. J. 2005. Enhancing sustainability of the international trade in seahorses with a single minimum size limit. *Conservation Biology* **19**: 1044-1050.
Garcia, S. M. and Newton, C. 1997. Current situation, trends, and prospects in world capture fisheries. In *Global Trends: Fisheries Management*, eds. E. K. Pikitch, D. D. Huppert, and M. P. Sissenwine. Bethesda, MD: American Fisheries Society, pp. 3-27.
Giles, B. G., Truong, S. K., Do, H. H. and Vincent, A. C. J. 2005. The catch and trade of seahorses in Vietnam. *Biodiversity and Conservation* DOI: 10.1007/s10531-005-2432-6
Goldsmith, E. 1996. Global trade and the environment. In *The Case against the Global Economy and for a Turn toward the Local*, ed. J. Mander and E. Goldsmith. San Francisco, CA: Sierra Club, pp. 78-91.
Gordon, H. S. 1954. The economic theory of common property resource: the fishery. *Journal of Political Economy* **62**: 124-143.
Goulder, H. and Kennedy, D. 1997. Valuing ecosystem services: philosophical bases and empirical methods. In *Nature's Services: Societal Dependence on Natural Ecosystems*, ed. G. C. Daily. Washington, DC: Island Press, pp. 23-48.
Hall, H. J. and Warmolts, D. 2003. The role of public aquaria in the conservation and sustainability of the marine ornamentals trade. In *Marine Ornamentals: Collection, Culture and Conservation*, eds. J. Cato and C. L. Brown. Ames, IA: Iowa State University Press, pp. 307-327.
Hall, M. A., Alverson, D. L., and Metuzals, K. I. 2000. By-catch: problems and solutions. *Marine Pollution Bulletin* **41**: 204-219.
Hannesson, R. 1996. Long-term industrial equilibrium in an ITQ managed fishery. *Environmental and Resource Economics* **8**: 63-74.
Hutchings, J. A. 2000. Collapse and recovery of marine fishes. *Nature* **406**: 882-885.
Iheduru, O. C. 1996. The political economy of Euro-African fishing agreements. *Journal of Developing Areas* **30**: 63-90.
IMF (International Monetary Fund). 2000a. *Globalization: Threat or Opportunity?* Washington, DC: International Monetary Fund.
IMF. 2000b. *World Economic Outlook: World Economic and Financial Surveys.* Washington, DC: International Monetary Fund.
International Telecommunications Union. 2003. International Telecommunications Union Website. Available online at www.itu.int
IUCN. 2003. *2003 IUCN Red List of Threatened Species.* Gland, Switzerland: World Conservation Union. Available online at www.iucnredlist.org
Jackson, J. B. C., Kirby, M. X., Berger, W. H., et al. 2001. Historical overfishing and the recent collapse of coastal ecosystems. *Science* **293**: 629-638.
Jennings, S. and Kaiser, M. J. 1998. The effects of fishing on marine ecosystems. In *Advances in Marine Biology*, vol. 34, eds. J. H. S. Blaxter, B. Douglas, and P. A. Tyler. London: Academic Press, pp. 201-352.
John, J. 1994. *Managing Redundancy in Overexploited Fisheries.* Washington, DC: International Bank for Reconstruction and Development (The World Bank).
Kaczynski, V. M. and Fluharty, D. L. 2002. European policies in West Africa: who benefits from fisheries agreements? *Marine Policy* **26**: 75-93.

Kaiser, M. J., Collie, J. S., Hall, S. J., Jennings, S., and Poiner, I. R. 2002. Modification of marine habitats by trawling activities: prognosis and solutions. *Fish and Fisheries* **3**: 114–136.

Kibel, P. S. 2001. The paper tiger awakens: North American environmental law after the Cozumel Reef case. *Columbia Journal of Transnational Law* **39**: 395–482.

Kuiter, R. H. 2000. *Seahorses, Pipefishes and their Relatives: A Comprehensive Guide to Syngnathiformes*. Chorleywood, UK: TMC Publishing.

Kuiter, R. H. 2001. Revision of the Australian seahorses of the genus *Hippocampus* (Syngnathioformes: Syngnathidae) with a description of nine new species. *Records of the Australian Museum* **53**: 293–340.

Larkin, S. L. and Degner, R. L. 2001. The U.S. wholesale market for marine ornamentals. *Aquarium Sciences and Conservation* **3**: 13–24.

Lintner, B. 1999. The Third Wave: a new generation of Chinese migrants fan across the globe. *Far Eastern Economic Review* **162**: 28–29.

Lourie, S. A. and Randall, J. E. 2003. A new pygmy seahorse, *Hippocampus denise* (Teleostei: Syngnathidae), from the Indo-Pacific. *Zoological Studies* **42**: 284–291.

Lourie, S. A., Vincent, A. C. J., and Hall, H. J. 1999. *Seahorses: An Identification Guide to the World's Species and their Conservation*. London: Project Seahorse.

Mace, P. M. 1997. Developing and sustaining world fisheries resources: the state of the science and management. In *Developing and Sustaining World Fisheries Resources: The State of the Science and Management, Proceedings of the 2nd World Fisheries Congress*, eds. D. H. Hancock, D. C. Smith, and J. Beumer. Collingwood, VIC: CSIRO Publishing, pp. 1–20.

Marine Aquarium Council. 2004. Marine Aquarium Council Website. Available online at www.aquariumcouncil.org

Marine Stewardship Council. 2004. Marine Stewardship Council Website. Available online at www.msc.org

Martin-Smith, K. M. and A. C. J. Vincent. 2006. Exploitation and trade of Australian seahorses, pipehorses, sea dragons and pipefishes (syngnathids). *Oryx* **40**: 141–151.

Martin-Smith, K. M., Samoilys, M. A., Meeuwig, J. J., and Vincent, A. C. J. 2004. Collaborative development of management options for an artisanal fishery: seahorses in the central Philippines. *Ocean and Coastal Management* **47**: 165–193.

McPhee, D. P. and Skilleter, G. A. 2002. Harvesting of intertidal animals for bait for use in a recreational fishing competition. *Proceedings of the Royal Society of Queensland* **110**: 93–101.

McPherson, J. M. and Vincent, A. C. J. 2004. Assessing East African trade in seahorse species as a basis for conservation under international controls. *Aquatic Conservation: Marine and Freshwater Ecosystems* **14**: 521–538.

Milazzo, M. 1998. *Subsidies in World Fisheries: A Re-Examination*, Technical Paper No. 406. Washington, DC: World Bank.

Munro, G. R. 1979. The optimal management of transboundary renewable resources. *Canadian Journal of Economics* **12**: 355–376.

Munro, G. R. and Sumaila, U. R. 2002. The impact of subsidies upon fisheries management and sustainability: the case of the North Atlantic. *Fish and Fisheries* **3**: 233–250.

Myers, R. A. and Worm, B. 2003. Rapid worldwide depletion of predatory fish communities. *Nature* **423**: 280–283.

Neumayer, E. 2000. Trade and the environment: a critical assessment and some suggestions for reconciliation. *Journal of Environment and Development* **9**: 138–159.

Nogales, S. F. 2002. The NAFTA environmental framework, Chapter 11 investment provisions and the environment. *Annual Survey of International and Comparative Law* **8**: 97–106.
Nordstrom, H. and Vaughan, S. 1999. *Trade and Environment*. Geneva, Switzerland: World Trade Organization.
OECD (Organisation for Economic Co-operation and Development). 1997. *Economic Globalisation and the Environment*. Paris: Organisation for Economic Co-operation and Development.
Pauly, D. 1990. On Malthusian overfishing. *Naga, the ICLARM Quarterly* **13**: 3–4.
Pauly, D. 1995. Anecdotes and the shifting baseline syndrome of fisheries. *Trends in Ecology and Evolution* **10**: 430.
Pauly, D. 1997. Small-scale fisheries in the tropics: marginality, marginalization, and some implications for fisheries management. In *Global Trends: Fisheries Management*, eds. E. K. Pikitch, D. D. Huppert, and M. P. Sissenwine. Seattle, WA: American Fisheries Society, pp. 1–10.
Pauly, D., Christensen, V., Dalsgaard, J., Froese, R., and Torres, F., Jr. 1998. Fishing down marine food webs. *Science* **279**: 860–863.
Pauly, D., Christensen, V., Guénette, S., *et al.* 2002. Towards sustainability in world fisheries. *Nature* **418**: 689–695.
Poston, J. D. L., Xinxiang Mao, M., and Yu, M.-Y. 1994. The global distribution of the overseas Chinese around 1990. *Population and Development Review* **20**: 631–645.
Project Seahorse. 2004. Project Seahorse Website. Available online at www.projectseahorse.org
Sadovy, Y. J. and Vincent, A. C. J. 2002. Ecological issues and the trades in live reef fishes. In *Coral Reef Fishes: Dynamics and Diversity in a Complex Ecosystem*, ed. P. F. Sale. Boston, MA: Academic Press, pp. 391–420.
Sheetz, J. and Seligmann, J. 2000. *The Availability of Seahorses in the U.S. for Use in Traditional Chinese Medicine*. Yarmouth Port, MA: International Fund for Animal Welfare.
Sraffa, P. and Dobb, M. H. 1951–1973. *The Works and Correspondence of David Ricardo*. Cambridge, UK: Cambridge University Press.
Stone, C. D. 1997. Too many fishing boats, too few fish: can trade laws trim subsidies and restore the balance in global fisheries? *Ecology Law Quarterly* **24**: 505–544.
Sumaila, U. R. and Watson, R. 2002. The rights to fish: a critique of ITQs. In *Ecosystem-Based Management for Marine Capture Fisheries: A Policy Direction*, eds. T. Ward, D. Tarte, E. Hegerl, and K. Short. Sydney, NSW: WWF, pp. 41–43.
Sumaila, U. R., Pitcher, T. J., Haggan, N., and Jones, R. 2002. Evaluating the benefits from restored ecosystems: a back to the future approach. In *Proceedings of the 10th International Conference of the International Institute of Fisheries Economics and Trade*, eds. R. S. Johnston and A. L. Shriver. Corvallis, OR (on CD-ROM).
Tseng, W. 1994. *Economic Reform in China: A New Phase*, Occasional Paper No. 107. Washington, DC: International Monetary Fund.
UNESCAP. 2003. United Nations Economic and Social Commission for Asia and the Pacific Website. Available online at www.unescap.org
Valley, P. J. 2004. Tension between the Cartagena Protocol and the WTO: the significance of recent WTO developments in an ongoing debate. *Chicago Journal of International Law* **5**: 369–378.
Vincent, A. C. J. 1995. Trade in seahorses for traditional Chinese medicines, aquarium fishes and curios. *TRAFFIC Bulletin* **15**: 125–128.

Vincent, A. C. J. 1996. *The International Trade in Seahorses*. Cambridge, UK: TRAFFIC International.

Watling, L. and Norse, E. A. 1998. Disturbance of the seabed by mobile fishing gear: a comparison to forest clearcutting. *Conservation Biology* **12**: 1180–1197.

WHO (World Health Organization). 2002. *Traditional Medicine Strategy 2002–2005*. Geneva, Switzerland: World Health Organization.

Wood, E. M. 2001. *Collection of Coral Reef Fish for Aqauria: Global Trade, Conservation Issues and Management Strategies*. Ross-on-Wye, UK: Marine Conservation Society.

World Tourism Organization. 2003. World Tourism Organization Website. Available online at www.world-tourism.org

WTO (World Trade Organization). 1994. *Uruguay Round Agreement on Subsidies and Countervailing Measures: Article 1–9*. Geneva, Switzerland: World Trade Organization. Available online at www.wto.org/english/docs_e/legal_e/24-scm_01_e.htm

Yu, D. W., Sutherland, W. J., and Clark, C. 2002. Trade versus environment. *Trends in Ecology and Evolution* **17**: 341–344.

8

Wronging rights and righting wrongs: some lessons on community viability from the colonial era in the Pacific

INTRODUCTION

The term "globalization" is used to describe an increase to a worldwide scale of the interconnectedness of cultural, environmental, and social phenomena. The term has been applied to commercial, ecological, economic, financial, organizational, religious, spiritual, and trade activities, among a great many other processes and structures. Although identified with various trends that developed largely during the last half-century, it can be argued that the substance of globalization does not connote anything particularly new. In contrast, the speed with which it is now occurring is new, and results from the intensity of modern communications. It is this awesome speed of contemporary communication combined with the portability, increasingly low cost, standardization, and integration of the required hardware and software that now enables the process of globalization to penetrate into the remotest corners of the world, and to tie hitherto isolated fishing communities, for example, into the mainstream of the world fish trade.

But it was not always thus. Although some would argue that the characteristics of globalization can be identified at other times in history, their assertions founder on the issue of communication speed. What can now be achieved with a PDA in every pocket certainly would not have been feasible in the days of sail-borne commerce. Whereas one might agree that the *outline* of Western commercial, military, and imperial globalization could be traced to the beginnings of the European rampage overseas in the fifteenth century, without lightning-fast communication this outline could not have evolved into globalization as we now understand it. Without the spokes of fast and

Globalization: Effects on Fisheries Resources, ed. William W. Taylor, Michael G. Schechter, and Lois G. Wolfson. Published by Cambridge University Press. © Cambridge University Press 2007.

interactive communication, the metropolitan hub could neither check on nor correct and control the various activities going on around the colonial rim.

My purpose in this brief essay is to illustrate that contention by using cases drawn from the management of fisheries in some Pacific territories controlled by the British. Under British Imperial rule traditional fishing rights were recognized as valid, based on a policy under which indigenous peoples were not divested of their lands, but, on the contrary, were protected and their rights secured by the colonial authorities. Despite this policy, in Oceania, as elsewhere, community-based marine resource management systems were massively affected by the impacts of colonialism and all that it implied (Ruddle 1994a, b, c, d, e). One of the massive if insidious impacts of both historical and contemporary globalization is the imposition of standard Western systems of resource management. In every respect this is the cultural equivalent of a major reduction in biodiversity. Coastal communities throughout the tropics experienced this early in the colonial era, when many communities were wrongly deprived of their traditional rights to fisheries and other resources. In some cases these have only recently been restored to them.

Based on that policy, under the Treaty of Waitangi (1840), the British Crown recognized the New Zealand Maori resource rights and the Crown's duty to protect Maori interests and their full, exclusive, and undisturbed possession of their lands and estates, forests, and fisheries. Similarly, in Fiji the British Crown expressly ordered that land and reef titles be held in trust for the Fijians. So in the far distant metropolitan "hub" of London orders were issued to protect fisheries rights for Maoris and Fijians, and it was assumed that this policy would be implemented in the distant colonies. But there was no way that the correct implementation of policy could be rapidly confirmed. With an increasing European settlement and deteriorating race relations and land wars, the social climate changed in New Zealand. Local legislation gradually eroded and then denied the existence of Maori marine resource rights, in utter disregard of the situation as understood in 1840. Somewhat differently, Fiji provides an excellent example of a blatant attempt by local colonial officials to destroy a traditional management system in favor of expatriate entrepreneurs and in defiance of the expressed wishes of the British Crown and the unambiguous orders of the metropolitan government. In both New Zealand and Fiji activities in the local parts of the colonial "rim" contradicted original policies made in the Imperial "hub."

Papua New Guinea provides a contrasting case. There the customary fishing and marine resources rights of indigenous peoples were recognized or protected by the various colonial administrations, in accordance with colonial policy, which in this country accorded with an uncompromising reality. In the Papua New Guinea of colonial times, as today, the sheer logistical and practical complexity of attempting to incorporate customary rights into a system of legal norms was almost overwhelming. An extraordinarily diverse society like Papua New Guinea, set in a formidable physical environment, makes it daunting if not impossible to consider formulating appropriate law and policies to embrace the specific customs of some 700 (if not more) distinct cultural groups.

At the colonial policy level in London, it is possible to accept the concept of colonial era globalization, based on a standard policy. But at the local level implementation diverged widely, and substantial differences arose in each location. At that level globalization disintegrated. Regardless of local differences in political and economic conditions, which were substantial, this must in large part be attributed to the inability of central colonial officials in London to confirm and repeatedly check that policies were being implemented as intended. Without a rapid means of communication, that could never be attained.

Whereas it is all too easy to attribute such a situation to negative factors, for example the greed of local expatriate entrepreneurs combined with compliant local colonial administrators, it should not be forgotten that it was a major shock to many colonial administrators to be confronted with the entirely alien concept of traditional community-based resource management systems. So not surprisingly in many cases, despite official policy, they set about undermining something which they both barely understood and saw as constraining Imperial governance and Western-style coastal–marine resource development.

The British, Dutch, French, German, Japanese, Portuguese, Spanish, and U.S. administrations in the Pacific Islands, and the post-colonial continuation by independent nations of the laws introduced and policies pursued by those regimes, have been a major "global" factor that either by *default* or *deliberately*, undermined customary law and community resource rights. Default was widespread and understandable. Despite noble policy, crafted no doubt by well-meaning officials in London who had but the faintest idea of the island realities, it never was the objective of colonial regimes to adapt metropolitan legal systems to indigenous systems and institutions. Rather the goal was that the latter should be displaced and native peoples educated to use Western systems and institutions. To have encouraged community-based management systems rooted in local systems of customary law would have been inimical to this objective. Rather, the objective would be attained by either legislating directly against community-based systems or allowing them to wither and become displaced during a gradual process of modernization and Westernization.

As a consequence, for fishing communities the principal impact of the colonial era in the Pacific Islands region is a strongly contradictory legal complexity; Western-based State law of the now independent nations, which essentially regards all waters below the high tide mark as being state property and open of access, is at odds with local customary law, which recognizes some form of marine property right. Worse, it is generally accepted by Westerners and those Western-trained that customary law, which locally legitimizes customary rights to resources, is invalid for upholding legal claims, because it is unwritten, not made by either a sovereign or legally constituted legislative body, and arises from societies lacking any notion of "law." As a consequence, in the Pacific Islands region, the relationship between the customary law that governs, or governed, community-based marine resource management and statutory law is highly varied and extremely complex.

THE CASE OF THE NEW ZEALAND MAORI FISHERY RIGHTS

British settlement of New Zealand and recognition of Maori resource rights was based on the Treaty of Waitangi (1840), which described the nature of the Crown's right to govern and the protection of Maori interests. In the English version of the Treaty, the Maori received full, exclusive, and undisturbed possession of their lands and estates, forests, and fisheries for as long as they wish to retain them.

But the Treaty of Waitangi lacked legal standing because it was neither ratified nor passed into law. As a result, it has been treated as a legal nullity. In addition, there is both an English and a Maori version of the Treaty, neither of which is a translation of the other! This, of course, added to the confusion and the scope for malfeasance. And all that was further compounded by interpretation problems of mutually incomprehensible concepts of property rights.

Not surprisingly, Maori rights were gradually eroded, a process accelerated by legislation aimed at dispossession. Conspicuous were the destruction of traditional Maori authority by the Native Land Act 1862, which individualized title to tribal lands (Kawhuru 1977), and land confiscation under the New Zealand Settlements Act 1863.

Fisheries laws followed. Although Maori fishing rights were provided for in law, in practice they were denied. Rights under the Treaty of Waitangi were acknowledged in general terms by the Fish Protection Act 1877 and by the Fisheries Conservation Act 1884. The Fisheries Amendment Act 1903 reintroduced a general and essentially meaningless provision (since it was left to administrators to interpret, and so basically ignored in practice) that "Nothing in this Act shall affect any existing Maori fishing rights." Conflicting interpretation of that general statement led to many court cases. But it was invariably held that it did not provide for fishing rights.

The Oyster Fisheries Act 1866, the first New Zealand fisheries legislation, was also the first statutory expression of erroneous assumptions regarding Maori fishing. By allowing only exclusive subsistence use of oyster beds near Maori villages, this Act implied that: (1) the Crown had an unencumbered right to dispose of foreshore fisheries because foreshore and the sea space beyond belonged to the Crown; (2) the Treaty of Waitangi could be ignored; (3) Maori fishing traditionally had no commercial component; (4) the Crown alone had the right to manage fisheries; and (5) only non-Maori people had the right to commercially exploit traditionally Maori inshore fisheries (Waitangi Tribunal 1988).

Thus proof of customary entitlement was no longer acceptable as evidence of a fishing right, as confirmed by the Larceny Act 1869, which made it an offence to take fish from private waters or from an area governed by a fishery right. This effectively demonstrated that unless specifically provided for, traditional Maori fishery rights lacked any status.

The Fish Protection Act 1877 illustrates typical legislative window dressing, by blithely assuming that Maori interests under the Treaty of Waitangi could be accommodated by a general statement "that nothing in the Act was to affect any of the provisions of the Treaty of Waitangi, or to take away Maori rights to any fishery secured by it" (Section 8). Such a statement is hardly convincing when everything else in the piece of legislation was clearly contrary to Treaty of Waitangi principles (Waitangi Tribunal 1988).

The social climate also changed drastically. In 1840 the intentions of the Crown were benign. Governors were directed "to honorably and scrupulously fulfil" the treaty conditions "as a question of honour and Justice no less than of policy" (Wards 1968:171). But policy changed as permanent settlers began to outnumber the Maori and the first Colonial parliament was formed, in 1855. Racial relations changed drastically as a consequence of the "Land Wars," in the 1860s.

But unlike the colonizers the Maori never entirely forgot their treaty-enshrined fishing rights and management systems; they were embedded in and transmitted via a comprehensive system of local knowledge and a rich oral culture. Preserved in that way, and together with the enormous injustice of their dispossession, the legacy remained as an *aide-mémoire* for posterity, until conditions were suitable to seek redress. Although there were occasional appeals for justice, substantive action had to await the general politicization that characterized the 1960s.

RIGHTING THE WRONGS: THE WAITANGI TRIBUNAL

In 1975 the Treaty of Waitangi Act established the Waitangi Tribunal to investigate Maori claims and make recommendations relating to the Treaty of Waitangi. Fisheries claims became a major focus in the resurgence of Maori ethnic politics. Whereas they had long since lost most of their lands, the Maori retained a residual interest in marine waters. The official perspective was that marine fisheries remained common property under the Crown, whereas versions of the Fisheries Act acknowledged some residual but unspecified Maori fishing rights. Thus fisheries issues pertained to a vital resource, and although acknowledged had never been specified.

Under the Waitangi Tribunal Maori claimants sought to make "Cultural Deprivation" actionable. In the first three fisheries claims (Motonui, Kaituna, and Manukau) they emphasized loss of reefs as a cultural resource. Since seafood and reef habitats are highly prized in Maori culture as *taonga katoa* ("treasures"), pollution damage is a cultural affront, by degrading the *mana* ("authority," "status," "prestige") of the tribes involved, and so protected by the Article 2 of the Treaty of Waitangi. In this way culture, demonstrated through local knowledge, became an instrument of Maori empowerment. The Waitangi Tribunal hearings on the three water rights cases show that simple issues of effluent pollution to food sources have been expanded into far wider ethnic demands (Levine 1987; Oliver 1991).

The Maori successfully demonstrated that traditional life was intimately shaped by the availability and sustainability of renewable natural resources. Water is regarded as so vital to life that it acquired a spirit (*wairua*). Traditionally, the Maori believed that life is derived from the waters of the womb of the Earthmother (*Papatuanuku*), and that water, the life-giving essence, must remain pure and unadulterated to ensure life for the following generations (Taylor and Patrick 1987).

Thus to pollute seriously diminishes "the life-force (*mauri*) of the water, demeans its *wairua* and thereby affects the *mana*, the prestige, of those who use it and its resources" (Taylor and Patrick 1987:22). The Maori relied heavily on marine and freshwater fish. Seafood (*kaimoana*) was of great cultural importance since much prestige and social standing accrued to groups that could provide a lavish feast at social and cultural events. This remains important to the Maori (Sandrey 1987). On those bases, according to the principles of the Treaty of Waitangi, cultural deprivation is actionable.

One of the keys in righting the wrong to Maori fishing rights was the use of local knowledge and historical documentation by early European visitors to refute erroneous perceptions. That the New Zealand Maori were historically expert fishers with a profound fisheries tradition was recognized during the earliest European contacts, accounts of which commented on the sizes and composition of catches, the gear employed, fisheries knowledge, long-distance fishing expeditions, large fishing fleet operations, trading in marine products, and traditional fisheries management systems (Ruddle 1995). Despite that documented tradition, through the social, economic, and legal processes described above, the heavy Maori involvement in fisheries characteristic of the early decades of the nineteenth century had so declined in the twentieth century that little credence of its former importance existed. Changed economic circumstances of the Maori as well as generally acculturation took their inevitable toll on the old ways. Thus it became widely believed that Maori fisheries never were commercial and always only a household subsistence activity done on a few local reefs and grounds (Waitangi Tribunal 1988).

The erroneous assumptions about Maori fishing embedded in early legislation became uncritically accepted. Maori fishing was assumed to traditionally have been:

(1) limited to few species;
(2) confined to limited inshore areas;
(3) for household subsistence; and
(4) rightfully managed by the Crown.

But the Waitangi Tribunal was able to demonstrate the falsehood of those perceptions by (Waitangi Tribunal 1988, 1992):

(1) systematically collecting evidence from Maori fishers to establish empirically the existence of a local knowledge base; and

(2) validating local knowledge bases from both historical records and scientific evidence of fisheries biologists and other specialists.

In this way, the historical wronging of New Zealand Maori fisheries rights has been corrected.

FISHERIES AND THE EARLY COLONIAL REGIME IN FIJI

The case of Fiji is well documented. It provides an excellent example of a blatant attempt by local colonial officials to destroy a traditional management system in favor of expatriate entrepreneurs and in defiance of the expressed wishes of the British Crown and the unambiguous orders of the metropolitan government.

At the time of cession of Fiji to the British Crown, in 1874, the question of customary resource rights was of major concern to the High Chiefs, most of whom wanted to attach conditions regarding their land and fishing grounds before agreeing to the cession of the country. But they were dissuaded from so doing during the final negotiations by Robinson, the British representative, who explained that Queen Victoria "was willing to accept the offer of cession ... but that conditions attached to it would hamper, and might even prevent, the good government of the country" (Derrick 1946:248). While the High Chiefs agreed with this, it was apparent that they expected to have their lands and waters returned, in accordance with Victoria's "generosity and good faith" (Derrick 1946: 248).

Detailed instructions regarding the verification and simplification of Fijian land titles of lands to be held in trust for the Fijians were given to the British Governor of Fiji by the Secretary of State for the Colonies (Despatch No. 1, March 4, 1875). But since no similar clear statement was forthcoming with respect to the reefs, the chiefs began to question the situation. They sent two letters to Victoria expressing their anxiety that their ownership of the reefs appeared to have passed from them.

In response, Kimberley, then Secretary of State for the Colonies, wrote to Des Voeux, Governor of Fiji, instructing him that he (Kimberley) was commanded by Victoria to inform the High Chiefs that Des Voeux was to investigate the entire matter, "and that it is Her Majesty's desire that neither they nor their people should be deprived of any rights which they have enjoyed under their own laws and custom" (Despatch No. 69, June 2, 1881). In another dispatch Kimberley further instructed Des Voeux to:

examine into the statements now advanced by the chiefs, and if you are satisfied that these reefs are the recognised property of native communities... (Despatch No. 71, June 2, 1881), or that they are required for the use and occupation of some Chiefs or tribe, you will take such measures as may be necessary to secure to the rightful owners the possession of their respective reefs and to effect the registration of them under the Ordinance relating to native lands; in the same way as other lands (not covered by water) which are the property of the different mataqali[1]... If there are any reefs not claimed as the property of any Native Chiefs or Community they will continue to be the property of the Crown together with the other lands which became vested in Her Majesty under the terms of the Deed of Cession.

Thus clearly it was both the policy and the intention of both Victoria and the British government that, according to customary law, the reefs and fishing grounds have Fijian owners in the same way that their lands did. In November of 1881, Des Voeux conveyed equally unambiguously the contents of those two dispatches during his opening address to the Council of Chiefs, and added that steps would be taken to ensure that the *mataqali* would obtain the reefs that belonged to them. This removed any doubts that the chiefs might have had (Proceedings of the Council of Chiefs held at Nailaga, Ba, November 1881, p. 32).

However, neither Royal Command nor the official British government policy was ever implemented. There is nothing to demonstrate that anything was ever done to follow up Des Voeux's opening address statement of November 1881. The Native Lands Commission was unable to devote time and personnel to marine matters. This reneging on royal wishes and official policy is exemplified by the behavior of Thurston, Acting Governor, who in 1886 wrote to the Secretary of State for the Colonies that "[i]t has been the habit of natives of this Colony to claim as absolute and exclusive, a proprietary right in the reefs... and in some cases this has led to pretensions that could not be recognised.... It is however inconsistent with the altered conditions of the country that any exclusive rights of the nature indicated can be enjoyed by one class only of Her Majesty's subjects" (Despatch No. 24, February 17, 1886).

In 1886, Thurston also opened the bêche-de-mer fishery to non-Fijians, in the interests of the export economy and under strong

---

[1] A *mataqali* is "an agnatically related social unit – usually a lineage of the larger clan" (*yavusa*) (Ravuvu 1983:119).

pressure from the colonists. The chiefs accepted this as only a temporary measure, in that only the outer reefs would be thus opened. But in 1887 the new Governor, Mitchell, opened all reefs to bêche-de-mer fishing, in the interests of the economy (Despatch No. 87, June 13, 1887).

Further, the Rivers and Streams Ordinance 1882 was now being interpreted as meaning that the private fishing rights of Fijians in all rivers and steams were abolished and that they now belonged to the Crown (Colonial Secretary's Office 3114/1891). In addition, colonial officials held the view that there were no longer exclusive tribal fishing grounds (Colonial Secretary's Office 1304/1893).

FISHERIES AND THE COLONIAL REGIME IN PAPUA NEW GUINEA

In Papua New Guinea, the attitude of the colonial authorities toward community-based marine resource management was quite unlike the case of Fiji, New Zealand, or most other colonies in the Pacific Islands region. In Papua New Guinea the customary fishing and marine resources rights of indigenous peoples were recognized or protected by the various colonial administrations. The contemporary recognition of traditional fishing rights is based on colonial policy under which the indigenous peoples were not divested of their lands, but, on the contrary, were protected and secured by the colonial authorities. Hence, today some 97 percent of the land area is under customary tenure and customary law. Similarly, colonial sovereignty of marginal seas did not displace traditional community-based marine tenure, including fishing rights, which were statutorily protected early in the colonial era. For example, the Fisheries Ordinance of 1922 (New Guinea) provided: "This ordinance shall not apply to any native fishing in waters in which by native custom he has any right of fishing" [Section 2A].

As a result, in most traditional coastal communities of Papua New Guinea traditional fishing rights still regulate activities. Colonial policy was motivated by the need to ensure a steady supply of marine fish to the coastal population. Further, many fishers lacked access to alternative resources from which to make a living; their livelihood and incomes depended on the continued recognition of their traditional fishing rights. For example, in 1884, in the Territory of Papua, just before establishment of the British Protectorate, commercial exploitation of pearl oysters, pearl shells, trochus shells, and bêche-de-mer by expatriates had become widespread. In 1891 the administration enacted licensing and other ordinances to regulate these activities. In some areas, like the Trobriand Islands, the Pearl Shell and Bêche-de-mer Fishery Ordinance of 1894 (Papua) was used in 1903 and 1910 to close the fishery to expatriates. However, indigenous peoples, such as the Trobriand Islanders, were permitted to exploit these resources commercially (Tom'tavala 1990).

Indigenous peoples' fishing rights in waters adjacent to a landholding were explicitly recognized in 1952. Pursuant to the Pearl, Pearl-Shell and Bêche-de-mer Ordinance 1911–1932 (Papua), a proclamation was made for the "Protection of Fishing Rights in Waters Adjacent to Property." Outsiders ("any person other than the [land] owner, lessee or occupier") were prohibited from taking pearl oyster shell, trochus shell, or bêche-de-mer within 800 m of the high tide mark on the foreshore and within a sea area bounded by the seaward projection of the lateral boundaries of the landholding. The most important commercial marine fishery products were thereby reserved for the indigenous landowners. Fish, crustacea, oysters, other shellfish, and all forms of marine animal life other than whales were included in 1953.

CONCLUSIONS AND POLICY IMPLICATIONS

The example of the New Zealand Maori demonstrates how validated local knowledge can be applied to restore fisheries property rights, particularly in cases where a treaty exists, and especially at a time when resurgent ethnic pride coincides with a Western liberalist trend to right past wrongs. In particular, the example demonstrates a simple data collection and validation methodology and a culturally sensitive "business environment" that is replicable in other contexts. It is likely to be particularly effective, of course, in cases where historical documentary evidence exists, as it does usually in Western colonial archives.

The main policy implication goes far beyond local knowledge, *sensu stricto*. The imputation is that local knowledge, based on generations of praxis, demonstrates entitlement via prior established and continuous usage – "credentials of ownership" as in the case of the Torres Strait Islands (Nietschmann 1989) – and therefore a property right upheld in customary law. The prime policy issue then becomes that of accepting local knowledge *qua* customary law within the framework of Western-based legal systems.

In customary law, as exemplified by that of the Kiriwinan Islanders of the Trobriand Islands, Papua New Guinea, traditional claims are substantiated by records preserved in lore, legend, song, and dance. As observed by Williamson (1989:31–32), himself a Western-trained lawyer, "Traditions and customary usage are important in resolving disputes relating to maritime claims. Folklores, legends, songs and dances to a Kiriwinan are like principles of the English Common Law to judges in common law jurisdiction."

In the context of economic and social change, during which rights to resources increase in value, groups may attempt to obtain codification of their customary rights. This occurred in Papua New Guinea (1990). The most compelling reason for codification is to restore to local communities the authority to protect their rights. One principal reason why traditional community-based management systems have been undermined is that the quality and security of rights has been eroded: "Traditional [local] authority has been usurped, replaced by the ephemeral authority of central governments; the institutions in which [traditional management systems were] previously 'informally codified' have collapsed" (Graham 1992:36). Thus modern codification is required to reinstitute local authority to protect rights.

This linkage between the preservation of systems of traditional authority and the preservation of traditional resource management systems has been recognized in the constitutions of some Pacific Island nations, particularly in Cook Islands, Samoa, Fiji, Vanuatu, Marshall Islands, the Federated States of Micronesia, and Palau (Pulea 1993). This enables traditional leaders to influence national and lower-level marine resource management, and in particular to assist in the reconciliation of the ambiguities between statutory and customary law.

Co-management is the outcome of these approaches: national government sets rules and principles, simultaneously recognizing traditional rights and allowing local government to manage locally within this national legislative framework. It can be argued that local "title to" resources implies an obligation to manage them effectively. But this is problematical because it goes beyond fisheries legislation to include political issues of local autonomy, national policy, hereditary claims and rights, and other highly contentious factors.

Nevertheless, the reality is that in a great many nations in the Pacific Islands region, particularly in the far-flung archipelagic states where the central government lacks the capacity to management fisheries comprehensively, *de facto* co-management has long existed in practice. It remains now for legislation to formalize this, and for central governments to shoulder a larger, complementary share of the task. Practical management considerations make it likely that this trend will continue worldwide.

The New Zealand Maori case provides both a solid example of what can be achieved, given the political will, and a commonsense methodology for substantiating and upholding historical property rights claims grounded in non-Western legal concepts and systems of law. It is high time to finally right the wrong of the earlier phase of the process of globalization. In doing this it may well be that a countervailing process of "localization" assumes global proportions.

References

Derrick, R. A. 1946. *A History of Fiji*. Suva, Fiji: Government Printer.

Graham, T. 1992. The application of traditional rights-based fishing systems to contemporary problems in fisheries management: a focus on the Pacific Basin. M.S. paper, Marine Resource Management Program, College of Oceanography, Oregon State University, Corvallis, OR.

Kawhuru, I. K. 1977. *Maori Land Tenure*. Oxford, UK: Clarendon Press.

Levine, H. B. 1987. The cultural politics of Maori fishing: an anthropological perspective on the first three significant Waitangi Tribunal hearings. *Journal of the Polynesian Society* **96**: 421–443.

Nietschmann, B. 1989. Traditional sea territories, resources and rights in Torres Strait. In *A Sea of Small Boats: Customary Law and Territoriality in the World of Inshore Fishing*, Report No. 62, ed. J. C. Cordell. Cambridge, MA: Cultural Survival, pp. 60–93.

Oliver, W. H. 1991. *Claims to the Waitangi Tribunal*. Wellington: Department of Justice, Waitangi Tribunal Division.

Pulea, M. 1993. *An Overview of Constitutional and Legal Provisions Relevant to Customary Marine Tenure and Management Systems in the South Pacific*, Report No. 93/23. Honiara, Solomon Islands: Forum Fisheries Agency.

Ravuvu, A. 1983. *The Fijian Way of Life*. Suva, Fiji: Institute of Pacific Studies, University of the South Pacific.

Ruddle, K. 1994a. Local knowledge in the future management of inshore tropical marine resources and environments. *Nature and Resources* **30**: 28–37.

Ruddle, K. 1994b. Local knowledge in the folk management of fisheries and coastal marine environments. In *Folk Management in the World's Fisheries: Lessons for Modern Fisheries Management*, eds. C. L. Dyer and J. R. McGoodwin. Niwot, CO: University Press of Colorado, pp. 161–206.

Ruddle, K. 1994c. External forces and change in traditional community-based fishery management systems in the Asia–Pacific Region. *Maritime Anthropological Studies* **6**: 1–37.

Ruddle, K. 1994d. Marine tenure in the 90s. Pages 6–45. In *Traditional Marine Tenure and Sustainable Management of Marine Resources in Asia and the Pacific, Proceedings of the International Workshop*, July 4–8, 1994, eds. G. R. South, D. Goulet, S. Tuquiri and M. Church. Suva, Fiji: International Ocean Institute, South Pacific and Marine Studies Programme, University of the South Pacific.

Ruddle, K. 1994e. *A Guide to the Literature on Traditional Community-Based Fishery Management in the Asia–Pacific Tropics*, Fisheries Circular No. 869. Rome: Food and Agriculture Organization of the United Nations.

Ruddle, K. 1995. The role of validated local knowledge in the restoration of fisheries property rights: the example of the New Zealand Maori. In *Property Rights in a Social and Ecological Context, Part 2, Case Studies and Design Applications*, eds. S. Hanna and M. Munasinghe. Stockholm: Beijer International Institute of Ecological Economics; and Washington, DC.: World Bank, pp. 111-119.

Sandrey, R. A. 1987. Maori fishing rights in New Zealand: an economic perspective. *Proceedings of the International Conference on Fisheries*, August 10-15, 1986, Université du Quebec à Rimouski, Rimouski, Canada, pp. 499-503.

Taylor, A. and Patrick, M. 1987. Looking at water through different eyes: the Maori perspective. *Soil and Water* **Summer**: 22-24.

Tom'tavala, D. Y. 1990. National law, international law and traditional marine claims: a case study of the Trobriand Islands, Papua New Guinea. M.A. thesis, Department of Law, Dalhousie University, Halifax, Nova Scotia.

Waitangi Tribunal. 1988. *Muriwhenua Fishing Report (WAI 22)*. Wellington: Department of Justice.

Waitangi Tribunal. 1992. *Ngai Tahu Sea Fisheries Report (WAI 27)*. Wellington: Brooker and Friend.

Wards, I. 1968. *The Shadow of the Land: A Study of British Policy and Racial Conflict in New Zealand*. Wellington: Government Printer.

Williamson, H. R. 1989. Conflicting claims to the gardens of the sea: the traditional ownership of resources in the Trobriand Islands of Papua New Guinea. *Melanesian Law Journal* **17**: 26-42.

9

Cooperation and conflict between large- and small-scale fisheries: a Southeast Asian example

INTRODUCTION

As globalization has affected the market for marine products, increasing demand and prices have induced entrepreneurs to invest in more expensive, larger-scale fishing operations. These differences in scale range along a continuum from individual fishers using unmechanized gear to large factory trawlers employing tens of fishers, processors, and others. Many of the larger-scale operations coexist with small-scale operators, sometimes harvesting the same species in the same area. Differences in scale have existed throughout the history of fishery development (Thompson 1983; Sider 2003), but they have become more pronounced today, especially in developing economies.

Most discussions involving interactions between these large-scale and small-scale fishers focus on competition and conflict between the two (e.g., Bailey 1987a; McGoodwin 1990; Payne 2000),[1] as well as impacts on increases in social stratification and inequity (Bailey 1984). For example, McGoodwin (1990:18) notes "many small scale fishers now find themselves increasingly losing competitive struggles with industrialized fishers from urban ports in their own country, or with fishers who have come from distant lands." There is ample evidence to support the perspective that conflicts between large- and small-scale fishers result from the perception that the large catches and gears of the former reduce the number of fish available to small-scale, coastal fishers (e.g., Bailey *et al.* 1987; CEP 1989; Masalu 2000; Johnson 2002). In some cases, large-scale operations have resulted in destruction of

---

[1] These are categorized as type III conflicts in Bennett *et al.* (2001). In their survey of 62 villages in Ghana, ten in Bangladesh, and three of the six inhabited Turks and Caicos Islands, issues between different scales of fishers were mentioned only in Ghana.

Globalization: Effects on Fisheries Resources, ed. William W. Taylor, Michael G. Schechter, and Lois G. Wolfson. Published by Cambridge University Press. © Cambridge University Press 2007.

the small-scale fishers' gear (Bailey 1987b). Some small-scale fishers have placed artificial reefs and/or fish aggregating devices (FADs), which function to attract fish as well as both mark and "claim" their traditional fishing grounds and deter trawls, which could become entangled in the gear (Guillen et al. 1994; Jahara Yahaya 1994; Pramokchutima and Vadhanakul 1994). In some cases, the conflicts have led to violence, as in Indonesia, the Philippines and Malaysia (Bailey 1987b), and India (Bavinck 2000).

Today, with increasing numbers of reports on the potential collapse of the world's fisheries (e.g., Pauly et al. 1998), it is easy to reduce this conflict to a struggle between the small-therefore-good and large-therefore-bad industrial fisheries. To some, the big industrial trawlers, indiscriminately scouring the floor of the ocean, and taking tonnes of marine organisms, are the "bad guys" while the small-scale fishers in their sail- or small-motor-powered boats are the "good guys" having little impact on their prey (Kurian 2002).[2] This dichotomy into "bad guys" and "good guys" focuses attention on big versus little and neglects the fact that there is a continuum connecting the two, and that in some cases the little guys admire the big guys and aspire to become big guys themselves. It also neglects the fact that there are some cases in which there is a symbiotic relationship between the two, where one helps the other and vice versa. In cases such as these, management efforts directed at the big, bad guy may in fact hurt the little guy, a point we will return to in the conclusion.

LESSONS FROM THE VILLAGES

Though not directly researching interrelationships between large- and small-scale fishers, the author, nevertheless, made observations concerning these relationships while investigating other aspects of human adaptations to the coast. For example, during an assessment of small-scale fisher villages along the coast of Ecuador in the mid 1980s, fishers in some communities noted that they bought bycatch from shrimp trawlers operating in the coastal waters. This bycatch was composed of fish species that were sold in the local market. The small-scale fishers bought the fish at a relatively low price and were able to make a profit

---

[2] This oversimplifies Kurian's well-reasoned arguments, which go beyond the large- and small-scale fishery debate. Nevertheless, it forms part of his argument. For a viewpoint that recognizes that small-scale fishers can also harm the environment see McGoodwin (2002).

by selling them at a higher price in the local market. Similar interactions have been observed among small- and large-scale fishers along the coast of West Africa where industrial vessels in Senegal and Nigeria sometimes sell bycatch at a low price to canoe fishermen, who bring it to shore for traditional processing and distribution (Haakonsen 1992). Haakonsen also points out that small-scale fishers sometimes use industrial shore-side installations such as freezing plants.

Interactions, either positive or negative, between fishers of different scales were not brought up as being significant by community members in any of the villages where the author worked until the mid 1990s. While conducting a brief assessment of the human ecology of the coral reefs in Atulayan Bay, the Philippines, the author learned that small scale hook and line fishers would fish around FADs deployed just outside the bay by larger-scale ring-net (*kalansisi*) fishers (Gorospe and Pollnac 1997). The large-scale fishers requested only that the small-scale fishers report when sufficient aggregations of target fish appeared. As a result of this relationship, the large-scale fishers did not waste time and fuel going to an FAD that had insufficient target fish to justify deployment of a net, and the small-scale fishers could hook and line catch other fish that aggregated at various levels around the FAD. This symbiotic relationship between small- and large-scale fishers with regard to FADs had also been reported previously in the literature (Jahara Yahaya 1994; Pollnac and Poggie 1997). What had not been reported, however, was the fact that local small-scale fishers sometimes deployed FADs, allowing other small-scale fishers to fish around them for free unless they caught a yellowfin tuna, in which case they would pay the FAD owner 10 pesos. If a ring-net fisher fished the FAD, however, he was required to give the owner one-third of the catch. Hence, the local, small-scale fisher who did not have capital to invest in ring-net gear and boat could take advantage of a local resource with minimal investment. In turn, the ring-netter had access to another FAD where he could profitably set his net. This does not mean that there were no conflicts between Atulayan Bay small-scale fishers and the larger-scale ring-netters. Many objected strongly to deployment of ring-nets within Atulayan Bay, which was closed to this type of gear by both local and national laws (Gorospe and Pollnac 1997).

More interaction between large- and small-scale fishers was noted during a rapid assessment of coastal villages in the Minahasa Regency, North Sulawesi, Indonesia, in early 1997 (see Fig. 9.1 for locations of villages in Minahasa). During interviews in villages along the Sulawesi

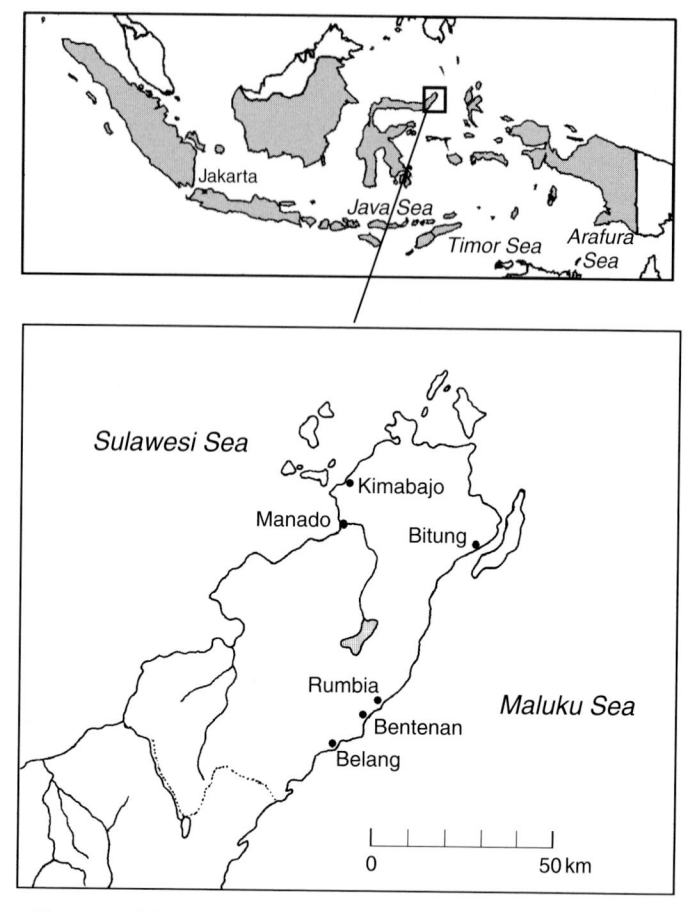

Figure 9.1 Minahasa Regency, North Sulawesi, Indonesia, with places discussed in text.

Sea coast, small-scale hook-and-line fishers noted that they used FADs placed by large-scale fishers in the offshore waters. In the recent past the FADs were installed by fishers from the Philippines, but when the interviews were conducted, they belonged to Indonesian fishers. Small-scale fishers were allowed to fish for free around the FADs as long as they did not use nets. This situation was observed again in late 2002. Complaints concerning the large-scale fishers were voiced in only one village along this coastline. In Kimabajo, fishers noted that before the offshore FADs were placed, the pelagic fish (tunas) would come into their bay to feed. They claimed that numbers were so large that the fishers could take fish from the water with their bare hands. After the placement of FADs, they reported that only a few pelagics come to feed in the bay. They complain that the now essential trips to the offshore FADs are dangerous, sometimes resulting in fishers being temporarily lost at sea or being forced to land at distant places after being blown astray in a storm.

On the Maluku Sea coast of North Sulawesi, another type of relationship between the large- and small-scale fishers was observed in the village of Bentenan. It was similar to those described thus far in that small-scale fishers could freely fish with hook and line around FADs placed by large-scale fishers, but small-scale fishers became more involved in building and deploying FADs. Some also own light boats, which function as FADs. Their ownership of FADs and light boats provides them with a greater share of the fish resources near their village for minimal cost. The following section provides a detailed examination of this relationship between large- and small-scale fishers in Bentenan.

LARGE- AND SMALL-SCALE FISHER COOPERATION ALONG THE MALUKU COAST OF MINAHASA

The capture fishery in Bentenan

The capture fishery plays a significant role in the life of the people of Bentenan. The beaches are lined with fishing vessels, and some sort of fishing activity is going on at all hours of the day and night, as evidenced by the departure and arrival of boats and their activities in the inshore and offshore areas. In 1997, the occupation of fishing contributed to the income of 83 percent of the households in the coastal subvillages of Bentenan (Pollnac *et al.* 1997). By 2002, it had changed little, decreasing slightly to 75 percent of households depending on fishing for at least part of their income. Inland subvillagers concentrate on farming.

Several types of boats are used by the fishers of Bentenan. The *londe* is a beautifully carved double-outrigger dugout, with gracefully curved projections at the base of the bow and stern. The projection at the base of the bow can be anywhere from 35 to 65 cm long, 8 to 10 cm high, and 6 cm thick. The projection at the stern is shorter. *Londe* are rarely motorized. The most common vessel in Bentenan is the *pelang*. The *pelang* is also a double-outrigger dugout, but it lacks the graceful carving and projections at the base of the bow and stern that characterize the *londe*. Many *pelang*, especially the larger ones, have plank extensions to increase the depth of the dugout hull. The *pelang* also encompasses a wider range of sizes (e.g., from 2.5 to 12 or more meters in length), with the larger ones frequently motorized. Another type of boat used by the fishers of Bentenan is the *pajeko*. The term *pajeko* also refers to a mini purse-seine net. The *pajeko* is the largest fishing craft used by the fishers of Bentenan, averaging about 16 to 20 m long, 4 m wide, and 2 m deep. They are usually powered by two to three 40-horse power (hp) outboard motors. There were only three in the village in 1997. The number had increased to eight by 2002.

A wide range of gear types is used in the Bentenan fishery. Perhaps most common and most widespread is the hand line. Hand lines are usually deployed from a *londe* or *pelang* but can be deployed from any type of boat or the shoreline. Hook size and number depend on target species. In most cases some form of bait is used, but lures designed for specific target fish are also deployed. Some of these lures are carved from wood; others are made of bits of frayed, colored plastic line.

Harpoons and spearguns are also used. The *tombak* is either a two- or three-pronged barbed device attached to a 1.5 to 2-m shaft. Fish are speared from the surface, either from a boat or while the fisher is standing in the shallows. There are also spearguns (*jubi*) that are used under water by divers. The gun is carved from wood and looks and handles like a slender rifle with a trigger. The power is provided by a length of rubber cut from an inner tube. Spears are steel rods approximately 0.8 cm in diameter with a toggle barb made from a bent nail inserted through a slot cut into the spear. A notch near the base of the spear engages the trigger mechanism. Spears are of varying length (1 to 2 m), depending on the target fish. Spear-fishers usually dive from *londe* or *pelang*, but they can swim out to the reef from the shoreline. Goggles, carved from wood with glass eyepieces, are used to improve underwater vision.

Both blast fishing (using explosives to stun or kill fish and/or extract them from hiding places in reefs) and poison fishing (used to stun fish and/or extract them from the reefs) were present in the Bentenan area in 1997. There has been a remarkable decrease in these types of fishing near Bentenan. In 2002, they were considered rare (Pollnac et al. 2003).

The gill-net is also commonly used in small-scale fishing. In the Bentenan area, the general term usually applied to the gill-net is *pukat kalenda*. The gear consists of one or several pieces of monofilament nylon netting. If several pieces are used, they are sewn together to form a net that can be longer, deeper, or both. The size of each piece is related to mesh size, which is related to target species. Piece sizes range from 25 × 2.25 m to 35 × 4 m, and mesh sizes used range from 3 to

7 cm. Floats are attached to the top of the net and weights to the bottom. A piece of net 30 × 3 m requires about 250 floats (5 to 7 cm in diameter) cut from the same material used to make the soles of relatively cheap sandals (flip-flops) and 5 kg of sinkers. Large stones are used as the main weights (e.g., at each end). The gear can be operated in many ways, depending on the target species. It can be deployed without the use of a boat in the nearshore waters, over the seagrass beds or near the coral reef flats. With a boat it can be deployed next to seaweed plantings (which act like an FAD), adjacent to coral reef structures or anywhere that target fish are known to school or move about. The gear can be set at the surface, drifting or fixed, mid-water or at the sea bottom. The net can also be used actively to encircle schooling fish at or near the surface. The fishers can then frighten the fish into the net by slapping the water with sticks, oars, or their hands, or diving into the water and herding the fish while swimming around and making noise. The technique involving scaring fish into the net is often referred to as *soma paka paka*.

Purse-seines, although not as numerous as hand lines and gillnets, are important because of their larger catches and the number of people employed per unit of gear. *Soma giop*, an older form of purse-seine, is being replaced by the *pajeko* in North Sulawesi (see Mantjoro and Yamao 1995). There were only two *giop* operated by fishers in Bentenan in 1997, and three *pajeko*, which increased to eight by 2002. The nets were hauled by hand in 1997, but by 2002, five of the *pajeko* used power winches for hauling nets (Pollnac et al. 2003). Both the *pajeko* and *giop* nets have a total length of about 300 meters. The *giop*, however, is much shallower – about 20 m against the usual approximately 60-meter depth of the *pajeko*. The small number of purse-seines belies their impact. The crew for a *giop* can be between 9 and 15 (12 is ideal), and for a *pajeko* between 15 and 20, so the boats provide employment for a large number of fishers. It is significant to note that installation of power winches did not result in reduction of crew size. We will return to this observation in the discussion section below.

Giop are deployed from large, usually motorized *pelang* (around 10 to 12 m long and 1 m or more wide). The *pelang* usually goes to sea and searches for schools of target fish. Where areas of migration and schooling are known, the *pelang* will sit in the water and wait until a school appears. When one appears, the net is deployed across the line of movement and long lengths of bamboo with colored (usually white, but sometimes blue or pink) plastic streamers are shaken over and on the water to herd the school of fish into the net.

Although the *pajeko* can be set around any school of fish in water of appropriate depth, the *pajeko* of Bentenan usually fish schools of fish that have been aggregated by light boats. Light boats are *pelang* with six to ten pressure lanterns (ideally with reflectors) that go out to sea at night ahead of the *pajeko*. When a school of fish has been aggregated, the light boat signals by blinking its lights, and the *pajeko* comes to set the net around the fish. It can also set its net around schools of fish aggregated by FADs, locally referred to as either *rumpon* or *rakit*, which are deployed only during a limited time period at Bentenan. These FADs consist of a bamboo raft with an attractant made of a line with palm fronds attached, which is suspended in the water below the raft. The raft is anchored to the bottom. The anchor can be a large stone, old engine block, or other heavy object. The anchoring cable is a multi-filament synthetic (polyethylene) rope.

Besides the FADs deployed in the deeper offshore waters by *pajeko* owners, FADs are constructed and deployed by small-scale fishers from Bentenan in the nearshore waters off Bentenan up the coast to Rumbia at depths of 30 to 60 *depa* (1 *depa* = approximately 1.6 meters) during July and August. These FADs are deployed mainly for the *pajeko* fleets from the villages between Bitung and Belang (Fig. 9.1). *Pajeko* fishers from villages along this strip of coastline congregate at the FADs off Bentenan, waiting for aggregations. One resident reported that it looks like a bus stop, with 30 to 40 boats, lights blazing, waiting offshore.

Large- and small-scale fishery cooperation

For purposes of the discussion presented here, the mini-purse-seine (*pajeko*) fleet that fishes along the Maluku Sea coast of Minahasa Regency will be considered the large-scale fishery. Although it is not as "large-scale" as an industrial trawler, its cost, harvesting ability, and crew size are much greater than those associated with the small-scale fishers in this area. Hook-and-line fishers, fishing from *londe* or *pelang*, and non-*pajeko* owners from Bentenan who own light boats or FADs are the small-scale fishers.[3] Interviews conducted to determine the distribution of benefits from fishing indicated that the FAD or the light boat (whichever was used to aggregate the fish for the *pajeko*) receives a full third of the catch. Another third of the catch goes to the boat and net,

---

[3] Many *pajeko* owners also own a light boat. Some deploy their own FADs. FADs and light boats deployed by small-scale fishers, however, increase the chances there will be an aggregation of fish available when the *pajeko* are searching for fish.

and the final third is distributed among the crew.[4] Hook-and-line fishers can fish freely around these fish aggregators until a *pajeko* wants to set its net. The share of the catch apportioned to the FAD or light boat seems excessive, especially in light of the cost of a *pajeko* in contrast to the fish aggregating gear.

The cost of the *pajeko* is extremely high. In 1997, a new, top-of-the-line net reportedly cost 30 million Rupiah (Rp) ($12 000 in 1997; $1 = approximately 2500 Rp), a new boat 15 million ($6000), and three 40-hp motors between 18 million and 20 million Rp ($7200–8000). Some boats could be purchased used for less than 5 million Rp and run with two motors, one of them smaller than 40 hp, reducing the costs somewhat. In contrast, a fully equipped light boat (new *pelang* with lights and motor) reportedly cost about 5 million Rp ($2000), and the cost of a nearshore FAD, including mooring lines, is around 200 000 Rp ($80). Maintenance for the FAD includes replacement of palm frond fish attractors as they become broken by wave action. Labor collecting and replacing the palm fronds is carried out by members of the FAD owner's household. The FADs blow away in storms that begin in November and December, but one good catch from an FAD can bring a profit to the fisher who deploys the device.

The deployment of the nearshore FADs and light boats by Bentenan small-scale fishers is an interesting example of these fishers taking advantage of a periodically available local resource where local capital resources are insufficient to provide the most effective harvesting techniques. The principal target for aggregation is *ekor kuning* (yellow tail scad, *Atule mate*), which reportedly begin aggregating just off the coast of Bentenan in August. Fishers also report that *ekor kuning* are spawning and gravid at this time. The FADs also aggregate *deho* (mackerel), *cakalang* (skipjack tuna), and *malalugis* (other scad). The FADs are fished by *pajeko* from other communities between Bitung and Belang. According to local fishers, they come to Bentenan waters because there are fewer fish in the other areas at the time that fish are aggregating off Bentenan. During this time, the non-local *pajekos* are based in Bentenan, where crews may stay with and cook in friends' or relatives' houses during the day.

The fisher who owns the FAD receives one-third of the market value of the harvest. At night, the FAD owner uses a *pelang* with lights to attract fish to the FAD. The *pajeko* encircles the fish around the FAD,

---

[4] This is a generalization of the share system, which varies in its details. For details see Pollnac *et al.* (1997, 2003).

day or night, after obtaining permission or confirmation from the FAD's owner. If the owner does not attend the FAD, the *pajeko* crew must find him and obtain permission. Catching fish around someone's FAD without permission is unlikely because one's reputation can be easily ruined when word of this transgression spreads along the coastal community of fishers. This system is based, in part, on mutual trust between the fishers of Bentenan and the outside fishers who use their FADs. Hand-line fishers are allowed to catch the fish around FADs day or night, even if the owner is not present. Hand-line fishers are not required to share the catch with the owner. If the FAD owner cannot go to the sea to aggregate fish around his FAD with a light boat, other fishers may make a deal to use their light boats to aggregate fish and offer them to a *pajeko*. In this case, the FAD owner receives one-half of the one-third share obtained from the *pajeko*. A boat cannot provide the share at the time of capture. It must first be taken to market and sold, the costs deducted, and then the shares calculated. The fisher from Bentenan watches the *pajeko* sail back to its home port (or some other market) and trusts that the owner will send him his share. The owner can go with them to witness the sale and receive the payment, but this is not necessary. It was reported that they always pay the share – that they would not be able to return to Bentenan if they did not.

DISCUSSION AND IMPLICATIONS FOR MANAGEMENT

Granting permission for small-scale hook-and-line fishers to fish around FADs in exchange for information on fish aggregations was reported in several locales. Large-scale fishers waste less time and fuel if they go to the FAD only when fish are available, and the only other way they could acquire this information is to post a lookout at the FAD.[5] The small-scale fishers also keep a lookout for unauthorized fishers, another task that would require posting a guard. Hence, this type of cooperation between large- and small-scale fishers makes economic sense for all parties. The relatively large share of the *pajeko* catch that goes to the FAD or light boat, however, seems out of proportion to the relative costs of the respective gear.

Initially, when queries were made concerning this disproportionate share, fishers replied that it takes a lot of skill to locate and aggregate

---

[5] There are modern FADs with fish finders that radio information back to the fishers, either on their boats or at their home port. This modern type of FAD was not observed off the coast of North Sulawesi.

fish, then move the light boat and fish into the proper position for setting the purse-seine. This interpretation suggests that the light boat owner/operator is being reimbursed, at least in part, for a needed skill. When lights are used to attract fish to an FAD at night, this explanation also seems to hold. But what about cases where the fish simply aggregate around the FAD without the use of lights, as takes place during daylight with the fish sometimes remaining at the device after sunset? The FAD's share is still one-third. Perhaps some other sociocultural factors, such as notions of fairness, are at work here.

Scott (1976) and others (e.g., Bailey 1983, 1991) have referred to an economy based on cultural values of generosity and fairness as a "moral economy." Bailey (1991:20) writes:

> The existence of a moral economy assumes: (1) the existence of commonly accepted cultural values of fairness and generosity, and that (2) these values are translated into behavioral norms governing asymmetrical relationships (e.g., patron–client, landowner–tenant, or rich–poor). Subsistence rights – the right to survive – are "socially experienced as a pattern of moral rights and expectations" (Scott 1976:6) enforced by informal social sanctions at the community level.

This concept provides a fairly good explanation for observed behavior between *pajeko* owners and the small-scale fishers and other community members in Bentenan. In the coastal subvillages of Bentenan, the asymmetrical relationships between the large- and small-scale fishers, owners and crew members, and between the rich and the poor are ameliorated in many ways that reduce potential sources of conflict. For example, the arrival and unloading of a *pajeko* is a community event marked by sharing of the catch not only with crew but also with the numerous community members who come to the landing and provide some assistance. The beach is crowded with villagers. The atmosphere is festive. Adults and children help unload the vessel, and help carry the boxes of fish to shore, across the beach, and onto the waiting transport vehicles. In exchange they receive a fish or two. The boat's crew could accomplish the transfer of fish, but the existing norm of behavior spreads the bounty of the catch widely throughout the village. Further evidence of this norm is provided by the fact that, after introduction of the power winch, crew sizes remained constant even though large crews were no longer necessary to retrieve the net. Established relationships between vessel owners and crew members were too strong, too bound by notions of fairness, to succumb to Western notions of economic efficiency.[6] It also seems that we can use the concept of the moral economy to understand the seemingly economically indefensible large share of the catch that is allocated to the light boat or FAD owner.

The same share system also holds for vessels from other communities along the Maluku Sea coast of Minahasa Regency when they use Bentenan FADs and light boats. For the most part, the immediate coastline of Minahasa Regency differs socially and culturally from the adjacent inland areas. In many villages, coastal residents are descendants of Islamic immigrants from other parts of North Sulawesi and elsewhere. Ethnically, many identify themselves as Bajo, Bolaangmongondo, Gorontalo, Sangir, and other coastal ethnic groups, in contrast to the non-coastal people who identify themselves as Minahasan and Christian. In Bentenan, most of the Islamic fishers live within 50 to 100 meters of the coast. Just beyond that line we find Christian Minahasans. These coastal dwellers have closer contacts with other fishers in villages along the coast than they do with the Christian farmers who live in their own village. Some fish off the coastlines of these other villages, and fishers from elsewhere along the coast fish the waters, moor their boats along the shore, sleep in friends' or relatives' houses, and pray in the mosque in Bentenan. In a sense, the fishers of Bentenan belong to a larger coastal "village" or community of fishers that stretches all along the Maluku Sea coast of Minahasa and perhaps farther. Notions of fairness in the moral economy extend throughout this extended coastal community; hence, the extension of the share system for light boats and FADs to boats from other parts of Maluku Sea coast of Minahasa.

Now, what does all this have to do with fishery management? First, the belief that large-scale fishers are pillaging stocks fished by small-scale fishers, thereby lowering the little guy's catch and income, may be either true or false, depending on the location and the fishery. Today, with many of the world's fisheries overexploited, there is a widely held belief among the environmentally conscious that "small is beautiful," as popularized by Schumacher (1973) over a quarter of a century ago. This positive evaluation of the relatively small, local fisher in contrast to the larger, industrialized fisher is implicit in the writings of many observers of the fishery (e.g., Cordell 1983; McGoodwin 1990; Binkley 2002; Playfair 2003). Hence, there is a tendency to accept the negative perceptions of large-scale fishers and target them for control

---

[6] See Pollnac (1982) for a similar response to the introduction of power winches into a Malaysian fishery.

to help the smaller, more environmentally friendly fishers. In reality, as presented in this chapter, there are situations where large- and small-scale fishers are interdependent, and attempts to reduce or stop the large-scale activities will inadvertently have a negative impact on the small-scale fishers. It is therefore important to investigate the relationships between various types of fishers to understand fully the potential impacts of management changes. Because these types of relationships can vary from place to place and time to time, it will be necessary to conduct rapid assessments across a range of fishing communities in the targeted areas.

It is also important to analyze these relationships with an open mind. Some tend to view intracommunity social stratification as a situation that invariably leads to tension and conflict. Notions of fairness and generosity that characterize the "moral economy" as discussed by Scott (1976) and Bailey (1991) reduce the chances for such conflict. The "moral economy" also results in economic relations of a type not easily analyzed using traditional Western economic theory. Hence, cost–benefit analyses of proposed management schemes may not reflect the reality of the villages involved. Failure to take into account these various relationships can result in unanticipated impacts that could influence community reactions to management efforts. If the community of fishers views the management effort as somehow unfair, it will be difficult to obtain their cooperation, and such cooperation is necessary to manage a fishery effectively.

References

Bailey, C. 1983. *The Sociology of Production in Rural Malay Society*. Kuala Lumpur: Oxford University Press.

Bailey, C. 1984. Managing an open access resource. In *People Centered Development: Contributions toward Theory and Planning Frameworks*, eds. D. C. Korten and R. Klauss. West Hartford, CT: Kumarian Press, pp. 97–103.

Bailey, C. 1987a. Social consequences of excess fishing effort. In *Proceedings, Symposium on the Exploitation and Management of Marine Fishery Resources in Southeast Asia*, Darwin, Australia, February 16–19. Bangkok: UNFAO Regional Office for Asia and the Pacific, pp. 170–181.

Bailey, C. 1987b. Government protection of traditional resource use rights: the case of Indonesian fisheries. In *Community Management: Asian Experience and Perspective*, ed. D. C. Korten. West Hartford, CT: Kumarian Press, pp. 292–308.

Bailey, C. 1991. Social relations of production in rural Malay society: comparative case studies of rice farming, rubber tapping, and fishing communities. In *Small-Scale Fishery Development: Sociocultural Perspectives*, eds. J. J. Poggie and R. B. Pollnac. Kingston, RI: International Center for Marine Resource Development, University of Rhode Island, pp. 19–41.

Bailey, C., Dwiponggo, A., and Marahudin, F. 1987. *Indonesian Marine Capture Fisheries*, ICLARM Studies and Reviews No. 10. Manila: International Center for Living Aquatic Resources Management.

Bavinck, M. 2000. The twilight zone. *Samudra* **27**: 16–19.

Bennett, E., Neiland, A., Anang, E., *et al.* 2001. Towards a better understanding of conflict management in tropical fisheries: evidence from Ghana, Bangladesh and the Caribbean. *Marine Policy* **25**: 365–376.

Binkley, M. 2002. *Set Adrift: Fishing Families*. Toronto, Ontario: University of Toronto Press.

CEP (Caribbean Environment Programme). 1989. *Regional Overview of Environmental Problems and Priorities Affecting the Coastal and Marine Resources of the Wider Caribbean*, Caribbean Environment Programme Technical Report No. 2. Kingston, Jamaica: United Nations Environmental Program.

Cordell, J. (ed.) 1983. *A Sea of Small Boats*. Cambridge, MA: Cultural Survival.

Gorospe, M. L. G. and Pollnac, R. B. 1997. The Tabao of Atulayan: communal use of private property in the Philippines. In *Fish Aggregating Devices in Developing Countries: Problems and Perspectives*, eds. R. B. Pollnac and J. J. Poggie. Kingston, RI: International Center for Marine Resource Development, pp. 13–26.

Guillen, J. E., Ramos, A., Martinez, L., and Sanchez Lizaso, J. L. 1994. Anti-trawling reefs and the protection of *Posidonia oceanica* (L.) Delile Meadows in the Western Mediterranean Sea: demand and aims. *Bulletin of Marine Science* **55**: 645–650.

Haakonsen, J. M. 1992. Industrial vs. artisanal fisheries in West Africa: the lessons to be learnt. In *Fishing for Development: Small-Scale Fisheries in Africa*, eds. I. Tvedten and B. Hersoug. Uppsala, Sweden: Nordiska Afrikainstitutet, pp. 34–53.

Jahara Yahaya. 1994. Fish aggregating devices (FADs) and community based fisheries management in Malaysia. In *Coastal Fisheries Management, Proceedings of the IPFC Symposium*. Bangkok: UNFAO Regional Office for Asia and the Pacific.

Johnson, D. 2002. Fishy comparisons or valid comparisons? Reflections on a comparative approach to the current global fisheries malaise, with reference to Indian and Canadian cases. *Maritime Studies* **1**: 103–121.

Kurian, J. 2002. People and the sea: a "tropical-majority" world perspective. *Maritime Studies* **1**: 9–26.

Mantjoro, E. and Yamao, M. 1995. Fish marketing systems in North Sulawesi: the development of commercial fisheries and its impact on distribution of fisheries produce. *Bulletin of Fisheries Economy* **38**: 101–117.

Masalu, D. C. P. 2000. Coastal and marine resource use conflicts and sustainable development in Tanzania. *Ocean and Coastal Management* **43**: 475–494.

McGoodwin, J. R. 1990. *Crisis in the World's Fisheries: People, Problems, and Politics*. Stanford, CA: Stanford University Press.

McGoodwin, J. R. 2002. Better yet, a global perspective? Reflections and commentary on John Kurian's essay. *Maritime Studies* **1**: 31–42.

Payne, I. 2000. *The Changing Role of Fisheries in Development Policy*, Natural Resource Perspectives No. 59. London: Overseas Development Institute.

Pauly, D., Christensen, V., Dalsgaard, J., Froese, R., and Torres, F., Jr. 1998. Fishing down marine food webs. *Science* **279**: 860–863.

Playfair, S. R. 2003. *Vanishing Species: Saving the Fish, Sacrificing the Fisherman*. Lebanon, NH: University Press of New England.

Pollnac, R. B. 1982. Sociocultural aspects of technological and institutional change among small-scale fishermen. In *Modernization and Marine Fisheries*

Policy, eds. J. R. Maiolo and M. K. Orbach. Ann Arbor, MI: Ann Arbor Science Publishers, pp. 225–247.

Pollnac, R. B. and Poggie, J. J. (eds.) 1997. *Fish Aggregating Devices in Developing Countries: Problems and Perspectives*. Kingston, RI: International Center for Marine Resource Development.

Pollnac, R. B., Sondita, R., Crawford, B., *et al*. 1997. *Socioeconomic Aspects of Resource Use in Bentenan and Tumbak*. Narragansett, RI: Coastal Resources Center, University of Rhode Island.

Pollnac, R. B., Crawford, B. R., and Rotinsulu, C. 2003. *A Final Assessment of the Coastal Resources Management Project Community-Based Sites of Talise, Blongko, Bentenan and Tumbak in the District of Minahasa, North Sulawesi Province, Indonesia*, Technical Report No. TE-03/02-E. Narragansett, RI: Coastal Resources Center, University of Rhode Island.

Pramokchutima, S. and Vadhanakul, S. 1994. The use of artificial reefs as a tool for fisheries management in Thailand. In *Proceedings, Symposium on the Exploitation and Management of Marine Fisheries Resources in Southeast Asia*. Bangkok: UNFAO Regional Office for Asia and the Pacific, pp. 442–448.

Scott, J. C. 1976. *The Moral Economy of the Peasant: Rebellion and Subsistence in Southeast Asia*. New Haven, CT: Yale University Press.

Schumacher, E. F. 1973. *Small Is Beautiful: Economics as if People Mattered*. New York: Harper & Row.

Sider, G. 2003. *Between History and Tomorrow: Making and Breaking Everyday Life in Rural Newfoundland*. Petersborough, Ontario: Broadview Press.

Thompson, P. R. 1983. *Living the Fishing*. London: Routledge & Kegan Paul.

ROSAMOND NAYLOR, JOSH EAGLE, AND WHITNEY SMITH

10

Response of Alaskan fishermen to aquaculture and the salmon crisis

INTRODUCTION

The rapid rise of salmon netpen aquaculture (referred to as "aquaculture" or "salmon farming" in this chapter)[1] has transformed global salmon markets. Since 1990, global farm salmon production has increased five-fold, and farms recently surpassed commercial fisheries as the largest source of marketed salmon (FAO 2003). Global salmon output, including fishery catch, has grown from less than 800 000 tonnes to more than 2 million tonnes during the past 15 years. Virtually all of the increase has come from farms. The global aquaculture industry, including salmon aquaculture, currently contributes over one-third of total world fish supplies (FAO 2003).

In this chapter we examine the growth in global farm salmon production and its economic consequences for fishermen in Alaska, where salmon netpen aquaculture is prohibited. Featured in this paper are the results of a survey of Alaskan salmon fishermen that we conducted in 2002–03. The survey results illustrate the economic impacts of the aquaculture industry on individual fishermen, the fishermen's adjustments to changing economic conditions in the fishery, and their views on the causes and possible solutions to the current "crisis." We also describe how policy in Alaska has influenced the efficiency of fishing activities and discuss how the survey results might be used to inform the political debate on restructuring the state's salmon fishery.

[1] We use the terms "aquaculture" and "salmon farming" to refer to netpen production as opposed to hatchery production of salmon. Hatchery operations for salmon are widespread in Alaska but are not the focus of this chapter.

Globalization: Effects on Fisheries Resources, ed. William W. Taylor, Michael G. Schechter, and Lois G. Wolfson. Published by Cambridge University Press. © Cambridge University Press 2007.

BACKGROUND

Salmon netpen farming originated in Norway in the early 1970s and expanded into Scotland, Japan, Chile, Canada, and the United States in the 1980s (Anderson 1997). Between 1980 and 1987, salmon aquaculture production increased worldwide by 1300 percent, and production spread into new countries, including Ireland, New Zealand, Australia, and the Faroe Islands (FAO 2003). By 1988, aquaculture production dominated the fresh and frozen salmon market in Europe, and U.S. imports of farm salmon accelerated (Anderson 1997; Sylvia et al. 2000). Despite strong salmon runs in Alaska – many of which are supplemented by hatcheries – the state's contribution to the global salmon market declined from 40 to 50 percent in the early 1980s to 17 percent in 2001 (Knapp 2002) (Fig. 10.1). Alaska's declining share in total production is due in large part to increased netpen production of salmon worldwide.

The salmon aquaculture industry is dominated by a small group of multinational companies, mostly from Europe, which distribute consistent salmon products, such as fresh fillets and steaks, to global markets year round (Naylor et al. 2003; Eagle et al. 2004). The industry has thrived with the globalization of the world economy. In particular, it has benefited from the rapid expansion in seafood trade; the decreased cost of transporting fresh products around the world; more information, via the Internet, on fish stocks and markets; a strong

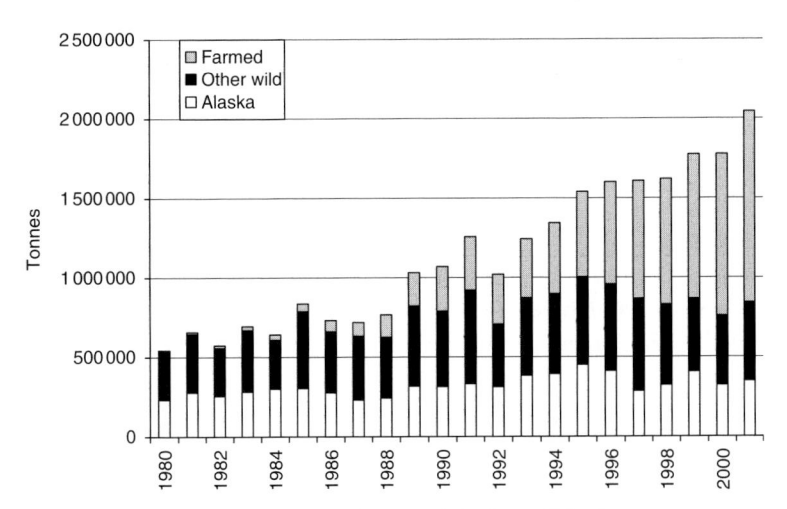

Figure 10.1 Change in contribution of Alaskan wild salmon capture to global farm and wild salmon supplies, 1980–2001. (*Source*: Knapp 2002.)

market demand for homogeneous, made-to-order products; and Web-based, business-to-business interactions (Knapp 2002). What has emerged is an industry dominated by a half dozen multinational firms, most of which produce a diversity of aquaculture and agriculture products. The salmon fishing industry is made up of many small businesses that operate at arm's length from processing corporations, but the farming industry is made up of companies with corporate affiliations. It is typical for an aquaculture multinational to have subsidiaries that include feed, hatchery, grow-out, distribution, and value-added processing companies. Most of the aquaculture multinationals are also involved in the farm production of other species, including trout, halibut, cod, turbot, bluefin tuna, sturgeon (for caviar), and sea bream. The diversity of activities and production locations provides some buffering during sectoral downturns.

Market competition

Fishermen now operate in a changed economic and political environment, with farm salmon outcompeting fishery salmon in the marketplace (Eagle *et al.* 2004). Global markets favor consistency and predictability of production (Knapp 2002). Salmon farmers have far greater control over the timing, consistency, and quantity of production than do fishermen. The fishing industry is limited to catching salmon that are migrating back to spawning rivers between June and September, and these fish can be caught only during short regulatory "openings." Catches are unpredictable over the long term because run sizes vary from year to year. Run sizes are determined by a host of factors, such as the life cycle of individual salmon populations, fishing effort, climatic and habitat conditions, and ecological factors such as food web dynamics and disease (Miller and Fluharty 1992).[2] When fish are caught during openings, they arrive in pulses by the millions, in varied condition, on the docks of processing plants. The fish must be processed as quickly as possible (the "sell it or smell it" doctrine) before the next load arrives. It is for this reason that a significant share of Alaska salmon is still canned, particularly pink salmon (ADFG 2003a). For other species, the largest share of production is headed, gutted,

---

[2] Climatic shifts (e.g., the Pacific Decadal Oscillation) create long-term unpredictability in the size of fish stocks; for example, salmon catches in Alaska varied fivefold (from 30 million to 150 million fish) in a 25-year period preceding the 1990s. During the 1990s, average catch in Alaska was over 175 million fish (ADFG 2002).

frozen, and then shipped to distant plants for further processing into fillets and steaks. Only a relatively small share of wild salmon caught in Alaska is sold fresh.

The aquaculture industry, on the other hand, can produce a consistent quality of fresh salmon – specified to order by size and cut – at any time during the year. Each salmon farming company stocks a calculated number of smolts in netpens on the basis of an estimate of market conditions 2 years hence, when the fish will be ready for market. Although actual production on any given farm may be affected by a number of factors, such as disease, storms, and marine mammal predation, the operation of multiple sites in various countries generally results in an even and predictable flow of production worldwide.

A drop in salmon prices

Prices for both farm and fishery salmon have fallen in line with the growth in global salmon supplies fueled by farms and hatcheries (Table 10.1).[3] Prices of farm Atlantic salmon have dropped 61 percent between 1988 and 2002. Prices of the five species of salmon caught in Alaska's commercial fisheries dropped 54 to 92 percent from 1988 (an exceptionally high price year) and 2002, and 36 to 82 percent from a price averaged over the 1984 to 1992 period to the 2002 price (ADFG 2003b). As a consequence, the ex-vessel value of the Alaskan salmon fisheries has declined from more than $700 million in 1988 to just over $200 million in 2002. Asset values have also plummeted (Table 10.2). The sale price for limited-entry salmon permits in Bristol Bay fell by 76 to 90 percent between 1993 and 2002, and prices for limited-entry permits in some other lucrative salmon fisheries declined by roughly 50 percent or more during the same period.[4] Despite these sharp

---

[3] An estimated 4.4 billion salmon fry were released by hatcheries from Alaska, Japan, Russia, and Canada in 2001, adding to global salmon supplies. Despite extremely low survival, hatchery fish currently account for one-third of total salmon capture in Alaska (averaged across all species) and virtually all chum capture in Japan (Goldburg and Naylor 2005).

[4] For example, average selling prices declined from $273 000 to $20 000 for Prince William Sound purse-seine permits, from $216 000 to $20 000 for Bristol Bay drift gill-net permits, and from $110 000 to $23 000 for southeast Alaska purse-seine permits between 1990 and 2002 (see CFEC 2003).

Table 10.1 *Change in ex-vessel prices for Alaskan salmon, 1984–2002 (U.S. dollars, nominal)*

	Ex-vessel price per pound			Percent change in price at 2002	
Species	1984–1992 average	1988[a]	2002	From 1984–1992 average	From 1988
Chinook	1.93	2.69	1.23	−36	−54
Chum	0.45	0.86	0.16	−64	−81
Coho	1.02	1.72	0.37	−64	−78
Pink	0.34	0.79	0.06	−82	−92
Sockeye	1.33	2.37	0.55	−59	−77
Farm Atlantic	NA	3.11	1.21	−61	

[a] 1988 was a peak price year.
Source: Alaska Department Fish and Game (2002). Ex-vessel prices are the prices fishermen receive at the dock.

Table 10.2 *Change in permit values for the selected Alaskan salmon fisheries, 1993–2002 (U.S. dollars, nominal)*

	Average sale price		Percent change in price from 1993 to 2002
Permit	1993	2002	
Bristol Bay set gill-net	49 100	11 900	−76
Bristol Bay drift gill-net	199 600	19 700	−90
Southeast drift gill-net	82 200	27 900	−66
Kodiak set gill-net	111 900	56 800	−49
Lower Yukon set gill-net	31 400	12 700	−60

Source: Commercial Fisheries Entry Commission (2003).

economic declines, however, Alaskan salmon catches have remained essentially level (Fig. 10.1).

Salmon price declines have had a severe impact on rural incomes in Alaska. At current levels of production, each 10 cent per pound decline in salmon prices translates to $66 million in lost income for Alaskan fishermen (Eagle et al. 2004). Moreover, since revenues from commercial salmon fishing finance subsistence fishing and hunting activities of many rural Alaskans, the scale of these activities has declined in some areas. A large number of people who have become dependent on the salmon industry live in isolated areas where other employment opportunities are not readily available. It has been estimated that commercial fisheries provided about 20 000 jobs in fishing and processing and another 15 000 related jobs in Alaska at the turn of the century (Colt 2001). Salmon fisheries represent roughly 50 percent of total direct and indirect statewide employment in the commercial fishing industry (DCED 1997).

Response by fishermen

The change in market conditions raises a number of questions related to capture fisheries. Is the growing aquaculture industry putting fishermen out of business? How much have fishermen's incomes decreased? How are fishermen adjusting to decreased prices and incomes? Will fishermen, a set of fiercely independent individuals, cooperate to find solutions for their ailing industry? These questions are frequently debated, often without reference to empirical evidence. Answers to these questions are important for understanding the economic behavior of fishing communities and its implications for proposed policies and solutions.

To observe and analyze some of the local impacts of price declines on fishing communities and individuals, we conducted a survey of fishermen in Alaska and the Puget Sound who own limited-entry permits in various Alaskan salmon fisheries. The survey provides a good illustration of the attitudes and behavior of salmon fishermen in the region and shows the differences in responses by fishing location and gear type. Because the 27 salmon fisheries in Alaska vary widely in average catch, catch trends, permit prices, permit holder participation, permit participation in other fisheries, history, and fishing attitudes, our survey does not capture all of the dynamics in fishing behavior in the state. Instead, it covers a spectrum ranging from high-valued, hatchery-supplemented fisheries in southeastern Alaska, where many gear types are used, to low-valued set- and drift-net fishing in western Alaska, where hatcheries are absent and run sizes are low. Our main objectives in looking at this spectrum are to show how Alaskan permit holders are responding generally to the changed economic conditions and to illuminate some of the similarities and differences between the widely diverse salmon fisheries within the state.

SURVEY METHODS

We conducted a survey of 91 individuals holding Alaskan salmon fisheries permits between November 2002 and May 2003.[5] Interviews were conducted in person with permit holders in southeastern Alaska (Juneau, Sitka, and Petersburg), Bristol Bay (Dillingham, Naknek, and King Salmon), and the Yukon–Kuskokwim Delta (Bethel area and Quinhagak). Some additional interviews were conducted randomly with permit holders in Prince William Sound (Cordova), Kodiak Island, Cook Inlet, and Chignik. The sample for each of the three main regions (southeastern Alaska, Bristol Bay, and western Alaska) was selected randomly across gear types. Respondents volunteered their time following town fisheries meetings, at the Fish Expo in Seattle, and by invitation through personal contacts in each region. All four major gear types – troll, purse-seine, drift gill-net, set net[6] – were represented in the survey. Each interview lasted about 20 to 30 minutes, and the response rate was 100 percent (all fishermen approached for the survey agreed to participate). The interviews were conducted mainly in the off-season between November and May, when most respondents were engaged in other employment, fisheries politics, or preparation for the upcoming fishing season. The sample distribution and the survey instrument are provided in the Appendix (see pp. 265–268).

In the interviews, we asked fishermen about their views on changing market conditions for salmon, about their adjustments to the change, and about their views on policy options for the salmon fishing industry.[7] The first section of the survey contained questions about the demographic characteristics of the respondents, such as age,

---

[5] Some of the Alaskan permit holders lived in Washington during the winter but were identified with the region of their permits.

[6] Drift and set netters fish with a net that is kept vertical in the water by a combination of floats and weights. Fishermen either anchor the net near shore or allow it to float freely for a short time before recollecting it. As its name suggests, the net catches fish by their gills as they attempt to swim through it. Purse-seiners use a much larger net, of finer mesh, that is set out in a circle around schools of fish and then tightened. Trollers catch fish with multiple hooks suspended by lines from rigging on their vessels. For a more detailed description of these gear types and their operation, see McMullan (1987).

[7] As a separate but complementary effort, the Commercial Fisheries Entry Commission conducted a survey of Bristol Bay drift gill-netters in 2002 (Carlson 2002). Their purpose was to collect data needed to determine the optimum number of permits in the fishery. The survey included some questions similar to those in the survey presented here.

residence, and income sources. The next two sections included a list of questions pertaining to subsistence and commercial salmon fishing activities, participation in other fisheries, relationships with processors, and attempts to market fish directly. The final sections of the survey sought fishermen's perceptions of problems within Alaska's salmon fisheries and policies that might be implemented to help the fishing industry. The problems were ranked on a scale of 0 to 5 (0 being no problem and 5 being the most serious problem), and the distance between rankings (0 to 5) was assumed to be equal for statistical purposes. Fishermen were also asked whether they were in favor of or opposed to harvesting cooperatives, an individual quota system, government disaster relief, government buyback programs, and government funding for quality or marketing programs. Finally, fishermen were asked their opinions on salmon farming and about its effects on wild salmon stocks and the environment.

SURVEY RESULTS

Our random sample of respondents, stratified across regions, included 83 men and eight women ranging in age from 22 to 66 years. The median age was 48 years. Ten respondents lived in Washington during the off-season when they were not fishing in Alaska, and 81 respondents lived in Alaska year round. Only three of the 91 respondents had sold their permits prior to our survey. For other respondents not wanting to fish, emergency transfers to family members or friends had been arranged on a temporary basis. Because such transfers involve a complex set of rules, almost all respondents in our survey had renewed their permits. Overall, the sampling method selected for fishermen who were still in the salmon fishing business.

Salmon fishing activity

As shown in Table 10.1, prices for Pacific salmon (all five of the species caught in Alaskan fisheries) have dropped precipitously, causing fishing revenues to decline. Fishermen were asked about changes in their salmon fishing incomes. Respondents estimated, on average, that their incomes had dropped 47 percent from peak levels in the late 1980s and early 1990s to current levels in 2002–03. Almost two-thirds (61 percent) of respondents had other employment outside of fishing. Teaching, construction, and tourism-related work were the most common outside jobs. Fifty-six percent of respondents participated in other fisheries, such as crab, halibut, sablefish, and herring. Many fishermen reported that their earnings from these other fisheries were becoming increasingly important for their livelihood and, in some cases, more important than salmon. Some expressed a desire to invest more in these fisheries. These results suggest that, though salmon farming may reduce the number individuals fishing for salmon in Alaska, fishing effort may be transferred to other species of fish.

Interestingly, although 84 percent of respondents thought that the salmon fishing industry was in crisis, 97 percent of all those surveyed planned to continue salmon fishing in the future.[8] Of the 84 percent who believed that salmon fishing was in crisis, 70 percent were optimistic and thought the situation would improve with time.

Ranking of problems

Fishermen were also asked to rank five phenomena in order of their detrimental impact on the salmon fishing industry: low prices, salmon farming, overcapitalization (too many boats or excessively sophisticated gear for the size of the fishery), fisheries management (including hatcheries), and run size. Overall, respondents across the diverse set of regions and fisheries viewed low prices and salmon farming as having the greatest impact on the industry (Fig. 10.2). Averaging across all regions, respondents ranked fisheries management and run size as less significant problems. In addition, fishermen generally did not rank overcapitalization as a major problem, even in Bristol Bay, where such conditions are notorious (BBEDC 2003).

We then analyzed the survey results to see if subgroups of fishermen felt differently about the cause of the "salmon crisis." Figure 10.3 shows the variation in mean rankings for each problem among fishermen in southeastern Alaska, Bristol Bay, and the Yukon–Kuskokwim Delta. An analysis of variance showed that there is significant variation among the regions in opinions about run size ($p = 0.0002$), management ($p = 0.0015$), and low prices ($p = 0.0017$). The

---

[8] This figure represents the number of fishermen who fished with their permits in 2002 and who planned to continue fishing with their permits in 2003 and beyond. According to data collected by the Commercial Fisheries Entry Commission (CFEC 2003), the actual number of permits fished relative to the number of permits renewed for various salmon fisheries in the state varies widely but is lower overall than our survey numbers suggest.

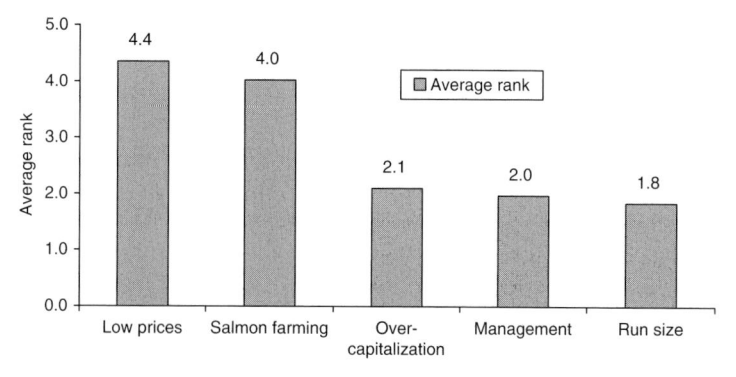

Figure 10.2 Survey results showing fishermen's perceptions of problems affecting their industry (average for all respondents in the survey). Rankings are from 0 (no problem) to 5 (very significant problem).

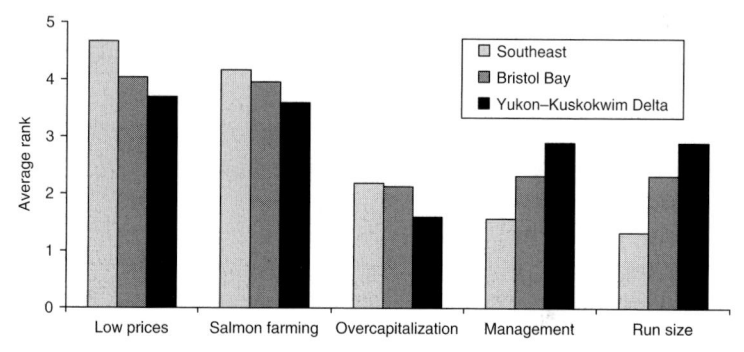

Figure 10.3 Survey results showing how members of various salmon fisheries (defined by region) perceived industry problems. Rankings are from 0 (no problem) to 5 (very significant problem).

size of salmon runs was not considered a major problem by most respondents. The exception is those in the Yukon–Kuskokwim Delta, where runs of some species, such as chum, have been low in recent years. Respondents in the Yukon–Kuskokwim Delta also placed blame on the Department of Fish and Game for poor management, while those in southeastern Alaska did not feel that management contributed to their problems. This difference might be explained by the fact that nearly all of the state-supported salmon hatcheries are in the southeastern part of the state. In other parts of the state, the hatchery program is viewed as detrimental to salmon ecology and economics. Finally, respondents in the Yukon–Kuskokwim Delta were less concerned about low prices than those in the southeast and Bristol Bay.

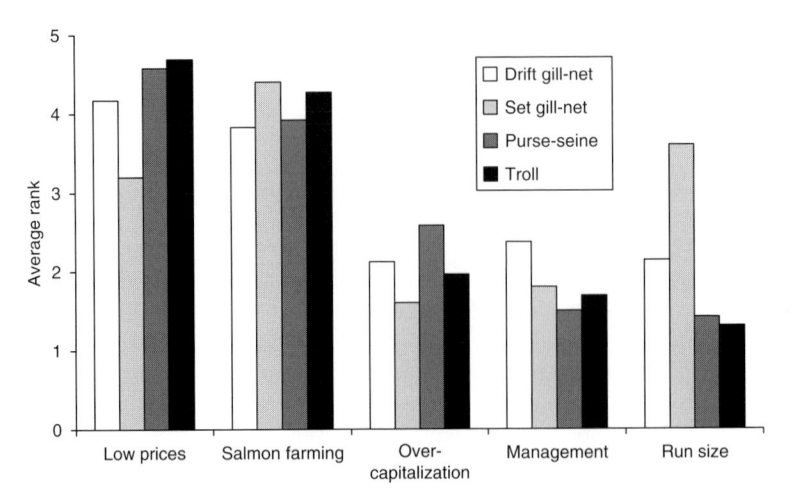

Figure 10.4 Survey results showing how members of various salmon fisheries (defined by gear type) perceived industry problems. Rankings are from 0 (no problem) to 5 (very significant problem).

Their main concern was low run size and the closure of several commercial fishing areas for much (and in some cases all) of the season.

We also used analysis of variance to see whether the type of fishing gear used by individual respondents could help explain the variation in rankings. Figure 10.4 shows mean rankings of fishermen using the four main gear types in the Alaskan salmon fishery. There was significant variation in responses between gear types for the rankings of low prices ($p = 0.0037$) and run size ($p = 0.0005$). Perhaps because theirs is a fishery with low capital requirements, set netters did not consider low prices to be as large a problem as did trollers, purse-seiners and drift gill-netters. Trollers and purse-seiners, who gave the highest rankings to low prices, have much higher fixed and variable costs than set netters and are thus more affected by fluctuations in prices.[9] On the other hand, set netters ranked low run sizes as a major problem, whereas other fishermen ranked low run size much lower on the scale. In recent years, set netters in both the Bristol Bay region and the Yukon–Kuskokwim Delta, unlike fishermen in other parts of the

---

[9] This result did not correlate well across species; i.e., permit holders in troll fisheries (catching a greater percentage of high-valued chinook and coho) and purse-seines (catching a greater percentage of lower-valued pinks and chums) both ranked low prices as a major problem.

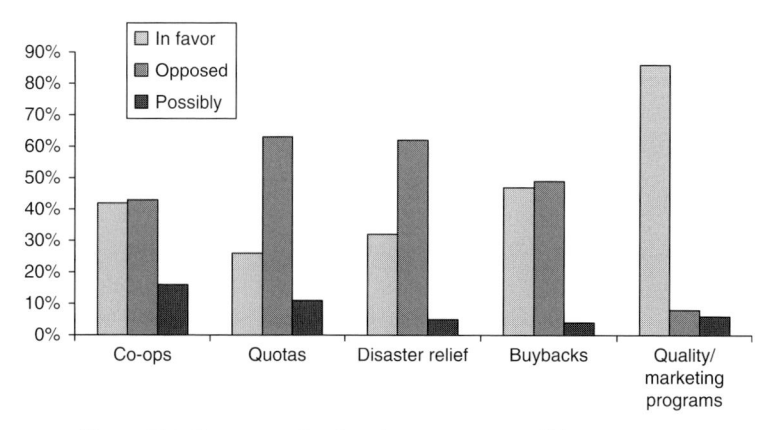

Figure 10.5 Survey results showing support, possible support, or opposition to various policy options for improving the economic condition of the Alaska salmon industry (average for all respondents in the survey).

state, have seen their fishing opportunities and catches decline dramatically. There was no significant variation in the other rankings, although responses on management showed fairly high variation ($p = 0.0599$). Drift gill-netters in overcrowded Bristol Bay gave a higher ranking to management problems than did other fishermen.

Policy preferences

When asked about policies to help the salmon fishery, 86 percent of respondents favored the development of quality and marketing programs to improve prices and build markets.[10] Despite the feeling that such efforts would help, however, only 29 percent of respondents had tried direct marketing of their catch in the past. Many respondents had contemplated direct marketing of their fish, but few followed through. Those who had taken on direct marketing and custom processing were beginning to learn the challenges and risks that processors have traditionally faced (e.g., insurance, shipping, up-front costs and market development). As shown in Fig. 10.5, views were mixed among salmon

---

[10] The Alaska Seafood Marketing Institute (ASMI) was established in 1981 to address marketing concerns of the fishing industry. Fishermen currently pay a 1 percent tax on their output to support ASMI. There is no clear indication that ASMI has been successful to date, and, therefore, most fishermen would like to see additional or improved programs in place.

permit holders on other policy options, such as cooperative fishing programs, quotas, government buyback programs, and disaster relief programs. More fishermen opposed an individual quota system than any other option.

There was some variation in attitudes toward policies among different regions and gear types. For example, 82 percent of respondents from the Yukon–Kuskokwim region were in favor of more disaster relief money from the federal or state government, whereas the majority of respondents in southeastern Alaska and Bristol Bay (67 percent and 60 percent, respectively) were opposed to it (Fig. 10.6A). Seventy-seven percent of Bristol Bay respondents expressed support for a buyback program, while only 30 to 35 percent of respondents in the other regions supported the idea (Fig. 10.6B). Again, set netters showed the most variation and difference from other gear types in the survey.

The survey also revealed – not surprisingly – that fishermen generally favor trade restrictions on farm fish and labeling laws that require the source of the fish to be identified. They recognized that Alaska's remoteness contributes to high costs and inconsistent quality. Transportation is expensive and unreliable, making it difficult to market fresh wild Alaskan salmon or even frozen products. Given the impediment of transportation costs, most respondents felt that making Alaskan salmon competitive with farm fish in the large retail market will be a challenge.

Salmon farming

In the last set of questions, we asked fishermen whether they were opposed to salmon farming and what they knew about its environmental impacts. Ninety-four percent were opposed to salmon farming, and 93 percent were aware of some ecological consequences of salmon farming.[11] Respondents mentioned escaped Atlantic salmon and disease transfer most often when asked about environmental impacts. Ninety-eight percent of respondents indicated that they believed that

---

[11] The ecological impacts from salmon aquaculture in this region include the transmission to and amplification of diseases and parasites in wild fish; the establishment and possible invasion of escaped farm fish in wild fish habitat; the release of untreated nutrients, chemicals, and pharmaceuticals from open netpens into marine ecosystems; the killing of marine mammals that prey on netpens; and food web effects associated with the use of small pelagic fish for feed (for more details, see Naylor et al. 2003).

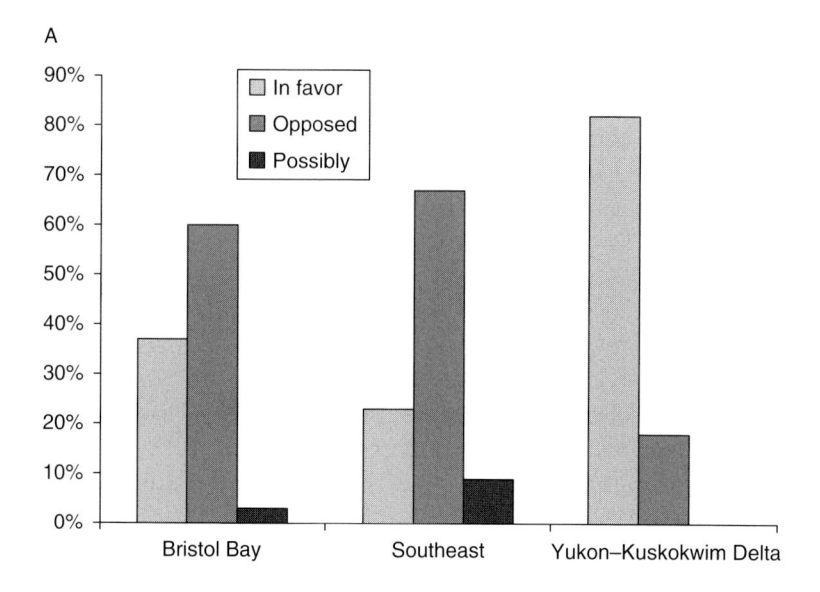

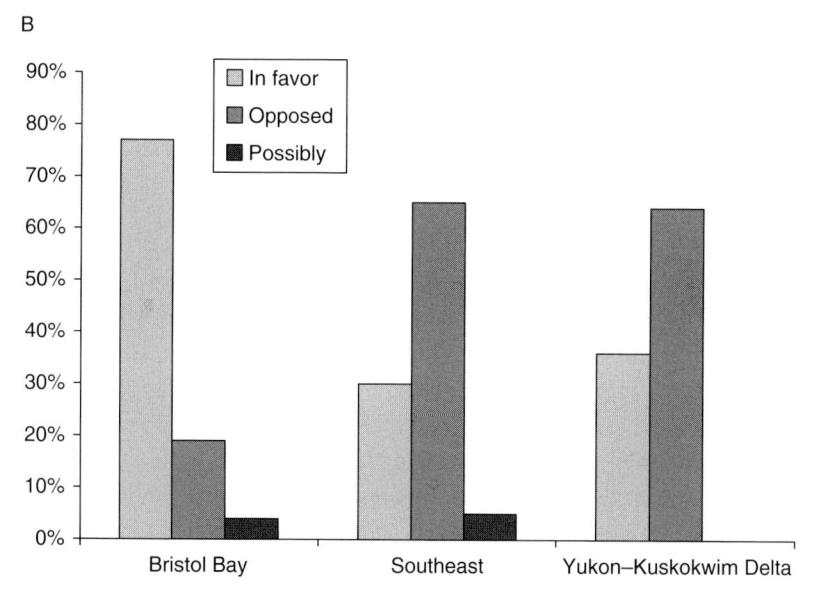

Figure 10.6 Survey results showing support, possible support, or opposition in Bristol Bay, southeastern Alaska, and the Yukon-Kuskokwim Delta to (A) disaster relief and (B) government buy backs.

salmon farming was at least somewhat responsible for the current low market prices for fishery salmon. Finally, and perhaps most surprisingly, there was also widespread acknowledgment that the aquaculture industry had increased the size of the U.S. consumer market for salmon and that, as a result, sales of wild-caught salmon were likely to rise in the future.

Several conclusions can be drawn from these results. There was widespread agreement that salmon farming had driven market prices for salmon down and reduced profits and incomes in the fishing industry. Though fishermen were aware of the relationship between farming and price declines, they also acknowledged that other factors contribute to current economic conditions. Many fishermen were making adjustments in effort, expenditures, and investments to ride out this difficult period, but almost all of the respondents planned to stay in the fishery. Alaska landings have remained fairly constant up to the present, although an increasing percentage of landings originate in hatcheries (Eagle et al. 2004). Recognizing the severity of the economic situation, fishermen appeared willing to engage in policy discussions and to consider substantial changes in the way salmon fishing is regulated.

DISCUSSION

The most severe problem facing the major salmon fisheries, according to fishermen in our survey, is the low price they receive for their fish. It is difficult for fishermen to compete with aquaculture producers, particularly because farm fish are produced year round and provide consistent, high-quality products to large retail chains. For both salmon fishermen and salmon farmers, a cost-cutting strategy is required to boost profits in a low-price environment. This strategy is especially difficult for Alaska salmon fishermen because the state's fishing laws have resulted in significantly higher costs than would occur under an optimal system (Eagle et al. 2004).

Policies perpetuating the salmon "crisis"

Alaska's salmon fishing laws are the result of a complex, dynamic political economy and years of legislative struggles between ethnic groups, residents and non-residents, fishermen and fish processors, commercial and sport fishermen, urban and rural fishermen, and fishermen and environmentalists. Conservation was often achieved not by reducing the number of fishermen but instead by requiring the use of less efficient gear. The main cause of high costs is the derby system created by Alaska's limited-entry laws. A derby system means that permit holders in a given fishery may catch as many fish as possible within the time limits of an opening. Because it is a race for volume, a derby fishery gives permit holders every incentive to invest in faster boats with larger storage capacity.

The legislated inefficiency of the current regulatory structure exacerbates the natural disadvantages of salmon fisheries with respect to salmon farming. Derby-style salmon fisheries motivate fishermen to catch and unload fish as quickly as possible before the opening closes. Fishermen often compromise quality in their haste, and fish of varied quality are mixed in tenders and at the processors, thus removing any reward for treating catches with care. Because all the fish arrive at the dock at the same time, prices paid to fishermen are lower than if the supply were restricted or spaced out over time. Finally, processors must shape their approach to suit the derby. Only those methods that are capable of preserving large amounts of fish quickly, such as canning, are viable. Yet the traditional product forms have diminishing appeal for modern consumers. Fresh farm salmon has redefined consumer preferences and expectations for salmon.

Our survey did not directly seek fishermen's views on fundamental regulatory problems for Alaska's salmon industry. We did not list "fisheries laws" as a potential problem. Likewise, evaluation of fisheries management assessed the degree of success of the Alaska Department of Fish and Game in maintaining salmon runs and maximum yield for fisheries rather than the efficiency of derby-style fisheries. Some fishermen felt that management was a problem, and a few even mentioned the structure of openings and the race for fish, but most did not. Views on overcapitalization were mixed, and the average ranking was also low. Nonetheless, fishermen's policy preferences revealed at least some recognition of legislated inefficiency in Alaskan salmon fisheries. Some fishermen were in favor of cooperatives, quota systems, and buybacks, all of which are cost-reducing measures that would alter income distribution within the industry. The majority of fishermen in our survey, however, were opposed to these restructuring options.

The policy reform debate

What structural changes can be made in the fishery – and what changes are fishermen and other groups in Alaska likely to support – to make it more competitive in the future? To date, the Alaskan and federal governments have tried to lessen the economic impacts on the fishing industry by establishing subsidies in the form of disaster relief (distributed principally in western Alaska) and government purchases of canned pink salmon. These approaches might be useful when the industry faces short-term problems such as natural disasters (Eagle et al. 2004). In the context of long-term changes, however, these subsidies do not improve economic conditions in the fishery because they allow the industry to ignore fundamental reforms that could increase efficiency and create more resilient businesses. Moreover, our survey showed that most fishermen, with the exception of those in the Yukon–Kuskokwim Delta (where commercial fishing often supports subsistence activities), did not benefit from disaster relief and did not see the need to continue in this policy direction.

In addition to disaster relief, policy discussions in recent years have focused on marketing. Many people within the salmon fisheries agree that the wild salmon industry must distinguish its product from the farm salmon product – in taste, nutritional value and marketing of a "wild" (and, by association, healthy) product.[12] The survey results showed that fishermen support these efforts. It is easy to understand why quality and marketing programs are attractive to fishermen. Though Alaskan salmon fishermen already pay a 1 percent marketing tax, additional funding and implementation of these programs are currently in the public sector domain. Bolstering these programs would thus require no substantial change within the industry itself, except a commitment to quality.

Much less political attention has focused on the more controversial idea of legal restructuring aimed at lowering costs and improving quality. Allowing new programs such as fishing cooperatives, quotas, permit buybacks, or even fish traps and wheels[13] would slow the pace

---

[12] There is a large debate on the relative health benefits of wild versus farm salmon. Some studies have shown that farm salmon have a higher fat content and a different, less beneficial fatty acid composition than wild salmon (van Vliet and Katan 1990; George and Bhopal 1995). Limited tests have also shown that farm salmon contain more dangerous chemical substances than fish that feed in the wild (Hites et al. 2004). Substantial research on these issues is ongoing.

[13] Use of wheels and traps would greatly reduce labor costs and would likely lead to enhanced fish quality because fish remain alive in traps and wheels until they are removed by fishermen. The fish trap is a floating or fixed device positioned across the migration paths of salmon on their spawning runs and designed to lead salmon into a holding section from which escape is virtually impossible. The

of fishing and provide a mechanism to improve quality and competitiveness. These programs would reduce effort and production costs (Eagle et al. 2004). A slower pace of fishing would mean higher dock prices because delivery would be spread over time, and higher quality, thus higher market prices. Slower fishing would also lead to easier, more customized processing because processors would not have to accommodate extreme pulses in delivery. A major downside to these programs, however, would be the loss of direct and indirect fishing jobs, which would affect both individuals and communities, especially small and remote communities.

Any type of restructuring program for the Alaskan salmon fisheries will undoubtedly have differential impacts by region and gear type. The results of our survey demonstrated the strongest effects by region, with southeastern Alaska, Bristol Bay, and the Yukon-Kuskokwim Delta – three very different points along a wide spectrum of commercial salmon fishing in Alaska – facing quite different economic and biological conditions within their fisheries. Southeastern Alaska has large hatchery-supplemented salmon runs, while the Yukon-Kuskokwim Delta does not rely on hatcheries and has very low runs for some salmon species. Likewise, the commercial salmon fisheries are more lucrative in southeastern Alaska and Bristol Bay than in the Yukon-Kuskokwim Delta, where commercial fishing largely helps to support subsistence activities. In addition, gear type is an important economic and political feature of these regional salmon fisheries. Our survey results showed some significant variation in perceptions of the salmon "crisis" by gear type, although all groups felt that salmon farming and low prices were the largest problems. The political debate over restructuring will be complex, as is currently demonstrated in the Bristol Bay salmon fisheries (BBEDC 2003).

Resistance to change

The political feasibility of industry restructuring remains in question. Barring a worsening of the economic situation, our survey results suggest that most fishermen are likely to oppose large-scale cooperative, quota, or government buyback programs in Alaska. Although

> trap can be opened to permit escape as desired and can be used to hold fish for a short period of time before processing. Fish wheels consist of two large baskets that turn on an axle. They are rotated by the river current and scoop up passing fish as they turn. Captured fish slide down a chute into a holding box that is emptied several times a day (Cooley 1963; Colt 1999).

77 percent of our respondents in Bristol Bay showed support for a government buyback program, a study by the Alaska Department of Fish and Game showed that Bristol Bay permit holders would be willing to accept such a program only if the permit value was in the order of $100 000.[14] The general message is that fishermen do not want to pay out of their own pockets for a solution to the crisis.

Why are Alaskan fishermen generally resisting structural change? The first reason is that they may not see the "crisis" as permanent. Our survey shows a high degree of optimism among fishermen. Most of them believe that the market for their products will improve over time, and most of them intend to keep fishing salmon in the future despite the decline in prices and fishing incomes. This optimism may be fueled by the cyclical nature of the fishing industry; in the past, downturns have always been reversed. In addition, fishermen have become accustomed to the political and economic support they receive in Alaska, particularly with Ted Stevens in the Senate.

Another reason for resisting change is that most fishermen believe that the problem lies outside of their own industry – the culprit is the salmon aquaculture industry, not inefficient policy within the fishing industry. As noted above, fishermen in our survey mainly blamed low prices and salmon farming for the current problems. Very few fishermen thought that the problem resulted from fishing costs being too high. In some regions, such as the Yukon–Kuskokwim Delta, fishermen also blamed nature (low run sizes) and management for their problems. Both of these factors lie outside of direct industry control.

The combination of long-term optimism and blaming others for the problem robs fishermen of the motivation needed to address the fundamental causes of the decline in the fishing industry. This behavior has been documented in other situations of common property resource use (Thompson 2000). The motivation for supporting change is also diminished by the fact that many fishermen, as noted in our survey, have found outside employment to supplement their incomes.

CONCLUSION AND PROJECTIONS FOR THE FUTURE

The global fishing sector is undergoing major structural adjustments with the rise in aquaculture, and Alaskan salmon fishermen are feeling

---

[14] Average permit value for drift gill-netters in Bristol Bay in 2002 was $20 000 (see note 4 above).

the stresses of an altered economic landscape. Like other industries that have faced economic revolutions in manufacturing or trade, such as the automobile and steel industries, the fishing industry must now accept change and adapt in ways that will improve competitiveness. Yet there is widespread resistance among fishermen to the types of structural changes that will accomplish this goal. Most fishermen are in the business because they like their independence, their ability to earn large incomes occasionally (even if these periods are balanced with low-income years), and their freedom to be out of doors fishing. The future of the Alaskan salmon industry is uncertain, but one fact remains clear from our survey: many salmon fishermen want to keep fishing.

How will Alaska's salmon fisheries operate 20 years from now? It would be surprising to see fish traps and fish wheels as the dominant form of salmon capture because many fishing constituents would object to the redistribution of income that would likely result from such capture systems (Eagle et al. 2004). It is also highly unlikely that the salmon fishermen of Alaska will become salmon farmers during the period. It is far more likely that a number of cooperatives will emerge within the Alaska's salmon industry, some of which will rely on boats and others on traps and wheels. Buyback programs might also be implemented in certain areas, such as Bristol Bay, but with little likely effect on fishing volume or quality.

Regardless of the restructuring methods in play, new fishing programs are likely to be regionally based with specific attention to the politics of gear types, processors and markets. Our survey suggests that a one-size-fits-all policy change is unlikely to be politically successful in the Alaskan salmon fishery industry. Unless some change in the direction of policy occurs, however – whether it be piecemeal or wholesale – fewer people will have the luxury of remaining in the commercial salmon fishing industry in Alaska in the long run.

Acknowledgments

The authors thank Gunnar Knapp, Walter Falcon, Ashley Dean, Ellen McCollough, Marshall Burke, and one anonymous reviewer for their helpful comments and assistance on the manuscript; the David and Lucille Packard Foundation for research funding; Paula Terrel, Grant Trask, Pat Kehoe, Anne Mosness, and Chase Hansel for providing survey respondent contacts; and fishermen in our survey for their time and effort.

References

ADFG (Alaska Department of Fish and Game). 2002. *Alaska Commercial Salmon Harvests 1970–2001*. Available online at www.cf.adfg.state.ak.us/geninfo/finfish/salmon/catchval/history/1970-2001s.htm

ADFG. 2003a. *Commercial Operator Annual Reports*. Alaska Department of Fish and Game (available by request).

ADFG. 2003b. *Ex-Vessel Price per Pound: Time Series by Species*. Available online at www.cf.adfg.state.ak.us/geninfo/finfish/salmon/CATCHVAL/BLUSHEET/84-02exvl.pdf

ADFG. 2004. *Salmon Fisheries in Alaska*. Available online at www.cf.adfg.state.ak.us/geninfo/finfish/salmon/salmonhome.php

Anderson, J. L. 1997. The growth of salmon aquaculture and the emerging new world order of the salmon industry. In *Global Trends: Fisheries Management*, eds. E. K. Pikitch, D. D. Hubbert, and M. P. Sissinwine. Bethesda, MD: American Fisheries Society, pp. 175–184.

BBEDC (Bristol Bay Economic Development Corporation). 2003. *An Analysis of Options to Restructure the Bristol Bay Salmon Fishery*. Available online at bbsalmon.com/FinalReport.pdf

Carlson, S. 2002. *Survey of Bristol Bay Salmon Drift Gillnet Fishery Permit Holders: Preliminary Summary of Responses*, Report No. 02-4 N. Juneau, AK: Commercial Fisheries Entry Commission. Available online at www.cfec.state.ak.us/RESEARCH/02_4N/BBAYSURV.PDF

CFEC (Commercial Fisheries Entry Commission). 2003. *Permit Values*, Available online at www.cfec.state.ak.us/pmtvalue/mnusalm.htm

Colt, S. 1999. *Salmon Fish Traps in Alaska*. Anchorage, AK: Institute of Social and Economic Research, University of Alaska Anchorage. Available online at www.iser.uaa.alaska.edu/publications/fishrep/fishtrap.pdf

Colt, S. 2001. *The Economic Importance of Healthy Alaska Ecosystems*. Anchorage, AK: Institute of Social and Economic Research, University of Alaska Anchorage. Avalible online at http://hosting.uaa.alaska.edu/afsgc/healthy_ecosystems.pdf

Cooley, R. A. 1963. *Politics and Conservation: The Decline of Alaska Salmon*. New York: Harper & Row.

DCED (Alaska Department of Community and Economic Development). 1997. *The Alaska Seafood Industry*. Available online at www.dced.state.ak.us/cbd/seafood/pub/seafood.pdf

Eagle, J., R. Naylor, and W. Smith. 2004. Why farm salmon outcompete fishery salmon in the market. *Marine Policy* **28**: 259–270.

FAO (Food and Agriculture Organization). 2003. *Fishery Statistical Databases*. Available online at www.fao.org/fi/statist/FISOFT/FISHPLUS.asp

George, R. and Bhopal, R. 1995. Fat composition of free-living and farmed species: implications for human diet and sea-farming techniques. *British Food Journal* **97**(8): 19–22.

Goldburg, R. and Naylor, R. 2005. Future seascapes, fishing, and fish farming. *Frontiers in Ecology and the Environment*, **3**: 21–28.

Hites, R. A., Foran, J. A., Carpenter, D. O., et al. 2004. Global assessment of organic contaminants in farmed salmon. *Science* **303**: 226–229.

Knapp, G. 2002. *Challenges and Strategies for the Alaska Salmon Industry*. Anchorage, AK: Institute of Social and Economic Research, University of Alaska Anchorage. Available online at www.iser.uaa.alaska.edu/iser/people/knapp/Knapp%20Salmon%20Presentation%2001.pdf

McMullan, J. 1987. The organization of the fisheries: an introduction. In *Uncommon Property*, eds. P. Marchak N. Guppy, and J. Mc Mullan. Toronto, Ontanio: Methuen.

Miller, K. A. and Fluharty, D. L. 1992. El Niño and variability in the northeastern Pacific salmon fishery: implications for coping with climate change. In *Climate Variability, Climate Change, and Fisheries*, ed. M. Glantz. Cambridge, UK: Cambridge University Press, pp. 49–88.

Naylor, R. L., Eagle, J., and Smith, W. L. 2003. Salmon aquaculture in the Pacific Northwest: a global industry with local impacts. *Environment* **45**(8): 18–39.

Sylvia, G., Anderson, J. L., and Hanson, E. 2000. The new world order in global salmon markets and aquaculture development: implications for watershed management in the Pacific Northwest. In *Sustainable Fisheries Management: Pacific Salmon*, eds. E. E. Knudsen, C. R. Steward, D. D. MacDonald, J. E. Williams, and D. W. Reiser. Boca Raton, FL: Lewis Publishers, pp. 393–405.

Thompson, B. H. 2000. Tragically difficult: the obstacles to governing the commons. *Environmental Law* **30**: 241–287.

van Vliet, T. and Katan, M. 1990. Lower ratio of n-3 to n-6 fatty acids in cultured than in wild fish. *American Journal of Clinical Nutrition* **51**: 1–2.

Appendix: Survey of owners of Alaskan salmon fishing permits

The survey of fishermen described in the text included limited-entry permit holders in the following fisheries: Bristol Bay drift gill-net and set net; southeast Alaska drift gill-net, purse-seine, and troll; lower Yukon and lower Kuskokwim drift gill-net and set net; Cook Inlet set net; Prince William Sound purse-seine and drift gill-net; Chignik purse-seine; and Kodiak purse-seine. We interviewed 91 individuals, a small number relative to the 12 000 salmon permit holders in the state (Table 10.A1). Because of high travel expenses within Alaska and limited resources for conducting the survey, we focused our attention on southeastern Alaska and Bristol Bay (82 percent of our respondents). Together these regions account for roughly half of the state's total salmon permit holders and about half of the total value of salmon catch, as shown in Table 10.A1 below. We also interviewed fishermen in the Yukon–Kuskokwim Delta region (12 percent of our respondents), a region subject to much of the recent political attention regarding disaster relief programs. Though the Yukon–Kuskokwim Delta accounts for over 12 percent of the state's total salmon permit holders, it is insignificant in terms of total catch volume or value. Finally, we surveyed a few individuals who hold permits in other regions (6 percent of our respondents), but these interviews were not systematic. Given that the number of respondents from the "other regions" was small, these interviews were used only in the description of statewide trends but were not included in cross-region statistical comparisons.

The number of survey respondents for each region is shown in Table 10.A1 in relation to the role that each region plays in the state's salmon fishing industry. The survey instrument is provided in Table 10.A2.

Table 10.A1 *Number of survey respondents in relation to total salmon permit holders, production volume, and production value by region*[a]

	Alaska total	Southeast Alaska[b]	Bristol Bay	Yukon–Kuskokwim Delta[c]	Other regions[d]
Number of interviewees in our survey	91	44 (48.4%)	31 (34.1%)	11 (12.1%)	5 (5.5%)
Number of salmon permit holders[e]	12 693	4033 (31.8%)	2920 (23.0%)	1555 (12.3%)	2950 (23.2%)
Total catch volume (thousands of pounds)[e]	773 336	264 168 (34.2%)	100 096 (12.9%)	4849 (.6%)	346 304 (44.8%)
Total catch value (thousands of U.S. dollars)	$194 816	$51 124 (26.2%)	$47 692 (24.5%)	$2839 (1.5%)	$83 738 (43.0%)

[a] The table aggregates values for all fisheries from a given region, e.g., purse-seine, drift gill-net, set net.
[b] Includes troll fisheries.
[c] Does not include respondents in the upper Yukon fisheries.
[d] Includes respondents from Cook Inlet, Prince William Sound, Chignik, and Kodiak Island, but does not include respondents from the Alaska Peninsula/Aleutian Islands, Norton Sound, and Kotzebue regions.
[e] Alaska Department of Fish and Game 2004; Commercial Fisheries Entry Commission 2003.

Table 10.A2 *Survey instrument used in research*

Demographics
What is your name, age, (and gender)?
What is your permanent place of residence?
Do you have other employment beyond fishing?
If so, what is it?
What is your percent annual income from salmon fishing?
What was the percentage at your peak of salmon fishing?

Table 10.A2 (*cont.*)

Subsistence
Do you do any subsistence fishing?
If so, what are the relative amounts of time you spend for commercial vs. subsistence?
What are the uses of subsistence fish?
Do you have a fish camp?

Salmon fishing activity
What gear type do you fish?
What is your permit location?
Are you still (currently) using your permit to fish (e.g., in past and upcoming year)?
If not, when did you last fish with your permit?
Did you sell your permit, or has it been renewed but not fished?
When did you purchase your permit?
Do you own or fish other salmon permits? If so, which?
Do you have permits for other fisheries?
If so, what permits?
Do you own the boat that you fish?
When did you purchase it?
Do you have other boats? What type?
What are your target species?
What processor do you use?
Is there any value adding for your fish?
Have you done any direct marketing? If so, when did you start?

Future plans and views of fishing
Do you think there is currently a crisis in salmon fishing?
If so, do you think it is a permanent situation?
Please rank the following commonly discussed problems for salmon fisheries from 0 (no problem) to 5 (very significant problem):
    Low prices
    Management (including hatcheries)
    Salmon farming
    Overcapitalization
    Run size
Do you plan to stay in fishing?
Do you plan to make any new investments in fishing gear or activities?
Do you have outstanding loans for your fisheries permits or gear?

Policy response
Please indicate whether you would support, possibly support, or oppose the following commonly discussed policy responses to the economic decline in the salmon fishing industry:

Table 10.A2 (cont.)

>Co-ops (fishing, not marketing)
>Quota system (for individuals)
>Disaster relief (government funded)
>(Have you received disaster relief funds previously?)
>Buybacks (government funded)
>Quality/marketing programs

Personal views on salmon farming

Do you think salmon farming affects wild stocks?
If so, how?
Are you opposed to salmon farming in Washington, British Columbia, and Alaska?
Are you opposed to farming elsewhere?
Do you think salmon farming has contributed to the decline in salmon prices?
If given the income or opportunity, would you invest in a salmon farm?

JOSEPH J. MOLNAR AND WILLIAM H. DANIELS

11

Tilapia: a fish with global reach

INTRODUCTION

Tilapia is the most widely produced fish in global export aquaculture and second only to carps as the most widely farmed freshwater fishes in the world (Naylor *et al.* 2000).[1] The world harvest of farm-raised tilapia surpasses 800 000 tonnes (FAO 2004). Tilapia is grown in more than 75 countries, and China is the leading producer with 706 585 tonnes in 2002, or 47 percent of total world production (FAO 2004). Although a freshwater fish, tilapia can tolerate some salinity and so is hardier than many other breeds. This increases the range of possibilities for culture. Depicted on the walls of Egyptian tombs, tilapia in Biblical times was known as *musht*, Arabic for "comb." More recently known as "St. Peter's fish," it is understood that tilapia (*Tilapia galilaea*) from the Sea of Galilee were used to miraculously feed the multitude.[2] Some attribute

---

[1] Until the late 1970s, the tilapias were all classified into a single genus, *Tilapia*, but most taxonomists now classify them into three genera, *Tilapia, Saratherodon*, and *Oreochromis*, according to their breeding behavior. In this chapter, "tilapia" will centrally refer to *O. niloticus*; other subspecies will be mentioned as specifically identified by cited sources. There is a broad diversity of species, subspecies, hybrids, crosses, and commercial varieties of tilapia cichlids reared for food fish. Examples that include combinations of species and commercial names include: red tilapia (hybrid of *O. niloticus* × *O. aureus*), Florida red tilapia (*O. urolepia hornorum* × *O. mossambicus* hybrid), and many others.

[2] First distributing loaves and fishes – sometimes claimed to be tilapia – to 5000 (Matthew 14:15–21), Jesus later fed 4000 from "a few small fish" and seven loaves of bread (Matthew 15:32–38). It was from the ranks of fishermen that Jesus Christ called the first apostles (Mark 1:16–20), including Peter and John. Fish and fishing were often associated with his ministry and later were used as a symbol of it. Jesus miraculously calmed the storm from a fishing boat (Matthew 8:23–26) and spoke many parables to the crowds while standing in a fishing boat (Matthew 13:1–58); once miraculously paid taxes with a coin taken from inside a fish (Matthew 17:27); once had the disciples make a catch of fish so great that their nets miraculously

---

Globalization: Effects on Fisheries Resources, ed. William W. Taylor, Michael G. Schechter, and Lois G. Wolfson. Published by Cambridge University Press. © Cambridge University Press 2007.

the naming of tilapia to Aristotle, from Greek for "distant," a fitting etymology for a globalized fish.

Globalization – understood broadly as a process resulting from the growing integration of product, labor and capital markets, common technologies, increasingly similar patterns of food consumption, and changes in the international trade regime – has become a major restructuring force for food systems in the developed and developing world (Steeten 2001; Roth 2002). It implies longer production chains that link distant production centers to centers of consumption (Steeten 2001). The process of globalizing a commodity involves human actors who engage to create, legitimize, and maintain uniform characteristics in things outside the national boundaries of their origin (Tanaka and Busch 2003). A globalized commodity is one where the constraints of geography on the social and cultural identity on a food item have receded. Others view globalization as a dark force to be resisted, with new concerns about the national origin, traceability, and desires for knowledge about sources of food (Bonanno et al. 1994; Waters 1995).

Globalization – as one source of social and economic restructuring – is having significant impacts on aquaculture industries, as well as the nations and locales where fish are grown. Globalization also implies rapid and widespread diffusion of information about cultural items. It also suggests some degree of standardization of food items, as the nomenclature and product identity requirements of large European Union and U.S. markets tend to impose a template on tilapia producers in the dispersed periphery. Handling, packaging, and labeling standards structure the way the commodity is produced and managed in distant locales. Access to export markets is tied to meeting the phytosanitary, cold chain, food handling, and safety standards of those markets.

Tilapia has become a globalized fish in three central ways. First, introductions of the organism have made it part of ecosystems across the planet. Tilapia are a truly globalized fish because of the widespread introductions of the organism and its various subspecies in the tropics and, increasingly, in temperate areas. This aspect of its globalization is not without controversy.

Second, through globalization, tilapia has become a widely known restaurant and supermarket item in the developed world and a central aquaculture crop and food source in the developing world. The fish is becoming a globally available commodity because

> were filled to overflowing (John 21:1–14); and ate a piece of broiled fish with the disciples after resurrection from the tomb (Luke 24:42–43).

of industrialized production and processing facilities that make tilapia a reliably supplied restaurant and consumer item.

Global increases in consumption of food fish will take place predominantly in the developing countries, where population is growing and higher incomes are allowing purchase of high-value fisheries items for the first time by many people (Costa-Pierce et al. 2003). Most recently, tilapia culture has grown because industrialized shrimp producers experienced devastating losses from viruses. Tilapia became an alternative rotational crop to counter the impacts of disease. Fish production is a key element of food security in least developed countries and a critical area where innovative programs are needed to increase production.

Finally, an international network of institutions, firms, and people moves across the planet introducing production systems, management strategies, and genetic material in ways that accelerate the previous two trends but also may have certain consequences and impacts on global development (Egna and Boyd 1997). A transnational class of production technicians moves in a labor market for skilled facility managers who understand the disciplines of global supply chains and markets.

The purpose of this chapter is to examine some implications of the rise of tilapia as a cultured fish in the developing world and a widely accepted consumer item in developed nations. We argue that tilapia's integration into the world system has been largely beneficial, although the realization of its potential as an enterprise for small- and medium-scale commercial farming is just beginning to accelerate.

TILAPIA AS A GLOBAL ORGANISM

Tilapia: the species

Tilapias are members of the Cichlidae family, which numbers approximately 1300 species and 105 genera (Stickney 1988, 2002). Possibly 900 cichlid species occur in Africa and the Middle East, including the native tilapia, which number approximately 100 species in three genera (Balarin 1979).

Tilapia is common to the warm, weedy waters of sluggish streams, canals, irrigation ditches, ponds, and small lakes. Most tilapias are strictly freshwater fish, but some have adapted to brackish or saltwater environments, and some can tolerate environments with an extremely high temperature and very low oxygen. In fresh water, they are primarily algae and plant feeders. Many are mouthbrooders, although some build spawning nests, which they guard after the eggs hatch. Most are small, although some reportedly can grow as large as 10 kg, and they are schooling species. Despite their abundance, tilapia have little to no sportfishing value in some areas where they have been introduced, including North America, but some species are pursued by anglers in their native range, especially in southern Africa. As non-predatory fish, they do not respond to most lures and casting presentations but are caught with coarse fishing methods.

Large-scale commercial culture of tilapia is limited almost exclusively to the culture of three species: Nile tilapia (*Oreochromis niloticus*), known for its high yield; blue tilapia (*O. aureus*), a cold-resistant strain; and Java tilapia (*O. mossambica*), which, when hybridized, produces reddish fish (Stickney 2002). Although Nile tilapia is potentially the most profitable of the species, it is also the least tolerant of cold water conditions. They grow maximally at 29 °C (85 degrees Fahrenheit) with a lower lethal temperature of 11.6 °C (53 degrees Fahrenheit). Therefore, they are primarily cultured only where warm water is naturally available or can be artificially supplied in a cost-effective manner.

Of the three tilapia species with recognized aquaculture potential, the Nile tilapia is the most commonly used species in fish farming. Grow-out strategies for tilapia range from the simple to the very complex. Simple strategies are characterized by little control over water quality, food supply, and sex composition of stocked fish – and by low fish yields. As greater control over water quality, fish nutrition, and reproduction is imposed, the production cost and fish yield per unit area increase (Gur 1997). Across this spectrum, there is a progression from low to high management intensity.

Israel, the United States, and Belgium have been centers of technology development for the tilapia. Israeli firms and consultants dominate the organization and management of industrialized production of tilapia, but Asian production and innovation centers – China, Taiwan, and Thailand – now drive much of the industry. Belgian scientists have played important roles in introducing tilapia and other species for culture in central Africa and other locales, a record that is not without criticism from environmentalists. U.S. development projects have fostered the culture of tilapia as an enterprise for small- and medium-scale commercial farms around the world (Molnar *et al.* 1996).[3]

---

[3] The Pond Dynamics/Aquaculture Collaborative Research Support Program (PD/A CRSP) is a global research network to generate basic science that may be used to advance aquaculture development. One of a family of research programs funded by the U.S. Agency for International Development (USAID), the CRSP focuses on

New varieties

In the 1990s, an effort was made to breed a hybrid of high-yielding species that would be faster-growing, more robust, and more efficient as a feed/food converter than traditional strains. The Asian Development Bank funded the World Fish Center (formerly the International Center for Living Aquatic Resource Management – ICLARM) to evaluate and disseminate the results of the Genetic Improvement of Farmed Tilapia (GIFT) project (Eknath and Acosra 1998). As a highly visible public sector effort to identify superior genetic stock, the project fell short of expectations. After some delays, GIFT fish became available, but it became apparent that the hybrid stocks required high maintenance to maintain superior performance. This was due to the polyploidy of the various performance traits, i.e., natural genetic drift in a short-generation organism and the general tendency of regression to the mean that affects any selected trait.

Commercial and industrial tilapia production requires all-male tilapia fingerlings so it can exploit the superior growth rates of male fish and avoid reproduction in grow-out populations. *Oreochromis mossambicus* and other species, especially *O. aureus* and *O. urolepis hornorum*, continue to be important in aquaculture because of factors such as cold tolerance, development of red lines of tilapia, and hybridization to produce all-male offspring (Popma and Phelps 1998). Currently, there are three main paths to achieving all-male grow-out stocks: hand sexing, sex reversal of newly hatched fish using 17α-methyltestosterone, and YY chromosome technology which employs hormone treatment of the broodstock to induce all-male offspring for grow-out (Mair 2001).

Commercial fish breeders work continuously to develop and introduce products that have various new performance traits. A few notable examples are, a number of brand identities of red tilapia for fillet color and new strains of white tilapia for cold-water tolerance. Operators of commercial and industrial farms take a good deal of care and expense to identify hatchery broodstock, often developing their own proprietary lines. Public sector evaluations of these branded broodstocks are not commonly available to evaluate the firms' performance claims.

> improving the efficiency of aquaculture systems. The PD/A CRSP began work in 1982 in Thailand; there were also projects in the Philippines, Honduras, United States, Indonesia, and Panama, and, until recently, Rwanda. At all the sites, the goal is the same: to identify constraints to aquaculture production and to design responses that are environmentally and culturally appropriate. The research network's global experiment has focused on the Nile tilapia, although some sites have devoted attention to marine shrimp and other locally significant species.

Transgenic fish

Identifying and introducing genes linked to specific performance traits is widely believed to be an imminent source of rapid and dramatic improvements in productivity for tilapia and other cultured species. According to Smith (2003), transgenic fish, including Atlantic salmon and tilapia, with improved growth rates have been produced, and comparative growth trials with non-transgenic fish have shown the benefits of transgenic technology to both researchers and producers. Other potential applications of transgenic change concern the production of disease-resistant fish or fish with an improved carbohydrate metabolism that accelerates growth. However, the application of gene transfer technology to commercial fish species raises major concerns among consumers and environmentalists about the benefit of their use in aquaculture and the possible environmental and human health risks associated with genetically modified fish.

Smith (2003) summarizes concerns linked to the possibility of transgenic fish interbreeding with wild native stocks, leading to undesirable ecological impacts and dilution of the wild fish genetic pool. He notes that the escape of farmed fish into the sea is relatively common and can induce gene diffusion to wild fish. The risks of interbreeding apply to introduced strains as much as genetically engineered transgenic fish, but Smith (2003) notes that public concern is much greater when genetically engineered fish may be involved. The possible transfer of transgenes to wild fish and the possibility of transgenic fish establishing themselves as permanent residents of an environmental ecosystem are the most important negative considerations in applying this technology to fish culture. Therefore, the enormous commercial potential benefits of transgenic fish technology in research and in aquaculture will not be achieved without effective isolation of genetically modified fish from the wild fish genetic pool. In considering transgenic research and its applications, the possibility of access by genetically modified animals to the environment and their interaction with the environment must be considered.

Tilapia as an alien invader

The risk of tilapia escaping from fish farms to the environment is difficult to evaluate. Tilapia is a warm-water fish that will not likely survive winter water temperatures in most temperate locations. Feral tilapias that have escaped from aquaculture and aquarist operations have been reported captured from the wild in a number of U.S. states but exist in sufficient abundance to support viable fisheries only in two locales (Costa-Pierce 2003). California's Salton Sea has a dense population of tilapia (*O. mossambicus*), and six shallow lakes in Polk County, Florida, are reported to have established populations (Costa-Pierce and Riedel 2000). However, in cooling reservoirs or locations where warm water is discharged to streams and rivers, tilapia may be able to find a temporary thermal refuge.

In the United States, diverse regulations govern the introduction and culture of tilapia, and many contradictions exist. For example, outdoor tilapia culture is entirely outlawed in Florida (one of 16 states), and Louisiana farmers can grow tilapia only in aboveground tanks, but Alabama and Georgia producers can grow tilapia in earthen ponds without issue. Many U.S. states require special permits to possess and farm tilapia. In California, farmers are restricted by permit to tilapia subspecies previously introduced into state waters, such as *T. zillii*. In Alabama, where tilapia has been cultured since the 1960s, no evidence of displacement of native species has been recorded.

Internationally, the regulation of tilapia ranges from severe to schizophrenic and non-existent to actively promoted. As will be discussed further below, some Australian states make serious efforts to prevent and eradicate any species of tilapia. In Malawi and some other African Great Lakes nations, tilapia culture is restricted to native species to protect the complex ecological structure of the many native cichlids present in those water bodies. On the urging of environmentalists, Peru passed a law in the 1980s prohibiting the culture of tilapia in the Amazon. Yet previous donor-supported government efforts had already established a thriving network of tilapia producers in the Alta Selva that continues to this day. Others have expressed concern that the Amazonian species bocachico (*Prochilodus* sp.) and tilapia, specifically *O. mossambicus*, occupy similar ecological niches – to the detriment of the bocachico (AUPEC 2003). Tilapia was introduced into Colombia's department of Valle del Cauca in 1962, where its much simpler reproductive process is a key advantage of the still-water breeding tilapia over the riverine bocachico.

In most developing countries, tilapia culture is viewed as an enterprise to be encouraged and made productive for its potential to provide income and food security. Coates (1997) provides a prototypical example of the introduction of tilapia into a non-native environment for otherwise well-intentioned purposes. Papua New Guinea is the world's largest tropical island and has extensive areas of freshwater habitat. In 1983, the introduction of more species of fish into the Sepik Ramu basin was proposed as a way of improving the fishery. The first fish imported, introduced, and stocked was *T. rendalli*. It was selected because of the ease of obtaining fingerlings and an anticipated short establishment time that would lead to quick tangible benefits. Little mention is made of the ecological consequences of the new species or the extent to which the expected benefits of the introduction were realized. Similarly, Singapore residents view tilapia as a feral species that has established itself in freshwater habitats. *Oreochromis mossambicus* was introduced locally during the World War II by the Japanese and today has the status of a minor sportfish on the island.

Others connect tilapia introduction to some negative impacts on the array of species in a water body. South African environmentalists work to remove *O. mossambicus* from Groenvlei Lake near Sedgefield (Cape Nature Conservation 2003). Groenvlei is the only freshwater lake in the Wilderness Lakes system and has no in-flowing rivers and no link to the sea. Several thousand years ago, the water body was cut off from Swartvlei, the largest of the Wilderness Lakes, by windblown sand. The waters of Groenvlei gradually lost their salinity, and today the algae that grow in the freshwater environment of Groenvlei give the lake a greenish tint (Cape Nature Conservation 2003). Largemouth black bass (*Micropterus salmoides*) and tilapia (*O. mossambicus*) have been introduced, but carp (*Cyprinus carpio*) seems to be the invasive fish that centrally threatens Groenvlei's indigenous fish. Tilapia per se are not the sole threat to this lake; rather it is one of several conditions undermining the unique array of species found there that have become a conservation objective.

One of the most consistent sources of opposition to tilapia introduction and propagation is in Australia. Tilapia are declared noxious in Queensland, making it illegal to possess, rear, sell, or buy tilapia. Fishermen are cautioned that it is important that tilapia are not moved from one water body to another because this is the main way that they are spread. The Australian Department of Primary Industries and Fisheries (DPI) instructs fishermen that tilapia are not to be used as bait, live or dead. If caught, tilapia fish are to be killed humanely and disposed of away from the water body. It is a citable offense to release tilapia into Queensland waterways or to use them as bait, live or dead, and penalties up to A$150 000 apply (DPI 2001, 2003).

The Central Queensland Department of Fisheries' efforts focus on one of several aquarium species, *T. mariae* (DPI 2001, 2003).[4]

---

[4] Introduced in the 1970s as aquarium fish, black mangrove cichlids are found in northern Queensland waters around the Cairns region. This species of tilapia is less tolerant of cool temperatures than *O. mossambicus* and, therefore has a

Although there has been very little research done on the effects of tilapia on the ecology of native fish, Australian (DPI 1997, 2003) concerns center on the potential for tilapia to compete for space and food with native species (Merrick and Schmida 1984).

The effect of tilapia on the ecology of species takes place in the context of other introductions, and the interactions create changes that create new niches or empty others held by pre-existing species. Tilapia was introduced as prey for predator bass (*M. salmoides*) to support sport-fishing in Kenya's Lake Naivasha,[5] but native species were displaced. Water conflicts from a dramatic rise in population and development along its shores caused major problems for local communities.

A different experience is reported for another African Great Lake by Njiru (2003). *Oreochromis niloticus* was introduced to Lake Victoria in the 1950s. It remained relatively uncommon until 1965, when its numbers began to increase dramatically. It is now the third commercially important fish after the Nile perch (*Lates niloticus*) and dagaa (*Rastrineobola argentea*) (Ogutu-Ohwayo and Balirwa 2004). Njiru (2003) observed that tilapia could be filling niches previously occupied by cichlid and non-cichlid fish that are no longer suppressed by predators that no longer exist. Thus, tilapia introductions are alternatively viewed as building sport fisheries, as enriching underpopulated water bodies with a productive item for human consumption, as destructive competitors for native species, or as opportunistic outsiders that expand to fill niches vacated by other interventions or circumstances.

In Kerala, the southernmost state in India, Ramachandran (2000) reports excessive replacement of the ecological niches of native fish species by tilapia (*O. mossambicus*) in the forested watersheds of major river systems in the state. The Central Marine Fisheries Research Institute, Mandapam, brought the first consignment to India from Bangkok in August 1952. The second consignment was brought in the same year from Sri Lanka. Thereafter, it was introduced in other southern India states to augment farm-based fish production.

---

narrower range (Grant 1997; Holloway and Hamlyn 2001). *Tilapia mariae* has vertical stripes on the head and body and varies in color from dark olive green to light yellowish green. Larger fish have less distinct stripes. They grow to around 25 to 30 cm, reaching sexual maturity at around 19 to 20 cm. They are substrate spawners and prefer to attach their eggs to hard surfaces. They do not build nests but do look after their eggs and young. In Queensland, they spawn between September and March (DPI 2003).

[5] The name "Naivasha" is derived from the Maasai word *enaiposha*, which means "receding water."

Ramachandran's (2000) study endeavored to assess the potential of a native ornamental fishery that had good demand in national and international markets. The native fishes had dwindled wherever tilapia were found in large numbers. While concluding that the introduction was irreversible, Ramachandran (2000) called for regulation of the import of exotic species in India.

The Kerala experience reflects the second thoughts that some countries have about tilapia introductions that took place, as in this case, more than 50 years ago. Previously unnoticed declines in other species are attributed to tilapia introductions, often in the context of other ecological changes and alterations in the pattern of human impact. Ignorance or willful neglect of international guidelines (e.g., ICES 1998) can result in the escape of exotic species and animal pathogens into the environment with a potential for unfortunate impacts on native aquatic species.

TILAPIA AS A GLOBAL COMMODITY

Commodification is a globalization process associated with mass production in an industrialized context. Commodification means that products are traded on international markets on the basis of the expectation of future fluctuations in supply and demand, regardless of provenance or the local value systems of the farmers and societies that produced them (Nuffield Council 1999).

Commodification is a combination of technical and market shifts that alter the role and function of a dietary item in global food systems. The process may be characterized as follows. A previously wild-caught fish is discovered to have desirable consumption properties. At first it becomes an exotic restaurant menu item, but its popularity and desirability grow among a broader set of consumers. Technical advances then enable the reproduction and rearing of the fish on a commercial scale. The identification of a food item with standardized features and qualities facilitates its production and exchange in the global marketing system. As production increases, supermarkets begin to promote the item, more producers enter the market, prices fall, and the species – in this case, tilapia – becomes the "aquatic chicken" (Little 1998).

Commodification facilitates the marketing and distribution of an item because multiple sources can meet the needs of multiple destinations. In the evolution of product markets, Indonesian tilapia farms now supply U.S. markets with flash-frozen whole tilapia, benefiting consumers who need inexpensive fish protein. Advances in post-harvest technology now focus on enhancing the shelf-life and appearance of fresh tilapia fillets by closely trimming the fat layer or, more controversially, treating first with carbon monoxide to retard color changes.

Consumer item

Costa-Pierce and his colleagues (Costa-Pierce et al. 2003) recently summarized some major trends in fish consumption. The reported production of fish for direct human consumption doubled between 1950 and 1970 and has stabilized since then at an average of 9 to 10 kg of fish per capita per year, notwithstanding world population growth. Fish consumption per person is expected to continue to rise. Supply will probably be limited by environmental factors, and a likely range for annual demand is 150 million to 160 million tonnes, or between 19 and 20 kg per person in 2030. Global increases in consumption of food fish will take place predominantly in the developing countries, where population is growing and higher incomes are allowing purchase of high-value fisheries items for the first time by many people. In least developed countries fish production is a key element of food security. On the other hand, fish protein is needed to prevent malnutrition and is a critical area where innovative programs are needed to increase production (Costa-Pierce et al. 2003).

Tilapia is a good source of protein. Globally, fish provide about 16 percent of the animal protein consumed by humans and are a valuable source of minerals and essential fatty acids. Fish is the primary source of omega-3 fatty acids in the human diet; these are critical nutrients for normal brain and eye development of infants, and they have preventative roles in a number of human illnesses, such as cardiovascular disease, lupus, and depression and other mental illnesses (Costa-Pierce et al. 2003). A 100-g serving of tilapia has 82 calories, 50 mg of cholesterol, and 34 mg of sodium, plus 0.14 gram of omega-3 fatty acids (Silvers and Scott 2002). Recent scientific findings about the beneficial consequences of weekly fish consumption for reduced heart disease risk (Hu et al. 2002) and, most recently, reduced incidence of Alzheimer's disease have expanded overall demand for fish as a food item (Morris et al. 2003).

Tilapia also has a number of culinary attributes valued by food preparers and consumers (Young and Muir 1998). Flesh color, texture, flavor, and size specifications can all be realized within a short growing cycle that produces a mild, soft, lean white fish fillet with tender flakes and a slightly sweet taste. Tilapia, usually priced at levels between chicken and salmon, is a widely available item in supermarkets in developed countries.

Consumer awareness of tilapia is still low compared with awareness of established species such as salmon or tuna, and there is considerable room for expanding tilapia markets (AquaSol 2003). The central product variants are live, whole, fillets, fresh, and frozen. Though Taiwan is the major U.S. source of imports in total volume, Western Hemisphere producers dominate fresh tilapia imports into the United States. Costa Rica, in particular, is the clear leader in the shipment of fresh fillets to the United States at the present time. Taiwan, Thailand, and Indonesia are the largest providers of frozen fillets to the United States. In the domestic markets of developing countries, tilapia are important food fish that are harvested from cultured ponds and from the wild for human consumption. For example, tilapia are netted in Mexican lakes where they were introduced, and some find their way to U.S. markets.

Part of the commodification process is the evolution of postharvest product forms that add value for both the consumer and the distribution system. More novel product forms such as surimi and all-but-cooked platters for home consumption increase demand for tilapia and increase its presence in high-value niches in the food system. This is particularly likely given food preparation trends, consumer concerns about safety, and other competing substitute aquaculture products.

The cost structure of production of tilapia is highly competitive, as they have among the most favorable ratios of food conversion and growth rates. From an environmental perspective, tilapia also have advantageous synergies with shrimp production for industrial-scale producers which should enhance international diffusion of the product. Tilapia also can be grown with feed containing little or no fishmeal. This is a clear advantage over salmon and other carnivorous species that currently require expensive and environmentally unsustainable feed formulations (Naylor et al. 2000), although there are now diets for these carnivorous species that contain less fishmeal.

THREATS TO TILAPIA AS A COMMODITY

The future of tilapia is bright; as a cultured product it is at least as safe and wholesome as wild-caught species. However, in addition to the consumer hazards listed above, Garrett et al. (1997) identify some less obvious ("shrouded") public and animal health hazards associated with ignorance, abuse, and neglect of aquaculture technology. These concerns apply to tilapia in several ways.

Technology ignorance

Planning for new freshwater and marine aquaculture sites should include discussions of the potential effect of large or small impoundments on such issues as disease transmission, water supply, irrigation, and power generation. Garrett et al. (1997) cite the common practice of creating numerous small fishpond impoundments, which often occurs when aquaculture becomes popular in a locale. This approach, however, may pose more risks to human health than the creation of a single large impoundment. Small impoundments greatly increase the overall aggregate shoreline of ponds, causing higher densities of mosquito larvae and cercaria, which can increase the incidence and prevalence of diseases such as lymphatic filariasis and schistosomiasis, respectively (Garrett et al. 1997). Practitioners and even some of those offering technical assistance may not be aware of the potential for microbial disease transmission. Ironically, tilapias are sometimes introduced into ponds and impoundments to control mosquitoes and dengue.

The most advanced level of environmental management is the development and enforcement of rules and regulations about how activities should be conducted to provide environmental protection. In pond aquaculture, contamination of natural waters with nutrients, organic matter, and suspended solids in effluents usually is the major environmental concern. Pond aquaculture normally cannot be conducted without effluents. The most advanced aquaculture operations reduce concentrations and loads of pollutants in effluents to levels that will not cause deterioration of water quality in receiving waters. Many nations, however, do not have or do not enforce environmental regulations for aquaculture (Boyd and Wood 2001).

For centuries, food growers have cultured species in wastewater-fed ponds and grown secondary vegetable crops in wastewater and sediment material in integrated aquaculture operations (Garrett et al. 1997:454). Fishery aquaculturists, however, rarely consider the potential for transmission of human pathogens to cultured species and secondary vegetable crops. Although aquaculture researchers working in developing countries infrequently refer to the potential human health implications of aquaculture, such risks often seem minuscule relative to the background risks from other sources of disease transmission, malnutrition, poverty, and food insecurity.

Technology abuse

Technology abuse includes the willful misuse of therapeutic drugs, chemicals, fertilizers, and natural fishery habitat areas (Garrett et al. 1997). The widespread use and misuse of antibiotics to control diseases in aquaculture species is worldwide, and Garrett et al. (1997) indicate that it will probably increase as aquaculturists move towards more intensive animal husbandry – feed-based rearing techniques and higher stocking densities. Garrett et al. (1997) cite the example of illegal use of chloramphenicol in shrimp culture to control diseases, which may result in unacceptable levels in the harvested product. Similarly, they note that improper or illegal use of chemicals (e.g., tributyl tin) to control pond pests such as snails can also result in human health hazards. The abuse and misuse of raw chicken manure as pond fertilizer may result in the transmission of *Salmonella* from manure to the cultured product.

The destruction of mangrove areas to build aquaculture ponds can have a drastic impact on the survival of wild aquatic species through the degradation of essential fish habitats and nurseries (Garrett et al. 1997:455). Although mangrove destruction is primarily associated with shrimp pond production, tilapia does play a role as a rotational crop. Mangrove loss continues to occur, but shrimp farming is not the only pressure on this forest resource.

Abuse of antibiotics can have negative consequences for fish farmers and human populations (Hough 2002). Increased production of fish from aquaculture has occurred primarily as a result of increasing feed inputs into ponds and other production systems, thereby increasing yields per hectare by an order of magnitude compared with extensive production systems in which rearing water is fertilized only (Costa-Pierce et al. 2003). Dense populations of genetically similar fish are more vulnerable to rapidly spreading disease, but prophylactic use of antibiotics can breed resistant bacteria with deleterious consequences for human and animal health. To date, the hardiness of tilapia has made the need for extensive use of antibiotics moot.

Production technology is available to achieve high levels of fish production, but high levels of inputs used in such systems have implications for the sources of feed protein and the environment that receives fish waste (Costa-Pierce et al. 2003). Higher inputs mean two things to the aquaculture feed industry: more feed and higher-quality feed. Currently, global feed production for farmed fish and crustaceans is approximately 13 million tonnes per year, and predictions are for feed production to increase to more than 37 million tonnes by the end of the decade (Costa-Pierce et al. 2003). Feeds for salmonids and marine fish have always been complete feeds – i.e., ones that supply all of the nutritional needs of the fish. Pond-reared tilapias, in contrast, obtain a significant proportion of their nutritional needs from pond biota, particularly when reared at lower stocking densities.

The degree to which feeds must supply essential nutrients to pond-reared fish increases as rearing densities increase beyond the capacity of natural foods in ponds to supply them. Fish farmers around the world have found that as they increase feed inputs, the biomass and economic yields from ponds increase as well. Thus, great areas of low-input, pond-based aquaculture mainly in Southeast Asia and China are being converted from low-input to high-input systems. Care must be taken to ensure that increased aquaculture production is not associated with higher harvest rates of forage fish species used to produce fish meals and oils, which would lead to a net loss of fish production – i.e., the capture for fishmeals and oils would exceed the amount of fish produced for consumption (Costa-Pierce et al. 2003). Such systems depend upon high-quality feeds to supply an increasing proportion of nutrients used by the fish and are sustainable insofar as they are economically viable and the trash species used as protein inputs are not depleted.

Experiments with all-vegetable feeds indicate that fishmeal can be completely replaced by supplementing the feed with mineral phosphorus. Feeding trials suggested an advantage to using feed with a 35 percent protein level and a lysine-to-protein ratio of 4.6 to 4.8 percent. Vitamin supplementation in feeds for tilapia during the fattening stage had a positive effect on survival but did not seem to affect fish growth (Bureau et al. 2002).

Technology neglect

The final "shrouded" hazard that Garrett et al. (1997:455) associate with aquaculture involves technology neglect, which includes such events as the abandonment of small aquaculture ponds in tropical countries, leading to increased mosquito habitats and concomitant increases in malaria. On the other hand, there is some evidence that tilapias are efficient consumers of such larvae, and small tilapia are sometimes stocked in Chinese water supplies for just such purposes. Production processes can exhibit technology neglect if employees are not trained in the proper use and application of therapeutics and chemicals, for example. Similarly, post-harvest risks connect to the management of processing facilities and maintenance of the cold chain after live fish become an item for human consumption.

Environmental managers are concerned about the possible transmission of exotic pathogens into the environment from tilapia processing plant wastewater discharge and solid waste material landfill leakage (Garrett *et al.* 1997). Tilapia processing plant Hazard Analysis and Critical Control Points (HACCP) include unload/receive, de-ice/wash, thaw, fillet, wash, re-ice, de-ice/wash, re-ice and dip/glaze. Application of HACCP principles at aquaculture sites and processing plant locations has the potential to control transmission of exotic human and animal pathogens (Hoskin 1993; FDA 2001).

Ignorance of the microbial profile of aquaculture products can also affect human health, as evidenced by reports of transmission of streptococcal infections from tilapia to humans, which resulted in several meningitis cases in Canadian fish processors. A change in marketing strategies to sell live fish in small containers instead of ice packs resulted in human vibrio infections from live tilapia in Israel in 1996. Such bacteria can be present in other aquacultured and wild-caught species in addition to tilapia. Ignorance of the hazards associated with the use of untreated animal or human waste in aquaculture ponds to increase production also has human health implications (Garrett *et al.* 1997).

The bacterium *Streptococcus iniae*, which can be found on the surface of the fish, including the head, spine and fins, is not contracted by eating cooked tilapia. However, these bacteria can enter a person's bloodstream through a cut in the skin during handling of the raw fish. The bacteria cause cellulitis, a fast-spreading infection of the tissue between the skin and the muscle. The sufferer typically notices redness and swelling at the injury site, then ends up in the hospital emergency room with a high fever and chills and must be treated with antibiotics. The most severe of the Toronto cases resulted in meningitis and heart valve infection. The *Streptococcus iniae* bacterium has been associated with meningitis in farmed fish since the 1970s. Such bacteria can be present in other aquacultured and wild-caught species in addition to tilapia (FDA 2001). Human failure – large and small – can have unpleasant and costly consequences from food-borne illness. Although technological neglect is a problem for aquaculture products in general, it may be a somewhat greater concern for tilapia because this species is predominantly sourced from farms in developing countries in the tropics where heat, humidity, and sanitation are salient issues.

USE OF HORMONES IN REPRODUCTION

Consumer perceptions of the quality of tilapia may be affected by the fact that hormones are sometimes used to produce all-male fingerlings. The most serious barrier to large-scale commercial tilapia farming was overpopulation and stunting resulting from reproduction before fish reached a commercially acceptable size (Popma and Masser 1999). Mixed-sex culture is commercially feasible in cages and in polyculture with a predator species, but the slower growth of females has stimulated the commercial farming of all-male populations. All-male fingerlings were first obtained on a commercial basis by the manual separation of the sexes by visual examination of the genital papilla (Popma and Phelps 1998).

Certain hybrid crosses can produce all-male offspring, but the oral administration of androgens to recently hatched fry is the method currently used to produce most monosex fingerlings for commercial-scale farming. Typically, fry are stocked in the first week at 100/m^2 and then transferred and stocked at 750 to 1000/m^2 for the last 2 weeks. The fry are fed a commercial diet containing 30 parts per million (ppm) of 17α-methyltestosterone (MT) for 21 days (Guerrero and Guerrero 1988). MT technology dominates commercial production systems because of its relative ease and effectiveness, but it poses some consumer acceptance issues despite the complete absence of MT in adult fish when procedures are followed correctly. Many methods have been used to control undesirable tilapia reproduction, but hand-sexing combined with co-stocking of a predator species is the most desirable method for subsistence production where simple, durable technologies are preferred.

An alternative approach for commercial production or public sector hatcheries is the use of males with YY chromosomes. The possibility of producing all-male offspring without oral administration of hormones to the fry has stimulated research on the development of "supermale" broodfish that produce all male offspring when crossed with normal females. This may be accomplished by feminization and selected crosses, by feminization and gynogenesis, and by feminization and androgenesis (Popma and Phelps 1998). Kocher (2001) notes that the process produces all-male fish by a breeding scheme that includes sex-reversal of XY males to phenotypic females. Subsequent gynogenesis, or crossing the XY females with normal XY males, will produce some YY "supermales." When mated to normal XX females, these YY males produce nearly 100 percent XY male progeny. Mair and his colleagues (Mair *et al.* 1997; Mair 2001) have articulated the central advantages of this approach, i.e., that it does not involve genetic engineering and no hormones are used in the fish reared for human consumption.

CONCLUSION: TILAPIA AS A GLOBAL LIVELIHOOD

In developing countries, 85 percent of aquaculture is for local consumption. Fish such as tilapia and carp are bred, for example, in village fishponds for consumption within the community or for sale in nearby markets. This type of fish farming is relatively low cost. Although global carp production still exceeds global tilapia production, tilapia is increasing at a much faster rate. A key aspect of the future of aquaculture is that China's role in world fisheries issues cannot be ignored. Even allowing for large margins of error, it is clear that the rate of continued aquaculture development in China and its diffusion to other developing countries are key variables affecting fisheries (Costa-Pierce *et al.* 2003).

A poverty focus would suggest concentrating on aquaculture in developing countries that produce low-value food fish (Edwards 1999; Edwards *et al.* 2002). However, the rosy outlook for high-value aquaculture items such as crustaceans and mollusks in developing country urban markets also suggests the importance of finding ways to keep poor fishers involved in these key sectors (Little *et al.* 2000). While export-oriented industrial and commercial aquaculture practices bring much-needed foreign exchange, revenue, and employment, more extensive forms of aquaculture benefit the livelihoods of the poor through improved food supply, reduced vulnerability, employment, and increased income (Costa-Pierce *et al.* 2003).

Introductions of tilapia may slow, but its already established presence in most tropical nations bodes well for its continued advance as a global commodity. The fact that so many farmers repeatedly elect to use tilapia as a crop in their farming systems suggests that the properties of small- and medium-scale production systems are sufficiently advantageous to motivate sustained practice of the enterprise.

The global food system has awakened to the advantages and prospects of tilapia as a food item. Large-scale industrialized aquaculture operations supply international markets with a standard product processed in ways that ensure quality and safety for the consumer, as well as protecting the reputation and standing of tilapia in the marketplace.

A network of international institutions exists to maintain and advance the culture of tilapia. Innovations in genetics, culture techniques, and feed composition continue to increase the productivity and sustainability of tilapia culture. Although many individuals and private firms are developing broodstock and production systems that advance the tilapia industry, the institutions that provide research, training, and technical assistance to farmers and those working with farmers provide an important leveling function. They ensure that the benefits of aquaculture reach the rural sector in ways that ensure food security and poverty alleviation, things that tilapia have been shown to be efficacious in accomplishing.

Finally, some forces resist globalization as a threat to local autonomy and a source of environmental destruction. A middle ground seeks to ensure that world institutions guide the globalization process to ensure that the developing world maintains access to developed-nation markets, technology, and investment capital. Tilapia is but one stream in the ocean of globalization, albeit one that is meeting both the food security and poverty alleviation needs of growing numbers, as well as contributing to economic growth and international trade for nations desperately in need of an export niche in the world system.

Acknowledgments

This chapter is a revised version of an invited paper presented to the American Fisheries Society symposium *Globalization: Effects on Fisheries Resources*, August 12, 2003, Quebec, Canada. The research was supported by USAID through the Pond Dynamics/Aquaculture CRSP and the Alabama Agricultural Experiment Station, Auburn University. We thank the anonymous reviewers for helpful comments.

References

AquaSol. 2003. *Tilapia Farming*. Available online at www.fishfarming.com/tilapia.html
AUPEC (Agencia Universitaria de Periodismo Científico). 2003. *Bocachico y tilapia, guerra a muerte*. Cali, Colombia: Agencia Universitaria de Periodismo Científico.
Balarin, J. D. 1979. *Tilapia: A Guide to their Biology and Culture in Africa*. Stirling, UK: University of Stirling.
Bonanno, A., Lawrence, B., Friedland, W., Gouveia, L., and Mingione, E. (eds). 1994. *From Columbus to ConAgra: The Globalization of Agriculture and Food*. Lawrence, KS: University Press of Kansas.
Boyd, C. E. and Wood, C. W. 2001. *Aquaculture Best Management Practices as a Possible Focus for Future PD/A CRSP Research*, Mimeo. Auburn, AL: International Center for Aquaculture and Aquatic Environments, Auburn University.

Bureau, D. P., Gibson, J., and El-Mowafi, A. 2002. Use of animal fats in aquaculture feeds. In *Avances en Nutricion Acuicola V, Memorias del VI Simposium Internacional de Nutricion Acuicola*, eds. L. E. Cruz-Suarez, D. Ricque-Marie, M. Tapia-Salazar, and R. Civera-Cerecedo. September 3-7, 2002, Cancún, Quintana Roo, Mexico.

Cape Nature Conservation. 2003. *Clearing Groenvlei Lake of Alien Fish*. Available online at www.capenature.org.za/index.php?fArticleId=31

Coates, D. 1997. Experiences with exotic introduction and stocking in Papua New Guinea. In *Report of the Technical Consultation on Species for Small Reservoir Fisheries and Aquaculture in Southern Africa*, ALCOM Report No. 19, eds. H. W. van der Mheen and B. A. Haight, November 7-11, 1994, Livingstone, Zambia, pp. 36-37.

Costa-Pierce, B. A. 2003. Rapid evolution of an established feral tilapia (*Oreochromis* spp.): the need to incorporate invasion science into regulatory structures. *Biological Invasions* **5**: 71-84. Available online at www.uri.edu/cels/favs/invasionab.html

Costa-Pierce, B. A. and Riedel, R. 2000. Fisheries ecology of the tilapias in the subtropical lakes of the United States. In *Tilapia Aquaculture in the Americas*, vol. 2, eds. B. A. Costa-Pierce and J. Rakocy. Baton Rouge, LA: World Aquaculture Society Books, pp. 1-20.

Costa-Pierce, B. A., Hardy, R., and Kapetsky, J. 2003. *Review of the Status, Trends and Issues in Global Fisheries and Aquaculture, with Recommendations for USAID Investments*, Report of the USAID SPARE Fisheries and Aquaculture Panel (Dr. Barry Costa-Pierce, Panel Chair), Mimeo. Washington, DC: USAID. Available online at http://pdacrsp.oregonstate.edu/miscellaneous/F&A_Subsector_Final_Rpt.pdf

DPI (Department of Primary Industries and Fisheries). 1997. *Fish Guide: Saltwater, Freshwater and Noxious Species*. Brisbane, QLD: Great Outdoors Publications.

DPI. 2001. *Control of Exotic Pest Fishes: An Operational Strategy for Queensland Freshwaters*. Brisbane, QLD: Department of Primary Industries and Fisheries. Available online at www2.dpi.qld.gov.au/fishweb/9091.html

DPI. 2003. *Tilapia* (Tilapia mariae), DPI Note No. f0115 Brisbane, QLD: EDFISH (Fisheries and Wetlands Education Project), Fisheries Communication Unit of Queensland Fisheries Service.

Edwards, P. 1999. Aquaculture and poverty: past, present and future prospects of impact, a discussion paper prepared for the *5th Fisheries Development Donor Consultation*, February 22-24, 1999, Rome, Italy. Available online at www.aqua.ait.ac.th/modules/library/singlefile.php?cid=1&lid=294

Edwards, P., Little, D. C., and Demain, H. 2002. *Rural Aquaculture*. New York: CAB International.

Egna, H. S. and Boyd, C. E. (eds.) 1997. *Dynamics of Pond Aquaculture*. New York: CRC Press.

Eknath, A. E. and Acosra, B. O. 1998. *Genetic Improvement of Farmed Tilapias (GIFT) project*, Final Report. Manila: ICLARM.

FAO (Food and Agriculture Organization). 2004. *Fishery Statistical Database*. Available online at www.fao.org/fi/statist/fisoft/FISHPLUS.asp

FDA (U.S. Food and Drug Administration). 2001. *Fish and Fisheries Products Hazards and Controls Guidance*, 3rd edn. June 2001. Washington, DC: U.S. Food and Drug Administration Center for Food Safety and Applied Nutrition. Available online at www.cfsan.fda.gov/~comm/haccp4.html

Garrett, E. S., dos Santos, C. L., and Jahncke, M. L. 1997. *Public, Animal, and Environmental Health Implications of Aquaculture: Emerging Infectious Diseases*. Atlanta, GA: National Center for Infectious Diseases, Centers for Disease

Control and Prevention. Available online at www.cdc.gov/ncidod/eid/vol3no4/garrett.htm

Grant, E. M. 1997. *Grant's Guide to Fishes*. Brisbane, QLD: Grant Pty Limited.

Guerrero, R. D. and Guerrero, L. A. 1988. Methyltestosterone and the production of all-male fingerlings. In *Proceedings of the 2nd International Symposium on Tilapia in Aquaculture*, eds. T. Bhukaswan, K. Tonguthai, and J. L. Maclean, March 16-20, 1987, Bangkok, Thailand, pp. 183-186.

Gur, N. 1997. Innovations in tilapia nutrition in Israel. *Israeli Journal of Aquaculture/Bamidgeh* **49**(3): 151-159.

Holloway, M. and Hamlyn, A. 2001. *Freshwater Fishing in Queensland: A Guide to Stocked Waters*, 2nd edn. Brisbane, QLD: Department of Primary Industries.

Hoskin, G. P. 1993. *FDA and Aquaculture*, a speech presented by George P. Hoskin to the Joint Committee on Aquaculture, Fish Net Conference Bulletins FN93-06-4/07/93. Washington, DC: U.S. Food and Drug Administration Center for Food Safety and Applied Nutrition. Available online at http://vm.cfsan.fda.gov/~ear/FDA-AQUA.html

Hough, C. 2002. *The Role of Codes of Conduct in Aquaculture*. Copenhagen: Federation of European Aquaculture Producers. Available online at www.aquachallenge.org/workshop_materials/Hough.pdf

Hu, F. B., Bronner, L., Willett, W. C., et al. 2002. Fish and omega-3 fatty acid intake and risk of coronary heart disease in women. *Journal of the American Medical Association*. **287**: 1815-1821.

ICES (International Council for the Exploration of the Seas). 1998. *Code of Practice on the Introduction and Transfer of Marine Organisms*. Copenhagen: International Council for the Exploration of the Seas.

Kocher, T. D. 2001. Tilapia genomics and applications. In *Biotechnology-Aquaculture Interface: The Site of Maximum Impact Workshop*, March 5-7, Shepherdstown, WV. Available online at www.nps.ars.usda.gov/static/arsoibiotecws2001/contributions/Kocher.htm

Little, D. 1998. Options in the development of the "aquatic chicken." *Thailand Fish Farmer* July/August. Available online at www.aquafind.com/articles/opt.php

Little, D., Murray, F., and Kodithuwakkub, S. 2000. Understanding demand: how the poor benefit from tilapia production in the Northwest dry zone of Sri Lanka. *Proceedings of the Department for International Development (DFID) Southeast Asia Aquatic Resources Management Programme Email Conference*, June 2000. Available online at www.dfid.stir.ac.uk/Afgrp/greylit/PS020.pdf

Lopez, M., Chu, P., Reimshuessel, R., Serfling, S., and Geiseker, C. 2004. Determination of 17-alpha-methyltestosterone in tilapia, poster presented at the *10th FDA Science Forum*, Washington, DC, no. A-27. Available online at www.cfsan.fda.gov/~frf/forum04/A-27.htm

Mair, G. 2001. Genes and fish: topical issues in genetic diversity and breeding. *Aquaculture Asia* **6**(4): 16-17.

Mair, G. C., Abucay, J. S., Skibinski, D. O. F., Abella, T. A., and Beardmore, J. A. 1997. Genetic manipulation of sex ratio for the large scale production of all-male tilapia *Oreochromis niloticus* L. *Canadian Journal of Fisheries and Aquatic Sciences* **54**: 396-404. Available online at www.fishgen.com/yysexdet-cjf1.pdf

Merrick, J. R. and Schmida, G. E. 1984. *Australian Freshwater Fishes: Biology and Management*. Brisbane, QLD: Griffin Press.

Molnar, J., Hanson, T., Lovshin, L., and Circa, A. 1996. A global experiment on tilapia aquaculture: impacts of the Pond Dynamics/Aquaculture CRSP in Rwanda, Honduras, the Philippines and Thailand. *Naga, The ICLARM Quantity* **19**(2): 12-17.

Morris, M. C., Evans, D. A., Bienias, J. L., *et al.* 2003. Consumption of fish and ω-3 fatty acids and risk of incident Alzheimer disease. *Archives of Neurology* **60**: 940–946.

Muchiri, M. 2001. Ban on lake fishing makes a lot of sense. *Daily Nation*, February 13, 2001. Available online at www.nationaudio.com/News/DailyNation/13022001/Comment/Comment3.html.

Naylor, R., Goldburg, R., Primavera, J., *et al.* 2000. Effect of aquaculture on world fish supplies. *Nature* **405**: 1017–1024.

Njiru, M. 2003. *Changes in Feeding Biology of Nile Tilapia*; Oreochromis niloticus *(L.) after Invasion of Water Hyacinth in Lake Victoria, Kenya*. Kisumu, Kenya: Kenya Marine and Fisheries Research Institute. Available online at http://filaman.ifm-geomar.de/Training/Abstracts/Kenya/Murithi%20%20Njiru.html

Nuffield Council. 1999. Genetically modified crops: the ethical and social issues. London: Nuffield Council on Bioethics. Available online at www.nuffieldbioethics.org/publications/gmcrops/rep0000000115.asp

Ogutu-Ohwayo, R. and Balirwa, J. S. 2004. *Management Challenges of Freshwater Fisheries in Africa*. Jinja, Uganda: Lake Victoria Fisheries Organisation. Available online at www.worldlakes.org/uploads/AfricanFisheriesManagement_2.3.04.pdf

Popma, T. and Masser, M. 1999. *Tilapia: Life History and Biology*, SRAC Publication No. 283. Stoneville, MS: Southern Region Aquaculture Center. Available online at http://aquanic.org/publicat/usda_rac/efs/srac/283fs.pdf

Popma, T. J. and Phelps, R. P. 1998. Status report to commercial tilapia producers on monosex fingerling production techniques. In *Simpósio Sul Americano de Aquicultura, Aquicultura Brasil* 10, Recife, Brazil, pp. 127–145.

Ramachandran, A. 2000. *Tilapia and Kerala Water Systems*, Report to the Marine Products Export Development Authority. Cochin, Kerala, India: School of Industrial Fisheries of the Cochin University Science and Technology. Available online at www.blonnet.com/businessline/2000/08/04/stories/070405y1.htm

Roth, E. 2002. *Does Globalization of Markets Influence the Aquaculture Challenge in Asia?* Copenhagen: Federation of European Aquaculture Producers. Available online at www.aquachallenge.org/workshop_materials/Roth.pdf

Silvers, K. M. and Scott, K. M. 2002. Fish consumption and self-reported physical and mental health status. *Public Health Nutrition* **5**: 427–432.

Smith, T. 2003. *Transgenic Fish Stay in the Pond: EC-Sponsored Research on Safety of Genetically Modified Organisms*. Brussels: European Commission. Available online at http://europa.eu.int/comm/research/quality-of-life/gmo/07-fish/07-index.html

Steeten, P. 2001. *Globalisation: Threat or Opportunity?* Copenhagen: Copenhagen Business School Press.

Stickney, R. R. 1988. Tilapia. In *Culture of Nonsalmonid Freshwater Fishes*, ed. R. L. Stickney. Boca Raton, FL: CRC Press, pp. 81–116.

Stickney, R. R. 2002. Tilapia update 2001. *World Aquaculture* **33**(2): 12–17.

Tanaka, K. and Busch, L. 2003. Standardization as a means for globalizing a commodity: the case of rapeseed in China. *Rural Sociology* **68**: 25–46.

Waters, M. 1995. *Globalization*. London: Routledge.

Young, J. A. and Muir, J. F. 1998. Tilapia: can the aquatic chicken fly? In *Proceedings of the 9th Biennial Conference of the International Institute of Fisheries Economics and Trade*, eds. B. Aarset and T. Vaasdal, Tromso, Norway, pp. 88–99. Abstract available online at www.nfh.uit.no/iifet98.asp?id=2

WILLIAM W. TAYLOR AND NANCY J. LEONARD

12

The influence of globalization on the sustainability of the North American Pacific salmon fisheries

INTRODUCTION

Throughout history, people in every corner of the world have been harvesting fish for nourishment and economic well-being. These reasons for fishing have persisted, and global fishing activity continues to support people's social, biological, and economic welfare. Fisheries have evolved over time into complex and often interlocking supply chains involving wild and captive production, harvesting, processing, marketing, and distribution systems. This evolution has been largely due to increases in the global consumption of fish products; the development of policies, agreements, and other political tools that facilitate trade between countries; and increases in the financial investments and subsidies for the development of fish production, location, capture, and processing technologies (Arbo and Hersoug 1997; Stone 1997; Harris 1999; Cole 2003; and see Taylor *et al.*, Chapter 1). These technological and governance advancements increase the geographical distribution and intensity of fishing, and the resulting fishing activity has frequently outpaced the natural production of fish biomass. Therefore, many regions of the world are now experiencing the consequences of commercial overfishing, such as fish stock depletion, socioeconomic hardships in fishing communities, and political intervention (Hanna 1997).

The term "globalization" implies that local events affect and are affected by events in other regions. This concept applies to several complex and interrelated aspects of commercial fisheries. For example, ecological globalization of fisheries results when habitat modification and overfishing in one region can significantly affect the population status of highly migratory fish stocks such as salmon throughout their

Globalization: Effects on Fisheries Resources, ed. William W. Taylor, Michael G. Schechter, and Lois G. Wolfson. Published by Cambridge University Press. © Cambridge University Press 2007.

range. Economic globalization of fisheries recognizes that fish market prices, which affect demand and fishing effort, can be influenced by the supply of substitute fish products from other regions of the world. Politically, globalization of fisheries refers to international treaties, subsidies, and trade policies that can influence the activities of both local and distant fishing operations. The cumulative impact of the globalization of the fisheries resource is magnified at the local level through significant alterations of the social fabric of fishing communities (see Frank *et al.*, Chapter 16).

Each of these concepts of globalization will be discussed further in this chapter which examines issues surrounding the sustainability of North Pacific salmon fisheries. Though these salmon fisheries are shared primarily by the United States and Canada, events in other regions of the world significantly affect the sustainability of these fisheries. In turn, the status of North Pacific salmon stocks can influence fishing policies and activities in other parts of the world. The objective of this chapter is to achieve greater understanding of the globally mediated aspects of salmon fisheries management. The rational, equitable, and sustainable harvest of North Pacific salmon must be enacted using a global framework.

NORTH PACIFIC SALMON FISHERIES

North Pacific salmon fisheries are permeated with complex and interrelated ecological, social, economic, and political issues. Current systems of salmon management, production, and harvesting must overcome a fundamental incompatibility: limited resource production and increasing societal demand. Pacific salmon have long played integral ecological and social roles, supporting ecosystem balance, food systems, economic activity, and cultural identity. Indeed, they are among the most economically and culturally significant fishes in North America. Salmon are also among the most vulnerable species; the survival of many salmon stocks is increasingly threatened by diverse human activities, including habitat destruction and overfishing (Bodi 1996). In response to these threats, many genetically distinct populations of Pacific salmon populations – including sockeye (*Oncorhynchus nerka*), chinook (*O. tshawytscha*), coho (*O. kisutch*), and steelhead (*O. mykiss*) salmon species – are currently listed under the 1973 U.S. Endangered Species Act as either endangered (five distinct populations), or threatened (21) or as candidates for listing (four) (NOAA Fisheries 2004).

Globalization and salmon ecology

Salmon have a unique life history (National Research Council 1996), evolving with innate systems of navigation and the ability to live in both fresh water and salt water. Their life cycle takes them through numerous types of habitat and ecological communities (Miller 1996). Salmon hatch from eggs laid on inland stream and river bottoms and migrate to open water. Depending on the species, these fish spend the next 2 to 7 years in open water, continuing to grow and mature. Toward the end of this period, salmon reach sexual maturity and return to the river in which they hatched. After migrating upstream to the approximate location of their birth, salmon reproduce. The long migration and intense reproductive activity consumes nearly all of their energy, and they die. As salmon carcasses decay, they release nutrients and contribute to ecosystem productivity (Miller 1996; Knudsen 2002).

The complex life cycle of salmon intersects a diversity of human activities, many of which threaten the survival of these populations, either directly by fishing pressure or indirectly by reducing the productive capacity of their habitats (Bodi 1996; Knudsen 2002). For example, hydropower dams block migration routes and alter habitat types. Logging operations increase erosion and sedimentation, destroying the pebble substrate in which salmon reproduce. Urban pollutants also affect river water quality and ecosystem balance. The ocean habitat in which salmon spend most their lives can also be degraded by pollution, which further reduces their productivity. Though many of these impacts are localized, some are not. For instance, it has been shown that global climate change and the intensity of El Niño events have significantly influenced salmon production in the eastern Pacific Ocean (Kruse 1998).

In addition to pressures from human land development, salmon are also exposed to various ecological and fishing stressors throughout their life history (Miller 1996). While in oceans, salmon are targeted by domestic, foreign, and multinational commercial fishing fleets. These fleets have become and continue to become more efficient, using sophisticated tracking and harvesting technology, including satellite imagery, aerial tracking, global positioning systems, sophisticated sonar, and vessel-based processing facilities (Fairlie et al. 1995; Hanna 1997). During their upstream migration, salmon are largely targeted by recreational and subsistence fishers. When the forces of overfishing, habitat destruction, and pollution are combined, their cumulative impacts can significantly hinder the productivity and survival of salmon. For example, more than 200 genetic stocks in the U.S. Pacific Northwest are currently considered threatened (Bodi 1996).

While salmon harvests are diminishing in many locations, the production of remaining salmon is shifting because of global climatic fluctuations that have influenced the distribution and production of salmon regionally and globally (Hanna 1997). Since the mid 1970s, sea surface temperatures have increased near Alaska, resulting in increased upwelling and production of zooplankton biomass. Fishery scientists believe these conditions favor the productivity of Alaska's salmon fishery. Indeed, Alaska has landed record harvests in recent years, accounting for 97 percent of the U.S. commercial harvest (Miller 1996). Harvests in British Columbia, Washington, and Oregon as well as in California have in turn decreased substantially. Ironically, rivers in these states and in the province of British Columbia produce most of the salmon that are later harvested by Alaskan fisheries after they migrate north to the more productive waters. In response, some commercial fishers in Canada have maximized their effort to prevent any fish from making it to Alaska (Wood 1995). As a result, intense international conflicts surrounding overfishing, stock interception, and equitable harvests are increasing.

Salmon ecology in North America is highly dependent on local and global influences that affect their population dynamics and the integrity and productivity of their habitats (Hanna 1997; Roberts 1997; Knudsen 2002). Salmon depend on a broad range of habitat types that are distributed over a large area to complete their life history. Stocks that originate in streams of the Pacific Northwest are significantly affected by changes in each of these habitats. An evaluation of the Pacific salmon fisheries has shown that the sustainability of these unique fishes and their fisheries is directly related to the quantity and quality of their habitats, which are related to human activities including land and water development projects, and pollution, and magnitude of fishing intensity.

Globalization and the politics of salmon fishing

Many countries active in the commercial fishing enterprise recognize the need to promote sustainable harvest and equitable allocations, as well as to reduce conflicts among stakeholders. Several international treaties have been proposed and/or ratified, many of which have substantially altered the patterns and rates of fishing (Colson 1995). The United Nations has been instrumental in many of these treaties, such as the 1982 United Nations Convention on the Law of the Sea, which established exclusive economic zones (EEZ), which give ocean-bordering countries rights to regulate and harvest ocean resources up to 200 miles (320 km) from their shores (Marsh 1997; Tahindro 1997). The 1995 U.N. Conference on Straddling Fish Stocks and Highly Migratory Fish Stocks is a multinational agreement that uses population indicators of EEZ-straddling fish, such as salmon, as a precautionary conservation tool in management (Colson 1995; Tahindro 1997).

Throughout the world, other international agreements have been developed to enhance the effectiveness of regional fisheries management. In the case of North Pacific salmon, Canada and the United States have recognized the need for international cooperation in managing salmon fisheries for more than 50 years. During this time, the two governments explored numerous management options and conducted a number of tense negotiations that ultimately led to the development and approval of the 1985 Pacific Salmon Treaty (PST). This landmark agreement, which formalized the international sharing and management of salmon (Miller 1996), contained two main principles: the *conservation principle*, which obliges the parties to prevent overfishing; and the *equity principle*, which allows each country to receive benefits proportional to the quantity of salmon produced in its rivers. In addition to these principles, the PST also formed the Pacific Salmon Commission, a forum for facilitating binational and Native American cooperation and collaboration on related management and research issues. Once fully implemented, the PST was expected to promote significant advances toward sustainability. The equity principle under the PST has yet to be implemented, however, despite extensive negotiations of annual management plans between the parties. This failure of implementation may be related to the diversity of stakeholders, with competing agendas, represented by managers and negotiators. This diversity results in disagreements about the scientific validity of the stock assessment data and hence its interpretation (Ebbin 1996). In turn, this has led to uncertainty in setting allocations and harvest levels (Miller 1996). Citing this uncertainty, the parties have generally been unwilling to agree to any harvest limits under the PST (Wood 1995).

Although a powerful ideological management tool, the PST suffers from an important limitation: it cannot impose any penalties for failure by parties to implement the principles. Therefore, it has done little to redirect the distribution of intercepted salmon. In fact, the disagreements over equity increased greatly when the expiration of the original PST in 1992 left Canada and the United States without a binational framework for managing the shared salmon stocks (Macdonald *et al.* 1999). As a result, the international "salmon war" between Canada and the United States escalated during the late 1990s, culminating in intense debates between the nations on renegotiating the PST as well as intense disputes between rival fishing operations (Wood 1995). These disputes climaxed in the summer of 1997, when Canadian fishers in Prince Edward Sound blockaded a U.S. ferry in protest of U.S. fishing and interception rates (Wood 1997). This tension continued until 1999, when Canada and the United States finally reached a new agreement under the PST that helped resolve some of the disagreements, including a more equitable sharing of the salmon resource (MacDonald *et al.* 1999). The salmon conflicts between U.S. and Canadian commercial operators have had profound impacts on the salmon, including overfishing and depletion of many genetically diverse stocks, which today limit the ability of local and regional wild fish operations to compete with the increasing presence of alternative sources of salmon. This decrease in regional salmon stocks productivity and harvest could ultimately undermine the ecological, economic, and social fabric of North Pacific fishing communities in Canada and the United States, including Native tribes.

Globalization and salmon markets

According to basic supply and demand principles in microeconomics, the demand for salmon in fish markets is inversely related to the supply of salmon and other fish, although other influences, such as marketing can influence this relationship. In general, however, when salmon and other fish are abundant in fish markets, the demand and price for additional salmon are diminished. Likewise, the intensity of Pacific salmon fishing activity is largely determined by their regional market price, which will be low when other fish are available or high as other fish become scarce.

Globalization of fishing markets has intensified the effects of the supply and demand model. With the establishment of 200-mile (320-km) EEZs, the distribution of fishing rights among countries shifted dramatically, closing off fishing areas for some countries (e.g., Spain, Portugal, and Japan) while expanding the fishing potential in other countries (e.g., countries in Latin America, Asia, and Africa) (Arbo and Hersoug 1997; Marsh 1997). As a result, many fish-consuming countries increasingly became dependent on imports at the same time that countries with newly developed fishing industries became important exporters. As explained by Arbo and Hersoug (1997:122), "What a country actually produces is no longer necessarily related to its domestic supply of resources."

Such international exchange has been facilitated by other macroeconomic global trends, such as the general liberalization and harmonization of global trade, industry subsidies for capitalization and technology, and the rise of large multinational fishing corporations (Arbo and Hersoug 1997; Stone 1997). Trade agreements, such as the North American Free Trade Agreement (NAFTA) and the General Agreement on Tariffs and Trade (GATT), were created to reduce barriers to trade and promote free-market conditions. These conditions have enabled large fishing corporations to pursue new multinational mergers and investment strategies to increase their competitive advantage, often forcing smaller-scale fishing operations to less productive areas or out of production (Fairlie *et al.* 1995; see Pollnac, Chapter 9).

In the case of salmon fisheries, this economic globalization has introduced new competitors in the fish markets. Large foreign and multinational operations, able to attract the necessary investment for acquiring new harvesting technologies, have expanded their presence in fish markets once dominated by local and regionally produced salmon. For example, U.S. imports of salmon species grew to approximately 221 959 tonnes in 2003 (up from approximately 47 471 tonnes in 1989), while exports decreased from about 179 933 tonnes in 1989 to 148 378 tonnes in 2003 (NOAA 2004). The tremendous growth in U.S. importation of salmon products indicates a shift in trade balance in favor of imports. Trends in salmon production also indicate that future salmon markets will be increasingly influenced by competition from aquaculture operations (see Naylor *et al.*, Chapter 10) – more than half of all U.S. salmon imported in 2003 (73 percent or 161 756 tonnes) came from salmon farms. At the same time, U.S. exports of farmed salmon rarely exceed 11 percent (2041 tonnes) (NOAA 2004).

What does this mean to fishing operations that target wild salmon populations in the North Pacific? We believe that unless Canadian, Native, and U.S. fishers can overcome their conflicts and adapt to more equitable systems of access and allocation, their salmon fisheries will continue to drop in importance at fish markets, being unable to compete with fish produced in other regions of the world. As long as globalized systems of salmon production and distribution maintain and increase their presence, American, Canadian, and Native peoples from regions of the North Pacific will become increasingly vulnerable to ecological, economic, and social hardships. To be competitive in the global salmon market, North American local, regional, and national communities will need to mitigate significant habitat losses and agree on equitable allocation and access systems to ensure sustainable production of salmon stocks and competitive advantage in the world marketplace for salmon products.

GLOBALIZATION AND SUSTAINABILITY OF SALMON FISHERIES

To mitigate the threats to salmon in the North Pacific, Canadian, U.S., and Native managers and stakeholders must pursue sustainable harvesting and allocation systems. This will require innovation, coordination, and cooperation throughout their management planning and implementation processes. If North Pacific salmon fisheries are to be sustainable, salmon fishers and managers must redefine their relationships with the salmon resources, as well as with competing fishing groups and nations.

The ecological, economic, and political dimensions of North Pacific salmon conflicts are obvious. Globalized habitat changes have shifted the distribution of salmon on the high seas, resulting in distributional shifts and allocation conflicts, economic inequity, and international political intervention. The global ramifications and resulting conflicts are replete with issues related to the sustainability of salmon fisheries, such as risk of overfishing, uncertainty in stock assessments, equitable allocation, trade and subsidy policies, investments in technology, and authority to impose (and ignore) international agreements. Yet the ever-expanding presence of domestic, foreign, and multinational operations, both in aquaculture and capture fisheries, will forever influence the market value and demand for salmon products from the North Pacific. This, in turn, will have permanent affects on the ecological, economic, and social welfare of fishery-dependent stakeholders in the region.

Industrialized commercial fishing operations have evolved into a globalized web of highly efficient – and often destructive – fish harvest and production systems. Though certain people throughout the world are benefiting from their transnational investments and increased distribution of fish products, the globalized system has often resulted in devastating effects on the resource base and dependent communities in many regions of the world (Marsh 1997; Stone 1997). One such region becoming increasingly vulnerable to these global influences is the North Pacific. As in many fishery-dependent regions throughout the world, the sustainability of these fisheries measured by the long-term ability to maintain fish production for ecological, economic, and social welfare (Dixon and Fallon 1989) will increasingly be determined by trends in the globalized fishing industry, which often occurs far from their local communities.

The long-term consequences of unrestricted "free-market" global investments and technologies in fish production, harvest, and distribution may result in irreparable ecological, economic, and social alterations in historically fishery-dependent regions of the North Pacific. Conventional economic theories that promote investment, subsidies, and maximized returns and government decisions to invoke subsidies for fisheries activities must take into consideration the unique nature of renewable natural resources such as fisheries. The production and distribution of such resources are inherently tied to the health of complex fisheries ecosystems and can sustain only a defined and relatively limited level of harvest (Marsh 1997; Roberts 1997; Stone 1997). To surpass this sustainable level will permanently alter the distribution and abundance of salmon. In North America, for healthy, sustainable salmon stocks to occur, it will require diverse stakeholders, from the historic commercial (including Native peoples) to the burgeoning recreational industries, to view salmon ecology and management in a holistic manner, recognizing that the biological capacity of salmon is limited. Those in fishing industries and governments who influence the globalized system of fish production and harvesting must consider these facts and cooperate in shifting to more sustainable strategies (Hanna 1997; Marsh 1997; Stone 1997). Only if this happens will future generations enjoy the benefits provided by this unique and valuable fishery resource.

Acknowledgment

The authors of this chapter benefited greatly from the insights and dedicated efforts of Dr. Kristine D. Lynch to understand the impact of globalization on fisheries resources. Dr. Lynch shared her knowledge on the variety of salmon management principles and practices across disciplines and regions of North America (Lynch *et al.* 2002). She was instrumental to our thought processes and provided the needed encouragement to bring this chapter to conclusion. We also appreciate the efforts of Dr. Michael Schechter, who started the authors thinking about the role of globalization as a driving force in fisheries governance, particularly related to interjurisdictional management.

References

Arbo, P. and Hersoug, B. 1997. The globalization of the fishing industry and the case of Finnmark. *Marine Policy* **21**: 121–142.

Bodi, F. L. 1996. Saving salmon in the Northwest. *Forum for Applied Research and Public Policy* **11**: 46–49.

Cole, H. 2003. Contemporary challenges: globalization, global interconnectedness and that "there are not plenty more fish in the sea": fisheries, governance and globalization – is there a relationship? *Ocean and Coastal Management* **46**: 77–102.

Colson, D. A. 1995. Current issues in international fishery conservation and management. *U.S. Department of State Dispatch* **6**(7): 100–105.

Dixon, J. A. and Fallon, L. A. 1989. The concept of sustainability: origins, extensions, and usefulness for policy. *Society and Natural Resources* **2**: 73–84.

Ebbin, S. A. 1996. The stock concept: constructing tools for pacific salmon management. *Coastal Management* **24**: 355–364.

Fairlie, S., Hagler, M., and O'Riordan, B. 1995. The politics of overfishing. *The Ecologist* **25**(2–3): 46–74.

Hanna, S. S. 1997. The new frontier of American fisheries governance. *Ecological Economics* **20**: 221–233.

Harris, M. 1999. *The Lament for an Ocean: The Collapse of the Atlantic Cod Fishery: A True Crime Story*. Toronto, Ontario: McClelland & Stewart.

Knudsen, E. E. 2002. Ecological perspectives on Pacific salmon: can we sustain biodiversity and fisheries? In *Sustaining North American Salmon: Perspectives across Regions and Disciplines*, eds. K. D. Lynch, M. L. Jones, and W. W. Taylor. Bethesda, MD: American Fisheries Society, pp. 277–320.

Kruse, G. H. 1998. Salmon run failures in 1997–1998: a link to anomalous ocean conditions? *Alaska Fishery Research Bulletin* **5**: 55–63.

Lynch, K. D., Jones, M. L., and Taylor, W. W. (eds.) 2002. *Sustaining North American Salmon: Perspectives across Regions and Disciplines*. Bethesda, MD: American Fisheries Society.

MacDonald, D. D., Hanacek, M., and Genn, L. 1999. Salmon, society, and politics: moving toward ecosystem-based management for Pacific salmon. In *Sustaining North American Salmon: Perspectives across Regions and Disciplines*, eds. K. D. Lynch, M. L. Jones, and W. W. Taylor. Bethesda, MD: American Fisheries Society, pp. 321–368.

Marsh, J. B. 1997. North Pacific fisheries environment: international issues. *Contemporary Economic Policy* **15**: 44–51.

Miller, K. A. 1996. Salmon stock variability and the political economy of the Pacific Salmon Treaty. *Contemporary Economic Policy* **14**: 112–129.

National Research Council. 1996. *Upstream: Salmon and Society in the Pacific Northwest*. Washington, DC: National Academy Press.

NOAA. 2004. *Annual Product Data by Country/Association: NOAA-Foreign Trade Information*. Available online at www.st.nmfs.gov/st1/trade/trade_prdct_cntry.html

NOAA Fisheries. 2004. *Endangered Species Act Status of West Coast Salmon and Steelhead*. Available online at www.nwr.noaa.gov/1salmon/salmesa/pubs/1pgr.pdf

Roberts, C. M. 1997. Ecological advice for the global fisheries crisis. *Trends in Ecology and Evolution* **12**: 35–38.

Stone, C. D. 1997. Too many fishing boats, too few fish: can trade laws trim subsidies and restore the balance in global fisheries? *Ecology Law Quarterly* **24**: 505–544.

Tahindro, A. 1997. Conservation and management of transboundary fish stocks: comments in light of the adoption of the 1995 Agreement for the Conservation and Management of Straddling Fish Stocks and Highly Migratory Fish Stocks. *Ocean Development and International Law* **28**: 1-58.

Wood, C. 1995. Northern defiance: tempers flare in Pacific salmon war. *Maclean's* **108**(30): 12-14.

Wood, C. 1997. Scaling down the B.C. salmon wars. *Maclean's* **110**(32): 16-17.

Part III Governance and multilevel management systems

GRANT FOLLAND AND MICHAEL G. SCHECHTER

13

Great Lakes fisheries as a bellwether of global governance

INTRODUCTION

The purpose of this chapter is both straightforward and unique: to apply a global governance approach (Brühl and Rittberger 2001)[1] to the study of Great Lakes (and particularly Michigan) fisheries. Case studies in global governance are rare; case studies focused in the United States are even rarer. Although students of international relations have done some work on fisheries (Peterson 1993) and international law scholars much more,[2] little of that is recent enough to apply and contribute to the insights of the global governance literature.[3]

---

[1] Three major models of global governance have been distinguished: authoritative coordination by a world state; governance under a hegemonic umbrella; and order as a result of horizontal self-coordination, governance without government (as with the classic balance of power). In the third model, which mostly accords with Great Lakes fisheries management, "the coordination of international activities is affected by states agreeing, for their mutual benefit, upon norms and rules to guide their future behaviour and to create mechanisms which make compliance with these rules and norms possible (i.e. in each actor's self-interest)." Kaye (2000) has argued that the Pacific "tuna war" in the 1980s between the United States and the Pacific island micro-states "provides strong evidence to suggest an absence of an identifiable hegemon" in the context of marine living resource management.

[2] There has been a recent mini explosion of literature that tries to break down distinctions between international relations and international law. Though that is a welcome trend, it has not yet focused attention to the global governance literature, in general, or fisheries governance, in particular. Much of it has focused on the lawyers' interest in the sources and impact of "soft law" (the non-legally binding rules of international institutions) and international relations scholars' interest in the related notion of norms, including their creation and impact. See, for example, Beck et al. (1996) and Slaughter et al. (1998).

[3] Barkin and DeSombre (2000) are an interesting exception, but given the thrust of their article, they do not build specifically on the global governance literature.

Globalization: Effects on Fisheries Resources, ed. William W. Taylor, Michael G. Schechter, and Lois G. Wolfson. Published by Cambridge University Press. © Cambridge University Press 2007.

Moreover, an inquiry such as this one seems in keeping with Francis' notion of what is needed to garner further insights for "realistic applications of the 'ecosystem approach'" to the Great Lakes:

> More attention needs to be given to the study of these Great Lakes actor systems and to their "dynamics" over time. Such study could provide a common focal point for the diverse disciplines and specialties that comprise the social sciences and the humanities. The subject is well within their domain, in a way that the Great Lakes viewed only as hydrological–biological phenomena never was. (Francis 1987:235)[4]

Thus this paper's purpose is to investigate "the processes and institutions, both formal and informal, that guide and restrain the collectivities of a group" (Keohane and Nye 2002) to begin to answer the question "of how the various institutions and processes of global society can be meshed more effectively in a way that would be regarded as legitimate by attentive publics controlling access to key resources" (Keohane 2002) (of the Great Lakes, in this instance).

The thesis of this chapter is equally straightforward and bold, namely, that the phenomena that have come to characterize global governance characterize Great Lakes and Michigan fisheries and, in many instances, actually were foreshadowed by Great Lakes and Michigan fishery management. We understand global governance not as global government but as a framework for the crystallization of norms and more formal rules necessary to address global problems, articulated and implemented by a set of public and private institutions. Thus "[g]lobal governance refers to multilevel governance which includes not only levels of policy-making beyond the nation-state but also the subnational (i.e., regional or local) levels" (O'Brien *et al.* 2000; Brühl and Rittberger 2001). It is a policy-making process wherein the influence of non-state actors (especially non-governmental organizations, NGOs) has been on the incline and which is characterized by increased openness, accountability, and transparency, and debates over the proper levels, degree, and forms of management.

In at least five areas, Great Lakes and Michigan fisheries and fisheries policy have embodied or anticipated trends that have come to be articulated under the rubric of global governance: (1) global norms such as biodiversity and the ecosystems approach were voiced

---

[4] Such an approach also presents a challenge to international legal scholars because of their statecenteredness, whereas ecosystems, as exemplified by the Great Lakes basin, rarely correspond to state boundaries (Kaye 2000).

in the Great Lakes and Michigan discourse before they were codified in a more global one; (2) the rise, increased influence, strategies, and functions of Michigan NGOs foreshadowed those operating through a global governance framework; (3) the increased openness, accountability, and transparency – in the argot of globalization – of policy processes to non-governmental, grassroots, and subnational actors were presaged by the more diffuse and open to stakeholder pressure processes in the Michigan Department of Natural Resources (MDNR), the Great Lake Fishery Commission (GLFC), and the International Joint Commission (IJC); (4) Michigan fisheries policy has long been, or has attempted to be, integrated into a larger framework of shared sovereignty;[5] and (5) Great Lakes governance, including the IJC and the GLFC, has also been wrestling with what globalization scholars refer to as subsidiarity (Begg *et al.* 1993; CDLR 1994; Knight 2000; Swaine 2000).[6] That is, Great Lakes fisheries policy-makers, analysts, and advocates have debated over the optimal level or site of instruments and institutions to regulate fishing activities and issues of access and influence of public citizens (Becker 1993). In Francis' perspective, the answer seems to be found in the "flexibility" of governance, i.e., how quickly it is susceptible to modification or change (Francis 1990), whereas global governance scholars and policy-makers focused on issues of subsidiarity usually conclude that different levels of policy-making are better suited for different sorts of issues, as there are costs

---

[5] This is not to suggest that unilateral actions are entirely absent in Great Lakes fisheries policy and practice any more than they are in global politics. Among the most noteworthy was Michigan's 1964 decision to introduce salmonids into the Great Lakes, a decision at odds with the preferences of the U.S. Bureau of Commercial Fisheries (Tanner and Tody 2002).

[6] Subsidiarity, a principle originally derived from Catholic theology, is most frequently connected with governance in the European Union. It is defined as the constant search for a decision-making level as close to the citizen as possible. It can be contrasted with decentralization, in which power is delegated down. Subsidiarity assumes that there are inherent powers at the lower level and thus power can be allocated upward as well as downwards, although it incorporates a presumption in favor of allocation downward in case of doubt. Subsidiarity raises issues similar to those raised in discussions of federalism, but federalism is normally thought of as limited to governmental participation, whereas subsidiarity's emphasis on citizens opens up space for NGOs. Moreover, subsidiarity calls for decisions to be made closest to the citizen to establish greater democratic control, whereas federalism argues for greater localism because that is expected to better reflect the population's interests.

and benefits to both centralization and highly democratic decision-making modes (Begg et al. 1993).

As will be seen, for the most part changes in norms, policies, and policy processes in the Great Lakes stem from responses to the degradation and overexploitation of the environment, which led ultimately to a collapse of commercial fishery and a shift toward a recreational one. Thus, somewhat akin to the scholars of successful regional economic institutions, we argue that the evolution of global institutions as they relate to Great Lakes fisheries is largely a consequence of a set of exogenous events, not all of which many would characterize as positive, most notably the task expansion of the IJC pursuant to the negotiation of the Great Lakes Water Quality Agreement, which can be understood as at least a "partial response to Lake Erie's much publicized decline" and the establishment of the GLFC as a response to the sea lamprey predation of the early 1950s.[7] As Chiarappa and Szylvian explain:

> With rates of occupation attrition running high, Lake Michigan commercial fishers turned to the federal government for aid, and tempered their longstanding opposition to international co-management schemes with Canada by supporting the 1955 establishment of the Great Lakes Fishery Commission under Pres. Dwight D. Eisenhower. More specifically, commercial fishers who found their traditional management systems both marginalized and incapacitated by the sea lamprey problem could do little but watch as state and federal fisheries managers assumed an increasingly aggressive stance in guiding the future of Lake Michigan's fisheries. Finding themselves forced to coexist and, at times, totally capitulate to fisheries managers in a manner that was unprecedented on the Great Lakes, commercial fishers saw the beginning of a fractured relationship with government regulators that culminated in the planting of Pacific salmon. (Chiarappa and Szylvian 2003:117–118)

The shift in power within the GLFC from commercial fishers to sport fishermen followed in due course (Chiarappa and Szylvian 2003).

---

[7] The Convention on Great Lakes Fisheries between the United States and Canada, which established the GLFC, came in the aftermath of a number of failed attempts at creating an international institution for the protection and perpetuation of Great Lakes fisheries (GLFC 2000; Inscho and Durfee 1995). The GLFC had two missions from the outset: eliminating the sea lamprey and restoring self-sustaining populations of lake trout. Though it has made considerable progress in controlling the sea lamprey, the lake trout population is not self-sustaining, except in Lake Superior.

GLOBAL GOVERNANCE, LAWS, INTERGOVERNMENTAL INSTITUTIONS, AND FISHERIES MANAGEMENT

Globally, fisheries are governed by a constellation of hard[8] and soft[9] law (Franck 1990; Edeson 1999; Hillgenberg 1999; Abbott and Snidal 2000), states, regional, and universal intergovernmental and nongovernmental organizations. Hard law gives form to the international fisheries management regime – described by scholars as the "most complex" environmental regime (Porter *et al.* 2000) – chiefly through the 1982 Third Law of the Sea Convention (LOS III).[10] The convention, arrived at by governments negotiating through the aegis of the United Nations Conference on the Law of the Sea (UNCLOS), limits the "varying degrees of sovereignty and jurisdiction along with rights and responsibilities" (Kimball 2001) that these states have in regard to fisheries and toward the seas in general. Moreover, LOS III serves as a "unifying framework for a growing number of more detailed international agreements ... [for] the management of marine resources" (Kimball 2001). Hard law governing global fisheries was elaborated upon with the 1995 Agreement Relating to the Conservation and Management of Straddling Fish Stocks and Highly Migratory Fish Stocks, or more simply, the Fish Stocks Agreement (FSA). The FSA

---

[8] Hard law is generally understood as including treaties (universal, regional, or bilateral), customary international law, and general principles of international law. They are "hard" because they are legally binding. Examples include the Agreement to Promote Compliance with International Conservation and Management Measures by Fishing Vessels on the High Seas, 1993; the Agreement for the Implementation of the Provision of the United Nations Convention on the Law of the Sea of 10 December 1982, relating to the Conservation and Management of Straddling Fish Stocks and Highly Migratory Fish Stocks, 1995; and the Great Lakes Water Quality Agreement as Amended, 1987.

[9] Soft law is taken to include international norms, General Assembly resolutions, and similar documents (such as the Rio Declaration, Agenda 21, the Rome Consensus on World Fisheries adopted by the Food and Agriculture Organization's Ministerial Conference on Fisheries, the Rome Declaration on the Implementation of the Code of Conduct for Responsible Fisheries, and the Declaration of the International Conference on Responsible Fishing, the so-called Declaration of Cancún) issued by intergovernmental institutions. Such documents, although not legally binding, can, over time, have considerable "compliance pull." Although increasingly important, soft law as it relates to regimes governing fishing has been "only marginally covered in legal literature."

[10] Although the United States has not ratified LOS III, it considers most of it customary international law, including those elements relating to fisheries.

seeks to guide the implementation of LOS III, with special care that states exploiting stocks within their 200-mile (320-km) exclusive economic zones granted by the Convention do not overexploit highly migratory stocks. Such stocks include the commercially lucrative tuna, whose migration patterns reveal an utter contempt for norms of state sovereignty. In addition to this legally binding hard law there are two key soft law accords: the 1995 Code of Conduct for Responsible Fisheries and the 1999 Food and Agriculture Organization International Plan of Action for Management of Fishing Capacity. Each articulates recommendatory norms for the sustainable exploitation of fish stocks, what Bratspies suggestively refers to as "environmental stewardship" (Bratspies 2001).

The legal framework characterizing global fisheries governance is at once the cause and the effect of states and non-state actors working through both regional and universal fisheries organizations. Such regional organizations include the Northwest Atlantic Fisheries Organization. The sole universal intergovernmental organization charged with management of fish stocks is the Food and Agriculture Organization's (FAO) Committee on Fisheries (COFI). These bodies serve as both political and scientific forums where treaties may be negotiated and new scientific data generated, disseminated, and, we hope, incorporated into global fisheries governance. That is, such intergovernmental organizations are at once arenas for state activity and at the same time, semi-autonomous actors unto themselves. As commercial fisheries the world over have become strained to the point of collapse, COFI's secretariat has become proportionately more assertive in pushing for new and more comprehensive norms to sustainably exploit fish stocks (Porter *et al.* 2000).

GLOBAL GOVERNANCE AND GREAT LAKES AND MICHIGAN FISHERIES MANAGEMENT

The global fisheries governance regime may be the world's "most complex," but it has no *direct* effect on Michigan fisheries or fisheries policy. Coastal U.S. states are integrated into the international fisheries governance regime through participation in regional fisheries bodies such as the various Pacific and Atlantic fisheries commissions, which in turn have relationships with COFI, but the regional intergovernmental body in which Michigan's fishery is integrated – the GLFC – has no relationship with COFI. Moreover, the Great Lakes' chief soft law document, the Joint Strategic Plan for Management of Great Lakes Fisheries (or simply the Joint Strategic Plan) – proposed in December 1980, adopted in 1981, and most recently revised in 1997 – has found virtually no place in the global governance literature. Many states' fisheries are affected by the global fisheries regime insofar as it articulates norms and regulations regarding commercial fishing practices; Michigan's commercial fishery, as noted above, declined decades ago and has been reincarnated as a recreational one (Brown *et al.* 1999), a topic heretofore not studied by students of global governance, most of whom focus on issues of international political economy rather than a framework for fisheries activities such as sport angling or fly fishing. Even though the Great Lakes, including Michigan, have all of the elements of multi-jurisdictional governance that define the global governance literature, they are overlooked because that literature has focused on multilateral institutions to manage commercial harvest rather than the recreational fishery that has dominated Michigan fisheries at least since the global governance literature appeared.

All of this is not to say, however, that a global governance approach is analytically bankrupt in examining the case of Great Lakes and Michigan fisheries and fisheries policy. In fact, as noted at the beginning of this chapter, the experience of Michigan fisheries may be seen, in certain regards, as a bellwether for fisheries global governance, including the fact that global norms such as those of biodiversity and the ecosystems approach were voiced in the Michigan discourse before they were codified in a more global one.

THE EVOLUTION OF GLOBAL NORMS: GREAT LAKES AS A BELLWETHER OF THE ECOSYSTEM APPROACH

The emergence and codification of international norms have historically been motivated by exogenous, contingent crises among other things. The norm of sovereignty, as codified by the Peace of Westphalia in 1648, was motivated by the crisis of the Thirty Years War, just as humanitarian norms contravening sovereignty have developed in response to specific crises such as the Holocaust and ethnic cleansing campaigns in the Balkans. Less dramatically but still importantly, case studies of biological and climate change regimes show that norms and institutions coalesced around these issue areas in response to exogenous crises. Global norms and regimes in the aforementioned issue areas became codified largely in response to international crises of the late 1980s and 1990s, such as global warming, the growing hole in the ozone layer, and the destruction of an entire ecosystem, i.e., the Brazilian rainforest (Arts 1998).

The ecosystem approach to natural resources policy is generally defined as a "holistic orientation toward resource management" (MacKenzie 1996). That is, it recognizes the interdependencies of ecologies and thus seeks policies that are "comprehensive in scope" and "integrated in content" (Mackenzie 1996).[11] It is only relatively recently, however, that this norm entered the consciousness of the international natural resources policy framework, and perhaps only implicitly, via the familiar 1987 Brundtland Commission. Just 15 years later, ecosystem management is a norm on its way to crystallization as it has now come to be widely advocated by governments and agencies worldwide (Hartig *et al.* 1998). The appearance of this norm in the Michigan natural resources policy discourse and attempts to integrate it into policies predate its incorporation in the global discourse. Some scholars have suggested that an ecosystem approach to migratory animals – such as fish – "began to emerge in this part of the world around 1968" (Dobson *et al.* 2002). Perhaps it first functionally entered the Great Lakes and Michigan policy framework through the IJC's 1978 Great Lakes Water Quality Agreement (GLWQA).[12] The 1978 agreement "embraced ecosystem concerns as well as concerns for toxic substances per se and included expanded responsibilities" for the IJC (Ryder and Orendorff 1999). Toward the end of implementing this new strategy, 43 areas of concern (AOCs) were identified in the Great Lakes basin and remedial action plans (RAPs) were designed to mitigate them. The GLWQA's emphasis on the ecosystem approach was introduced into

---

[11] Francis relates the scope of Great Lakes governance to one's "interpretation of what an ecosystem approach must entail. At the very least it should embrace matters being dealt with under binational agreements that concern the Great Lakes ... In addition, there are different configurations for governance over major ecosystem components of the basin; i.e., the atmosphere (or 'atmospheric region of influence' over the basin, which can be of continental or even biospheric scale); the lakes and connecting channels (rivers); tributary rivers and watersheds; groundwater aquifers; and coastal waters. Arrangements are also organized around seven distinct water uses: commercial navigation; hydropower generation and cooling water; domestic and industrial water supply; effluent disposal; sport and commercial fisheries; wildlife; and water-based recreation other than hunting and fishing" (Francis 1990:197–199).

[12] Michigan is obviously not a party to the binational GLWQA, but of the 43 areas of concern noted in Annex 2 of the agreement, 14 are within Michigan's jurisdiction. Accordingly, communities across Michigan have taken action to restore their local environments under Michigan's Action of Concern Program, a multiagency, locally driven team approach under the overall guidance of the respective remedial action plans mandated by the agreement.

Michigan fisheries policy through the GLFC's 1981 Joint Strategic Plan, to which the Michigan Department of Natural Resources is a party. Here, ecosystem management pursuant to the GLWQA came to be codified as one of four strategies for fisheries management, providing for "stronger links between agencies and the Great Lakes Fishery Commission's Habitat Advisory board" (GLFC n.d.). The ecosystem approach was further incorporated into Michigan natural resources and fisheries policy through a November 1994 binational workshop between the U.S. Environmental Protection Agency (EPA) and Environment Canada under the aegis of the IJC and Wayne State University (Hartig *et al.* 1998). Most recently, the commission's Strategic Vision for the First Decade of the New Millennium – or simply, the Strategic Vision – described the ecosystem approach as its fundamental concept because it is "well suited to address complex problems with extensive linkages such as the introduction of unwanted non-native species; toxic chemicals in fish; and non-point pollution sources" (GLFC 2001). Such a normative shift in policy – from a single-species approach to management to a more holistic ecosystem approach – was precipitated by a material shift. It could not, of course, emerge *ex nihilo* divorced from context. It was the collapse of the Great Lakes salmon stocks two decades ago with the lack of "obvious single species solutions" that "forced the fisheries management community to move their management strategy away from a single species focus to an ecosystem focus" (Ferreri *et al.* 1999). Thus the ecosystem approach has long been part of the Michigan fisheries and natural resources policy framework but has only recently come to be articulated at the global level. Fisheries management at a global governance level, however, remains concerned primarily with single-species management or a range of target species (Kimball 2001). The context that precipitated the shift toward the ecosystem approach in the Michigan discourse suggests that perhaps the recent acceleration of collapsing fisheries globally explains why the norm now is gaining currency at the global level.

The norm of biodiversity – defined as "the variability among living organisms from all sources, including, *inter alia*, terrestrial, marine, and other aquatic ecosystems and the ecological complexes of which they are part" (Article 2, 1992 United Nations Framework Convention on Biological Diversity – see United Nations [1992]) – is a relatively recent accretion to the global normative milieu. In comprehensive case studies of the Climate and Biodiversity Conventions, Arts (1998) has persuasively argued that "[t]he notion of *biodiversity* became popular only recently" and that this norm has become codified in only the Convention on Biological Diversity. Though the 1972 United Nations Conference on the Human Environment (UNCHE) and resultant Stockholm Declaration have been called the "genesis of modern environmental law" (Bowman 1996), the Conference and Declaration make no mention of any biodiversity norm, nascent or otherwise (Arts 1998). The origin of the biodiversity treaty has been traced to recommendations of the 1984 General Assembly meeting of the IUCN (Arts 1998), but it was only in 1987, with the third principle of the Brundtland Report, that the norm of biodiversity was explicitly articulated in an official context. The way in which it was articulated, not as "biodiversity" but as the longhand "biological diversity," testifies to the newness, then, of the norm. Biodiversity as a norm subsequently crystallized at the 1992 United Nations Conference on the Environment and Development (UNCED) and Convention on Biological Diversity, opened for signature at Nairobi May 22, 1992, and entered into force just 18 months later following its thirteenth ratification.[13]

The norm of and the need for the conservation of biodiversity entered into the Michigan natural resources policy discourse and framework before the norm was globalized by the reports, conferences and conventions of the late 1980s and early 1990s. Biodiversity became a policy concern for Michigan natural resources policy-makers – among others – in the 1960s. It was in the 1960s that Great Lakes policy-makers began to acknowledge the need to protect indigenous species against the unintentional introduction of exotic species (Ryder and Orendorff 1999). In the Great Lakes context, biodiversity was elaborated upon with the GLWQA of 1978 and the expanded role of the IJC, where its new strategy of ecosystem based management "involved the measurement of biodiversity at different trophic levels in order to determine the health of the Great Lakes basin ecosystem" (Ryder and Orendorff 1999). One could even argue that, from its inception in 1954, the GLFC put a premium on the conservation of biological diversity – one of its key mandates has been the mitigation of the adverse effects of the sea lamprey on the biological diversity of the lakes. Indeed, it was the issue of biodiversity itself that led to the genesis of the GLFC. The loss of biodiversity vis-à-vis the loss of the lake trout in the Great Lakes was what led to the establishment of the GLFC in 1954 (GLFC 2001). Commitment to biodiversity by the GLFC continues to this day and is explicit in the first principle of its vision statement on healthy

---

[13] Asserting this is not meant to imply that all of the provisions of the Convention are customary international law or necessarily binding on third parties.

ecosystems, where it pledges to "conserve native biodiversity." Toward that end the GLFC promises that "native species will not be lost from any Great Lake" while it contines with its rehabilitation of lake trout and efforts at its natural reproduction (GLFC 2001).

THE ROLE OF NGOs IN GLOBAL GOVERNANCE AND IN THE GREAT LAKES AND MICHIGAN FISHERIES MANAGEMENT

The experience of Michigan natural resources policy vis-à-vis fisheries has been ahead of the global curve not only in anticipating certain international norms but also in giving NGOs roles in the policy-making process. Whereas most of the academic literature treats the emergence of NGOs, their proliferating numbers (especially development-related NGOs), and their global influence as a relatively recent phenomenon – often temporally coincident with economic globalization or the growth of the global women's, environmental, or human rights movements – NGOs and grassroots actors have long been present and have had an influential role in Michigan fisheries policy, and some have gone on to become national forces as change agents in natural resources management. Further, NGOs operating within the Michigan and Great Lakes context perform the same functions and strategies that NGOs operating in the global milieu have come to perform. NGOs operating in Michigan and at the international level – especially in matters relating to environmental policy – are known for "agenda setting"; pressuring, cajoling, and otherwise influencing government policy; participating in negotiations and overseeing implementation of policy; monitoring and verifying compliance with agreements; and influencing "values, social behavior and collective choice more generally among large groups of people – creating a form of 'world civil politics' in which state behavior becomes less central to collective choice" (Conca 1996:104; see also Victor *et al.* 1998; Wapner 1998; Porter *et al.* 2000).[14] Just as NGOs in the global governance framework

[14] In a study that includes NGOs operating in the Great Lakes fisheries arena, Born and Stairs (2002) note that NGOs are beginning to get involved in the planning aspect of watershed and aquatic resources. Their study underscores that this development is not simply "immature," but has been difficult in part because some bureaucrats' views that citizens have little to offer them and that NGOs are biased. This selfsame hesitation has been noted in the global governance literature. Indeed, hesitation has been greater by the International Monetary Fund, where almost all of the professionals are economists, than in the World Bank, where a wider variety of social scientists are employed (see O'Brien *et al.* 2000).

have been portrayed as coming to pressure policy-makers by wielding political and, especially, intellectual capital, so have Michigan-based NGOs come to constitute paradigmatic Haasian "epistemic communities" (i.e., groups or networks of specialists with recognized expertise in policy-relevant knowledge areas) (Haas 1992), influencing policy through the collection and dissemination of consensual scientific knowledge.[15] At times, their knowledge is solicited by those formally charged with making decisions. The role of the 1971 Symposium on Salmonid Communities in Oligotrophic Lakes (SCOL I) is a classic example: "an important stimulus leading to a broader thrust in thinking about fish and fisheries within the context of a lake ecosystem." Its relationship to policies affecting the Lake Ontario ecosystem has been recently recounted (Mills et al. 2003). The GLFC's commissioned research is a somewhat different but also relevant example.

Any treatment of the deep roots of grassroots NGO involvement in Michigan natural resources policy must include an analysis of the rise of the Michigan United Conservation Clubs (MUCC). MUCC was established on November 9, 1937, by an alliance of 35 conservation organizations (MUCC 2002). Since then, it has expanded to include 100 000 members as well as 500 affiliated clubs, making it the largest statewide conservation NGO in the United States (MUCC 2002). MUCC is able to mobilize these vast numbers through its grassroots organizational structure, made up of 20 statewide districts consolidated into four regional bodies. This grassroots strategy of political influence is accompanied by other strategies of overt political influence such as lobbying offices in Washington, DC, and Lansing, MI, as well as mounting court challenges and assisting in policy implementation. Notably, MUCC was an influential amicus in the Supreme Court battles of the 1980s. In *Michigan v. United States*, MUCC and the Michigan Department of Natural Resources (MDNR) took Native American tribes to court to contest their use of gill nets, which MUCC considered to be environmentally harmful as well as unfair to its members (Folland interview with Schroeder 2003). In addition to this political

---

[15] Great Lakes focused NGOs, like Great Lakes United (GLU), also perform the role of NGOs frequently connected to global ad hoc conferences, namely as consciousness-raisers. In 1986, 4 years after its founding, GLU held more than a dozen hearings around the Great Lakes to make people aware of the Great Lakes Quality Agreement and the IJC, to which periodic reports were to be submitted. Not only were the hearings well attended, but attention to and participation in the IJC's biennial meetings increased dramatically. More recently (in June 2003), GLU issued a Green Book, or citizens' guide, to actions that can be taken to restore the Great Lakes–St. Lawrence River ecosystem (GLU 2003).

"hard power" of grassroots mobilization, direct lobbying, and court challenges, MUCC also employs more subtle strategies to influence fisheries policy. For example, it hires scientific experts to act as liaisons between their sportfishing lay members and MDNR. This allows members to articulate their concerns to a proxy, who because of his or her expert knowledge is able to speak the same – in the words of one such expert – "biologist language" as MDNR officials and thus through such intellectual credibility gains entrée into the policy-making process (Folland interview with Schroeder 2003). The success of MUCC in this regard, as well as other NGOs to be considered, squares nicely with empirical research that attempts to explain the rise in influence of global NGOs, as for example, Arts finds regarding global NGO influence on the Biodiversity and Climate Conventions (Arts 1998:257–260).

MUCC, therefore, like many NGOs at the global level, trades in multiple currencies of influence, from political capital to intellectual capital. Its success in doing so, and its doing so ahead of the global governance curve, has been precipitated by the same material changes to the local political context that caused the normative shift and anticipation of the ecosystem approach in Michigan fisheries – that is, the effective collapse of Michigan's commercial fishery and its replacement by a recreation focused one in the 1960s and 1970s (Dobson et al. 2002). As the recreational fishery expanded, grassroots-based anglers, sportfishers and conservationists came to have more influence over fisheries policy than they did when Michigan's fishery was managed as a faltering commercial one. Nevertheless, the very rise and emergence of NGOs and their influence in the case of Michigan fisheries have also, paradoxically, led to a decrease in MUCC influence. MUCC in the 1960s and 1970s commanded perhaps as many as 137 000 members; today membership is substantially lower. Though this is a multivariate phenomenon (Schroeder interview 2003), one key reason for the dwindling numbers is a proliferation of NGOs in the Michigan arena. As greater numbers of conservationist NGOs have entered the fold, from Trout Unlimited to the Charter Boat Association[16] to the Steelheaders[17] and others, MUCC has had to compete for members.

---

[16] The Michigan Charter Boat Association (MCBA) is an educational NGO, focused on educating its members about issues affecting the charter industry.

[17] The Michigan Steelhead and Salmon Fishermen's Association (MSSFA), often simply referred to as Michigan Steelheaders, or simply Steelheaders, is an educational and advocacy NGO, aimed at promoting sportfishing in the Great Lakes and their tributary streams and rivers and the wise management of anadromous

A broad-based NGO such as MUCC finds it tough to compete in the face of a proliferation of specialized NGOs which are often believed better able to advance the goals of clienteles concerned with specific agendas or issue areas. Such NGOs are perceived as more legitimate lobbyists. Such a perception has been long understood as the most essential variable in explaining lobbyists' influence (Cohen 1959).

Trout Unlimited (TU) is another significant Michigan-based NGO which embodies the trend in global governance toward the emergence and increased influence of non-state actors. TU, like MUCC, has deep roots, and anticipated this trend in global governance. TU has been helping to shape Michigan fishery policy for more than four decades. Also as in the case of MUCC, TU combines the political capital of its grassroots mobilization of 125 000 volunteers and 500 chapters nationwide (Trout Unlimited 2002a), lobbying efforts in Washington and legal expertise with intellectual capital. TU collects and disseminates scientific knowledge in the spirit of a true "epistemic community" toward the end of developing and recommending policies that are, in the words of TU's first president, "substantially correct, both morally and biologically" (Trout Unlimited 2002a). TU is a more specialized NGO than MUCC, seeking the conservation of trout and salmon fisheries specifically. Ironically, this species-specific mandate led to TU's first successful policy to change its "put-and-take" policy of trout stocking and start managing for wild trout along the lines of a "healthy habitat" – or in other words – an ecosystem approach (Trout Unlimited 2002b). Here again we see the interrelatedness of the collapse of the commercial fishery, its replacement with a recreational focus, a shift in norms toward a respect for biodiversity, and the value of an ecosystem approach as well as the concomitant, perhaps resultant, perhaps causational, rise of NGOs.

That is not the end of TU's story, however, but merely its beginning. TU's success in Michigan policy-making in 1963 created a spillover effect, catalyzing the creation of TU chapters in Illinois, Wisconsin, New York, and Pennsylvania. By 1965 it had won its first national campaign in stopping the construction of the decidedly trout-unfriendly Reichle dam in Montana (Trout Unlimited 2002a). Again, it is interesting to note that in seeking policies to conserve a type of fish – salmonids – TU in turn advocates policies that manage for a healthy ecosystem at large. That is, single-species management and ecosystem

trout and salmon and the waters they inhabit. (For a comparison of Steelheaders and MUCC, see Zuverink interview in Chiarappa and Szylvian [2003].)

management are not dichotomous, as is often asserted, but interdependent. This is especially so in the case of trout. Trout and salmon species "are often viewed as indicators of overall environmental health" (Trout Unlimited 2002b) because of their pronounced sensitivity to environmental change. Insights such as this reflect the scientific knowledge and moral authority that have led to TU's intellectual credibility and influence over fisheries policy, underscoring that this NGO often performs roles identified with more prototypical epistemic communities. It is noteworthy that TU and Great Lakes United (GLU) were virtually the only NGOs at the September 1998 meeting of the Lake Michigan Lakewide Stocking Conference that argued for ecosystem restoration (i.e., against the stocking of non-native game fish from hatcheries) as a necessary precondition for real improvement to the Great Lakes aquatic ecosystem. In addition to launching an educational program publicizing this controversial position – one subsequently identified with GLU but only a minority of TU members – they launched a comprehensive review of current hatchery/stocking practices, bringing together specialists from state, federal, and academic institutions from both the United States and Canada (Frank 1998).

TRANSPARENCY IN GLOBAL GOVERNANCE AND IN THE GREAT LAKES AND MICHIGAN FISHERIES MANAGEMENT

Another hallmark of the trend toward global governance, and a logical extension of the increased numbers and influence of NGOs, is the increased openness, accountability, transparency, and responsiveness of policy-making structures to non-state, including substate actors.[18] Though there are still limitations on NGO access to the United Nations (UN), including to UN-sponsored global conferences, the trend toward greater access is obvious (Willetts 1996; Schechter 2001). Even the World Bank and International Monetary Fund (IMF) have not proven immune to calls for greater transparency and accountability. Their responsiveness, although certainly not adequate for all, is evident to

---

[18] On the relationship between public participation and transparency, see Wiser (2001). Note also that there is a growing literature that speaks to the seeming paradox of an increased transparency of key state and intergovernmental organization actors in the global policy-making process, at least resulting in part from pressure from NGOs, most of which suffer from lack of transparency and accountability themselves (see, for example, Nelson 1995).

anyone looking at their websites. For example, IMF loan agreements, heretofore secret, are now regularly posted on the fund's Website (Scholte and Schnabel 2002). Furthermore, students of global governance are increasingly monitoring the work of Transparency International, an NGO aimed at making the actions of multinational corporations more transparent and accountable (Galtung 2000). Again, the case of Michigan fisheries and fisheries policy-making is reflective if not anticipatory of this trend in the study and practice of global governance.

Michigan fisheries' policy-making structures were not always open or accountable to outside actors or stakeholders, however.[19] In the decades preceding the collapse of Michigan's commercial fishery, policy-making bodies were much more closed to the participation of stakeholders, including non-state grassroots pressure groups. However, as with many of the phenomena in Michigan fisheries policy-making that anticipated or reflected trends in global governance, here, too, the shift from a commercially stocked fishery to a recreationally stocked fishery was concomitant with an increased openness of policy-making authorities to outside actors. For example, in the mid 1980s, when salmon stocks were collapsing, the MDNR set up lake advisory councils (Ferreri *et al.* 1999). These councils opened political space in the fisheries policy arena by allowing the MDNR to receive input from interested publics as well as other agencies to shape the direction of fisheries management (Ferreri *et al.* 1999). This trend in incorporating of stakeholders and opening up the process has continued vis-à-vis MDNR hearings. The cold water regulations hearings held by the MDNR during the 1990s, for example, are noted for their incorporation of stakeholder input (Folland interview with Taylor 2003).

This increased openness of fisheries policy-making structures to pressure groups and stakeholders, however, is not limited to MDNR hearings or lake advisory committees, although these remain crucial as the states, the province of Ontario and the tribes retain primary management authority on the Great Lakes. The GLFC's processes are important, as well, because beginning in the late 1970s, the provincial and state jurisdictions asked the commission to facilitate the

---

[19] This is not, however, to suggest that issues of transparency are absent from the Michigan context. There have been accusations, for example, that the newest Tribal Fishing Consent Agreement was negotiated in secrecy. "When the consent agreement was announced, many felt deprived of an opportunity to make comments and considered on whose behalf the state was negotiating during the talks." Concerns focused on how tribal gill net fishing related to the interests of recreational fishing (Great Lakes Conservation Task Force 2002:72–3).

multijurisdictional management of the lakes; the tribes joined much more recently. And GLFC processes have been and are increasingly accessible to external actors. This accessibility includes the lake committees, which supplement individual jurisdictions' processes for receiving and considering public input. Through the annual meetings of lake committees – open to the public and attended by commercial and recreational interest groups (Dobson *et al.* 2002) – stakeholders and pressure groups offer input and seek to influence policy. Lake committees constitute the "action arm" of the aforementioned 1981 Joint Strategic Plan (GLFC n.d.). One of the four strategies enshrined in the Joint Strategic Plan is accountability and this strategy combined with another of the four – the ecosystem approach – augurs well for the inclusion of non-state actors and stakeholders in the policy process.

Perhaps unsurprisingly, the normative shift from single-species management to the ecosystem approach – an inherently more holistic approach to management – has resulted in an increased premium on the involvement of non-state actors. Dobson *et al.* (2002) have cogently related this latter phenomenon to the implementation of the ecosystem approach pursuant to the 1978 GLWQA.

Trends toward increased openness of policy-making processes regarding GLFC continued in recent years. Indeed, it is telling that, to its twin mandates of Great Lakes ecosystem management and integrated management of sea lamprey, the Commission has added a third in its strategic vision: institutional/stakeholder partnerships (GLFC 2001). As a third pillar to its *raison d'être*, GLFC seeks "strengthened and broadened partnerships" not only among fish management agencies and environmental agencies but also non-agency stakeholders. Toward that end, the GLFC strategic vision seeks to increase travel funds to commission advisors who act as liaisons to GLFC stakeholders and interest groups. The commission will also seek to establish a communications framework to help implement the Joint Strategic Plan by facilitating communication between agencies and non-agency stakeholders (GLFC 2001).

More generally, O'Gorman and Stewart (1999) make a persuasive case that the resolution of future conflicts, like those over alewife management in the Great Lakes and between recreational fishing objectives and those whose objectives include restarting indigenous fish species, requires better ecosystem management "mechanisms for integrating scientific knowledge ['epistemic communities' in the language of students of global governance], public consultation and decision making" (O'Gorman and Stewart 1999). The key question, however, is whether the evolving form of governance in the Great Lakes in general, and Michigan in particular, is up to the task. Though mechanisms exist for integrating scientific knowledge and for seeking public input, the vital question is whether the evolving governance structure effectively and appropriately balances those two necessary and desirable inputs in a way that results in optimal and legitimate decision-making.

SOVEREIGNTY IN GLOBAL GOVERNANCE AND GREAT LAKES FISHERIES

The changing nature of sovereignty in the international system has been seized upon as a major trend characterizing global governance and globalization in general. On the one hand, there is an increased diffusion of sovereignty as states have increasingly devolved authority and traditional state competencies to the substate level, a phenomenon connected with the notion of federalism and, more recently, subsidiarity. On the other hand, states have ceded sovereignty not only below but also above vis-à-vis political integration, or a "pooling" of sovereignty, with the European Union and the World Trade Organization being the most prominent examples. Again, both these phenomena have had long-standing precedents in Great Lakes and Michigan fisheries and fisheries policy-making.

In one sense, the Michigan fisheries policy-making process has always been the beneficiary of devolved sovereignty, if not actually subsidiarity, given the nature of the U.S. federal structure in which it is embedded.[20] That is, owing to Article X of the U.S. Constitution, sovereignty vis-à-vis fisheries policy is inherently devolved because it is counted among the "reserved powers" of state governments (Dochoda 1999). Though Article X is indeed qualified by the supremacy of federal law over state law and sovereignty ceded through international and tribal treaties, the prerogative to regulate the management and exploitation of fish stocks rests chiefly with state governments.[21]

---

[20] Interestingly, Born and Stairs (2002) credit the recently expanded devolution of decision-making and fiscal responsibility to federal decentralization and budget cutting. To the extent that these are contributing factors, they would be suggestive of the consequences of neo-liberal trends in the global political economy.

[21] Francis offers a fascinating parallel story relating to Great Lakes environmental protection policy. He writes of both the devolution of authority from the federal governments of the United States and Canada and from governments to nonstate actors. He credits much of this to a serious underestimation by both the U.S. and Canadian governments of "the strength of public concern about the lakes and public support for environmental measures" (Francis 1990:199–200).

If the nature of Michigan (and Canadian)[22] fisheries existing in a federal framework inherently devolves authority, the nature of Michigan fisheries existing within the context of a shared resource such as the Great Lakes implies a pooling of sovereignty as well. That Michigan fisheries include the Great Lakes inherently embeds it in a system of multijurisdictional as well as ecological interdependence. To be sure, it must share this resource with two federal governments, seven other states, a Canadian province, two intergovernmental organizations (the IJC and GLFC), and a number of tribal arrangements, not to mention countless municipal authorities and sundry aforementioned NGOs (Ferreri *et al.* 1999). If the absence of all these management authorities would lead to certain environmental stress, the presence of these myriad authorities certainly leads to a degree of what has been termed "jurisdictional stress" (Ferreri *et al.* 1999).

This jurisdictional stress is both the product of and has sought to be ameliorated by coordinated management and shared sovereignty. For example, efforts at coordinated fisheries management in the Great Lakes go back far before the vogue for neo-liberal institutionalism or shared sovereignty. There were countless efforts to create an international institution for the management of Great Lakes fisheries. Among these were a stillborn attempt to create a "joint board between Canada and the United States" in 1893, an aborted treaty agreement for coordinated management between the United States and Great Britain for fisheries in U.S. and Canadian waters in 1908 (it was terminated in 1915 because of the failure of the U.S. Congress to approve the regulations), and the Convention Between the United States of America and Canada for the Development, Protection and Conservation of the Fisheries of the Great Lakes which was dead in the water in 1946 because of opposition in the U.S. Congress to transferring regulatory authority from the states to an international body (Dochada 1999; GLFC 2000). Thus, despite consistent efforts from 1893 to 1954, an international body for the coordinated management of the Great Lakes failed, primarily because of "an unwillingness of the states and the province to cede ... authority to the U.S. or Canadian federal governments or to an international commission" (GLFC 2000) (though the IJC, established by the Boundary Waters Treaty of 1909, has significantly affected Great Lakes fisheries – not least of all through the

---

[22] As Dochoda (1999:94) notes, "Canadian federal laws take precedence over provincial laws for the Great Lakes fishery where these are in conflict, but only to the extent necessary to protect the fishery."

encouragement of a ban on phosphates – its focus was never explicitly on fisheries).[23] Therefore, when the 1954 Canada–U.S. Convention on Great Lakes Fisheries finally created an international body, the Great Lakes Fishery Commission, it should not be surprising that the Commission was circumscribed in its charge. That is, the GLFC was limited to a "study and advise" mandate (including, however, the important role in facilitating, funding, publishing, and disseminating fisheries research), recommendatory powers and regulatory powers only in the management of the invasive sea lamprey. It is the only U.S.–Canadian fishery commission without regulatory authority, primarily because of the "vehement opposition raised by the vested American interests – particularly by spokesmen of the Ohio commercial fishermen" (Dochoda 1999). Indeed, it is because of rather than in spite of such limited power that this Commission has succeeded where countless other attempts failed. As Margaret Dochoda puts it: "In retrospect it was perhaps the absence of regulatory authority in the Great Lakes Fishery Commission that allowed states to be comfortable in using its binational forum" (Dochoda 1999).

The relative success of the commission in this regard is significant. It testifies to the logic of functionalism, that strategy for regional integration that was at play at the same time across the Atlantic in the European Coal and Steel Community (ECSC). In both cases, both the GLFC and the ECSC give succor to the functionalist theories and practice of political integration of Jean Monnet, Robert Schumann, and David Mitrany, who maintained that the best way to integrate intractable policies is to start with narrow, relatively non-controversial issue areas (Cram 2001). GLFC has not led to the supranational regional integration of a continent in the way the ECSC has, but it has, nonetheless, engendered its own spillover effects. Its narrow mandate widened with the 1981 Joint Strategic Plan, encouraging closer coordination between management agencies and a paradigm shift toward an ecosystem-based approach.

SUBSIDIARITY: A CONTESTED CONCEPT IN GLOBAL GOVERNANCE AND GREAT LAKES FISHERIES MANAGEMENT

The United Nations, the European Union (EU), the IJC, and the GLFC are working on the same problem: how to bring about the best and most legitimate policy decisions in a world of overlapping and

---

[23] For an account of its role in the phosphates ban issue, see Munton (1980).

competing governance claims. All agree that this requires more public participation than anticipated when their constitutive documents were written, and that decisions require deference to knowledgeable individuals, including scientists as well as those most likely to be affected by the policies. All also agree that though subsidiarity maximizes participation, accountability, and transparency, there are trade-offs in time and, often, ease of coordination across jurisdictional lines. The authors of the proposed EU Constitution punted on this issue; early drafts of the Constitution worked hard to concretize subsidiarity (e.g., defining what sorts of issues should be handled in what decision-making forums), but the final document almost entirely avoided the issue.

The IJC for years wrestled with arguments that it had overdone subsidiarity or at least public input into water management decision-making. Indeed, it has been argued that the 1987 amendment of the 1978 GLWQA transferring some of the responsibilities previously conferred on the IJC back to the contracting parties, that is, the federal governments of the United States and Canada, was a consequence of widely shared frustration about the IJC's inability to act. The jury is still out on the effectiveness of this measure (Great Lakes Conservation Task Force 2002), but a couple of things are clear that are relevant to the ongoing debates in Europe, at the EU and relating to the Great Lakes.

Manno's study (1994) of the 1987 revision process found that "when governments undertake public participation and public consultative activities, they often do so in a manner that suggest[s] government's responsibility to strike a balance between competing stakeholder interests, as though all stakeholder interests were of equal value and each had equal power, ability, and motivation to articulate and defend its interests ... Yet stakeholder rights and interests are multidimensional and power is not equally distributed, nor are costs and benefits" (Manno 1994). More generally, it seems that though it may not be clear whether subsidiarity results in optimal decisions, policymakers recognize that the locus of decision-making matters. In addition, some empirical evidence suggests that stakeholders have also come to understand that the level and the arena of decision-making significantly affect fisheries policies and, especially, priorities among competing goals.[24] Though policy participants and policy analysts of Great Lakes fisheries management do not seem to have solved the challenges presented by advocates of subsidiarity in multijurisdictional environments,

---

[24] See, for example, Tables 1 and 3 in Born and Stairs (2002).

it seems that those involved in constitution writing in Europe would have done well to have paid some heed to the experience of those on the other side of the Atlantic.

CONCLUSIONS

Though the global governance regime of fisheries worldwide may not speak directly to the experience of Great Lakes and Michigan fisheries or fisheries policy-making, a global governance approach to conceptualizing Great Lakes and Michigan fisheries is not analytically bankrupt. The phenomena that have come to characterize global governance – crystallization of emergent norms; the rise of NGOs; increased openness, accountability, and transparency of policy processes; a trend toward diffused and shared sovereignty; and debates over the proper level of decision-making – have been well articulated in Michigan and Great Lakes fisheries and, in many instances, actually foreshadowed by Michigan and Great Lakes fisheries.

Why have Great Lakes and Michigan fisheries embodied or foreshadowed a global governance approach? For the most part, these changes in norms, policies, and policy processes stem from responses to the degradation and overexploitation of the environment which led ultimately to a collapse of the commercial fishery and a shift toward a recreational one. Only in recent years has attention been paid to collapsing and overexploited global fish stocks, and thus a global governance approach appealed to. A realization of degradation and overexploitation in Michigan fisheries and the Great Lakes occurred decades earlier. Moreover, Michigan fisheries vis-à-vis the Great Lakes have always been predisposed to the efficacy of a global governance approach, given Michigan's place within a federal context on the one hand and the context of an inherently shared and multi-jurisdictional resource in the Great Lakes on the other. Thus it seems past time for global governance scholars to pay attention to the case of Great Lakes fisheries and fisheries management and for those studying the Great Lakes to apply the insights of global governance approaches.

Acknowledgments

We wish to thank the two reviewers of this chapter, whose countless suggestions have helped us to clarify our thoughts and to make our account a more accurate one. We also wish to thank William W. Taylor for encouraging us to pursue this topic and the Michigan State

University Institute for Public Policy and Social Research for providing funds to assist in our research and the writing of this chapter.

References

Abbott, K. W. and Snidal, D. 2000. Hard and soft law in international governance. *International Organization* **54**: 421–56.

Arts, B. 1998. *The Political Influence of Global NGOs: Case Studies of the Climate and Biodiversity Conventions*. Utrecht, Netherlands: International Books.

Barkin, J. S. and DeSombre, E. R. 2000. Unilateralism and multilateralism in international fisheries management. *Global Governance* **6**: 339–60.

Beck, R. J., Arend, A. C., and Vander Lugt, R. D. (eds.) 1996. *International Rules: Approaches from International Law and International Relations*. New York: Oxford University Press.

Becker, M. L. 1993. The International Joint Commission and public participation: past experience, present challenges, future tasks. *Natural Resources Journal* **33**: 235–274.

Begg, D., Crémer, J., Danthine, J.-P. et al. 1993. *Making Sense of Subsidiarity: How Much Centralization for Europe? Monitoring European Integration*, CEPR Annual Report No. 4. London: Centre for Economic Policy Research.

Born, S. M. and Stairs, G. S. 2002. *An Assessment of State Planning for Coldwater Fisheries Management in the United States*. Trout Unlimited. Available online at: www.tu.org

Bowman, M. 1996. The nature, development and philosophical foundations of the biodiversity concept in international law. In *International Law and the Conservation of Biodiversity*, eds. M. Bowman and C. Redgwell. London: Kluwer Law International, pp. 5–49.

Bratspies, R. 2001. Finessing King Neptune: fisheries managements and the limits of international law. *Harvard Environmental Law Journal* **25**: 213–258.

Brown, R., Ebner, M., and Gorenflo, T. 1999. Great Lakes commercial fisheries: historical overview and prognosis for the future. In *Great Lakes Fisheries Policy and Management: A Binational Perspective*, eds. C. P. Ferreri and W. W. Taylor. East Lansing, MI: Michigan State University Press, pp. 361–362.

Brühl, T. and Rittberger, V. 2001. From international to global governance: actors, collective decision-making, and the United Nations in the world of the twenty-first century. In *Global Governance and the United Nations System*, ed. V. Rittberger. Tokyo: United Nations University Press, pp. 1–47.

CDLR (Council of Europe Steering Committee on Local and Regional Authorities). 1994. *Definition and Limits of the Principle of Subsidiarity*. Strasbourg, France: Council of Europe.

Chiarappa, M. J. and Szylvian, K. M. 2003. *Fish for All: An Oral History of Multiple Claims and Divided Sentiment on Lake Michigan*. East Lansing, MI: Michigan State University Press.

Cohen, B. C. 1959. *The Influence of Non-Governmental Groups on Foreign Policy Making*. Boston, MA: World Peace Foundation.

Conca, K. 1996. Greening the UN: environmental organisations and the UN system. In *NGOs, the UN and Global Governance*, eds. T. G. Weiss and L. Gordenker. Boulder, CO: Lynne Rienner, pp. 103–119.

Cram, L. 2001. Integration theory and the study of the European policy process: toward a synthesis of approaches. In *European Union: Power and Policy-Making*, 2nd edn, ed. J. Richardson. New York: Routledge, pp. 51–73.

Dobson, T., Regier, H. A., and Taylor, W. W. 2002. Governing human interactions with migratory animals, with a focus on humans interacting with fish in Lake Erie: then, now and in the future. *Canada–United States Law Journal* **28**: 389–446.

Dochoda, M. R. 1999. Authorities, responsibilities, and arrangements for managing fish and fisheries in the Great Lakes ecosystem. In *Great Lakes Fisheries Policy and Management: A Binational Perspective*, eds. C. P. Ferreri and W. W. Taylor. East Lansing, MI: Michigan State University Press, pp. 93–110.

Edeson, W. 1999. Closing the gap: the role of "soft" international instruments to control fishing. *Australian Year Book of International Law* **20**: 83–104.

Ferreri, C. P., Taylor, W. W., and Robertson, J. M. 1999. Great Lakes fisheries futures: balancing the demands of a multijurisdictional resource. In *Great Lakes Fisheries Policy and Management: A Binational Perspective*, eds. C. P. Ferreri and W. W. Taylor. East Lansing, MI: Michigan State University Press, pp. 539–548.

Francis, G. 1987. Editorial: Toward understanding Great Lakes organizational ecosystems. *Journal of Great Lakes Research* **31**: 233.

Francis, G. 1990. Flexible governance. In *An Ecosystem Approach to the Integrity of the Great Lakes in Turbulent Times*, eds. C. J. Edwards and H. A. Regier. Ann Arbor, MI: Great Lakes Fishery Commission, pp. 195–207.

Franck, T. 1990. *The Power of Legitimacy among Nations*. New York: Oxford University Press.

Frank, A. 1998. GLU and Trout Unlimited plan regional fish conferences. Available online at www.glu.org/english/information/newsletters/12_4-fall-1998/GLU-conferences.html

Galtung, F. 2000. A global network to curb corruption: the experience of transparency international. In *The Third Force: The Rise of Transnational Civil Society*, ed. A. Florini. Washington, DC: Brookings Institution, pp. 17–47.

GLFC (Great Lakes Fishery Commission). 2000. *The Great Lakes Fishery Commission Established by Treaty to Protect Our Fishery*, Fact Sheet 1. Ann Arbor, MI: Great Lakes Fishery Commission.

GLFC. 2001. *Strategic Vision of the Great Lakes Fishery Commission for the First Decade of the New Millennium*. Ann Arbor, MI: Great Lakes Fishery Commission.

GLFC. n.d. *A Joint Strategic Plan for Management of Great Lakes Fisheries*, Fact Sheet 10. Ann Arbor, MI: Great Lakes Fishery Commission. Available online at www.glfc.org

GLU (Great Lakes United). 2003. *Green Book*. Available online at www.glu.org/home.html

Great Lakes Conservation Task Force. 2002. *An Action Plan to Protect the Great Lakes*. Final Report, The Citizens' Agenda, Presented to Senate Majority Leader Dan L. DeGrow. Lansing, MI: Michigan State Senate and the Citizens of Michigan.

Haas, P. M. 1992. Epistemic communities and international policy coordination. *International Organization* **46**: 367–390.

Hartig, J. H., Zarull, M. A., and Law, N. L. 1998. An ecosystem approach to Great Lakes management: practical steps. *Journal of Great Lakes Research* **24**: 739–750.

Hillgenberg, H. 1999. A fresh look at soft law. *European Journal of International Law* **10**: 499–515.

Inscho, F. R. and Durfee, M. H. 1995. The troubled renewal of the Canada–Ontario agreement respecting Great Lakes water quality. *Publius: The Journal of Federalism*. **25**: 51–70.

Kaye, S. M. 2000. *International Fisheries Management*. Boston, MA: Kluwer Law International.

Keohane, R. O. (ed.) 2002. *Power and Governance in a Partially Globalized World*. London: Routledge.
Keohane, R. O. and Nye, J. S., Jr. 2002. Governance in a globalizing world. In *Power and Governance in a Partially Globalized World*, ed. Robert O. Keohane. London: Routledge, pp. 193–218.
Kimball, L. A. 2001. *International Ocean Governance: Using International Law and Organizations to Manage Marine Resources Sustainably*. Gland, Switzerland: World Conservation Union.
Knight, W. A. 2000. *A Changing United Nations: Multilateral Evolution and the Quest for Global Governance*. New York: Palgrave.
MacKenzie, S. H. 1996. *Integrated Resource Planning and Management: The Ecosystem Approach in the Great Lakes Basin*. Washington DC: Island Press.
Manno, J. P. 1994. Advocacy and diplomacy: NGOs and the Great Lakes Quality Agreement in environmental NGOs. In *World Politics: Linking the Local and the Global*, eds. T. Princen and M. Finger. New York: Routledge, pp. 69–120.
MUCC (Michigan United Conservation Clubs). 2002. *Inside MUCC*. Available online at www.mucc.org/inside_mucc/about.htm
Mills, E. L., Casselman, J. M., Dermott, R., *et al.* 2003. Lake Ontario: food web dynamics in a changing ecosystem (1970–2000). *Canadian Journal of Fisheries and Aquatic Sciences* **60**: 471–490.
Munton, D. 1980. Great Lakes water quality: a study in environmental politics and diplomacy. In *Resources and the Environment: Policy Perspectives for Canada*, ed. O. P. Dwivedi. Toronto, Ontario: McClelland & Stewart, pp. 153–178.
Nelson, P. 1995. *The World Bank and Non-Governmental Organizations: The Limits of Apolitical Development*. New York: St. Martin's Press.
O'Brien, R., Goetz, A. M., Scholte, J. A., and Williams, M. 2000. *Contesting Global Governance: Multilateral Economic Institutions and Global Social Movements*. Cambridge, UK: Cambridge University Press.
O'Gorman, R. and Stewart, T. J. 1999. Ascent, dominance, and the decline of the alewife in the Great Lakes. In *Great Lakes Fisheries Policy and Management: A Binational Perspective*, eds. C. P. Ferreri and W. W. Taylor. East Lansing, MI: Michigan State University Press, pp. 489–507.
Peterson, M. J. 1993. International fisheries management. In *Institutions for the Earth: Sources of Effective International Environmental Protection*, eds. P. M. Haas, R. O. Keohane, and M. C. Levy. Cambridge, MA: MIT Press.
Porter, G., Brown, J. W., and Chasek, P. S., 2000. *Global Environmental Politics*, 3 edn. Boulder, CO: Westview Press.
Rittberger, V. (ed.) 2001. *Forward in Global Governance and the United Nations System*. New York: United Nations University Press.
Ryder, R. A. and Orendorff, J. A. 1999. Embracing biodiversity in the Great Lakes ecosystem. In *Great Lakes Fisheries Policy and Management: A Binational Perspective*, eds. C. P. Ferreri and W. W. Taylor. East Lansing, MI: Michigan State University Press, pp. 113–143.
Schechter, M. G. 2001. Making meaningful UN-sponsored world conferences of the 1990s: NGOs to the rescue. In *United Nations-Sponsored World Conferences: Focus on Impact and Follow-Up*, ed. M. G. Schechter. Tokyo: United Nations University Press, pp. 189–217.
Scholte, J. A. and Schnabel, A. 2002. Introduction. In *Civil Society and Global Finance*, eds. J. A. Scholte and A. Schnabel. London: Routledge, pp. 1–12.
Slaughter, A.-M., Tulumello, A. S., and Wood, S. 1998. International law and international relations theory: a new generation of interdisciplinary scholarship. *American Journal of International Law* **92**: 367–397.

Swaine, E. T. 2000. Subsidiarity and self interest: federalism at the European court of justice. *Harvard International Law Journal* **41**: 1-128.

Tanner, H. A. and Tody, W. H. 2002. History of Great Lakes salmon fishery: a Michigan perspective. In *Sustaining North American Salmon: Perspectives across Regions and Disciplines*, eds. K. D. Lynch, M. L. Jones, and W. W. Taylor. Bethesda, MD: American Fisheries Society, pp. 145-158.

Trout Unlimited. 2002a. *Trout Unlimited Today*. Available online at www.tu.org/about_tu/tu_mission.html

Trout Unlimited. 2002b. *Trout 101*. Available online at www.tu.org/campaigns/trout101.html

United Nations, 1992. *United Nations Framework Convention on Biological Diversity*, Document DPI/130/7. Reprinted in 31 International Legal Materials 818.

Victor, D. G., Raustiala, K., and Skolnikoff, E. B. (eds.) 1998. *The Implementation and Effectiveness of International Environmental Commitments: Theory and Practice*. Cambridge, MA: MIT Press.

Wapner, P. 1998. Reorienting state sovereignty: rights and responsibilities in the environmental age. In *The Greening of Sovereignty in World Politics*, ed. K. T. Liftin. Cambridge, MA: MIT Press, pp. 275-297.

Willetts, P. 1996. Consultative status for NGOs at the United Nations. In *The Conscience of the World: The Influence of Non-Governmental Organisations in the U.N. System*, ed. P. Willetts. Washington, DC: Brookings Institution, pp. 31-62.

Wiser, G. M. 2001. Transparency in 21st century fishery management: options for public participation to enhance conservation and management of international fish stocks. *Journal of International Wildlife Law and Policy* **4**: 95-129.

Interviews

Folland interview with Brandon Schroeder, former fisheries policy official for MUCC, February 21, 2003, East Lansing, MI.

Folland interview with William W. Taylor, chairperson, Michigan State University Department of Fisheries and Wildlife, February 6, 2003, East Lansing, MI.

DEAN BAVINGTON AND JAMES KAY

14

Ecosystem-based insights on northwest Atlantic fisheries in an age of globalization

FISHERIES SYSTEMS IN AN AGE OF GLOBALIZATION: THE NEED FOR NEW HEURISTICS TO COPE WITH COMPLEXITY

In an age of consolidated vertically integrated seafood corporations, industrial capture and culture fisheries, declining wild fish populations, shifting government policies, and unpredictable biophysical changes operating on scales ranging from local to planetary, the complexity and uncertainty of fisheries systems have never been greater. In the face of global stock collapses, budgetary restraints, and extreme overfishing of species thought to be managed effectively, national fisheries managers are increasingly left to cope with and adapt to failures rather than taking a confident proactive stance toward fisheries under their managerial control (Ludwig et al. 1993; Thompson and Trisoglio 1997; Finlayson and McCay 1998; Ludwig 2001; Bavington 2002; McCay 2002). Human expropriation of up to 35 percent of marine primary productivity is overwhelmingly driven by the capitalist mode of production with its focus on capital accumulation and profitability through the generation of exchange value. Economic growth continues to be achieved in marine fisheries by adopting advanced industrial technologies in harvesting, maintaining government subsidies, and expanding demand for marine biomass in global markets. The aim is fulfilling the desires of wealthy consumers for wild, cultured, and terrestrial animal products (Pauly and Christensen 1995; Naylor et al. 2000; Pauly et al. 2001).[1] The global fisheries situation at local, regional, national, and transnational scales raises troubling ecological and equity issues central to sustainability,

---

[1] Wild seafood is removed from oceans for direct human consumption as well as use in aquaculture and terrestrial industrial farming through reduction fisheries to produce fishmeal for animal feeds and fertilizers.

Globalization: Effects on Fisheries Resources, ed. William W. Taylor, Michael G. Schechter, and Lois G. Wolfson. Published by Cambridge University Press. © Cambridge University Press 2007.

and involves interconnected biophysical, social, and political complexity (Caddy and Regier 2002; Barange 2003; Bavington et al. 2004).

As in previous decades with smaller freshwater fisheries, attempts to address the complexity within fisheries management in the open ocean have revolved around calls to move from single species to ecosystem-based management with an emphasis on modeling a wider number of variables beyond commercially relevant fish populations, and integrating a wider diversity of knowledge outside traditional fisheries biology and economics, such as the experiential knowledge of fishers, interdisciplinary science, and the concerns of environmentalists (Neis 1992; Pitcher et al. 1998; Neis et al. 1999; Neis and Felt 2000; Charles 2001; DFO 2002a; McCay 2002; Berkes 2003; Busch et al. 2003; FAO 2003a, b; Garcia et al. 2003; Latour et al. 2003) (see Table 14.1).

The need to address irreducible uncertainty with respect to scientific knowledge (Kay and Schneider 1994), regime shifts (Steele 1998; Scheffer et al. 2001), fishing down theory (Pauly et al. 1998; Caddy and Garibaldi 2000), hierarchical nesting from the local to the global across system types and temporal scales (Boyle et al. 2002), and the democratic inclusion of a range of knowledge types and forms of stakeholder participation in the management process (Neis and Felt 2000; McCay 2002) has raised continuing challenges for the implementation of ecosystem-based fisheries management (FRCC 1997; Link 2002; Garcia et al. 2003). Challenges facing fisheries managers include a complex array of interconnected biophysical, socioeconomic, legal, and political systems flowing from a diverse set of historical circumstances, and operating at a variety of spatial and temporal scales. For example, the weak interpretation and application of the precautionary principle (Garcia 1994; FAO 1995; DFO 2002b) by fisheries regulatory agencies has permitted economic globalization, corporate interests, and technological innovations to outpace and overcome efforts at sustainable fisheries use and control.[2]

---

[2] The precautionary *principle* has been adopted by a number of international and national fisheries management organizations, but has mainly been implemented as the precautionary *approach* (DFO 2001, 2002b). The precautionary *principle* places the burden of proof on development proponents to prove that any harm resulting from their operations will be reversible; the precautionary *approach* places the burden of proof on regulators to show that restrictions on marine development will not irrevocably or unnecessarily harm the economy. The precautionary *principle* implies a deontological ethical framework that places strict limits on economic development; the shift in language toward precautionary *approaches* subtly promotes a permissive utilitarian ethical stance focused on maximizing the greatest good for the greatest number of economic actors (VanDeVeer and Pierce 1998; Coward et al. 2000).

Table 14.1 Comparison between single-species and ecosystem-based fisheries management for marine fisheries

Criteria	Single-species fisheries management	Ecosystem-based fisheries management
Management	Sector-based. Control oriented: aimed at target commercial species and harvesters.	Area-based. Coping oriented: focusing on habitats, ecosystem integrity, and facilitating the self-control of people vs. optimizing or maximizing landings.
Governance objectives	Not always coherent or transparent. "Optimal" system output focused on commercial fish production.	A desired state of the ecosystem (health, integrity). Conservation, rebuilding of depleted stocks and risk management.
Scientific input	Formalized (particularly in regional commissions). Variable impact. Population statistics (large number systems).	Less formalized. Less operational. Often insufficient. Stronger role of advocacy, post-normal science and local ecological knowledge. SOHO systems (middle number systems).
Decision-making	Most often top-down. Strongly influenced by industry lobbying.	Promotes bottom-up, participatory approaches. Strongly influenced by environmental lobbies. Stronger use of tribunals and councils.
Role of the media/PR	Historically limited. Growing as fisheries crisis spreads.	Stronger use of the media, public relations and facilitation techniques.
Local, regional, and global institutions	Central role of the Food and Agriculture Organization of the United Nations, regional fishery bodies (e.g., North Atlantic Fisheries Organization), and national fishery departments (e.g., Department of Fisheries and Oceans Canada).	Central role of United Nations Environment Program (UNEP), the Regional Seas Conventions, and partnerships between various levels of government, including municipal scale, self-managing professional fish harvester organizations, and community-based organizations.

Table 14.1 (cont.)

Criteria	Single-species fisheries management	Ecosystem-based fisheries management
Geographical basis	A process of overlapping and cascading subdivision of the oceans for allocation of resources and responsibilities (e.g., NAFO subdivisions 2J3KL).	A progressive consideration of larger-scale ecosystems for more comprehensive integrated management (e.g., from specific areas to entire coastal zones and large marine ecosystems [LME]).
Stakeholder and political base	Narrow. Essentially fishing industry stakeholders.	Much broader. Society wide. Often with support from subsistence, recreational and small-scale fisheries.
Global instruments	1982 Law of the Sea Convention, UN Fish Stock Agreement, and FAO Code of Conduct.	Ramsar Convention, UN Conference on Environment and Development and 1992 Agenda 21, Convention on Biological Diversity and Jakarta Mandate.
Measures	Regulation of fishing activity outputs (removals, quotas) or inputs (gear, effort, capacity) to maintain resource productivity and trade benefits.	Protection of specified areas and habitats, including limitation or exclusion of extractive human activities. Total or partial ban of some human activities. A focus on controlling human interaction with marine resources vs. the resources themselves.

Source: Adapted from Busch *et al.* (2003), Garcia *et al.* (2003), and Bavington (2002).

The task of monitoring, managing, and governing these systems presents enormous challenges that require innovative ways of thinking and acting. In recent years, ecosystem-based approaches of various types have been presented as alternatives to single-species models which – despite their many identified failings – continue to influence how monitoring and management are perceived and implemented in fisheries systems throughout the world (FRCC 1997; Caddy 1999; Caddy and Cochrane 2001; Caddy and Regier 2002; Link 2002; Pauly *et al.* 2002; Barange 2003; Garcia *et al.* 2003).

This chapter presents an adaptive ecosystem-based approach that represents fisheries systems as self-organizing, holarchic, open (SOHO) systems as a way to characterize the complexity of globalized fisheries systems and begin to help frame what can be done at this moment in history. Building on the work of others in the field of ecosystem-based fisheries management, the chapter argues that SOHO systems require new approaches to monitoring, management, and governance that differ from and improve upon approaches flowing from the single-species perspective. The chapter begins with a brief description of the SOHO systems heuristic and complex systems theory in general, and proceeds to explore the implications of these perspectives for the monitoring and management of globalized fisheries systems. The case of the collapsed northern cod fishery in Newfoundland and Labrador, Canada, is used to provide both a concrete illustration of the benefits offered and challenges posed by the SOHO ecosystem approach, and to raise larger questions surrounding the human–marine relationship.

COPING WITH COMPLEXITY: THE SELF-ORGANIZING HOLARCHIC OPEN (SOHO) SYSTEMS HEURISTIC

Self-organizing holarchic open (SOHO) systems models are complex systems heuristics used to help understand and deal with complexity and sustainability issues (Kay *et al.* 1999; Boyle *et al.* 2002) (Fig. 14.1). SOHO systems descriptions are useful in situations that involve high levels of uncertainty, and conflict, and large decision stakes (Ravetz 1999).[3] SOHO systems understanding builds on the tradition of

---

[3] Many have argued that these situations call for a new type of science that moves beyond the expert-based reductionist science of the Enlightenment and the modern period. Ravetz and Funtowicz (1999) have proposed "post-normal science" (PNS) for situations that lie outside the controlled environments of the laboratory, situations where stakes are high, certainty is low, and the need for "extended

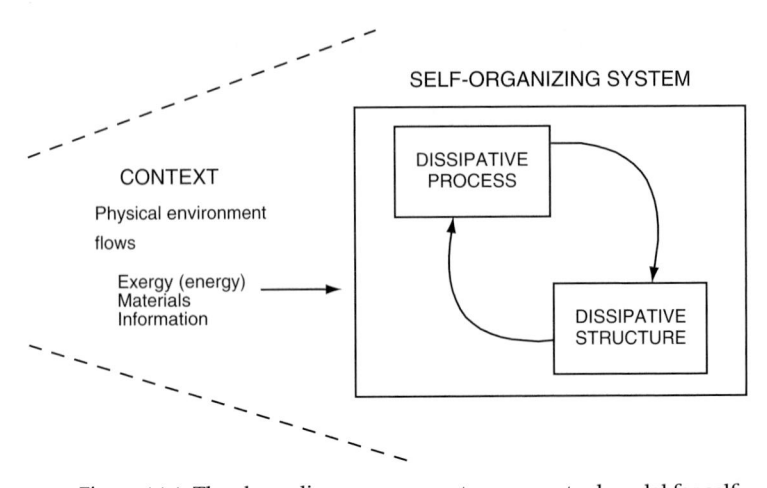

Figure 14.1 The above diagram represents a conceptual model for self-organizing systems as dissipative process/structures. "Self-organizing dissipative processes emerge whenever sufficient exergy is available to support them. Dissipative processes restructure the available raw materials in order to dissipate the exergy. Through catalysis, the information present enables and promotes some processes to the disadvantage of others. The physical environment will favor certain processes. The interplay of these factors defines the context for (i.e., constraints) the set of processes which may emerge. Once a dissipative process emerges and becomes established it manifests itself as a structure. These structures provide a new context, nested within which new processes can emerge, which in turn beget new structures, nested within which ... Thus a SOHO system emerges, a nested constellation of self-organizing dissipative process/structures organized about a particular set of sources of exergy, materials and information, embedded in a physical environment" (Kay et al. 1999:724). In a marine ecological setting, examples of *structures* could include individuals of a particular species, stocks of fish or breeding populations, coral, eelgrass, etc. The ecological *processes* would take in reproduction, metabolism, primary productivity, etc. The *context* would be determined by the available set of nutrients and energy sources in the physical environment, and the information would consist of the biodiversity (Kay et al. 1999).

facts" and an "extended peer community," beyond that of narrow disciplinary scientific expertise, demand a focus on the quality of the scientific process as opposed to the singular expert drive for universal Truth (Ravetz and Funtowitcz 1999). As an applied science involving high levels of uncertainty surrounding measurements, large decision stakes for the communities involved, and conflicts around how to proceed, fisheries science and management seem to be good candidates for the post-normal science approach.

Bertalanffy's general systems theory (1950) and Koestler's notion of holons and holarchy (1978),[4] with an emphasis on nested hierarchy theory, self-organization, and the openness of social and ecological systems to energy, materials, and information (Kay et al. 1999).[5] Applied to fisheries, SOHO systems models focus attention on the hierarchical nature of fisheries systems by considering issues of scale (temporal and spatial), system type (physical, biological, societal), and the bounding and nesting of fishing activities within these systems. Additionally, SOHO systems models draw attention to the dissipative structures and processes that can evolve in marine ecosystems out of specific physical environments that provide high-quality energy (exergy), material, and information permitting the emergence of self-organization (Kay and Schneider 1994; Kay et al. 1999) (Fig. 14.1).[6]

From the nested hierarchical perspective of SOHO systems modeling, marine ecosystems provide the context for the emergence of complex societal fishing systems, or fishing societies, which exhibit self-organizing dissipative structures and processes contingent on the extraction of structure or biomass from the marine ecological system. In addition to framing the nested hierarchical nature of fishing activities, SOHO systems models focus attention on the feedbacks that exist between fishing activities, ecological structure, and the physical environmental context out of which both ecological and societal self-organizing systems emerge (Kay et al. 1999).

SOHO fisheries systems can exist in a number of stable states around attractors[7] (sometimes referred to as ecological regimes) and can resist movement away from them (Kay et al. 1999). This resistance to change is accomplished by feedback loops in the SOHO system that

---

[4] A holon is a whole/part entity or system that exists contextually in a nested network of other holons forming a holarchy (Kay et al. 1999).

[5] The hierarchical nature of complex systems requires that they be studied from different types of perspectives and at different scales. With SOHO systems descriptions there is never only one correct perspective; rather a diversity of views is required for understanding (Boyle et al. 2002).

[6] High-quality energy is referred to as *exergy* and is a reflection of how organized or useful energy is in its ability to do work (see Kay 1991; Kay and Schneider 1992; Kay et al. 1999).

[7] "A SOHO system exhibits a set of behaviors that are coherent and organized, within limits. The nexus of this organization at any given time is referred to as an attractor" (Kay et al. 1999:725).

serve to maintain the system's current state, but when critical thresholds are breached and the SOHO system moves beyond the domain of an attractor, system change tends to be erratic and catastrophic, flipping from one regime/attractor into a new one (Holling and Meffe 1996; Kay and Regier 1999; Kay et al. 1999; Boyle et al. 2002). Flips in aquatic systems have been observed in Lake Erie involving a benthic and pelagic attractor (Regier and Kay 1996; Kay and Regier 1999) and regime shifts have been postulated for marine systems, including changes from ground fish to crustacean regimes driven by modifications in water temperature and salinity in the northwest Atlantic off Newfoundland and Labrador, Canada (Rose 2003), and others throughout the world's oceans (Steele 1998; Scheffer et al. 2001). The precise time when flips occur, the domain-space of the new attractor, and the exact state that the system will change into are generally not predictable (Kay et al. 1999). This is so because in any given SOHO fishery system, often several are possible. The state that the system arrives in is a function of its history, and there is never a universally "correct" state for a system to be in, in spite of the fact that specific groups of people will prefer certain system states (Kay et al. 1999; Boyle et al. 2002; Wilson 2002). For example, cod fishers in Newfoundland and Labrador prefer ground fish regimes with healthy stocks of capelin, and other cod prey, and low numbers of cod predators such as harp seals; crab and shrimp harvesters depend on crustacean regimes with appropriate environmental and trophic dynamics to maintain their fishery (Gray 2000; Wilson 2002). The preferences of fishers and other marine stakeholders for particular marine regimes or attractors will be mediated by their values, perspectives, linkages to other systems, historical attachments, understanding, interests, and a multitude of other factors. For instance, market demand can influence how fish harvesters and processors interpret marine regime shifts. The direction of the regime change relative to trophic levels and biodiversity might alter how marine biologists interpret new system states, perceiving them either as signs of ecological recovery or further evidence of fishing down sequences (Pauly et al. 1998; Pauly and MacLean 2003; Bavington et al. 2004).

The SOHO systems model represents a unique understanding of the complexity of fisheries systems and suggests a different role for science, monitoring, management, governance, and the quest for sustainability. It removes the idea of an objective, context-free assessment of changes in fisheries systems, and focuses on the importance of *who* is making monitoring and management decisions and *what* is motivating

```
R
A
N
D
O
M
N
E
S
S
```
↑

Unorganized complexity (aggregates)
Large number systems (linear and equilibrium thermodynamics)
Tool: statistics (probabilistic rules)
Order out of chaos or complexity (take the average of behavior)
Parts are similar and interact in just a few ways
Statistical control

- -

Organized complexity (SOHO systems)
Middle number systems (Neither over- nor under-connected)
Tool: post-normal science (complex systems rules)
Too many parts for deterministic analysis,
not enough for probabilistic analysis
Many parts with diverse ways of interacting
Non-linear thermodynamics and dynamic self-organization
Limited control/coping and adapting

Organized simplicity
(Newtonian mechanics)
Small number systems
Tool: calculus
(deterministic rules)
Linearity and equilibrium
analysis/machines
Direct control

COMPLEXITY →

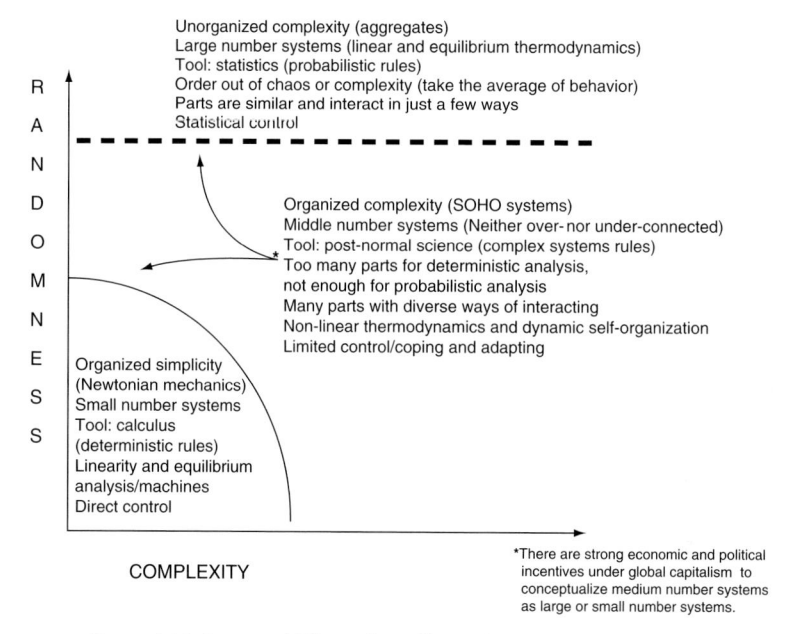

*There are strong economic and political incentives under global capitalism to conceptualize medium number systems as large or small number systems.

Figure 14.2 Large, middle, and small number systems. (*Source:* adapted from Kay and Foster 1999 and Weinberg 1975.)

their preferences and perspectives. The hierarchical nature of SOHO fisheries systems must be understood through a consideration of issues involving system scale and type, and the bounding and nesting of the system, which inevitably involve identifying important processes, structures, feedbacks, and contexts (Boyle *et al.* 2002). What constitutes an important process, structure, feedback, or context will be characterized by the position of the fishery system observer and can never capture everything – "importance" will always involve values and interests that require multiple observers to capture a diversity of perspectives on the SOHO fishery system (Kay *et al.* 1999). In addition, though possibilities of future system states can be offered, exact predictions of fishery system dynamics are not possible when fisheries systems are understood with SOHO systems models. This is because self-organizing systems display "middle number" behavior (Fig. 14.2) that can be described and understood by exploring *possible* attractors accessible to the system, the feedbacks that *may* maintain the system at the attractors, the external influences that *could* define the context for a specific attractor (or regime), and the conditions under which flips between attractors (or regimes) are *likely* (Kay and Foster 1999; Kay *et al.* 1999; Boyle *et al.* 2002). In the middle number systems that characterize most of the environmental and social issues associated with marine fisheries today, the assumptions of large number systems, where statistics can be applied, and of small number systems, where deterministic mathematical methods such as calculus can be used, are of limited applicability (Fig. 14.2).

Fisheries issues exhibit complex nested system dynamics that contain too many diverse elements for deterministic analysis and not enough elements of identical and average type to apply statistics (Weinberg 1975; Kay and Foster 1999). They involve citizens who, as long as they are allowed political freedom to deliberate on fisheries issues, defy deterministic explanation and exhibit non-linear dynamics and adaptive self-organization. The decision stakes surrounding fisheries issues are often extremely high, with associated levels of uncertainty that move beyond the scope of normal science (Funtowitz and Ravetz 1993). Typical managerial responses to fisheries systems involve attempts to absorb complexity and uncertainty – problems are recast as being amenable to direct or statistical control (Ludwig *et al.* 1993; Torgerson 1999). Strong socioeconomic actors and institutionalized attractors seek certainty in fisheries systems to allow for profit maximization, planning, and management over the short term. The demand for stability and certainty in fisheries systems tends to displace fishery problems from the messy middle number sphere into more controllable and economically useful articulations (Bella 1997). The practice of displacing complexity and uncertainty in the short term, however, can perversely lead to increasing overall uncertainty and complexity in fisheries systems over the long term (Bella 1997).

The application of the SOHO systems model to fisheries provides an example of an ecosystem approach that raises different questions, data needs, monitoring requirements, and managerial outlook from those offered by the reductive single-species approach (Slocombe 1993, 1998; Kay *et al.* 1999; Link 2002; Busch *et al.* 2003). Framing fisheries systems as self-organizing, holarchic, and open focuses attention on situations that are irreducibly complex and requires participatory approaches toward a much broader range of monitoring, management, and governance techniques (Jessop 1997, 2002). It is appropriate to refer to globalized fishery systems as complex, as opposed to complicated, because they involve many types of interacting systems (societal, biological, physical) displaying non-linear dynamics, and operating at a variety of temporal and spatial scales with multiple possible attractors and self-organizing regimes, and irreducible uncertainties associated with their observation and a variety of legitimate perspectives on their description (Table 14.2).[8]

SOHO systems descriptions allow managers and interested stakeholders to ask a series of new questions that go beyond the traditional focus on annual single-species stock assessments, total allowable

Table 14.2 *Properties of complex systems to keep in mind when thinking about SOHO fisheries system descriptions*

Non-linear	Behave as a whole, as fishery *systems*. Cannot be understood by taking them apart into pieces that can be added or multiplied together in a linear fashion.
Hierarchical	Are *holarchically nested*. The fishery system is nested within a system and is made up of other systems. The "control" exercised by a holon of a specific level always involves a balance of internal or self-control and external, shared, reciprocating controls involving other holons in a mutual causal way that transcends the old selfish–altruistic polarizing designations. Such nesting cannot be understood by focusing on one hierarchical level (holon) alone. Understanding comes from multiple perspectives of various *types* and *scales*.
Internal causality	Non-Newtonian: not a mechanism but rather is *self-organizing*. Characterized by goals, positive and negative feedback, autocatalysis, emergent properties and surprise.
Window of vitality	SOHO systems must have enough complexity but not too much. There is a range within which self-organization can occur. Complex systems strive for *optimum*, not minimum or maximum conditions.
Non-equilibrium/ dynamic stability (far from equilibrium)	Equilibrium points may *not* exist for the fishery system, but fishery systems may be characterized by dynamic flux and change within the constraints imposed by the physical environmental context.

[8] "If a system – despite the fact that it may consist of a huge number of components – can be given a complete description in terms of its individual constituents, such a system is merely *complicated*. Things like jumbo jets or computers are complicated. In a *complex* system, on the other hand, the interaction among constituents of the system and the interaction between the system and its environment, are of such a nature that the system as a whole cannot be fully understood simply by analyzing its components. Moreover, these relationships are not fixed, but shift and change, often as a result of self-organization" (Cilliers 1998: viii).

Table 14.2 (*cont.*)

Multiple steady states/ regimes	There is *not* necessarily a unique preferred system state in a given fishery. *Multiple attractors* can be possible in a fishery, and the current system state may be as much a function of historical accidents as anything else. For example, regime shifts in fishery systems may be a result of the combination of particular fishing pressures, changes in the physical context (such as ocean currents, water temperature, and salinity), and shifting ecological structures and dynamics. There may be multiple regimes, which are valued differently by various actors in the system.
Catastrophic behavior	The norm. Bifurcations: moments of unpredictable behavior. Flips: sudden, discontinuous, rapid change. Holling four-box cycle/shifting steady-state mosaic with exploitation, conservation, release and reorganization stages.
Chaotic behavior	Our ability to predict and forecast fishery systems is always limited. For example, with weather forecasts, the limit is between 5 and 10 days regardless of how sophisticated the computers are or how much information is available, because of the sensitivity to measurement conditions and unavoidable measurement errors. Stock assessments, therefore, cannot be used for anything like future predictions over multiple years, and we need to establish safe biological limits with reference to particular regimes or attractors rather than precise total allowable catch (TAC) levels.

Source: Adapted from Kay *et al.* (1999:727), Boyle *et al.* (2001), Gunderson and Holling (2002), and Ottino 2004.

catch (TAC) levels, and quota allocation in fisheries science monitoring and management. The new questions focus on the ecological and physical contexts within which fishing takes place and the feedbacks within and between societal systems, ecological systems, and the physical environmental context (Table 14.3, Fig. 14.3).

Table 14.3 *Questions that flow from the SOHO systems heuristic applied to fisheries*

(1) What are the elements (structures and processes) of the societal fishing system that you wish to maintain?
(2) What is the ecological context necessary to maintain the processes and structures of the societal fishing system (Fig. 14.3A)?
(3) What are the ecological structures and processes that provide the context for the societal fishing system?
(4) What is the physical environmental context necessary to maintain the processes and structures of the ecological system (Fig. 14.3D)?
(5) What structural changes does the societal fishing system make to the ecological system (Fig. 14.3B)?
(6) How does the societal system alter the physical environmental context (Fig. 14.3C)?

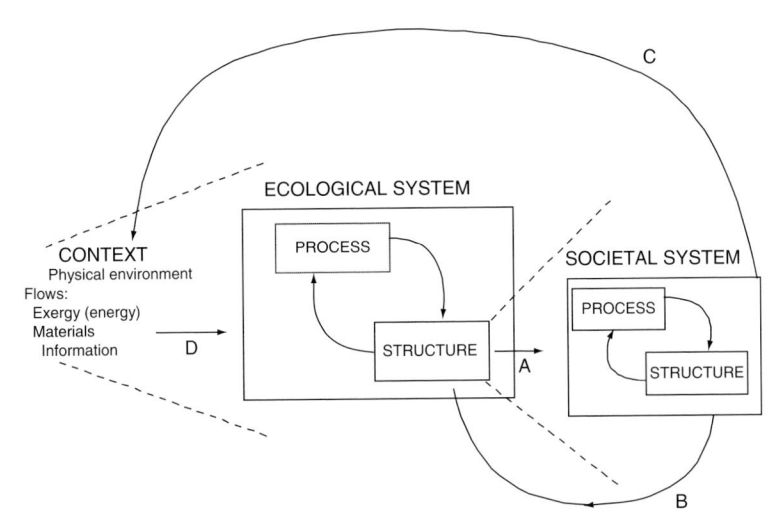

Figure 14.3 SOHO systems model. This model is used to illustrate Newfoundland and Labrador fishing dynamics by focusing on the physical environmental context (D), the self-organizing structures and processes of the ecological and societal systems, feedbacks between them (B and C), and the transfer of ecological structure (biomass) into the societal system (A).

SOCIOECOLOGICAL SOHO DESCRIPTION OF THE NEWFOUNDLAND AND LABRADOR COD FISHERY: FROM COD ABUNDANCE AND COLLAPSE TO THE CRUSTACEAN REGIME

In 1992, the northern cod (*Gadus morhua*) fishery off the northeast coast of Newfoundland and Labrador, Canada (North Atlantic Fisheries Organization [NAFO] divisions 2J3KL), was placed under moratorium after being fished to commercial extinction (Fig. 14.4).

Up until the moratorium, the northern cod fishery had been commercially exploited since the 1500s and was recognized as the richest ground fishery in the world, with total landings of more than 100 million tonnes of codfish up to 1992 (Kurlansky 1997; Rose 2003). Half of the 100 million tonnes of codfish was slowly captured over the 400-year period from 1500 to 1900 using preindustrial technology in the absence of scientific fisheries management; the other half was rapidly removed between 1900 and 1992 as fishing became both industrialized and increasingly managed. The convergence of high-tech bottom-trawlers from more than 20 nations targeting spawning aggregations of cod on offshore banks occurred as fisheries management matured into an applied science. In the period since World War II, fisheries science, along with other resource management sciences, focused on the efficient control of single-species population growth and productivity in the context of pressure to maximize profitability (Smith 1994; Hutchings and Myers 1995; Holm 1996). Northern cod fisheries management was initially implemented in the 1950s through the application of single-species population models and total allowable catch quota allocations among nations, administered through the International Commission for the Northwest Atlantic Fisheries (ICNAF), the North Atlantic Fisheries Organization (NAFO), and later, when coastal jurisdiction was extended out to 200 miles (320 km) in 1977, the Canadian Department of Fisheries and Oceans (DFO) (Lear and Parsons 1993; Hutchings *et al.* 2002).

Northern cod had been massively overfished when Canada nationalized the fishery within the 200-mile (320-km) limit in 1977 (Lear and Parsons 1993). The DFO implemented what was recognized at the time as a world-class single-species scientific monitoring and management system with the stated goal of rebuilding and conserving offshore cod stocks (Hutchings *et al.* 2002; McCay 2002). By the early 1980s, the DFO and the offshore fishing industry were confident that cod populations had been rebuilt or were well on their way to recovery. Fisheries management quotas were increased to give Canadian fishing

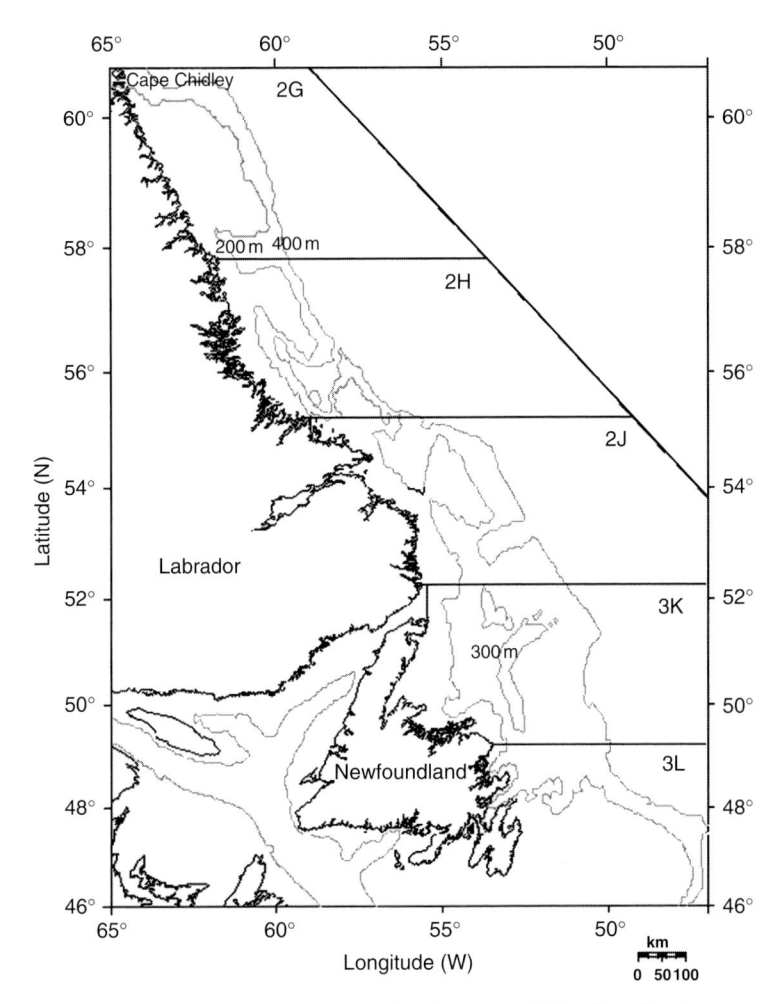

Figure 14.4 The northwestern Atlantic Ocean off Newfoundland and Labrador, Canada, illustrating NAFO fisheries management divisions 2GHJ3KL. The northern cod stocks are generally referred to as those stocks encompassing NAFO divisions 2J3KL.

companies access to the nationalized fish stocks (Lear and Parsons 1993; Rose 2003). However, inshore cod fishers complained of declining catches during this same period, doubting the confident scientific advice voiced by DFO managers and the self-interested opinions of the offshore fleet (Keats *et al.* 1986; Neis and Felt 2000; Hutchings *et al.* 2002; Bavington *et al.* 2004). In spite of several expert-based cod-fisheries reports and inquires into the state of the cod-fishing industry and scientific stock assessments throughout the 1980s and early 1990s

(Kirby 1983; Keats *et al.* 1986; Harris 1989; Finlayson 1994), quotas for northern cod remained high up until 1992, when the Canadian offshore trawler fleet was unable to find cod to catch on the offshore banks and a stock collapse crisis was declared (Steele *et al.* 1992; Walters and Maguire 1996). The resulting cod-fishing moratorium led to the largest single lay-off in Canadian history, massive social disruption in rural Newfoundland and Labrador, and billions of dollars of emergency federal assistance for unemployed cod-fishery workers (Millich 1999; Bavington 2001; Rice *et al.* 2003; Bavington *et al.* 2004).

The collapse of the northern cod has become a classic case of the perils associated with single-species fisheries management (Finlayson and McCay 1998; Newell and Ommer 1999). A belief that fishing mortality was the only relevant variable determining stock size and the associated assumptions that environmental variables remained relatively constant and favorable to cod population growth and that fishing did not significantly alter the natural growth rates and productive dynamics of cod populations permitted managerial confidence that controlling fishing effort would allow for accurate quantitative predictions of stock structure and ultimately the realization of recovery and conservation goals (Rice *et al.* 2003; Rose 2003). Based on the assumptions of single-species fisheries management, the initial cod population growth estimates claimed that cod would recover quickly (within 2 to 5 years) with the cessation of directed commercial fishing (Rice *et al.* 2003; Rose 2003; Bavington *et al.* 2004). This optimistic belief in rapid recovery was based on the assumption that ecological, environmental, and behavioral conditions associated with cod productivity had not significantly altered since the original rebuilding effort in the late 1970s, and that legal bycatch, illegal fishing, and scientific survey fisheries were inconsequential for rebuilding the cod population (Rice *et al.* 2003; Rose 2003).

After more than a decade since the moratorium, single-species management assumptions applied to the northern cod appear to have been overly optimistic and simplistic. Since 1992, fisheries scientists have found that cod have failed to recover across their range, and their spawning biomass has been all but eliminated – reduced by 99.9 percent of its historic maximum (Hutchings 2004). Other ground fish stocks have also collapsed, resulting in the complete closure of the cod fishery and the scientific recommendation that northern cod be listed as an endangered species (COSEWIC 2003). Increasingly, fisheries scientists argue that regional-scale regime shifts and fishing down sequences have occurred in the northwestern Atlantic as higher trophic level ground fish species decline and jellyfish, scavenger species (e.g., sculpins), crustacean populations (e.g. snow crab [*Chionocetes opilio*] and northern shrimp [*Pandalus borealis*]) expand (Pauly and Maclean 2003; Rose 2003). The fishing industry, the provincial Department of Fisheries and Aquaculture (DFA) and rural development boards are responding to these regime shifts by focusing harvesting effort on crustaceans (northern shrimp and snow crab) and developing harvesting strategies for under-utilized species such as sea urchins, jellyfish, and sculpins (Curtis 2002; Harte 2002).

The collapse of the Newfoundland and Labrador northern cod fishery provides a good case study to apply the SOHO systems model to work through some of the structural changes that have occurred in biophysical and socioeconomic systems, and the energy, material, and informational flows and feedbacks that currently exist within and among them. In addition to building a more realistic picture of the dynamics of exploited fish populations (Busch *et al.* 2003), a SOHO socioecological systems model of the Newfoundland and Labrador fishery helps to highlight the interconnections that exist between social and ecological systems and their nested contexts (Bavington and Kay 2003). The SOHO systems model provides a way to understand and integrate the multiple sociological and ecological perspectives on what has happened in the Newfoundland and Labrador fishery since the cod moratorium in 1992, especially the reasons why cod stocks have failed to recover as predicted and the context for socioecological interventions and actions at this moment in history.

TRANSFERRING MARINE ECOLOGICAL STRUCTURE (BIOMASS) INTO THE SOCIETAL SYSTEM OF NEWFOUNDLAND AND LABRADOR

The socioeconomic and cultural system of Newfoundland and Labrador is extremely dependent on extracting marine ecological structure through fishing (Fig. 14.3A). The cod fishery was the reason for European interest in the island, and England, France, Spain, and the Basques vied for colonial control of the rich fishing grounds from the time of European "discovery" in 1497 (Innis 1954). Approximately 100 million tonnes of codfish were removed from the marine ecological system from 1500 to 1992 (Rose 2003).[9]

---

[9] Other species were also harvested, and by the turn of the twentieth century, some were extirpated by overharvesting, including the great auk, marine mammals, sea birds, and others (Pauly and Maclean 2003; Rose 2003).

In the SOHO systems model of the Newfoundland and Labrador fishery (represented in Fig. 14.3), the arrow marked A represents the movement of ecological structure (codfish) into the societal system. The abundance of codfish was predicated on favorable ecological structures and processes making up a ground fish dominated regime with capelin forming a critical prey source for cod. Half of the ecological structure (cod biomass) was fished seasonally over a 400-year period from 1500 to 1900 using pre-industrial fishing technology, primarily single baited hooks on hand lines. With industrialization and changes in fishing and processing technologies, the remaining 50 percent of the 100 million tonnes of codfish were rapidly removed between 1900 and 1992, with year-round fishing activities taking place on the offshore banks, often during spawning or on pre-spawning aggregations (Steele *et al.* 1992; Rose 2003), and reduction, bait, and roe fisheries targeting squid, capelin, and other prey species of cod. Industrialization led to a spatial and temporal scaling-up of the fishery and initiated a fishing down trajectory (Pauly and Maclean 2003; Bavington *et al.* 2004). Industrialization of the fishery was also associated with Newfoundland and Labrador joining Canada in 1949 and a general modernization of the societal system. This modernization altered social processes and structures, shifting the peasant–fisher–merchant society, based on the household production of dried salt cod, into a modern market society with wage labor and factory production of frozen seafood products in fish plants (Polanyi 1957; Wright 2001; Ommer 2002).

Since 1993, the species composition and the annual biomass of landings extracted from the ecological system have changed dramatically, and this has had profound effects on the processes and structures of the societal system (Bavington *et al.* 2004). The overall biomass of landings in the Newfoundland and Labrador fishery has been cut roughly in half compared with average landings in the last decades of the pre-1992 period, and the main composition of the landings has shifted from ground fish (northern cod) to crustaceans (northern shrimp and snow crab). Counter-intuitively, from the perspective of fishing down theory, the high market value of the lower trophic level crustaceans (mainly snow crab) has resulted in an extremely profitable fishery, exceeding the historic cod fishery to become the most profitable in Newfoundland and Labrador's history. The shift from a ground fishery to one focused on crustaceans has altered the structure of the fishing industry, rural fishing communities, and the societal processes that rely on and help to sustain it (Neis *et al.* 2001).

The crustacean fishery involves fewer people, and with fewer fishers capturing a higher-value product, wealth has become increasingly concentrated within fishing communities and regions of the province where crab and licenses to catch and process them are plentiful and those areas where they are scarce (Bavington et al. 2004). Processor profit margins and market demand for crab have allowed some fishers to finance expensive upgrades on their boats (increasing the overall fishing capacity of the fleet) through "trust" agreements with fish processors, who gain guaranteed access to the profitable crab resource without having to compete with other processing companies (FFAW 2004). This change has undermined the DFO's fleet separation policy and encouraged corporate vertical integration of the industry by allowing processors to obtain proxy ownership of crab licenses that are supposedly held by small, owner-operated harvesting enterprises (FFAW 2004). In addition, fewer processing jobs are associated with crustaceans, and the market preference for unprocessed snow crab-in-the-shell has led to a disproportionate impact on women, who are the main workers in the processing sector (Neis et al. 2001; Bavington et al. 2004). Processing workers have also been exposed to new occupational hazards, such as crab asthma, associated with handling large amounts of crab (Neis et al. 2001). The reassignment of shrimp quotas to the province of Prince Edward Island in recent years may also be contributing to an increase in the number of seasonal migrant plant workers from rural Newfoundland going to that province. The change in landings has also resulted in fishers having to travel longer distances to catch their quota and pressure to expand fishing effort and the size of boats, resulting in a greater number of marine accidents (Wiseman et al. 2001).

Since 1993, corporate fish processing companies such as Fisheries Products International (FPI) have begun to source their product globally rather than locally. They focus on value-added processing that combines wild seafood caught off Newfoundland and Labrador with other seafood sourced internationally from both wild and cultured fisheries (Rowe 2004). FPI, one of the largest seafood corporations in North America, operates plants in Newfoundland which import farmed salmon from Chile, cod from the Barents Sea, and farmed tilapia from the United States for use in value-added products destined mainly for sale to large restaurant chains in the United States (Rowe 2004). FPI also sells premium value-added brands such as President's Choice™ and Costco™ in Canada (Rowe 2004). The switch to value-added processing resulted in FPI becoming profitable in the midst of the cod moratorium after years of financial trouble and government subsidies, and effectively decoupled the corporation from dependence on raw material captured or cultivated in Newfoundland and Labrador. The company's delinkage from local raw material dependence and reliance on the U.S. market have exposed the corporation, and the fishing industry in general, to risks associated with American currency fluctuations. A strong U.S. dollar increases export profits and demand for Newfoundland and Labrador crab, shrimp, and other processed seafood products; a falling U.S. dollar has the opposite effect (Rowe 2004).

Changes in price and market demand (crustaceans are worth more and give higher profit margins per pound than ground fish) and the relative abundance of crustaceans compared with ground fish species in Newfoundland and Labrador's ecological system have resulted in complex dynamics within the post-1992 Newfoundland and Labrador societal fishing system. The gradual replacement of the resilient, pre-modern, low-profit inshore cod-fishing society with a highly profitable, less resilient, modern, crustacean-based market society raises questions of social and ecological sustainability. The present societal system is extremely reliant on harvesting a low-trophic-level species (snow crab vs. northern cod), employing relatively few fishers and processing workers (whose corporate employers are tied into a global capitalist economic system heavily dependent on continuing U.S. consumer demand), high crab prices, and the abundance of snow crab in the ecosystem.[10] The present Newfoundland and Labrador fishery requires an exploration of the feedbacks that exist between the societal system, the ecosystem, and the broader physical environmental context to gain a better understanding of the vulnerabilities and options for intervention and action available at this point in history.

FEEDBACKS FROM THE SOCIETAL TO THE ECOLOGICAL SYSTEM

Feedback loops represented in Fig. 14.3B can help to frame understanding of fisheries systems as complex socioecosystems, and draw attention to interactions between societal, ecological, and physical systems. The SOHO systems model represented in Fig. 14.3 draws attention to

---

[10] There are signs that snow crab populations are declining as fishing effort has increased in recent years, and quotas have been cut in some areas, most notably off the southeast coast of Labrador (Bavington *et al.* 2004).

the ecological and physical systems that form the contexts for societal fishing systems and the relationships that exist among all three.

When we are trying to understand relationships in the Newfoundland and Labrador fishery, it is important to consider not only the transfer of ecological structure used by the societal system (Fig. 14.3A) but also the overall impact that fishing activities have on ecological systems. Arrow B in Fig. 14.3 can be used to represent the total amount of ecological structures (species) removed from the ecological system or killed through fishing practices.

The total amount of species killed will exceed those used by the societal system because ecological structure (overall biomass and species composition) can be disturbed or destroyed through bycatch, discards, and high grading with impacts on marine mammals, fish, seabirds, and other marine species critical for ecological integrity and health. Ongoing fisheries deploying non-selective gear types have resulted in high levels of cod bycatch and some intentional targeting of cod bycatch up to legal landing limits (Rice *et al.* 2003; Winsor 2004). Even noise produced by sonar, boat engines, and seismic testing can affect fish behavior and, potentially, population dynamics and broader ecological structures (Popper 2003; Rowe and Hutchings 2004). Shifts to crustacean fisheries, especially the shrimp fishery, have been associated with higher levels of bycatch due to the trawling gear used. Recent studies of the crab fishery have shown high mortality rates associated with throwing back juvenile crabs that are caught in pots and pulled to the surface. The high mortality is thought to be due to damage inflicted to crabs during handling and increased predation pressures after release (Grant 2004). Structural changes to marine ecosystems induced by fishing gear such as otter trawls have resulted in claims from crab fishers that shrimp trawling damages the bodies of snow crabs, leading to increased incidence of diseases and physical harm to legs and the carapace. Many more structural changes could be highlighted in relation to fishing and other human activities conducted in the marine environment; the SOHO ecosystem approach helps to provide a conceptual framework to think through these structural changes and the feedback loops from the societal system that co-produces them.

Structural changes in the ecological system feed back into ecological processes, causing alterations that in turn feed back into ecological structures, often in unpredictable ways. Thinking through these feedbacks (Fig. 14.3B) and facilitating the participation of a group of diverse stakeholders with knowledge of them, can help managers integrate knowledge from a diversity of observers, identify knowledge gaps, and assess the overall impact of harvesting technologies critical for the application of the precautionary principle in fisheries management (FAO 1995; DFO 2001, 2002b; Garcia *et al.* 2003).

In addition to feedbacks from fishing activities, the SOHO systems model (Fig. 14.3) can be used to think through the connections between other societal activities and ecological systems. For example, exploratory seismic testing for petroleum resources and oil pollution at sea can cause increased mortality of larval fish, changes in the behavior of marine organisms, seabird kills, and additional interactions that alter ecological structures and processes. These changes in the ecological system ultimately loop back to influence what ecological structures are available for societal fishing systems (Fig. 14.3A), and the changes induced in ecological systems can affect societal systems that are spatially and temporally separated from the societal system that initiated the activity. The spatial and temporal separation of feedbacks (Fig. 14.3B) from the changes induced in the ecological structure available for a specific societal fishing system (Fig. 14.3A) are especially relevant to think through as capitalist economies become more globally integrated, industrial aquaculture expands, and human systems enhance their feedbacks onto the physical environmental context over increasing spatial and temporal scales (Fig. 14.3C).

FEEDBACKS FROM THE SOCIETAL SYSTEM TO THE PHYSICAL ENVIRONMENTAL CONTEXT

Feedback C in Fig. 14.3 can be used to think through the relationships between activities in the societal system and the physical environment. The physical environmental context shapes the self-organization that is expressed in ecological processes and structures by determining the available energy, material, and information (Kay *et al.* 1999). Temperature changes (especially those associated with the North Atlantic Oscillation, the Labrador Current, and the Gulf Stream), relative amounts of precipitation, salinity, nutrients, genetic information, and introductions of ice and freshwater runoff into marine ecosystems can influence spawning success, recruitment, migration, and other behavioral attributes of species that are of direct interest to the societal system and those that form important indirect trophic interactions with commercially relevant species. Human activities in the societal system can feed back onto the physical environment (Fig. 14.3C) that comprises the context for a particular ecological system, and changes in the physical environment can cascade to affect ecological structures available to the societal system, leading to changes in societal structures and processes (Steele 1998; Kay and Regier 1999; Scheffer et al. 2001). High levels of uncertainty and complexity will be associated with the identification and prediction of these feedbacks, but the SOHO systems model (Fig. 14.3) helps to produce an integrative picture or narrative of the general trends, possible relationships, cascades, and results of current feedbacks and interrelationships.

Feedback loops from the societal system to the physical environment (Fig. 14.3C) alter the context for the whole marine ecological system. These feedbacks may include human-induced climate change, damming projects that influence the amount and temperature of fresh water, and the number of nutrients entering the marine ecosystem,[11] along with many others, such as intensive industrial aquaculture, which can introduce (or remove, in the case of bivalves) large amounts of energy and materials (nutrients, therapeutics and other chemicals used on the farm) and non-native behaviors and genetic information through escapees (Bavington 2000, 2001; Pauly et al. 2001). Fishing practices can also feed back to influence the physical environment and, therefore, the ecological context. Structural changes induced by fishing gear types, such as otter trawls, disturb ocean habitat (Watling and Norse 1998) and have recently resulted in legal action against the government of Canada by environmental groups focused on the risks to fish habitat associated with this type of deleterious feedback (Winsor 2004).

Many more perspectives on the interactions between and internal dynamics of the physical context, ecological system, and societal systems (Fig. 14.3) of the Newfoundland and Labrador fishery could be offered.[12] The SOHO systems model places the context, various systems, and their feedbacks into a hierarchy of holons to conceptualize the scalar dynamics that exist in complex SOHO fishing systems. Figure 14.3 illustrates a model that helps identify and think through feedback loops, but it does not illustrate the hierarchy of spatial and temporal scales that are involved in socioecological systems. It is crucial that we consider scale and cross-scale issues when thinking about globalized fishery systems (Kay et al. 1999; Boyle et al. 2002). Dams, for

---

[11] For example, the Churchill Falls hydroelectric dam in Labrador may have nutrient, water temperature, and salinity effects on Groswater Bay in Labrador.

[12] The perspectives that have been offered in this chapter reflect the knowledge, interests, values, and beliefs of the authors, and though we have tried to include a diversity of views, SOHO systems theory points to the need to involve a diversity of actors to complete the systems narratives and emphasizes the contingent and partial nature of all system descriptions.

example, alter the physical context of entire regions and require large-scale societal resources to construct and maintain. Aquaculture sites tend to alter contexts more locally, but industrial aquaculture operations can exhibit cross-scale feedbacks when farms draw on fishmeal, therapeutics, and terrestrial agricultural products from around the world (Naylor et al. 2000; Bavington 2001; Pauly et al. 2001). The attention to nested scalar relations can be used to explore the differences between inshore and offshore fisheries that target fish at different life stages and often from different populations (bay stocks vs. offshore stocks, for example) (Hutchings et al. 2002). Debt, global trade rules, and food safety standards originating in societal systems (institutional structures) at various scales can feed back to influence where fishing and aquaculture activities take place and the practices they employ. These feedbacks ultimately loop back into ecological structures and the physical environmental context.

Ultimately, one needs to think feedbacks through from the global to the local spatial scale, including the varying temporal rates that are operating. Failure to do this can lead to simplistic explanations and understanding of fishing dynamics, monitoring, management, and governance activities that can result in overconfident scientific assessments and naive policy prescriptions, as illustrated in the case of the collapse of the Newfoundland and Labrador northern cod fishery.

ECOSYSTEM-BASED FISHERIES MANAGEMENT: AN OPEN AND CONTESTED FUTURE

Global stock collapses, the complexity of marine fishery systems, and the failure of single-species management all demand a radical rethinking of human relationships with the sea. The SOHO systems heuristic helps to change how we think about fisheries systems to reflect lessons learned from complexity science, ecosystem approaches, and the many failures of single-species fisheries management. Under the SOHO systems approach, fisheries monitoring shifts from a narrow focus on experts obtaining single-species population information to permit accurate stock assessments and TAC levels to a much broader range of indicators of ecological health and integrity that are defined through negotiation with interested stakeholders.[13] From this new perspective, fisheries

---

[13] It is important to note that the terms "health" and "integrity" have a diversity of definitions and do not have an agreed operational meaning in ecosystem-based fisheries management (Garcia et al. 2003).

management changes from the confident control of commercial fish populations for powerful economic interests to a more humble focus on coping with and adapting to ecological systems while attempting to facilitate the control of anthropogenic feedbacks that produce unsustainable ecological and environmental contexts for a broad range of human activities and marine species (Larkin 1988; Thompson and Trisoglio 1997; Coward *et al.* 2000; Bavington 2002).

If the SOHO systems approach and other ecosystem approaches to fisheries management are to be implemented successfully, many political, economic, ethical, and institutional barriers will have to be challenged and removed, or avoided. For instance, the shift in emphasis from managing wild fish populations to managing people and their institutions under ecosystem-based fisheries management frameworks is ironically occurring at a time when human control over domesticated fish is expanding, and state regulatory institutions are being rationalized and downsized in many countries. The industrial domestication of profitable carnivorous fish such as steelhead, salmon, and cod on the controlled environments of fish farms is perhaps the only context where the simplifying assumptions of single-species fisheries management are currently applied with relative impunity. However, though single-species management for maximum production may produce profits on the controlled, confined contexts of industrial fish farms, the global expansion of industrial aquaculture and other ocean-based industrial activities increases the complexity, uncertainty, and challenges associated with fisheries management while exacerbating food security and sovereignty challenges (Bavington 2000, 2001). Furthermore, government attention and rhetorical support for various types of ecosystem-based fisheries management is occurring at a time when fisheries management institutions are experiencing intense pressure to cut back on their expenses and offload responsibility and costs onto industry user groups, voluntary organizations, and local forms of government. Neo-liberal political logic associated with economic globalization favors individual responsibility, public–private partnerships, and the market as the trusted mechanisms to implement ecosystem-based fisheries management. In the present political climate, entrepreneurial fishers are increasingly expected to help fund and participate in scientific data gathering, monitoring and management activities. Though these initiatives have the *potential* to bring stakeholders together to learn from past managerial mistakes and antagonisms, the emphasis on cost-cutting and shifting responsibility from state agencies to fishers runs the risk of passing on immense responsibilities without the necessary resources to implement effective ecosystem-based fisheries management. Institutional change and support, along with a shift in how science and the economy are perceived and ultimately function, will be crucial to the successful implementation of ecosystem-based fisheries management (Rogers 1995; Garcia *et al.* 2003).

Additionally, two vastly different responses to ecosystem-based fisheries management are vying for support. One is aimed at assimilating "the ecosystem approach, like the precautionary approach, within the existing methodology of fisheries science … The second response is to abandon the existing methodology of fisheries science, and, ceasing to try to measure fish stocks quantitatively, instead seek to monitor indicators of ecosystem health" (Gray 2002:3). Neo-liberal reform versions of ecosystem-based fisheries management continue to entrench the control-oriented managerial status quo with an emphasis on developing new quantitative models with layers of complexity that include all relevant processes and structures that promise to enable managers to predict ecosystems and eventually control them (Gray 2002). Radical interpretations of ecosystem-based fisheries management, such as those suggested by the SOHO systems heuristic, call for fundamental changes in how we understand what science and management can deliver to managers and include a strong emphasis on institutional change, sustainability principles, and democratic deliberation rather than further development of managerial technique. At this moment in history, it appears that the ideology of neo-liberal governance is encouraging reform interpretations of the ecosystem approach. As long as modern economic and technological imperatives are allowed to unproblematically dictate human–marine relationships, it appears unlikely that ecosystem-based fisheries management will be any more successful than the single-species approach. Successful ecosystem-based fisheries management will have to "make the structures and processes of modern life sufficiently problematic" rather than promoting their expansion (Rogers 1995:142).

The failure of single-species fisheries management and the rise of globalized economies and neo-liberal regulation theories are resulting in increased calls to treat fishers as if they were entrepreneurial farmers, promoting the vision of a privatized ocean full of tradable goods and services "sustainably" allocated through the invisible hand of the market (Bavington *et al.* 2004). The ocean "is a resource that must be preserved and harvested," *The Economist* magazine proclaims. "To enhance its uses, the water must become ever more like the land, with owners, laws and limits. Fishermen must behave more like ranchers than hunters" (Carr 1998:S3).

Rather than looking to the invisible hand of the market for magical solutions, the challenge of fisheries management requires difficult choices and decisions by all coastal and marine citizens. The SOHO ecosystem approach to fisheries management offers an alternative way of understanding fisheries systems. The approach, however, is embedded within conceptual, political, economic, and ethical contexts that affect how people will understand and implement it. The meaning, application, and impact of the SOHO systems approach will be influenced by these nested contexts and the ecosocial systems in which they are embedded. It is our hope that the radical challenges posed by complexity science and the ecosystem approach will be seriously engaged with and motivate normative action, as opposed to reformed versions of the managerial status quo, the tragic outcome of which is so starkly illustrated in the story of Newfoundland and Labrador's northern cod fisheries.

Acknowledgments

We would like to thank the participants and organizers of the special *Globalization: Effects on Fisheries Resources* session of the 133rd meeting of the American Fisheries Society in 2003 for the opportunity to present our ideas and for their thoughtful engagement and helpful comments on the first draft of this paper. An extra special thank you to Henry Regier for his ongoing encouragement and helpful suggestions during the completion of the paper, to the post-normal science discussion group at the University of Waterloo for providing friends and a forum to develop and discuss many of the ideas contained in the paper, and to the anonymous reviewers whose comments helped to polish the presentation and sharpen the ideas contained in the final version. Dean Bavington would also like to extend a special thank you to his co-author, the late Dr. James Kay, for introducing him to post-normal science and providing a generous amount of encouragement, intellectual support, and friendship in addition to being an influential mentor. His presence and guidance are sorely missed.

References

Barange, M. 2003. Ecosystem science and the sustainable management of marine resources: from Rio to Johannesburg. *Frontiers in Ecology and the Environment* **1**(4): 190–196.

Bavington, D. 2000. *From Hunting to Farming: Exploring the Development of Industrial Aquaculture in Newfoundland and Labrador from a Complex Systems Perspective*, awarded the Canadian Policy Graduate Research Award. Ottawa, Ontario:

Federal Policy Research Initiative. Available online at: www.jameskay.ca/about/grad/aquac.pdf

Bavington, D. 2001. From jigging to farming. *Alternatives* **27**(4): 16–21.

Bavington, D. 2002. Managerial ecology and its discontents: exploring the complexities of control, careful use and coping in resource and environmental management. *Environments* **30**(3): 3–21.

Bavington, D. and Kay, J. 2003. Fisheries systems in an age of globalization: ecosystem-based insights on monitoring, management and governance challenges. *American Fisheries Society 133rd Annual Meeting*, August 10–14, Québec City, Québec, Canada.

Bavington, D., Grzetic, B., and Neis, B. 2004. The feminist political ecology of fishing down: reflections from Newfoundland and Labrador. *Studies in Political Economy* **73**: 141–162.

Bella, D. 1997. Organizational systems and the burden of proof. In *Pacific Salmon and Their Ecosystems*, eds. D. Stouder, P. Bisson, and R. Naimon. New York: Chapman & Hall, pp. 617–638.

Berkes, F. 2003. Alternatives to conventional management: lessons from small-scale fisheries. *Environments* **31**(1): 5–19.

Bertalanffy, L. 1950. An outline of general systems theory. *British Journal of Philosophy of Science* **1**: 134–165.

Boyle, M., Kay, J., and Pond, B. 2002. Monitoring in support of policy: an adaptive ecosystem approach. In *Encyclopedia of Global Environmental Change*, vol. 4, ed. M. K. Tolba. Chichester, UK: John Wiley, pp. 116–137.

Busch, W., Brown, B., and Mayer, G. (eds.) 2003. *Strategic Guidance for Implementing an Ecosystem-Based Approach to Fisheries Management*. Silver Springs, MD: National Marine Fisheries Service, National Oceanic and Atmospheric Administration.

Caddy, J. F. 1999. Fisheries management in the twenty-first century: will new paradigms apply? *Reviews in Fish Biology and Fisheries* **9**: 1–43.

Caddy, J. and Cochrane, K. 2001. A review of fisheries management: past present and some future perspectives for the third millennium. *Oceans and Coastal Management* **44**: 653–682.

Caddy, J. F. and Garibaldi, L. 2000. Apparent changes in the trophic composition of world marine harvests: the perspective from the FAO capture database. *Oceans and Coastal Management* **43**: 615–655.

Caddy, J. and Regier, H. 2002. Policies for sustainable and responsible fisheries. In *The Encyclopedia of Global Environmental Change*, vol. 4, ed. M. K. Tolba. Chichester, UK: John Wiley, pp. 343–351.

Carr, E. 1998. Survey: The sea – a second fall. *The Economist* **347**(8069): S3.

Charles, T. 2001. *Sustainable Fishery Systems*. Oxford, UK: Blackwell Science.

Cilliers, P. 1998. *Complexity and Postmodernism: Understanding Complex Systems*. New York: Routledge.

COSEWIC (Committee on the Status of Endangered Wildlife in Canada). 2003. *COSEWIC Status Assessments, May 2003*. Available online at www.cosewic.gc.ca/htmlDocuments/Detailed_Species_Assessment_e.htm

Coward, H., Rosemary, O., and Pitcher, T. 2000. *Just Fish: Ethics and Canadian Marine Fisheries*. St. John's, Newfoundland: ISER Books.

Curtis, D. 2002. Sculpin Project update. Presentation to *2002 Fisheries Forum: Fisheries Issues and Opportunities*, November 14, Renews, Newfoundland.

DFO (Department of Fisheries and Oceans Canada). 2001. *A Canadian Perspective on the Precautionary Approach/Principle*, discussion document. Available online at: www.dfo-mpo.gc.ca/cppa/HTML/discussion_e.htm

DFO. 2002a. *Canada's Ocean Strategy: Policy and Operational Framework for Integrated Management of Estuarine, Coastal and Marine environments in Canada*. Ottawa, Ontario: Fisheries and Oceans.

DFO. 2002b. *Proceedings of the DFO Workshop on Implementing the Precautionary Approach in Assessments and Advice*, Canadian Science Advisory Secretariat. Available online at www.dfo-mpo.gc.ca/csas/Csas/Proceedings/2002/PRO2002_009b.pdf

FAO (Food and Agriculture Organization). 1995. *Precautionary Approach to Fisheries, part 1, Guidelines on the Precautionary Approach to Capture Fisheries and Species Introductions*, Fisheries Technical Paper No. 350. Rome: Food and Agriculture Organization of the United Nations.

FAO. 2003a. *The Ecosystem Approach to Marine Capture Fisheries*, FAO Technical Guidelines for Responsible Fisheries No. 4(Suppl.2). Rome: Food and Agriculture Organization of the United Nations.

FAO. 2003b. *Report of the Expert Consultation on Ecosystem-Based Fisheries Management*, FAO Fisheries Report No. 690. Rome: Food and Agriculture Organization of the United Nations.

FFAW (Fisheries Food and Allied Workers Union). 2004. Preserving the independence of the inshore fleet in Canada's Atlantic fisheries, Presentation to the *DFO Consultation on the Atlantic Fisheries Policy Review* January 20, St. John's, Newfoundland.

Finlayson, A. C. 1994. *Fishing for Truth: A Sociological Analysis of Northern Cod Stock Assessments from 1977–1990*. St. John's, Newfoundland: ISER Books.

Finlayson, A. C. and McCay, B. 1998. Crossing the threshold of ecosystem resilience: the commercial extinction of northern cod. In *Linking Social and Ecological Systems*, eds. F.Berkes, C.Folke, and J.Colding. Cambridge, UK:Cambridge University Press, pp. 311–337.

FRCC (Fisheries Resource Conservation Council). 1997. Towards an ecosystem approach to fisheries management, *Report of the FRCC Environment and Ecology Workshop*, December 15–16, University of Moncton.

Funtowitz, S. and Ravetz, J. 1993. Science for the post-normal age. *Futures* September 1993: 739–755.

Garcia, S. M. 1994. The precautionary principle: its implications in capture fisheries management. *Oceans and Coastal Management* **22**: 99–125.

Garcia, S.M., Zerbi, A., Aliaume, C., Do Chi, T., and Lasserre, G. 2003. *The Ecosystem Approach to Fisheries: Issues, Terminology, Principles, Institutional Foundations, Implementation and Outlook*, FAO Fisheries Technical Paper No. 443. Rome: Food and Agriculture Organization of the United Nations.

Grant, S. M. 2004. The mortality of snow crab discarded from Newfoundland and Labrador's trap fishery: at sea experiments on the effect of drop height and air exposure duration, presentation to the *Canadian Conference on Fisheries Research and the Society of Canadian Limnologists Annual Meeting*, January 8–10, St. John's, Newfoundland.

Gray, J. 2000. Who needs cod anyway? *Globe and Mail Newspaper*. March 11, A18.

Gray, T. 2002. Fisheries science and fishers' knowledge, paper presented to ENSUS 2002, the International Conference on Marine Science and Technology for Environmental Sustainability. Available online at www.efep.org/TSGENSUS.pdf

Gunderson, L. and Holling, C. S. 2002. *Panarchy: Understanding Transformations in Human and Natural Systems*. Washington, DC: Island Press.

Harris, L. 1989. *Independent Review of the State of the Northern Cod Stock*. Ottawa, Ontario: Department of Fisheries and Oceans.

Harte, R. 2002. Emerging fisheries development (jelly fish), presentation to the *2002 Fisheries Forum: Fisheries Issues and Opportunities*, November 14, Renews, Newfoundland.

Holling, C. S. and Meffe, G. 1996. Command and control and the pathology of natural resource management. *Conservation Biology* **10**: 328-337.

Holm, P. 1996. Fisheries management and the domestication of nature. *Sociologia Ruralis* **36**: 177-188.

Hutchings, J. 2004. Life history consequences of over-exploitation to population recovery of northwest Atlantic cod, presentation to the *Canadian Conference on Fisheries Research and the Society of Canadian Limnologists Annual Meeting*, January 8-10, St. John's, Newfoundland.

Hutchings, J. A. and Myers, R. A. 1995. The biological collapse of Atlantic Cod off Newfoundland and Labrador: an exploration of historical changes in exploitation, harvesting technology, and management. In *The North Atlantic Fisheries: Successes, Failures, and Challenges*, eds. R. Arnason and L. Felt. Charlottetown, Prince Edward Island: Institute of Island Studies, pp. 37-39.

Hutchings, J., Neis, B., and Ripley, P. 2002. The "nature" of cod, *Gadus morhua*. In *The Resilient Outport: Ecology, Economy, and Society in Rural Newfoundland*, ed. R. Ommer. St. John's, Newfoundland: ISER Books, pp. 140-185.

Innis, H. 1954. *The Cod Fisheries: The History of an International Economy*. Toronto, Ontario: University of Toronto Press.

Jessop, B. 1997. The governance of complexity and the complexity of governance: preliminary remarks on some problems and limits of economic guidance. In *Beyond Markets and Hierarchy: Interactive Governance and Social Complexity*, eds. A. Amin and J. Hausner. Aldershot, UK: Edward Elgar, pp. 111-147.

Jessop, B. 2002. *The Future of the Capitalist State*. Cambridge, UK: Polity Press.

Kay, J. J. 1991. A nonequilibrium thermodynamic framework for discussing ecosystem integrity. *Environmental Management* **15**: 483-495.

Kay, J. J. and Foster, J. 1999. About teaching systems thinking. *In Proceedings of the HKK Conference*, eds. G. Savage and P. Roe, June 14-16, 1999, University of Waterloo, Ontario, pp. 165-172.

Kay, J. J. and Regier, H. A. 1999. An ecosystemic two-phase attractor approach to Lake Erie's ecology. In *State of Lake Erie (SOLE): Past, Present and Future*, eds. M. Munawar, T. Edsall, and I. F. Munawar. Leiden, The Netherlands: Backhuys, pp. 511-533.

Kay, J. J. and Schneider, E. D. 1992. Thermodynamics and measures of ecosystem integrity. In *Ecological Indicators: Proceedings on the International Symposium on Ecological Indicators, Fort Lauderdale, Florida*, eds. D. H. McKenzie, D. E. Hyatt, and V. J. McDonald. Amsterdam, The Netherlands: Elsevier, pp. 159-182.

Kay, J. J. and Schneider, E. D. 1994. Embracing complexity: the challenge of the ecosystem approach. *Alternatives* **20**(3): 32-38.

Kay, J., Regier, H., Boyle, M., and Francis, G. 1999. An ecosystem approach for sustainability: addressing the challenge of complexity. *Futures* **31**: 721-742.

Keats, D., Steele, D. H., and Green, J. M. 1986. *A Review of the Recent Status on the Northern Cod Stock (NAFO Divisions 2J, 3K and 3L) and the Declining Inshore Fishery*, a report to the Newfoundland Inshore Fisheries Association on Scientific Problems in the Northern Cod Controversy. St. John's, Newfoundland: Department of Biology, Memorial University of Newfoundland.

Kirby, M. 1983. *Navigating Troubled Waters: A New Policy for the Atlantic Fisheries*. Ottawa, Ontario: Canadian Government Publishing Centre.

Koestler, A. 1978. *Janus: A Summing Up*. London: Hutchinson.

Kurlansky, M. 1997. *Cod: A Biography of a Fish that Changed the World*. New York: Penguin Books.

Larkin, P. 1988. The future of fisheries management: managing the fisherman. *Fisheries* **13**(1): 3-9.

Latour, R., Brush, M., and Bonzek, C. 2003. Toward ecosystem-based fisheries management: strategies for multispecies modeling and associated data requirements. *Fisheries* **28**(9): 10-22.

Lear, W. H. and Parsons, L. S. 1993. History and management of the fishery for northern cod in NAFO divisions 2J,3K and 3L. In *Perspectives on Canadian Marine Fisheries Management*, eds. L. S. Parsons and W. H. Lear. Ottawa, Ontario: Canadian Bulletin of Fisheries and Aquatic Sciences, pp. 55-90.

Link, J. 2002. What does ecosystem-based fisheries management mean? *Fisheries* **27**(4): 18-21.

Ludwig, D. 2001. The era of management is over. *Ecosystems* **4**: 758-764

Ludwig, D., Hilborn, R., and Walters, C. 1993. Uncertainty, resource exploitation, and conservation: lessons from history. *Science* **260**: 17-18.

McCay, B. 2002. Co-management and crisis in fisheries science and management. In *Marine Resources: Property Rights, Economics and Environment*, vol. 14, eds. M. Falque, M. De Alessi, and H. Lamotte. New York: JAI Press, pp. 341-359.

Millich, L. 1999. Resource mismanagement verses sustainable livelihoods: the collapse of the Newfoundland cod fishery. *Society and Natural Resources* **12**: 625-642.

Naylor, R., Goldberg, R., Primavera, J., et al. 2000. Effect of aquaculture on world fish supplies. *Nature* **405**: 1017-1024.

Neis, B. 1992. Fishers' ecological knowledge and stock assessment in Newfoundland and Labrador. *Newfoundland Studies* **8**: 155-178.

Neis, B. and Felt, L. (eds.) 2000. *Finding Our Sea Legs: Linking Fishery People and Their Knowledge with Science and Management*. St. John's, Newfoundland: ISER Books.

Neis, B., Schneider, D. C., Felt, L., et al. 1999. Fisheries assessment: what can be learned from interviewing resource users? *Canadian Journal of Fisheries and Aquatic Sciences* **56**: 1949-1963.

Neis, B., Grzetic, B., and Pidgeon, M. 2001. *From Fishplant to Nickel Smelter: Health Determinants and the Health of Women Fish and Shellfish Processors in an Environment of Restructuring*. St. John's, Newfoundland: Memorial University of Newfoundland.

Newell, D. and Ommer, R. 1999. Introduction: Traditions and issues. In *Fishing Places, Fishing People*, eds. D. Newell and R. Ommer. Toronto, Ontario: University of Toronto Press, pp. 1-12.

Ommer, R. (ed.) 2002. *The Resilient Outport: Ecology, Economy, and Society in Rural Newfoundland*. St. John's, Newfoundland: ISER Books.

Ottino, J. M. 2004. Engineering complex systems. *Nature* **427**: 399.

Pauly, D. and Christensen, V. 1995. Primary production required to sustain global fisheries. *Nature* **374**: 255-257.

Pauly, D. and Maclean, J. 2003. *In a Perfect Ocean: The State of Fisheries and Ecosystems in the North Atlantic Ocean*. Washington, DC: Island Press.

Pauly, D., Christensen, V., Salsgard, J., Froese, R. and Torres, F., Jr. 1998. Fishing down marine food webs. *Science* **279**: 860-863.

Pauly, D., Tyedmers, P., Froese, R., and Liu, L. Y. 2001. Fishing down and farming up the food web. *Conservation Biology in Practice* **2**: 25.

Pauly, D., Christensen, V., Guénette, S., et al. 2002. Towards sustainability in world fisheries. *Nature* **418**: 689-695.

Pitcher, T., Hart, P., and Pauly, D. (eds.) 1998. *Reinventing Fisheries Management*. Boston, MA: Kluwer Academic Publishers.

Polanyi, K. 1957. *The Great Transformation: The Political and Economic Origins of our Time*. New York: Beacon Press.

Popper, A. 2003. Effects of anthropogenic sounds on fishes. *Fisheries* **28**(10): 24–31.

Ravetz, J. R. 1999. What is post-normal science? *Futures* **31**: 647–653.

Ravetz, J. and Funtowicz, S. 1999. Post-normal science: an insight now maturing. *Futures* **31**: 641–646.

Regier, H. A. and Kay, J. J. 1996. An heuristic model of transformations of the aquatic ecosystems of the Great Lakes–St. Lawrence River Basin. *Journal of Aquatic Ecosystem Health* **5**: 3–21.

Rice, J., Shelton, P., Rivard, D., Chouinard, G., and Fréchet, A. 2003. Recovering Canadian Atlantic cod stocks: the shape of things to come? presented at the *International Council for Exploration of the Sea Annual Conference: The Scope and Effectiveness of Stock Recovery Plans in Fisheries Management*, Tallin, Estonia, CM 2003/U:06. Available online at www.ices.dk/products/CMdocs/2003/U/U0603.pdf

Rogers, R. 1995. *The Oceans are Emptying: Fish Wars and Sustainability*. Montreal, Québec: Black Rose.

Rose, G. A. 2003. *Fisheries Resources and Science in Newfoundland and Labrador: An Independent Assessment*, Research Paper for the Royal Commission on Renewing and Strengthening Our Place in Canada. St. John's, Newfoundland: Government of Newfoundland and Labrador.

Rowe, D. 2004. D. Rowe (Fisheries Products International President) interviewed on CBC Radio's Fisheries Broadcast, January 14.

Rowe, S. and Hutchings, J. 2004. Implications of mating systems for the collapse and recovery of Atlantic cod, presentation to the *Canadian Conference on Fisheries Research and the Society of Canadian Limnologists Annual Meeting*, January 8–10, St. John's, Newfoundland.

Scheffer, M., Carpenter, S., Foley, J. A., Folke, C., and Walker, B. 2001. Catastrophic shifts in ecosystems. *Nature* **413**: 591–596.

Slocombe, S. 1993. Environmental planning, ecosystem science, and ecosystem approaches for integrating environment and development. *Environmental Management* **17**: 289–303.

Slocombe, S. 1998. Lessons from experience with ecosystem-based management. *Landscape and Urban Planning* **40**: 31–39.

Smith, T. 1994. *Scaling Fisheries: The Science of Measuring the Effects of Fishing, 1855–1955*. Cambridge, UK: Cambridge University Press.

Steele, D. H., Anderson, R., and Green, J. M. 1992. The managed commercial annihilation of northern cod. *Newfoundland Studies* **8**: 34–68.

Steele, J. H. 1998. Regime shifts in marine ecosystems. *Ecological Applications* **8**(Suppl. 1): S33–S36.

Thompson, M. and Trisoglio, A. 1997. Managing the unmanageable. In *Saving the Seas*, eds. L. Brooks and S. VanDever. College Park, MD: ISea Grant, pp. 107–127.

Torgerson, D. 1999. *The Promise of Green Politics*. Durham, NC: Duke University Press.

VanDeVeer, D. and Pierce, C. 1998. *The Environmental Ethics and Policy Book*, 2nd edn. New York: Wadsworth.

Walters, C. J. and Maguire, J. J. 1996. Lessons for stock assessment from the Northern Cod collapse. *Reviews in Fish Biology and Fisheries* **6**: 125–137.

Watling, L. and Norse, E. A. 1998. Disturbance of the seabed by mobile fishing gear: a comparison to forest clearcutting. *Conservation Biology* **12**: 1180–1197.

Weinberg, G. M. 1975. *An Introduction to General Systems Thinking*. New York: John Wiley.

Wilson, J. 2002. Scientific uncertainty, complex systems, and the design of common pool institutions. In *The Drama of the Commons*, eds. E. Ostrom, T. Dietz, N. Dolšak, P. Stern, S. Stonich, and E. Weber. Washington, DC: National Academy Press, pp. 327–360.

Winsor, F. 2004. Fred Winsor (Executive Director of the Fisheries Recovery Action Committee [FRAC]) interviewed on CBC Radio's Fisheries Broadcast, January 28.

Wiseman, M., Burge, M., and Burge, H. 2001. *Fishing Vessel Safety Review (less than 65 feet)*. Ottawa, Ontario: Coast Guard DFO Intra-Departmental Working Group, Department of Fisheries and Oceans.

Wright, M. 2001. *A Fishery for Modern Times*. Oxford, UK: Oxford University Press.

P. VINCENT HEGARTY

15
"Fishy" food laws

INTRODUCTION

"Fishy" is defined in dictionaries as: "of, like or full of fish; dubious, open to suspicion and unsafe." The word "fishy" in the title of this chapter was chosen deliberately to convey two important elements of food laws and regulations. The first part of the chapter deals with some current regulatory issues in the international trading of fish and fishery products. The latter part of the chapter deals with some food regulations that come, in the opinion of this author, under the slang usage of the word "fishy." Some of these regulations cover fish oils and omega-3 fatty acids.

Food laws and regulations at the national and international levels have five distinct but interrelated objectives:

- protect public health
- inform the consumer
- ensure fair trade practices
- protect against fraud
- protect the environment.

The discussion here deals mainly with the first three objectives listed. It must be emphasized that these topics are not independent entities – they are closely interrelated. For example, food poisoning from unsafe fish or fishery products results in negative attitudes among some consumers and a resultant reduction in fish trade. Likewise, when a national regulatory agency rejects local or imported fish as unsafe for human consumption, it gives needed protection to consumers. These rejections also have an impact on the fish trade, and perhaps on environmental issues, as well as protecting the consumer from the fraudulent sale of fish that could be unsafe, misbranded, or mislabeled or any combination of these factors (see Tables 15.1–15.3).

Globalization: Effects on Fisheries Resources, ed. William W. Taylor, Michael G. Schechter, and Lois G. Wolfson. Published by Cambridge University Press. © Cambridge University Press 2007.

Table 15.1 *Food poisoning outbreaks and cases of illness in the United States, 1990–2003*

Food	Number of outbreaks	Cases of illness
Seafood	723	8 071
Produce	432	25 823
Poultry	354	11 894
Beef	343	10 872
Eggs	309	10 750
Multi-ingredient foods	601	18 006

Source: Center for Science in the Public Interest (2004).

Table 15.2 *U.S. Food and Drug Administration refusals of imported seafood, July 2001–June 2002*

Number refused		Number of seafood import refusals by reason					
Total	Seafood	Filthy	*Salmonella*	*Listeria*	Histamine	Poison	Other
15 522	1684	817	427	49	27	37	436

Source: FAO (2003).

FISH REGULATIONS TO PROTECT PUBLIC HEALTH

Fish exports/imports

Documentation from many countries demonstrates that food safety is a major public health issue. All food categories, including seafood, contribute to the problems caused by unsafe food. Data from the United States (Table 15.1) indicate that seafood caused the most outbreaks of food poisoning during the period from 1990 to 2003 (Center for Science in the Public Interest 2004). However, the same study shows that seafood caused fewer cases of illness than did produce, poultry, beef, eggs, and multi-ingredient foods.

The safety of fish and fishery products in the United States is determined by the Food and Drug Administration (FDA). Details of the reasons for rejection of unsafe domestically produced and imported fish are available (FDA 2004a, b). Seafood rejected by the FDA accounted for 11 percent of total food rejections over a 1-year period between 2001 and 2002. "Filthy" and *Salmonella* were the reasons most often

Table 15.3 *U.S. Food and Drug Administration violations for detaining fishery/seafood products, 2001*

Violation code	Number violations	Percent of all violations	Number of countries
Total violations	6405	100	86
Adulteration	5356	84	
Salmonella	1832	29	36
Filthy	1460	23	62
No process	683	11	54
Insanitary	351	6	25
Needs acid/Needs FCE[a]	336	5	42
Poisonous	231	4	38
Listeria	170	3	11
Histamine	123	2	11
Unsafe color	41	0.6	14
Insanitary manufacturing, processing, or packaging			
Manufacture insanitary	130	2	27
Misbranding			
Nutrition label	200	3	33
Lacks firm	140	2.2	32
Usual name	136	2.1	28
List ingredients	87	1.4	29
Lacks n/c[b]	84	1.3	25
False	70	1.1	13
No English	47	0.7	21
Labeling	46	0.7	21
Sulfite label	40	0.6	4
All other violations	64	1.0	35

[a] It appears the manufacturer is not registered as a low acid canned food or acidified food manufacturer pursuant to 21 CFR 108.25 (c) (1) or 108.35 (c) (1). FCE, Food Canning Establishment number.

[b] The article is in packaged form and appears not to have a label containing an accurate statement of the quantity or the contents in terms of weight, measure, or numerical count, and no variations or exemptions have been prescribed by regulations (FDA 2004b).

Source: Economic Research Service, USDA (2003).

Table 15.4 *European Union rejection/detention of imported seafood (totals for January 1999–June 2002)*

Cause of detention/rejection	Number of detentions/rejections
Microbial	208
Chemicals/residues	220
Parasites	32
Others	42
Labeling	20
Sanitary certificate	5
Shelf-life	4
Interrupted cold chain	2
Insects	5
Import prohibited	1
Mixing of fish species	1
Uncertified establishment packaging	2
Not specified	2

Source: FAO (2003).

cited for rejection of seafood (Table 15.2). "Filthy" is defined by the FDA as: "the article appears to consist in whole or in part of a filthy, putrid, or decomposed substance or to be otherwise unfit for food."

A different perspective on fish and fishery products rejected by the FDA is seen in Table 15.3. Adulteration accounted for 84 percent of the rejections. The number of exporting countries who had fish rejected by the FDA is extensive. Similar statistics are produced by the European Union (EU) (Table 15.4). The greatest number of rejections or detentions of imported seafood was for microbial and chemicals/residues violations of EU regulations. In summary, fish and fishery products are important components in the international trade in food; some of these products do not meet regulatory safety requirements. These violations of the regulations can be attributed broadly to two categories – filthy/unsafe products and ignorance of regulations. Those in the latter category may be producing safe fish and fishery products but are ignorant of the branding and labeling requirements of the importing country (Tables 15.3 and 15.4). These companies need education programs so as to avoid further detentions.

The United States, EU, and some other large fish-importing countries send food safety and regulatory experts to fish-exporting countries to ensure improved regulatory compliance. The resulting report consists of an evaluation of the regulatory controls dealing with the safety of fish and fishery products. The following information is typical of the investigation, conclusions, and recommendations in the resulting report, published on the EU Website. A mission to Saudi Arabia assessed the conditions of production of fishery products intended for export to the EU (European Commission 2003). The mission visited the Competent Authority (CA) offices at the central, regional, and local levels, laboratories, aquaculture shrimp farms, food processing establishments, and a fish feed mill. The main findings covered the structure and competencies of the CA, standards used by the CA to export fishery products to the EU, CA performance, laboratory services, and food safety controls.

The report ends with conclusions and recommendations for the implementation of regulations dealing with the safety of fishery products. This is the general format in these country reports; there may be additional issues such as pesticide residues, etc. Complete details on the countries visited and the findings related to fishery products are available by searching the EU Website (European Commission 2007).

Regulatory difficulties for developing countries

Some developing countries including Bangladesh, Thailand, India, Vietnam, and Indonesia are large exporters of fishery products (Allshouse *et al.* 2003).

Increased difficulty faced by developing countries in meeting international standards for food safety is a recurring theme at international regulatory meetings, including Codex Alimentarius. Specific details on the difficulties in meeting regulatory costs are given by Cato and Lima dos Santos (2000) for the Bangladesh frozen shrimp processing sector. These include higher costs for implementing Hazard Analysis and Critical Control Points (HACCP). The cost of maintaining a HACCP plan in Bangladesh ranged from $0.0148 to $0.0408 per pound, or from 0.31 to 0.85 percent of the 1997 price received. Comparable figures for the United States were $0.0009 per pound.

In general, regulatory problems for developing countries were investigated by the FAO Committee on World Food Security (FAO 1999). The following issues were considered for the whole food system but apply equally well for the fishery industry:

- *Food systems are complex.* In developing countries, they are also highly fragmented and predominated by small producers. Large quantities of food pass through the many food handlers and

middlemen involved in the food production, processing, and storage and distribution chain, and control is difficult. There is a significant risk of exposing food to contamination or adulteration.
- *National food control strategy.* This lays down the role of government agencies, various sectors of the economy, and consumers in dealing with new or emerging issues in food safety. This is non-existent in some developing countries.
- *Food legislation.* Food legislation in many developing countries is outdated and in need of review.
- *Food control service.* A national food control service consists of a food inspectorate, laboratories and analysts, and managers and supervisors. Again, many developing countries are unable to attain high standards in these areas because of costs and lack of adequately trained employees.
- *Compliance policies.* This is an official statement or group of statements that establishes specific or general limits to which products, processes, or conditions must comply and be in accordance with relevant laws and regulations.
- *Infrastructure development.* This requires money and resources that many developing countries do not have. Infrastructure includes adequate refrigeration, packaging, storage, and distribution facilities.

Assistance has been given to some developing countries that are major fish exporters. A good example is the assistance given to the South Pacific region to meet new fish import regulations (FAO 1998). This assistance included a needs assessment and collection of existing legislation, the drafting of a model law, the preparation of HACCP plans, and a regional workshop on the implementation of quality assurance.

In summary, considerable challenges face some fish-exporting developing countries in meeting regulatory requirements.

Hazard Analysis and Critical Control Points applied to seafood

Seafood HACCP is a science-based system of preventive controls for food safety that commercial seafood processors develop and operate to identify potential problems and keep them from occurring. The FDA HACCP program was designed to increase the margin of safety for U.S.

consumers and to reduce those illnesses that do occur to the lowest possible levels (FDA 2002). The seafood HACCP program has raised the standard for compliance much higher than it had been for the seafood industry. Now a seafood processor must have a system in place that consists of several complex elements that collectively make it unlikely that contamination will occur. (Detailed information on seafood HACCP regulations in the United States is available [FDA 2001a].)

Implementation of a HACCP program, including seafood HACCP, involves seven procedures (FDA 2001b):

- analyze hazards
- identify critical control points
- establish preventive measures with critical limits for each control point
- establish procedures to monitor the critical control points
- establish corrective actions to be taken when monitoring shows that a critical limit has not been met
- establish procedures to verify that the system is working properly
- establish effective record-keeping to document the HACCP system.

Developing countries account for almost 50 percent of global fish exports. Yet the cost and complexity of implementing HACCP programs make it difficult for some developing countries to meet international food safety standards and to compete on the international fish market (Cato 1998).

Risk analysis

Risk analysis comprises three interrelated components: risk assessment, risk management, and risk communication. Risk assessment is divided into hazard identification, hazard characterization, exposure assessment, and risk characterization. Risk management is divided into risk evaluation, option assessment, option implementation, and monitoring and review.

Risk analysis is applied regularly in assessing the safety of food, including fish and fishery products. It is now considered an integral part of the decision-making process of Codex Alimentarius (WHO 2004). A document titled *Draft Working Principles for Risk Analysis for Application in the Framework of Codex Alimentarius* gives a good

Table 15.5 *U.S. FDA violations for* Salmonella *by seafood product, 2001*

Seafood product	Percent violations
Shrimp and prawns	58
Lobster, tilapia, milkfish, oysters, squid, catfish, eel	23
Other	19

Source: Allshouse et al. (2003).

overview of risk analysis (Codex Alimentarius 2003). A joint FAO/WHO Expert Consultation on the Application of Risk Analysis to Food Standards Issues (FAO/WHO 1995) recognized the increased scientific, legal, and political demands being made on standards, guidelines, and other recommendations elaborated by Codex. In a response to these increasing demands, it was considered essential to have a greater application of risk assessment in the Codex decision-making process.

In fish, Tables 15.2, 15.3, and 15.5 indicate that *Salmonella* is a significant risk. Details on the conduct of a microbiological risk assessment should be useful for all engaged in the fish production, processing, and distribution chain (Codex Alimentarius 1999).

FISH REGULATIONS TO INFORM CONSUMERS

The important topic of fish regulations to inform consumers is given only a cursory examination here because of lack of space. Two examples from Europe are worth noting.

The EU has established common rules concerning fish products and the health and safety of consumers. Within the EU the member states are responsible for ensuring that food manufacturers and traders comply with these rules. There are increased efforts also to inform consumers on food safety issues related to fish products (European Commission 1998).

The second example is of new fish regulations to inform English consumers. This is achieved by the Fish Labelling (England) Regulations 2003 (Food Standards Agency 2003). Certain fish products must, when sold to the consumer, be labeled with the following information:

- the commercial designation of the species (i.e., an agreed common name for the species of fish)

- the production method (i.e., whether caught at sea, caught in inland waters, or farmed)
- the catch area (i.e., whether the ocean area or, in the case of freshwater fish, the country in which it was caught or farmed).

Business sectors affected by the Fish Labelling (England) Regulations 2003 include:

- retailers
- fish product manufacturers (including those making prepacked branded products)
- fishmongers, market stalls, etc.
- wholesale fish suppliers
- fish auctions, trawlers, etc.

Catering establishments are not affected by the new labeling requirements because the requirements do not apply to fish products that are processed in some way (e.g., by cooking) or that are served with accompanying ingredients (e.g., sauce, salad, etc.).

The Department for the Environment, Food and Rural Affairs (DEFRA) in the United Kingdom estimated that U.K. consumers spent more than £2 billion on fish in 2000. The anticipated costs of the new labeling legislation represent about 0.1 percent of the revenue earned from selling fish to consumers.

FISH REGULATIONS TO ENSURE FAIR TRADE PRACTICES

The information in Tables 15.2, 15.3, and 15.4 demonstrates the impact of national regulations on ensuring that imported fish and fishery products are safe for human consumption. It is important that the laws and regulations are equitable and transparent.

Most countries in the Organisation for Economic Co-operation and Development (OECD) apply HACCP systems. The enforcement of hygiene and sanitary requirements takes place through point inspection, through dedicated/licensed importers, or through systems of approval of establishments (Schmidt 2003). There is a growing awareness of traceability as an important regulatory element in all aspects of the food industry. In discussing traceability as applied to seafood, Schmidt (2000) defines "traceability" as the procedure or process through which products or services are traced en route from the supplier to the demander (end or intermediary consumer) and the recording of this track of events. The route from the supplier to the demander is often referred to as the chain of custody. Verification at each stage whenever the product or service is being handed over along the chain of custody is a way to trace the product or service. It also provides a means of ensuring that what comes out at the end of the process corresponds to what was put into the chain. All parts of the chain of custody, from the fisher to the consumer of fish, must have a stake in the outcome of introducing tracing systems. If not, tracing will not achieve its full potential, and cheating will continue. Implementing traceability may involve a great deal of education and reasoning to help ensure that all fisheries stakeholders have an interest in such systems and play by the rules (Schmidt 2000).

There are several ways to manage quality and safety of food. These are listed below (complete details are presented in FAO [2003]):

- good hygienic practices (GHP) / good manufacturing practices (GMP) or sanitation operating procedures (SSOP) or prerequisite programs
- Hazard Analysis Critical Control Points (HACCP)
- quality control (QC)
- quality assurance (QA) / quality management (QM) – ISO standards
- quality systems
- total quality management.

The list of official standards for fish and fishery products developed by Codex Alimentarius (2004) is a useful reference point. Details of standards for various fish products that are fresh, canned, quick-frozen, salted, dried, smoked, minced, and battered and/or breaded are available. They include also guidelines on the levels of methylmercury in fish and the sensory evaluation of fish and shellfish in laboratories.

Reports from meetings of the Codex Committee on Fish and Fishery Products are useful reference points in noting developments in regulatory issues. The most recent meeting included a draft code of practice for fish and fishery products (aquaculture and quick-frozen coated fish products). This includes a useful section on definitions, on aquaculture production, and on the processing of quick-frozen coated fish products. Other sections of this report deal with a proposed draft code of practice for fish and fishery products and a proposed draft standard for smoked fish (Codex Committee on Fish and Fishery Products 2003).

These resources provide useful reference points in ensuring fair trade practices in fish and fishery products. They are all accessible on the Internet.

SOME "FISHY" FOOD LAWS

A disclaimer is necessary for this section: the slang version of the word "fishy" is applied solely on this author's assessment of the following regulations.

Fish oils, omega-3 fatty acids, and the Dietary Supplement Health and Education Act of 1994

Fish oils and omega-3 fatty acid supplements (some are from fish) in the United States come under the Dietary Supplement Health and Education Act of 1994 (FDA 1994, 1995). It is commonly referred to as DSHEA (pronounced "D'Shay"). This act deregulated dietary supplements and reduced the FDA's regulatory authority over both supplements and conventional foods. With DSHEA the Congress gave the American consumer the "freedom to choose" dietary supplements. This was in response to a well-organized campaign by the dietary supplement industry. It is considered by many to be, at best, a faulty piece of legislation (Barrett 2000).

DSHEA has provisions that prevent federal regulatory agencies from acting against the interest of the supplement industry (Nestle 2003). DSHEA extended the legal definition of dietary supplements beyond vitamin and mineral supplements to include botanical, herbal, and diet products. A legal "welcome" was thus extended to fish oils, omega-3 fatty acids derived from fish and other food sources, and many other supplements. The safety and human health aspects of this extended group of supplements were less well researched. Manufacturers of these supplements were not required to show that their products were safe before selling them to consumers. The FDA was required to prove that these products were unsafe before they could be removed from the market. Thus, DSHEA removed the FDA's independent authority to take products off the market.

Barrett (2000) states that the FDA never had enough resources to deal with the huge amount of deception in the supplement and health food marketplace. In his opinion, DSHEA has made this situation worse, and the FDA should drop any pretense of being able to protect the public. He states that unless the U.S. Congress provides an adequate law, the FDA cannot protect the public from the deceptive marketing of what DSHEA calls "dietary supplements."

Fish oils and omega-3 fatty acids from fish and other sources come under DSHEA. Omega-3 fatty acids have been shown in clinical and epidemiological trials to reduce the incidence of cardiovascular disease (Kris-Etherton *et at.* 2002). Research does indicate that individuals at risk for coronary heart disease benefit from the consumption of marine- and plant-derived omega-3 fatty acids, but the ideal intakes are at present unclear. The American Heart Association recommends the inclusion of two servings of fish per week (particularly fatty fish). This recommendation must be balanced with concerns about environmental pollutants, in particular polychlorinated biphenyls (PCBs) and methylmercury. Kris-Etherton *et al.* (2002) state that omega-3 fatty acid supplements can reduce cardiac events such as death, non-fatal myocardial infarction, and non-fatal stroke, and decrease the progression of atherosclerosis in coronary patients. But they emphasize that additional studies are needed to confirm and further define the health benefits of omega-3 fatty acids and supplements for both primary and secondary prevention. These recommendations appear to be valid and appropriate health warnings. Meanwhile, DSHEA allows the American consumer to use omega-3 fatty acid supplements and fish oils that are essentially unregulated.

Chloramphenicol and shrimp exports

The following information highlights the difficulties imposed on fish-exporting countries in the developed world. It illustrates difficulties created when regulations are initiated arbitrarily and without proper consultations.

The delegation from Indonesia brought to the attention of the Thirteenth Session of the FAO/WHO Regional Coordinating Committee for Asia (Codex Alimentarius 2002:22) problems facing shrimp exporters due to the detection of residues or traces of chloramphenicol. At issue is the manner in which the regulation for chloramphenicol residues has become stricter in recent years. This resulted in the implementation of a zero-tolerance approach by importing countries and a progressive reduction in the limit of analytical detection. The Indonesian delegation questioned the scientific basis for imposing a zero tolerance because neither JECFA (the Joint Expert Committee on Food Additives) nor the Codex Committee on Food Additives and Contaminants (CCFAC) had established maximum residue limits for chloramphenicol, especially in shrimp. Hence, there was an urgent need to establish maximum residue limits (MRL) for chloramphenicol in shrimp to avoid such technical barriers to trade. An ironic twist to this discussion came from the delegation of Indonesia, which stated that fish and shellfish caught in the open sea had chloramphenicol at low levels.

The delegation of Vietnam pointed out that fish-importing countries had initiated progressive reductions of the limit of analytical detection for chloramphenicol. These were due to new scientific techniques and equipment in the fish-importing countries. This was done frequently without giving adequate advice, forewarning, or technical assistance to exporting countries. Such abrupt changes in analytical methodology meant that expensive investments in training and in laboratory equipment in the fish-exporting countries were suddenly made valueless. The delegation of India stressed that this problem was not exclusively confined to chloramphenicol in shrimp. It concerned also other antibiotics and contaminants in other products.

It was pointed out that JECFA had evaluated chloramphenicol on a number of occasions, concluding on each occasion that no residues of chloramphenicol in foods are acceptable. Therefore, no MRL could be established.

Saving two in a billion

The regulatory issue discussed in this section does not concern fish and fishery products. It is included here as a precautionary example of what could happen when food safety regulations are changed without full evaluation of consequences. In other words, the fishery industry should be on the alert to prevent a similar situation occurring in that industry.

New, stricter EU standards on aflatoxins were estimated to reduce the health risk by approximately 1.4 deaths per billion per year in the EU. But a consequence of imposing a new standard that might save fewer than two people in a billion was a decrease in African exports of more than 60 percent ($670 million) compared with regulations based on an international standard (Otsuki and Wilson 2001). This raises a familiar question: how safe is "safe"?

"Fishy wars" at the World Trade Organization

In recent years, numerous disputes involving fish and fishery products, including salmon, sardines, scallops, shrimps, and swordfish, have been dealt with by the World Trade Organization (WTO) (WTO 2004). Disagreements over the trade description of fish and fishery products constituted many of these disputes. For example, a trade description dispute brought by Peru against the European Commission involved essentially a definition of what constituted a sardine. It is not the brief of this paper to pass critical judgment on the merits of each of these trade disputes. But it is regrettable that some pressing issues in food safety that result in trade disputes are not resolved either at the WTO or elsewhere.

Food safety regulations and transgenic fish

Several reports indicate that genetically modified foods currently on the market are safe for human consumption. The FAO/WHO Expert Consultation on Safety Assessment of Foods Derived from Genetically Modified Animals Including Fish (FAO/WHO 2003) examined several food safety concerns. These included possible risks of consuming transgenes, their resulting protein, the potential production of toxins by aquatic transgenic organisms, changes in the nutritional composition of foods, activation of viral sequences, and allergenicity of transgenic products. It was concluded that these risks have been analyzed and the majority of genetic modifications of foodstuffs will be safe. It was pointed out that allergenicity poses the greatest potential for risk and harm.

National and regional regulatory systems exist to examine the safety of genetically modified foods. However, significant differences in testing methods cause inconsistent outcomes of safety evaluations. The World Health Organization (WHO), FAO, and the Codex Alimentarius Commission are working on safety assessments of genetically modified organisms (GMOs) and on international rules on the handling of genetically modified foods (WHO/WTO 2002).

The above information is reassuring, but it must be emphasized that the regulation of the safety of transgenic fish presents some unique difficulties. These difficulties are highlighted by the Pew Foundation (2003) in its report *Future Fish: Issues in Science and Regulation of Transgenic Fish*. It states that only one animal, a genetically modified fish, is thought likely to come to market in the near future. The report presents the following cautious assessment of whether transgenic (genetically modified) fish will be effectively regulated in the United States.

Legal authority

The report states that it is unclear if the FDA has the comprehensive legal authority needed to address all the food safety and environmental issues associated with transgenic fish. In the United States, the FDA has jurisdiction over transgenic fish because the biological material used to transform a genetically modified fish and the product expressed by the fish's transformed genetic construct both come under the legal definition of a "new animal drug." The FDA has indicated that it will use the same process to regulate transgenic fish that is now used to review a new animal drug. The Pew Foundation report concludes that "it is uncertain whether the FDA interpretation of the law to include genetic modification as a new animal drug would withstand legal challenge" (Pew Foundation 2003).

Adequacy of risk management tools

Because transgenic fish are dealt with like a new animal drug, a developer must prove the safety of a product before it goes to market. Contrast this with the non-regulation of dietary supplements including fish oils and omega-3 fatty acids from fish sources (see above). The FDA can impose the following prior to sale: restrictions in the use of the product by labels, conditions of use, and post-approval monitoring. Developers must report also any adverse effects developed post-approval. If necessary, the FDA can stop the marketing of a product. The Pew Foundation (2003) report points out if the FDA's authority to regulate transgenic fish as a new animal drug is ever challenged successfully and results in the FDA adopting another approach, then the benefits of the new animal drug approach may be lost.

Transparency, clarity, and public participation

If consumers have access to and understand the information needed for new product approval, it will increase consumer confidence in a new product. However, the FDA's application and approval process for the approval of new animal drugs is totally confidential and closed until the FDA grants approval. This legal requirement protects trade secrets and confidential business information. What this means is that consumers cannot have input into the approval process. Questions from consumers on acceptable risk to human health or to the environment are excluded. This is seen as a serious limitation to public acceptance. Contrast this with the open public debates held in the United States, Europe, and elsewhere on crops derived from biotechnology.

Resources and expertise

Some question the FDA's access to adequate resources and to expertise to include the necessary environmental assessments on transgenic fish, and suggest that the Fish and Marine Wildlife Service and the National Marine Fisheries Service have expertise that the FDA may lack. Furthermore, resource limitations could hinder the FDA's ability to ensure that developers follow any special conditions under which it approves the use of a product.

Efficiency and coordination

Several U.S. agencies have expertise in food safety and environmental risks associated with transgenic fish. How these agencies will work together to create a regulatory system to review transgenic animals in an effective, efficient, and well-coordinated manner remains unclear (Pew Foundation 2003). In summary, most countries have much work to do on the regulation of transgenic fish.

One recent development with potential implications in the commercial development of transgenic fish for human consumption is the "Night Pearl" zebra fish. The British Broadcasting Corporation (BBC 2003) reported that "a Taiwanese company has created this genetically modified (GM) ornamental fish that glows in the dark." This zebra fish is the first gene-altered pet to go on sale to the public. A spokesperson for the Aquatic Ornamental Trade Association said that interfering with the genome was unnecessary and that people did not want animals to become fashion accessories. Public perception and input into the development of new regulations is vital these days. Developments like this one are seen by many as unhelpful in making the case for the commercialization of transgenic fish for human consumption.

Despite the heated public debate in Europe and elsewhere, little formal consideration of the health and safety aspects of GMOs has occurred in the WTO (WHO/WTO 2002). The most detailed discussions to date have occurred in the Technical Barriers to Trade Committee, where labeling of GMOs in various countries is under scrutiny. Issues related to food safety and GMOs come under the Sanitary and Phytosanitary Agreement (SPS). A need was expressed for transparency and for the development of international standards for GMOs (including fish). In conclusion, much work still remains to be done in regulating the safety of GMOs.

CONCLUSIONS

In summary, it is in the best interests of all associated with the fisheries industry to know and apply correct and pertinent food regulations. As a result, their fishery products will enhance public health, inform the consumer, ensure fair trade practices especially in incidences of international trade disputes, give protection against fraud, and, where applicable, protect the environment.

References

Allshouse, J., Buzby, J.C., Harvey, D., and Zorn, D. 2003. International trade and seafood safety. In *International Trade and Food Safety: Economic Theory and Case Studies*, Agricultural Economic Report No. (AER828) ed. J.C. Buzby. Washington, DC: Economic Research Service/U.S.Department of Agriculture, pp. 109–129. Available online at www.ers.usda.gov/publications/AER828

Barrett, S. 2000. How the dietary supplement Health and Education Act of 1994 weakened the FDA. Available online at www.quackwatch.org/02ConsumerProtection/dshea.html

BBC (British Broadcasting Corporation). 2003. GM fish glows in the bowl. Available online at http://news.bbc.co.uk/1/hi/sci/tech/3026104.stm

Cato, J.C. 1998. *Seafood Safety: Economics of Hazard Analysis and Critical Control Point (HACCP) Programmmes*, FAO Fisheries Technical Paper No. 381. Rome: Food and Agriculture Organization of the United Nations. Available online at www.fao.org/docrep/003/X0465E/X0465E00.htm

Cato, J.C. and Lima dos Santos, C.A. 2000. Costs to Upgrade the Bangladesh Frozen Shrimp Processing Sector to Adequate Technical and Sanitary Standards and to Maintain a HACCP Program. In *The Economics of HACCP: New Studies of Cost and Benefits*, ed. L. Unnevehr. St. Paul, MN: Eagan Press, pp. 385–402.

Center for Science in the Public Interest. 2004. *Contaminated Produce Tops Food Poisoning Culprit*. Available online at www.cspinet.org/new/200404011.html

Codex Alimentarius. 1999. *Principles and Guidelines for the Conduct of Microbiological Risk Assessment*. Available online at www.codexalimentarius.net/download/standards/357/CXG_030e.pdf

Codex Alimentarius. 2002. *The need for MRLs for chloramphenicol in shrimp*. p. 22. Available online at www.codexalimentarius.net/download/report/420/Al03_15e.pdf. P. 22.

Codex Alimentarius. 2003. *Principles for Risk Analysis for Application in the Framework of the Codex Alimentarius*. Available online at www.fao.org/docrep/meeting/005/X7101E/x7101e0g.htm

Codex Alimentarius. 2004. *Current Official Standards*. Available online at www.codexalimentarius.net/standard_list.asp

Codex Committee on Fish and Fishery Products. 2003. *Report of the 26th Session of the Codex Committee on Fish and Fishery Products*. Available online at www.codexalimentarius.net/download/report/609/a104_18e.pdf

European Commission. 1998. *The Growing International Dimension of Fisheries Management*. Brussels: European Union.

European Commission. 2003. *Final Report of a Mission Carried out in Saudi Arabia from 18 to 25 October 2003 Assessing the Conditions of Production of Fishery Products Intended for Export to the European Union*. Brussels: European Union.

European Commission. 2007. *Fishery Products*. Available online at http://europa.eu.int/comm/fisheries

FAO (Food and Agriculture Organization). 1998. *Assistance to the South Pacific Region to Meet New Fish Import Regulations*. Available online at www.fao.org/docrep/field/383969.htm

FAO. 1999. *The Importance of Food Quality and Safety for Developing Countries*. Available online at www.fao.org/docrep/meeting/X1845E.htm

FAO. 2003. *Assessment and Management of Seafood Safety and Quality*, Fisheries Technical Paper No. 444. Rome: Food and Agriculture Organization of the United Nations. Available online at ftp://ftp.fao.org/docrep/fao/006/y4743e/y4743e00.pdf

FAO/WHO (World Health Organization). 1995. *Application of Risk Analysis to Food Standard Issues*, Report of a Joint FAO/WHO Expert Consultation, WHO/FNU/FOS/95.3. Geneva, Switzerland: World Health Organization. Available online at www.who.int/foodsafety/publications/micro/march1995/en

FAO/WHO. 2003. *Biotechnology and Food Safety: FAO/WHO Expert Consultation on Safety Assessment of Foods Derived from Genetically Modified Animals, Including Fish*. Available online at www.fao.org/ag/agn/food/risk_biotech_animal_en.stm

FDA (Food and Drug Administration). 1994. *Dietary Supplement Health and Education Act of 1994*, Public Law 103–417, 103rd Congress. Available online at www.fda.gov/opacom/laws/dshea.html

FDA. 1995. Review of the *Dietary Supplement Health and Education Act of 1994*. Available online at http://vm.cfsan.fda.gov/~dms/dietsupp.html

FDA. 2001a. *Fish and Fisheries Products Hazard and Controls Guidance: Seafood HACCP Regulation*. Available online at www.cfsan.fda.gov/~comm/haccp4x8.html

FDA. 2001b. *HACCP: A State-of-the-Art Approach to Food Safety*. Available online at www.cfsan.fda.gov/~lrd/bghaccp.html

FDA. 2002. *FDA's Evaluation of the Seafood HACCP Program for Fiscal Years 2000/2001*. Available online at www.cfsan.fda.gov/~comm/seaeval2.html

FDA. 2004a. *Import Refusals Report by Product*. Available online at www.fda.gov/ora/oasis/ora_ref_prod.html

FDA. 2004b. *Violation Code Translation*. Available online at www.fda.gov/ora/oasis/ora_oasis_viol_rpt.html

Food Standards Agency. 2003. *Fish Labelling Regulations 2003*. Available online at www.food.gov.uk/multimedia/pdfs/fish_lab_reg2003gn.pdf

Kris-Etherton, P. M., Harris, W. S, and Appel, L. J. 2002. Fish consumption, fish oil omega-3 fatty acids, and cardiovascular disease. *Circulation* **106**: 2747. Available online at http://circ.ahajournals.org/cgi/content/full/106/21/2747

Nestle, M. 2003. *Food Politics*. Berkeley, CA: University of California Press.

Otsuki, T. and Wilson, J. S. 2001. *Global Trade and Food Safety: Winners and Losers in a Fragmented System*, World Bank Working Paper No. 2689. Washington, DC: World Bank. Available online at http://go.worldbank.org/TXUABOCJR0

Pew Foundation. 2003. *Future Fish: Issues in Science and Regulation of Transgenic Fish*. Philadelphia, PA: Pew Initiative on Food and Biotechnology. Available online at http://pewagbiotech.org/research/fish

Schmidt, C. C. 2000. *Traceability and Fisheries: Consumer Wants and Government Needs – Lessons Learned from Past Experiences*. Paris: Organisation for Economic Co-operation and Development (OECD).

Schmidt, C. C. 2003. *Globalization, Industry Structure, Market Power and Impact on Fish Trade Opportunities and Challenges for Developed (OECD) Countries.* Available online at www.oecd.org/dataoecd/8/12/25012071.pdf

WHO (World Health Organization). 2004. *General Information Related to Microbiological Risks in Food.* Available online at www.who.int/foodsafety/micro/general/en/print.html

WHO/WTO (World Trade Organization). 2002. *WTO Agreements and Public Health.* Available online at www.who.int/media/homepage/en/who_wto_e.pdf

WTO (World Trade Organization). 2004. *Index of Dispute Issues.* Available online at www.wto.org/english/tratop_e/dispu_e/dispu_subjects_index_e.htm

Part IV Ethical, economic, and policy implications

KENNETH A. FRANK, KATRINA MUELLER, ANN KRAUSE,
WILLIAM W. TAYLOR, AND NANCY J. LEONARD

16

The intersection of global trade, social networks, and fisheries

GLOBALIZATION AND NETWORKS

In this chapter, we explore globalization through networks. Of course, globalization can be described in terms of networks of trade between countries and as executed by multinational corporations (Breiger 1981; Chase-Dunn and Grimes 1995; Kim and Shin 2002). And fisheries ecosystems have long been characterized in terms of networks of predator and prey relationships between taxon or species (e.g., Cohen *et al.* 1993; Gaedke 1995; Krause *et al.* 2003). But here we will explore how human social networks mediate between global economic exchanges and the dynamics of aquatic ecosystems. Thus we extend critiques of the globalization literature for lack of attention to individual agency (e.g., Schechter 1999:62) by calling attention to the effects of human relationships in globalization. Ultimately, our focus allows us to integrate theories related to social networks (e.g., social capital) as well as inform policy and management of and research on fisheries and their associated ecosystems.

What is globalization?

We define globalization as an increase in the rate of exchange of resources and information across geographic regions and cultures. Though communities have been interdependent through trade as long as people have traversed the oceans, our current awareness of globalization suggests that we are increasingly globalized – that the resources and related actions in distant regions of the world have an unrivaled immediacy in the lives of most people (Harrison 1996; Kim and Shin 2002; One World 2007). That is, exchanges across vast regions increasingly are realized more rapidly, occur more frequently, and

Globalization: Effects on Fisheries Resources, ed. William W. Taylor, Michael G. Schechter, and Lois G. Wolfson. Published by Cambridge University Press. © Cambridge University Press 2007.

require fewer middlemen. Furthermore, actions affecting the environment in one region of the world have increasing implications for those in locales previously considered distant.

With globalization comes an increased distribution of fished aquatic species (e.g., cod, salmon, tropical fish). This is attributed to increased demand associated with population increases and recognition of aquatic species as a healthy dietary staple or valuable economic resource or status symbol, and with increased technology for preserving and shipping that reduces transportation costs. Furthermore, globalization is associated with increased global trade of non-aquatic resources which can then affect local aquatic ecosystems.

Globalization has generally been associated with changes in aquatic ecosystems (Safina 1998). Many have noted that, with globalization, habitats have been destroyed (e.g., the Pacific Northwest: Gilden 1999), exotic species have been introduced (e.g., zebra mussels in the Great Lakes: Vanderploeg et al. 2002), and fish stocks have been depleted (e.g., cod: Finlayson and McCay 1998). On the other hand, others argue that changes in ecosystems are primarily due to increased population, and that increases in technology and resource-sharing are benefits of globalization because they facilitate more efficient use of aquatic resources (Hardin 1968). In some sense, aquatic systems are no different from land resources, political governance, and other social and biological systems affected by global changes in economic landscapes. And yet our awareness of the link between globalization and aquatic systems is perhaps pronounced because the ships that transport globalized trade come in direct contact with the resource – water. For example, tankers carrying increased oil exports from the Caspian Sea contributing to globalization served as a vector for zebra mussel invasions that altered ecosystems in the Great Lakes (e.g., Ricciardi and Rasmussen 1998).

Humans as mediators between global forces and ecosystems

Critically, the theoretical link between globalization and aquatic ecosystems is strikingly distant and asocial. It is as though global forces directly penetrate aquatic ecosystems with little thought regarding how these processes are mediated by human action (Schechter 1999). Thus the zebra mussel invasion is attributed to ballast water and changes in shipping technology and patterns (Ricciardi and Rasmussen 1998), the decline of the cod industry is attributed to increased competition across the Atlantic (Finlayson and McCay 1998), and the decline of the habitats in the Pacific Northwest is attributed to increased demand for salmon

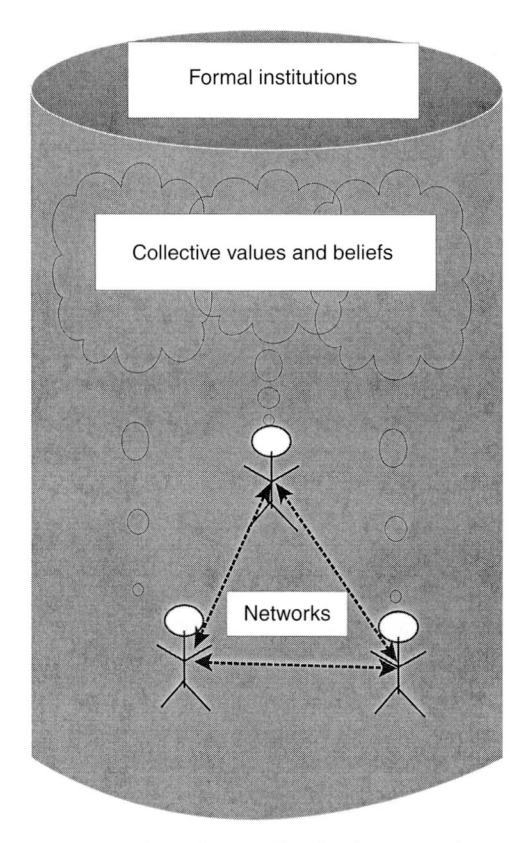

Figure 16.1 Mechanisms for coordinating human action to manage common resources.

and the competition of aquaculture (Finlayson and McCay 1998). In this asocial view, people at best are merely *Homo economicus*, rationally reacting en masse to global economic forces beyond their control.

The macroeconomic forces described in the preceding paragraph may well cause some changes in aquatic ecosystems. But even when they do so, their effects are transmitted, shaped, and mediated by people. How do the actions of shippers and official monitoring agents contribute to the zebra mussel invasion of the Great Lakes? In the face of global competition and demand, why do some salmon fishers of the Pacific Northwest resort to exploitative fishing techniques such as trawling (Gilden 1999; Vanderploeg *et al.* 2002) and cod fishers wipe out the fisheries on the North American eastern seaboard while the lobster fishers of Maine preserve their fisheries and livelihoods (Acheson 1988, 2003)?

The answers lie in human action, exchange, and relationships as depicted in Fig. 16.1. Some have identified formal political institutions that organized societies can implement to control fishing and preserve fisheries and their ecosystems. For example, the Canadian government engaged in extensive efforts to involve cod fishers in assessing the fish stocks and in educating them about the effects of their actions on the fish stocks (Finlayson and McCay 1998). These formal institutions are represented by the cylinder that contains human relations in Fig. 16.1. Others point to collective values and beliefs that govern engagement with aquatic resources. For example, Maine lobster fishers strongly agree with and adhere to the required release of gravid female lobsters or notched females (previously found to be gravid) to lower potential impact on lobster reproduction and sustainability. Similarly, the hunter's code of the Chisasibi Cree (of northern Ontario) speaks strongly against wastage, and tribal leaders discourage overhunting by declaring that it represents a lack of respect for caribou (Berkes 1998). Such collective values are represented by the "group thought bubble" in Fig. 16.1.

To these human elements we add networks. Granovetter (1985) argues that economic action is invariably embedded in social relations. This applies across a range of seemingly economic action, from the cozy gem trade embedded in Jewish enclaves in New York (Coleman 1988) to provision of financial support from kin and friends (Wellman and Wortley 1990) to expectedly arm's-length transactions such as those of French bankers (Frank and Yasumoto 1998). These networks are represented by the lines connecting the actors in Fig. 16.1. A key point here is that networks are ubiquitous, existing wherever a relation between two actors is possible, and, at the same time, they are not as tangible as the written documents of formal political institutions or the aggregate of perceptions defining collective values and beliefs.

If networks are ubiquitous and yet ephemeral, how can we use the metaphor and idea of networks to understand human action? A corollary of Granovetter's thesis is that the underlying structure of the social network will affect how a social system reacts to external changes and disturbances (e.g., Frank and Fahrbach 1999). As one example, internal fragmentation is likely to be exacerbated by an increased rate of external disturbances. As a particular example, existing social cleavages, based on location and type of gear (trawling vs. trolling) among fishers in the Pacific Northwest, were exacerbated by increased global competition (Gilden 1999). As a second example, highly centralized networks may effectively diffuse information but may be susceptible to extreme variability when central actors are exposed to frequent external changes. Most generally, the social structure of any system can be understood as one form of systemic response to forces that place uneven stress on the members of the system.

Social embeddedness and globalization

The embeddedness of economic transactions in interpersonal networks plays a particularly important role in light of globalization. When previously local exchange comes to include distant participants, the market is expanded beyond the realm of the local social network. As a result, information is critical to new transactions in the market (Granovetter 1985; Burt 2000). And yet information is particularly limited because of the distance between the trading parties and the lack of mediators of information. Furthermore, exchanges between distant parties may be unstable to the extent that there is not a long history of exchange and the parties are of different cultures and are exposed to different economic influences (Krempel and Plumper 2002).

When information is limited and conditions are unstable, exchange tends to become more embedded in social networks because others in the network can provide unique and valuable information or access to resources as a form of insurance in case of catastrophe (Granovetter 1973; Frank and Yasumoto 1998; Lin 2001). This suggests a paradox of globalization: globalization implies economic exchange between geographically, culturally, and socially distant parties, yet globalization accentuates the extent to which such exchanges are embedded in interpersonal social networks. This is not to say that exchange will be concentrated within local communities when external opportunities present themselves. Instead *how* local actors participate in globalized trade will be contingent on local networks.

The link between globalization and local networks follows naturally from the recognition that globalization and ecosystems can be described in network terms. That is, if globalization is a set of international trade and exchange relationships among countries or corporate actors, and if an ecosystem encompasses a set of predatory/competitive relationships among taxa as in a food web, then links between these two networks must be mediated by a set of actors who engage in the relationships associated with globalization and who also interact with the ecosystem. And these actors and their relationships typically consist of people who extract resources from aquatic ecosystems and then enter them into the system of potentially international trade. Thus the network approach can bridge the gap between ecological and social science (Hollingshead 1940). Pragmatically, the network approach informs management, for which there is an emerging recognition to understand and engage resource users in their social contexts (McDonough, *et al.* 1987; Lee 1993; Holling 1995; Berkes and Folke 1998; Blumenthal and Jannink 2000; Lal *et al.* 2001; Conway *et al.* 2002).

In the next section, we will explore in theory how small, local, human networks affect the way in which social systems use and affect aquatic resources in the face of globalization. We will then present three case studies that typify some of the processes we describe. We then present a set of analytic tools that might be applied to study how social capital is manifest through social networks of fishers. In the conclusion, we emphasize the need to understand human action in terms of human social networks and draw implications for fisheries managers and scientists.

SOCIAL CAPITAL OF HUMAN NETWORKS

We use social capital to explore how resources flow through human networks. Social capital can be defined as the potential to access resources through social relations (Bourdieu 1986; Coleman 1988; Portes 1998; Woolcock 1998; Putnam 2000; Lin 2001).[1] Thus a new lobster fisher may draw on his family's long-standing presence in the community to access important information to improve his catch, a fisher's wife may draw on her relationship for financial support or childcare while her husband is at sea (Conway *et al.* 2002), or a cod fisher may draw on kinship ties to share gear (Faris 1972).

Though stated at the level of the individual, the manifestation of social capital has implications for how communities manage common resources. When members of a community share resources through social relations, they reduce the need to invest in the infrastructure of other institutions (e.g., a legal system, formal organizations, and formal markets) that would otherwise be required to facilitate the flow of

---

[1] We are aware that social capital has been defined in multiple ways, from Coleman's (1990:303) "definition by function" to Putnam's (2000:19) link to "civic virtue." The result is that social capital has become one of the most ambiguous terms in the social sciences (Portes 1998) and may lose any distinctive meaning (Hirsh and Levin 1999). Our definition is consistent with the emerging consensus among sociologists, as typified by definitions offered by Portes (1998:7) and Lin (1999:30–31).

resources (Coleman 1990). This is especially important for small fishing communities that may have limited resources.

Because social capital applies at multiple levels of social organization (Bourdieu 1986; Coleman 1990; Gabbay and Leenders 1999), it has important potential for policy and management. On one hand, social capital attends to the rationality of resource allocations among individuals. But it also speaks to systemic properties that can facilitate such allocations. Thus though the theory of social capital has important value for basic research, in the last section we will draw on the social capital paradigm to help managers help members of communities manage common resources.

Manifestation of social capital also accentuates the value of social relationships. Thus, when members of a community access resources through social relations, they come to identify more strongly with others with whom they have social relationships and, potentially, the community in which the relationships are embedded (Lawler and Yoon 1998; Frank 2002). In turn, they are more inclined to extract natural resources with an appreciation for the general value of the resource to the community. Thus, communities that cultivate social capital may engage in what Hardin (1968) referred to as "mutual coercion, mutually agreed upon" action, which can reduce potential tragedies of the commons (Dietz et al. 2003; Pretty 2003).

Nested in their small communities, fishers are very aware of the functions of social capital, even if they do not call it that by name. For example, the Scarlet family of merchants was highly successful among the cod fishers of Cat Harbour because John Scarlet valued the social obligation of his customers as much as their immediate cash (Faris 1972:122–5). In particular, he would forgo immediate payment for equipment he lent in exchange for a future favor. This is a critical practice because the vicissitudes of the fishing industry sometimes leave fishers low on cash. As a second example, reflecting appreciation for the role of social capital in protecting the common resource, members of lobster gangs who cut the traps of encroaching gangs or individuals are at minimum condoned by others, who recognize that such behavior may help sustain the fishery and lobster stock (Acheson 1988).

Because social capital inheres in networks (indeed Burt [2000] refers to social capital as the "killer application" for social network analysis), it suggests we pay special attention to the network structures of the social systems we study. In particular, we will attend to the tendency for social systems to be made up of cohesive subgroups – sets of actors who interact frequently with one another.

Classic sociological theory describes people as establishing primary affiliations with members of their subunit while uniquely defining their roles through ties with members of other subunits (Durkheim 1933; Weber 1947; Simmel 1955; Nadel 1957; Granovetter 1973). Similarly, ecosystems have been conceptualized as divided into a set of integrated trophic levels or compartments (Bendix and Fisher 1961; May 1973; Cohen et al. 1993), which then mediate the effects of external perturbance. As examples in human social systems, kinship relations of cod fishermen are concentrated within crowds (Faris 1972), and long-standing social relationships among lobster fishermen are concentrated within gangs (Acheson 1988, 2003). Note, though, that cohesive subgroups represent the general sociological term defined exclusively in terms of the pattern of interaction, whereas crowds and gangs also derive from biological relationships and common land ownership.

Theories of social capital then suggest that the form of social capital will vary with distribution of relationships within and between cohesive subgroups (e.g., Frank and Yasumoto 1998; Lin 2001). Subgroups in which relationships are concentrated are most helpful in preventing catastrophic loss when disaster strikes. For example, a cod fisher caught in a storm is most likely to be rescued by members of his crowd (Faris 1972), or a lobster fisher who cannot fish because of personal illness or tragedy is most likely to receive economic help and emotional support from members of his gang (this may even include abstinence from fishing for a short period to protect the market [Acheson 1988]). Critically, social capital operates through the network – members of the crowd or gang who fail to provide help will themselves be ostracized. Portes and Sensenbrenner (1993) refer to social capital that inheres amidst dense social ties within subgroups as *enforceable trust*, while others refer to this as bonding social capital (Gittell and Vidal 1998; Putnam 2000).

In contrast to members of one's subgroup, members of a social system who are outside one's subgroup can provide unique information or resources that are critical to advancement (Granovetter 1973; Burt 1992; Frank and Yasumoto 1998; Lin 2001). As examples, John Scarlet expanded his business as a distributor of cod by taking over a store in a nearby town, thus gaining access to new customers (Faris 1972), and charter boat captains may communicate with captains not in their primary information-sharing subgroup to gain access to new information or gain favor with members of more successful fishing subgroups (Mueller 2004). Here, membership in a loose social network ensures that the boundary spanner will be treated reasonably fairly and will make moderate investments in the larger community, a form of social capital that Portes and Sensenbrenner (1993) describe as *reciprocity transactions*; the entrepreneur must give as much as he gets, although the network allows the reciprocity to be delayed over time. This is sometimes referred to as bridging social capital (Gittell and Vidal 1998; Putnam 2000).

Current theories of social capital suggest that people who have the most advantage and systems are the most efficient in taking advantage of opportunities and distributing resources when both forms of social capital are present (Frank and Yasumoto 1998; Gittell and Vidal 1998; Woolcock 1998; Putnam 2000; Lin 2001). Individuals can limit the effects of catastrophes by drawing on bonding social capital and can advance themselves by drawing on bridging social capital. At the system level, core components are sustained through bonding social capital; integration of the system as a whole is accomplished through bridging social capital. Thus, systems that contain both types of social capital have mechanisms for resisting the negative effects of external disturbances while taking advantage of the positive opportunities made possible by external change.

The theory of social capital pertains directly to resource flow within social systems. But for communities of fishers, the resources come directly from aquatic ecosystems for which there may be global competition. Furthermore, the resources may well then become part of a global exchange and are affected by global disturbances. In the next subsection, we draw on the theories of social capital to explore how human networks mediate between global exchange and aquatic ecosystems.

MEDIATION BETWEEN GLOBAL EXCHANGE AND AQUATIC ECOSYSTEMS THROUGH SOCIAL CAPITAL

The social capital that inheres in the networks of local fishing communities mediates between global events and aquatic ecosystems in three ways. First, the distribution of social capital affects how local fishing communities react to the increased competition that emerges with globalization. Second, the distribution of social capital affects how local fishers engage in exchange with members of the global marketplace. Third, the distribution of social capital shapes how fishers relate to a general increase in external disturbances – such as changes in climate, other economic conditions, or the rise in ecological concerns – generated by globalization. Below we characterize each of these mediating mechanisms through case studies. Importantly, the case studies feature a range of communities in size, technology, economic development, and aquatic environments. The critical feature in each of these studies is how these other factors may affect social networks and social capital, which then mediates between global events and aquatic ecosystems.

Globalization and increased competition

To describe how existing social structures affect the reaction of local communities to increased competition, we contrast the experiences of cod and lobster fishers on the eastern Atlantic seaboard of North America. The following is based on Mayo and O'Brien (2000) for the Northeast Fisheries Science Center. A key feature of the Atlantic cod (*Gadus morhua*), is its range and mobility. Atlantic cod that inhabit polar waters in the summer and autumn migrate to more southerly and deeper waters in winter and spring, while cod summering in the Nantucket Shoals region overwinter along the New Jersey coast. Some cod move considerable distances in search of food or in response to overcrowding at certain spawning grounds.

In an early example of globalization, improved technology (i.e., the factory freeze-trawler) in the 1960s made it possible for boats to weather the great storms and cold of the North Atlantic and thus to catch, process, and preserve cod caught off the shores of North America and distribute it to Europe and the Soviet Union (Finlayson and McCay 1998:316). Increased demand for cod and limited local supplies made it profitable to do so. Thus, large-scale economic forces induced competition for cod between North American fishers and the fishers of the world.

Though the increased competition to supply the market was a manifestation of globalization, the reaction of the local cod fishers was an expression of the local social network and institutions (Finlayson and McCay 1998). Consider the cod fishers of Cat Harbour, Maine, as one particular example (Faris 1972). Deep social divisions among these fishers emerged out of historical divisions between the French and the English and between the Catholic and the Protestant founders of the communities. These divisions translated into divisions between communities and even within communities between kin-based crowds or clans defined by proximity of their "gardens," or plots of land. Even the ties between members of a crowd or clan frayed when co-owned fishing gear was not required to fish. Thus the dense social relations required for bonding social capital were compromised.

Given their social structure, the cod fishers of Cat Harbour relied upon an unaffiliated merchant, John Scarlet, to negotiate exchange (this gave merchants extreme status in the community). Therefore, the fishers did not have a strong base of social capital on which to draw to combat global competition (although their cod economy positioned the merchants to take advantage of a global marketplace). Instead, the cod fishers generally relied on the Canadian government to protect the waters against outside forces (Finlayson and McCay 1998: 316–318). In response, the Canadian government enacted a 200-mile (320-km) zone of exclusive fisheries jurisdiction, which was then ostensibly supported by the European Community.

The difficulty of relying on the Canadian government to preserve the common resource was in enforcement and implementation. Partly because of the mobility of the cod, the boundary waters did not include all of the home range of the cod. Furthermore, the Canadian government did not sustain the interest and resources to ensure implementation of the boundaries and limits. The result was massive overfishing by all nations. When that was combined with political and fishery management miscalculations, the cod stock was depleted (Finlayson and McCay 1998).

Contrast the case of the cod fishers with that of the lobster fishers of Maine (Acheson 1988, 2003). The following is adapted from the summary by Idoine (2004) for the Northeast Fisheries Science Center. In contrast to cod, American lobsters have limited mobility and are concentrated in rocky areas where shelter is readily available, although occasional high densities occur in mud substrates suitable for burrowing. Because of their limited range, the principal fishing gear used to catch lobsters is the trap. Lobsters are also taken as bycatch with otter trawls. Before 1950, lobsters were taken offshore primarily as incidental trawl catches in demersal fisheries. Reported offshore lobster landings increased dramatically from about 400 tonnes during the 1950s to an average of more than 2000 tonnes in the 1960s. In 1969, technological advances permitted the introduction of trap fishing to deeper offshore areas, which helped to increase landings (trap landings increased from 50 tonnes in 1969 to 2900 tonnes in 1972). Total landings were steady from 1977 to 1986 ($\sim$17 600 tonnes/year). In the 1990s, with improved distribution and markets, the landings increased to approximately 32 000 tonne/year with a slight decline in 1992–93. Thus far, the lobster fishers have not experienced the drop-off in catches indicative of overfishing that the cod fishers have seen.

The question that arises is why the lobster fishers did not experience the same declines in their target species as the cod fishers. Though the answer is complex, involving a range of factors, one critical factor resides in the social relations of the lobster fishers (Acheson 1988, 2003). The social relations of Maine lobster fishers were extremely concentrated within harbor-based gangs. To defy the gang was to risk ostracization, a serious sanction because a fisher's success often depended on local and long-held fishing knowledge. Furthermore, because the gangs were defined by long-standing social and kin-based relationships, ostracization affected one's social standing in the community as well as one's standard of living.

The gangs drew on their social relationships to aggressively protect their territories. If a fisher placed traps in a harbor perceived to be the territory of a rival gang, the rival gang members would either first warn the intruder (verbally or by notching the traps) or, most likely, sabotage the perceived intruder's traps (e.g., cutting them, placing debris in them). The social structure of the gang was critical to perpetrating the sabotage. It often took several members to challenge a potential intruder, and all members of the gang were complicit in not revealing who sabotaged the traps. Although there were temptations to "free-ride" by not challenging rival fishers, ultimately the gangs were able to sustain territorial control.

The gangs ultimately drew on the social capital accumulated through establishing the territorial system of control to impose harbor-based trap and take limits lower than any externally imposed limits. Those who exceeded the limits were socially ostracized, leaving them to face the dangers of fishing and the vicissitudes of the market on their own. In fact, the formal authorities relied upon the sanctioning power of the gangs to impose legal limits once they were passed (Acheson 2003). Thus, the lobster fishers had strong stocks of social capital based on their gangs and well-defined norms of sabotaging outside competition to compete with fishers from other regions.

Interestingly, there are no legal restrictions to lobster fishing for people outside of the United States. It is possible for non-citizens to obtain commercial lobster fishing licenses as long as they are residents of Maine.[2] But just like the members of any outside gang, any lobster

---

[2] Personal communication, Helen Holt, supervisor of licensing, Department of Marine Resources, Licensing Office, Maine.

fisher outside the community or not following the socially accepted entry into the lobster fishing industry (i.e., having a Maine ancestry, working full time, involvement in the community, etc.) risked sabotage to his traps if he placed them in waters perceived to be the territory of a gang. Thus, though there are no formal or legal barriers to becoming a lobster fisher, the social capital among existing lobster fishers limits external entry into the harbor communities, ultimately giving long-standing members of the communities a competitive advantage.

Our emphasis on the effects of differences in social structures between the lobster and cod fishers on overfishing is perhaps too simplistic. The cod fishers did have their kinship-based crowds, which fishers drew on to limit losses in face of catastrophe and, therefore, potentially could have drawn on to resist overfishing. And social capital among the lobster fishermen was not uniformly strong – there were sometimes disagreements among the lobster fishers within a harbor that could have undermined the social restrictions on overfishing. Furthermore, Acheson argues that multiple factors contributed to the setting of trap limits (which were indicative of constraint through social structure): political entrepreneurs had to mobilize a following to win a distributional conflict against those who wanted to fish a large amount of gear; and a gang was inclined to enforce only territory that included a large amount of exclusive fishing area (Acheson 2003:75; Acheson and Gardner 2004). Acheson also notes that lobster gear is easy to monitor because lobsters are located closer inland than cod (10 to 15 miles [15 to 25 km] for lobster; cod can exceed the 200-mile [320-km] limit), and territories are more meaningful to the lobster fishers because lobsters are relatively sedentary compared with cod and, thus, have smaller home ranges (McCay 2001). Finally, the lobster fishery is less vulnerable to global competition because of the difficulties of shipping live lobster, although this has recently changed with increased fishing of lobsters in Mexico (*Los Angeles Times* 2002) and the advent of new processes for freezing and shipping lobster (Richardson 2003).

To further support our case that social capital does indeed operate to mediate the effects of globalization on aquatic resources, we consider a second comparison between the lobster fishers of Maine and fishers of the western Atlantic spiny lobster (*Panulirus argus*) of the Turks and Caicos Islands (TCI). We do not use the TCI comparison as our primary comparison because less is known about the TCI fishers than about either the cod or Maine lobster fishers.

Although entry into the TCI fishery by nationals ("belongers") is currently classified as open-access, Haitians, Dominicans, and other "non-belongers" are not allowed to fish commercially unless a belonger is aboard the fishing vessel with them at all times. Thus, the social model would seem to be similar to that of the gangs of Maine lobster fishers. But there is a general lack of enforcement of illegal fishing activities, and foreign poaching vessels correctly perceive this lack of enforcement and use it to their advantage. Though there is strong community pressure to enforce the law, the illegal activities continue. Critically, it appears that the TCI belongers are not able to garner social capital to sanction those who violate the rule. Thus, those who ride with non-belongers go unpunished, and few limits on catch rates are enforced.

The comparison of the Maine and spiny lobster fishers supports our emphasis on the importance of differences in social capital that ultimately affect how the Maine lobster and Atlantic cod fisheries were managed. For example, the U.S. government was able to draw on the social capital and norms of the lobster fishers to enforce harbor-based limits, whereas such an approach was considered but never implemented for cod fishers (Office of Technology Assessment 1977:321; Finlayson and McCay 1998). The lobster fishers also accepted limitations on their catch (e.g., the V-notch used to identify and limit the taking of fecund female lobsters) because they knew it would be imposed evenly and benefit the lobster fisher community in general (Acheson et al. 1998:400). In contrast, cod fishers ultimately accepted no such limit, likely because they recognized that such a limit could not be uniformly enforced and thus those who complied would be yielding their catch to other fishers.

Even with lobster communities, the distribution of social capital differentiates island based perimeter communities from harbor based nucleated communities (Acheson et al. 1998:79). Though all use the same equipment and fish in the same industry, the social networks of the island communities are denser than those of the harbor communities (owing, perhaps, to a sense of shared fate or what Portes and Sensenbrenner [1993] called "bounded solidarity"), and correspondingly, the ability to enforce sanctions to protect territories is greater in the perimeter defended areas (Acheson 2003:75).

It is impossible to say for sure whether the cod stock would have been preserved had the cod fishers traditionally relied on a more extensive and cohesive social network drawn from community membership to enforce fishing practices. But it is clear that the limits and laws of the fishery needed to be enforced on the micro level within small groups, and that greater stores of social capital deriving from stable social networks could have facilitated enforcement.

Though the lobster and cod fishers are just two of many types of fisheries and communities, in combination they represent a large population of fishers (e.g., McCay 2001), because general concerns regarding overfishing are ubiquitous (Pauly et al. 1998). But even the particulars of the lobster and cod fishers represent those of other fishing industries. The factory trawler is just one example of a technology that helped globalize a fishing industry (see Berkes and Folke 1998). The cooperation between lobster fishers and state government to enforce harvest limits is similar to that of some salmon fishers and Pacific Coast communities who have cooperated to restrict the amount, especially of young salmon, that are fished (Gilden 1999). The kin-based social structures of the cod fishers are similar to those used in Indian villages to preserve sacred tree groves (Gadgil et al. 1998), and the divisions that occurred within the cod and lobster fishers were similar to those experienced by the salmon fishers of the Pacific Northwest (Gilden 1999). Therefore, we believe the comparative analysis of the cod and lobster fisher networks and behaviors generally provides insight into how social structures of fishers mediate between global competition and the aquatic ecosystem.

We have characterized lobster and cod fishers as exposed to competition through globalization due to changes in transport, shipping, and communications. But improvements in technology have also increased the market opportunities for these fishers, in essence contributing to globalized competition experiences by others. The advent of the factory freeze-trawler in the 1930s enabled cod fishers to transport their catch throughout the world, while a new deep-freeze process has made it easier for lobster fishers to distribute their product across North America and overseas (Richardson 2003). In fact, more than half of the lobster catch within U.S. waters now is shipped outside the United States (Richardson 2003).[3]

Though lobster and cod fishers have expanded their market share by using technology to distribute and sell their products globally, these market effects of globalization are less pronounced for them than for others because the lobster and cod fishers have strong regional and national markets. Furthermore, the distribution to these markets is accomplished through well-established institutionalized channels

---

[3] Characteristically opportunistic, the lobster fishers of Maine are now integrating consumers into their community by tagging lobsters and having consumers record where a lobster was eaten and learn about the fisher who caught the lobster (Zezima 2003).

that differentiate little within local markets. That is, exchanges between wholesalers and distributors and fishers are not influenced strongly by personal social relationships and can more generally be considered to be market based. In contrast, in the next subsection we describe the distribution of ornamental fish and invertebrates from coral reefs, which has been dramatically altered by market opportunities that emerged with globalization.

Globalization and increased market opportunities

Currently, ornamental fish and invertebrates are collected on coral reefs generally within developing nations such as Indonesia, the Philippines, Kenya, the Maldives, and regions such as the Caribbean. These organisms are exported to wealthy nations such as the United States, European countries, and Japan, where they, depending on species, are placed into aquariums or eaten as a symbol of status (in Asian countries) (Wood 2001). The aquarium trade began in the early twentieth century, but the demand for ornamental fish increased in the 1950s with the expansion of the airline industry, and this demand has increased greatly in the past decade (Wood 2001). As mediated by human social networks that we will describe, this increase in demand has contributed to the degradation of the coral reef system, such as death of non-target species and loss of coral (Rubec et al. 2000), and altered the dynamics of the oceanic ecosystem, such as disrupting carbon cycling (McClanahan et al. 2002).[4]

As in other fish industries, fishers have responded to increased market demand for ornamental fish by using more aggressive, unselective fishing techniques. Historically, fishers who collected fish on coral reefs used cyanide. The cyanide immobilizes all of the fish in the area. Then collectors retrieve only the immobilized fish and invertebrates that are valuable to the aquarium trade (Rubec et al. 2000; Wood 2001). The collected fish are immediately placed in water without cyanide, which allows them to recover.

Though ornamental fishers use cyanide to be more expedient in the short run, the use of cyanide affects/harms the local ecosystem and its biota and is thus inefficient in the long run (Berkes and Folke 1998). Uncollected fish and invertebrates often die from overexposure

---

[4] Although coral reefs have been able to sustain themselves across a long geological time-frame, coral reefs are undergoing major declines worldwide (Hughes et al. 2003).

to cyanide (Rubec et al. 2000; Wood 2001), and cyanide impairs the photosynthetic process of zooxanthellae, the symbiotic algae of coral (Rubec et al. 2000). Additionally, because of its efficiency at stunning organisms, ornamental collectors using cyanide are vastly more able to overfish an ecosystem. For example, 25 percent of the species caught in the Maldives for the aquarium trade were either overexploited or close to overexploitation (Edwards and Shepherd 1992). The macroeconomic description is thus: through improved transportation and communication, globalization has generated a dramatic increase in exports for small coral-reef fishing communities, which has encouraged overfishing using non-selective harvest techniques that harm the ecosystem.

The description of the macroeconomic processes in the previous paragraph treats the members of the local fishing communities as interchangeable and members of the international community merely as "foreign" relative to the local community. Instead, there are several layers of variable and interacting networks that bring fish and invertebrates from their natural environment on the coral reef to an aquarium in a home. The basic layers are: collectors, who catch the animals on the coral reefs; exporters, who package and transport the animals to the affluent countries; dealers, who sell the animals at shops; and hobbyists, who buy the animals for their aquariums (Wood 2001).

As is natural in a chain of exchange, the actors along the chain are fairly segregated – hobbyists know only a few dealers, dealers know only a few exporters, etc. But the most critical disjuncture occurs between the collectors and exporters. Middlemen mediate between collectors and exporters, and in the process, they become key drivers in the economics of the ornamental fish trade (Christie and White 1997; Wood 2001). As the middlemen take their profit, they lower the price a collector receives for his animals. Furthermore, collectors often become heavily indebted to the middlemen, who act as creditors (Anonymous 1998), although collectors can have higher incomes than the other fisher folk within a community (Christie and White 1997). Collectors then resort to cyanide to reduce their immediate debt, with the cyanide often provided by the middleman (Anonymous 1998; Rubec et al. 2000; Wood 2001). Thus it is through the middleman that increased market demand may ultimately result in degradation of coral-reef ecosystems.

Market forces can also enforce positive environmental behaviors. A dealer whose shipment experiences higher mortality because the collector(s) used cyanide may stop buying from a given exporter and may tell other dealers to do the same (see Raub and Weesie [1990] for the effect of reputation in continued exchange). This pressure can be translated down to the collectors, although further research is needed to determine the sensitivity of the price of ornamental fish.

Supplementing market forces to use environmentally friendly techniques are various forms of social capital. Hobbyists who perceive themselves embedded in a large global community may choose to pay extra money for animals caught with non-lethal, highly selective methods. This is facilitated by the Marine Aquarium Council (MAC), which certifies chains of custody from "reef to retail," enabling consumers to identify and reward responsible businesses through their purchase of certified marine aquarium organisms, i.e., "those that were collected, handled and transported in a sustainable manner" (USAID Philippines 2006). Furthermore, the Marine Aquarium Council works to improve conservation at all the levels of trade (Holthus 1999), especially in the collector network by providing training in using nets (Wood 2001). Finally, exporters can facilitate conservation by establishing ties with collectors, bypassing the middlemen and their attendant economic pressures, by offering a stable method of transportation from the small village of the collector to the major city of the exporter.

It is worth considering how ornamental fishers might draw on the social capital within their local communities to limit catch rates and make their aquatic resources more resilient like the communities described by Berkes and Folke (1998) and Burger et al. (2001). But critically, collectors are often an ethnic minority or marginal to their communities (Christie et al. 1994; Anonymous 1998; Wood 2001). Collectors are also likely to be immigrants into their community (Edwards and Shepherd 1992; Christie and White 1997).

Because collectors typically are not embedded in long-standing social networks such as are the fishers of the New England seaboard, they have difficulty generating the social capital necessary to refrain from overfishing to preserve the common ecosystem. For example, on San Salvador Island within the Philippines, those who fish for food are the majority, while those who collect fishes and invertebrates for the aquarium trade are the minority and come from the Visayan region (Christie et al. 1994). Collectors' lack of social integration then restricts their social capital for two key types of actions. First, the collectors do not have the social capital to enforce restrictions on one another, and thus some use cyanide in spite of the potential threat to the ecosystem.[5] Second, because the collectors are not well embedded in the local community, members of the community assigned the transgressions of the few to the whole and thus did not fully address the needs of the collectors when setting up and protecting the boundaries and regulations for protected areas of the coastal region (Christie et al. 1994; Christie and White 1997). Although the collectors of aquarium fish have accepted and follow the regulations of the community, this lack of concern over their livelihood has further alienated them from the community (Christie and White 1997). (Outside international organizations also have come to the aid of these collectors by helping them form their own association to police the collectors who violate the regulations and explore alternative options for income).

The fragmented social capital among the collectors of ornamental fish also limits the potential for community management or co-management of resources among collectors, government agencies, and other members of the community (Christie and White 1997; Burger et al. 2001; Acheson 2003). Co-management includes training on sustainable methods of collection, such as net collection; enforcing regulations within the community and on outsiders who come within the resource boundaries; scientifically monitoring the resource; and participating in managing the resource (Christie et al. 1994; Christie and White 1997; Wood 2001). As an example of how co-management draws on social capital in contrast to more authoritative management techniques, consider the fact that the government of the Philippines was able to effectively regulate the aquarium trade in the Maldives with a quota system on exported animals. But when research suggested that there should be quotas on specific species to prevent overfishing (Edwards and Shepherd 1992), the government did not have the funds or manpower to enforce these regulations (Wood 2001). And it could not turn to local, informal enforcement because of the sparse social networks and lack of social capital among the collectors.

Though the evidence for collectors of ornamental fish does not permit exact description, the example of the coral reefs suggests how the distribution of social capital of small fishing communities mediates between the network of the global market and the network of an ecosystem (e.g., McClanahan et al. 2002). In the next subsection, we address how the distribution of social capital matters even when the

---

[5] In contrast, it is almost impossible to imagine a few lobster fishers drastically exceeding the trap limit without social consequence.

local network does not mediate directly between a global force and the local ecosystem.

Globalization and increased disturbances

Some aspects of globalization directly affect ecosystems, unmediated by human behavior. The most prominent of these is global warming, which affects water temperatures (Walker and Steffen 1997). In turn, water temperatures alter ecosystem dynamics, affecting everything from the reduction of the symbiotic algae in the coral reefs (coral bleaching) to the reproductive success of the Maine lobster (Acheson 2003). Though generated by human behavior, these effects of globalization are conveyed directly by the most common and ubiquitous of natural resources. This challenges the central thesis of this chapter because the effects of globalization are not directly conveyed by human social networks.

But human social networks affect how local communities respond to changes in the environment produced by globalization. Long-standing sociological theory argues that the more unstable the environment, the more complex the social structure generated to accomplish desired goals (Woodward 1965). Classically, this complexity can be accomplished through a refined division of labor which is then coordinated by a central bureaucracy (Durkheim 1933; Weber 1958). For example, schools that face rapid external changes in computer technology and student composition often create more elaborate divisions of labor, such as technology teachers or special education teachers (Zhao and Frank 2003).

More recently sociologists have identified complex networks for sharing information or resources as critical to systemic responses to changes in the environment (Abrahamson and Rosenkopf 1997; Frank and Fahrbach 1999). When conditions are unstable, people need new and varied information to accomplish their tasks efficiently. Because each human has a limited capacity to acquire and manage such information, humans rely on coordinated activity and social networks when tasks are complex and changing. Continuing the example, schools may respond to changes in computer technology or student composition not just with the introduction of new specialties and corresponding training but also by attending to how these specialized teachers coordinate with other teachers in their schools. Again, this returns us to the value of social capital – those who can access information and resources through their social networks will be better able to respond to external disturbances (Berkes and Folke 1998; Palsson 1998; Burt 2000; Frank *et al.* 2004).

Though the extant sociological literature typically focuses on well-defined formal organizations or firms such as a post office (Weber 1958) or school (Frank *et al.* 2004), there are examples in use of natural resources that are consistent with the theory regarding the importance of social capital in helping social systems react to change. Iceland increased interaction between policy-makers and commercial fishers to gauge changing fishing stocks via an annual trawling rally (Palsson 1998:59). As an example in wildlife management, the Chisasibi Cree elders coordinated a response to a decline in caribou by sharing stories of earlier declines. These stories reinforced the communal ties because they emphasized that members of the community shared a common fate. Furthermore, the elders symbolically and absolutely represented the kinship ties within the community. Members of the Chisasibi Cree then drew on the social ties to enforce restrictions and curtail exploitive hunting practices (see Sporrong 1998).

If one human response to instability is to coordinate action, then social capital becomes more salient in the turbulent times brought on by globalization. Just as an isolated teacher can be a good teacher (Hargreaves 1993) when conditions are stable, an isolated fisher can be effective when stocks and conditions are stable. This is because one who is isolated but experienced can draw on personal experience to solve most problems. But when conditions change, people seek new information and need to coordinate action (Frank and Fahrbach 1999). This will increase the importance of social capital – for example, as has been manifest in the hunting and fishing practices of many small, local cultures indigenous to North America.

We have made the argument that the distribution of social capital within a community affects how a community uses its natural resources in light of globalization. Then in the discussion we will argue that managers and researchers should attend to explicit network processes. As a bridge, in the next section we describe tools for graphically representing the structure of networks from which social capital emerges and for modeling the manifestation and benefits of social capital.

APPLICATIONS OF SOCIAL NETWORK TOOLS TO STUDY SOCIAL CAPITAL

The networks of the cod, lobster, and ornamental fishers in our case studies were essentially inferred. Cod fishers were expected to interact according to kinship and land ownership, lobster fishers were inferred to interact by gangs associated with their fishing port, and collectors of ornamental fish were generally described as disconnected from those who fished for subsistence food on the coral reef. There is no evidence to suggest that the inferences in these cases are incorrect, and the manifestations of social capital are certainly consistent with the inferences. But in other instances, network structure may not neatly align with geographic region or kinship, especially as global forces enter local communities. Thus we call upon researchers and managers to gather direct social network data (e.g., who talks to whom, who fishes with whom). This type of data has been used to study everything from electrical wiring rooms (Roethlisberger and Dickson 1941) to bankers (Stovel *et al.* 1996; Frank and Yasumoto 1998; Uzzi 1999) to tailors (Kapferer 1973) to teachers (Bidwell and Yasumoto 1999; Frank *et al.* 2004) to charter boat fishers (Mueller 2004).

Given our discussion of the distribution of social capital within and between cohesive subgroups, one could use explicit network data to identify subgroups and then link subgroup membership to the distribution of social capital. For example, Frank and Yasumoto (1998) identified cohesive subgroups based on friendships among the French financial elite. After using simulation to establish that the friendships were concentrated within subgroup boundaries at a rate that was unlikely to have occurred by chance alone, they embedded the boundaries in a sociogram (see Fig. 16.2). In this sociogram, each number indicates a member of the French financial elite (those who ran major public or private financial institutions), lines indicate a friendship between two people, and the circles represent subgroup boundaries. These subgroups substitute for the kinship boundaries assumed for the cod fishers or the geographic boundaries assumed for the lobster fishers. Distance between actors and subgroups are indicative of friendship strength, with shorter distances representing denser sets of friendships.

Frank and Yasumoto then showed that resource allocations varied with the subgroup boundaries. First, hostile actions, such as a corporate takeover, almost never occurred within subgroups, indicating that trust could be enforced via the dense friendships within subgroups. Second, supportive actions, as an example of reciprocity transactions, were more likely to occur amidst the sparse ties between subgroups. Frank and Yasumoto reasoned that members of the French financial elite supported others outside their subgroups because they already had social capital within their subgroups via enforceable trust – the dense social ties within subgroups increased the likelihood that

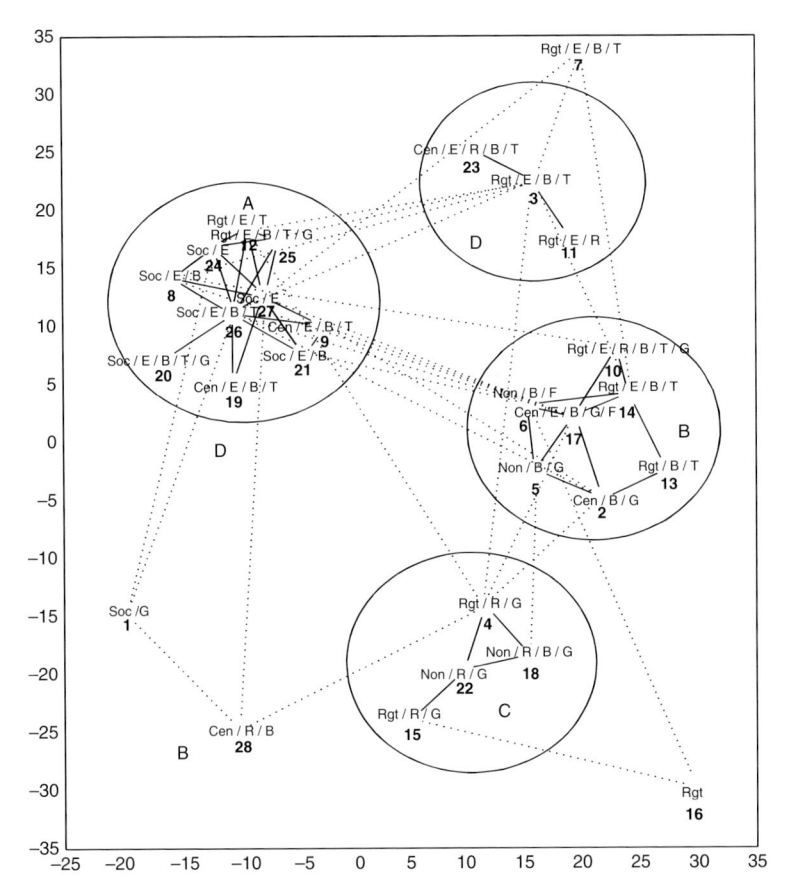

Figure 16.2 Crystallized sociogram: friendships among the French financial elite. Solid line within subgroups, dotted lines between. Scale = maximum weight / (density of exchange), expanded by 6 within subgroups. Party affiliation: Soc, Socialist; Cen, Central; Rgt, Right; Non, none/unknown. R, member of Résistance de socialisme; E, attended Ecole nationale d'administration (ENA); T, employed in the Treasury; B, employed in the banking industry; G, employed in the Grande Banque.

anyone who betrayed the trust of a subgroup member would be sanctioned, socially and otherwise, by the group as a whole. Thus, members of the French financial elite seek to engender new obligations and access new information and resources by helping those outside their subgroups.

The sociological forces affecting the French financial elite may seem far removed from those forces experienced by those who rely on natural resources for their livelihood. And yet both types of economies are generated by allocation and movement of what are potentially common resources, be they natural or fiscal. In fact, the allegorical "tragedy of the commons" generated as community members use public grazing land solely for their own good (Hardin 1968) could as easily apply as members of the French financial elite indirectly raid and manipulate the financial coffers of France with potential costs to the French economy.

Beyond the sociological phenomenon, the underlying theory of subgroups and graphical representation of social structure as in Fig. 16.2 have potential application to other systems. For example, Krause et al. (2003) used the same techniques to identify and represent food webs as compartments. Their analysis ultimately can inform how systems react to external disturbances depending on the entry point of the disturbance relative to the subgroup boundaries. The loss of a species that is a core member of a subgroup or a bridger between subgroups may have a more profound impact than the loss of a species that is only a peripheral member of one subgroup. For example, in a simulation exercise, two species core to their compartments and two species peripheral to their compartments were hypothetically removed from a food web of Lake Michigan. The loss of the core species had a larger impact on the overall structure of the food web and the structure of the compartment in which the core species was a member than the loss of the peripheral species did on the overall food web or their membership compartment (Krause et al. 2004).

To formalize the graphical representations, one could evaluate specific components of the theory of social capital using models incorporating network relations among humans. The key to this is to build network effects into standard regression type models used by social scientists and others (Doreian 2000). This moves network analysis past mere metaphor (Burt 2000) and allows network effects to be estimated controlling for, and compared with, alternative effects associated with individuals or formal organizations (Duke 1993; Frank 1998; Doreian 2000).

Focusing on the effect of social relations on an individual outcome, one could specify the competitive advantage an individual gained by accessing resources through social relations. For example, in a system such as the cod fishers network, define $kin_{ii'}$ to take a value of 1 if i and i' are kin, 0 otherwise. Next, define $gear_{i'}$ to represent the extent to which i' controls gear that might help a fisher be more successful. The term $\sum_{i=1}^{n} kin_{ii'} gear_{i'}$ then represents the total gear that a fisher could potentially access through kin, and the advantage that fishers gain as a result of accessing gear through kin is represented by β_1 in the following model:[6]

$$\text{catch rate}_i = \beta_0 + \beta_1 \sum_{i=1}^{n} (\text{kin}_{ii'}) \times (\text{gear}_{i'}). \qquad (16.1)$$

This model, known as a network effects model (Marsden and Friedkin 1994), can be generalized to include other types of social relations (e.g., members of a harbor gang) or resources (e.g., knowledge of lobsters). Furthermore, one can, of course, include other covariates in the model, such as the expertise of a fisher, equipment he owns, etc. Thus the model can generally be used to estimate the competitive advantage of social capital relative to the advantages of other characteristics (Burt 2000; Doreian 2000).

Researchers may also seek to model network relations as an outcome. For example, following social capital theory, one might model who allocates resources to whom to understand the factors that affect the flow of social capital in a system. In particular, define informs$_{ii'}$ to take a value of 1 if fisher i' provides information to fisher i, 0 otherwise. Then we can specify the effect of kinship on the likelihood that a fisher provides information to another as in the following logit model (Wasserman and Pattison 1996):

$$\log\left(\frac{p[\text{informs}_{ii'} = 1]}{1 - p[\text{informs}_{ii'} = 1]}\right) = \theta_0 + \theta_1 \text{kin}_{ii'}. \qquad (16.2)$$

Thus a large value of θ_1 would indicate that kin are more likely to provide information than non-kin. One can also add terms indicating the degree of similarity between i and i' on some characteristic, such as expertise, or seniority. These are known as homophily effects (Festinger *et al.* 1950; Homans 1950; Blau 1977; Feld 1981). Thus models such as (16.2) can be used to understand the factors that affect the rate and direction of flow of resources in a system, ultimately helping social scientists understand how resources come to be distributed in a system. This can help managers plan programs and anticipate how information, gear, etc. will diffuse among members of a community, which we address more in the discussion.[7]

---

[6] This model of influence can be estimated with the SAS program found at http://www.msu.edu/~kenfrank/software.htm

[7] Note that dependencies inherent in social network data pose important challenges for estimation of the parameters in models such as (16.2). We discuss these in the Appendix.

DISCUSSION

In this chapter, we have employed the theory of social capital to understand how human social networks mediate between the international exchange networks of globalization and aquatic ecosystems. When people can access resources through social relations, the network of relations affects how resources are extracted from aquatic ecosystems and how they are distributed across the globe. Thus social capital helps us understand how some fishers can limit the catch of others by threatening ostracization from the social network, and how entrepreneurial fishers and middlemen can draw on social relations to sell aquatic species globally.

As we focus on how resources are distributed through social networks, we must recognize that relationships are not uniformly distributed through a social system. Most systems are made up of integrated regions of highly concentrated relationships, such as the harbor gangs of lobster fishers or the kin-based crowds of cod fishers. Correspondingly, social capital has multiple forms, such as bridging and bonding, and is unevenly distributed across systems.

Individuals who have social capital can gain advantage over their competitors who may be less able to react to external changes (Burt 2000). Thus the merchant John Scarlet built his business through good standing within his local community and then expanded through new relations in a second community (Faris 1972). Similarly, coral-reef fishers have found new markets by building exchange relationships with middlemen who have access to the aquarium trade.

Some emphasize the advantages of social capital for the individual. But because social capital facilitates the flow of resources without reliance on more formal institutions, the distribution of social capital within a community can affect how members of the community manage their common resources. When there is adequate bonding social capital, members of a community need not resort to drastic, environmentally damaging behavior when faced with catastrophe under the stress of global competition. For example, the lobster fishers rely on the bonding social capital of their gangs to avoid overfishing. Bridging social capital helps members of neighboring communities manage their proximal resources, a proximity that is accentuated by globalization. For example, members of a small community in Bula, Fiji, drew on their social ties outside of the village to learn more about what damaged their coral reef and to coordinate action to preserve it (Coralfilm 2003).

Social capital, by definition, pertains to the resources actors can informally access through social relations. We recognize that those members of a community who have more formal authority or are perceived to be more legitimate than others may have more influence on communal action. But because globalization simultaneously increases the complexity of challenges to local communities and the exposure of general community members to external challenges, we anticipate that the capacity of those who have influence through formal authority, detached from networks, to make and implement effective decisions will be diminished. We also predict that globalization will make it necessary for effective decision-makers to engage in networks of their local communities. Furthermore, decision-makers must rely on networks to coordinate with those far removed from their immediate locales but whose actions may affect the local natural resource. Thus we anticipate that globalization will raise the demands on the networks of fisheries managers (see Dowding 2004), increasing the social transaction costs to ensure sustainable resource management. Indeed, reference to this phenomenon is already emerging in the governances and fisheries management literature, to which we turn in the next subsections.

Implications for governance

A key question is whether merely a more complex and dense network within existing communities can sustain the coordination necessary to respond to globalization. The existence of non-governmental organizations (NGOs) that have interests in issues affecting individuals and communities provides an example of a dense network within an existing community or between communities that can respond at local, regional, national, and global scales to the influences of globalization. Non-governmental organizations foster the formation of a dense subnetwork by providing a framework through which people with common concerns and interests can be brought together via global communication and transportation systems. They therefore facilitate the coordination of responses, creating a stronger, unified political voice at targeted levels of governance.

Non-governmental organizations have also served as a catalyst by which societal members attempt to influence governmental actors in relationship to issues related to fish, their environments, and fishing. But when external disturbances are increasingly initiated and moderated by people outside the community, perhaps the best communal response is to expand the network to coordinate with those who influence the availability of resources. Moreover, the composition of the networks may need to change to include government employees, environmentalists, politicians, and scientists (Dowding 2004). For example, the Marine Aquarium Council was formed to promote a sustainable fishery of aquarium organisms by certifying those organisms that are collected according to the goals of the council (Holthus 1999). This council is described as a global network made up of representatives from the aquarium industry, hobbyists, conservation organizations, government agencies, and public aquariums (Holthus 1999). Thus the increased rate of external disturbances may demand an extension of the relevant network needed for coordination, constituting a secondary effect of globalization (Dietz et al. 2003). This can ultimately lead to state and international coordination, such as the 1992 United Nations Framework Convention on Climate Change (part of the 1992 Rio Conventions) (United Nations 1992) or the 1997 Kyoto Protocol to the United Nations Framework Convention on Climate Change (United Nations 1997).

Implications for managers

Managing a fishery now requires much more than knowledge of the biological systems and manipulation of a select number of species in those systems. Managers must be aware of the multiple stakeholders who will be affected by and then can influence policy and action. Recognizing the human element, many have called for increased communication, awareness, and interaction between managers, policymakers, and stakeholders (McDonough et al. 1987; Lee 1993; Holling 1995; Berkes and Folke 1998; Blumenthal and Jannink 2000; Lal et al. 2001; Conway et al. 2002).

Here we extend this focus on the human element by calling attention to the underlying human networks. Rarely are groups as monolithic as outsiders perceive them to be (Freeman 1992). Thus when fishery managers interact with commercial fishers, they must be aware that there may be multiple groups of fishers and that to interact with one set is not to interact with all. Interaction with only a small group may lead to unintended consequences, potentially pitting one group against another with negative implications for the environment (Portes 2000). For example, coordinating policy with only one lobster gang would favor that gang over others and thus could intensify existing rivalries. Similarly, management decisions regarding coral reefs focused on the collectors do not attend to the economic pressures and social context in which the collectors operate and thus may merely exacerbate social cleavages between collectors and other members of their communities.

Even within the small communities in our case studies, there are discernible subgroups in the social structure. Correspondingly, social capital takes various forms and is not uniformly distributed throughout the system. This is accentuated when we consider the multiple stakeholders who can affect an ecosystem or who draw resources from an ecosystem (Gilden 1999; Lynch 2001). Stakeholders may organize by demographic region, industry, economic interdependency, etc., and managers must be aware of these networks as managers seek to engage the broader community in resource management.

We expect that an analysis of networks will be most important when trying to govern or constrain naturally competitive behavior.[8] For example, if the most economically advantageous fishing technique, trawling, were not the most harmful, then there might be little need to regulate the behavior (Gilden 1999). Correspondingly, policymakers could anticipate and accept that fishers would slowly adopt trawling as they gained access to the relevant knowledge and equipment. But because trawling is harmful to the environment or the fish population, we as a society may seek to regulate this behavior. And critical to implementing any regulation is to be aware of cleavages and even distrust among subsets of fishers. Therefore fishers cannot be appealed to as a single group, and they may have little underlying social structure on which to draw to self-regulate. Correspondingly, managers will have to engage each subgroup separately and will have to ensure that regulations are being enforced in economically equivalent ways.

An appreciation for network structure and the distribution of social capital is consistent with arguments that fisheries management strategies are best when tailored to the local social structure. For example, the U.S. federal government's strategy of localized enforcement of lobster trap limits drew on the extensive bonding social capital of the lobster fishers deriving from the well-marked social systems of the gangs. Absent such social structure and social capital, a similar enforcement strategy may not be successful. Thus government use of localized enforcement was limited on the coral reefs and failed

---

[8] Here, we assume that environmental norms and external actors either do not change or are not influential in light of economic pressures.

with the cod fishery. Where network structures are sparse, enforcement strategies based on market forces (e.g., individual transferable quotas: Burger et al. 2001; McCay 2001; Dietz et al. 2003) may be more effectively implemented.

Implications for research

The tensions we describe in our case studies reflect how the social structures of small communities mediate between competing demands of economic competition and ecological concern. Coral-reef fishers must choose between cyanide for immediate economic gain or techniques that are less destructive, but less economically beneficial in the short run. Similarly, the lobster and cod fishers must balance their immediate catch against degradation of the supply. Generally, individuals acting in their self-interest may choose the short-term economic benefit, calculating that they could fish in other areas should the current area become overfished.

Individuals embedded in the social structure of a community may be more likely to recognize that relocation will harm many members of the community and may compromise the social structure. Thus they have more of an interest in preserving the local resource for the long term. Though the tensions between economic competition and ecological concern are not new (Hardin 1968), they are exacerbated by globalization, which increases market pressures and opportunities while simultaneously heightening awareness of shared environments. Therefore, the demand on local communities to balance these tensions is increasing. Correspondingly, we argue that the classic tools of technology and policy used to negotiate competing tensions need to be augmented with more sophisticated ways of thinking about the interface between human and ecological networks. In particular, researchers need to attend to the causes and consequences of social relations among fishers as well as to the dynamics of food webs and the formation of legislation. Such knowledge could help managers understand how to develop and implement policy as well as inform policy-makers of possible unintended social consequences of their actions.

In fact, there has already been some interest in the link between human networks and ecology (McDonough et al. 1987; Maiolo and Johnson 1992; Gaedke 1995; Holling 1995; Lee 1999; Stepp et al. 2003). Though these have been important and valuable efforts, much of the literature reads as an attempt to apply generically the tools and metaphors of network analysis to study human social networks as they relate to ecology. As an extension, our theory of social capital draws explicitly on the social relations of humans to understand how resources flow without reliance on formal institutions. This should help researchers develop specific hypotheses beyond theoretical characterizations of whole systems as networks (e.g., Steins 2001). For example, researchers may hypothesize and test whether members of tightly knit social groups will provide extensive support to one another and be more inclined to cooperate to preserve a resource than members of loosely knit social groups.

The linkages between human networks and ecological networks could be extended by examining natural capital within ecological networks using similar methods to study social capital in human networks. Natural capital can be defined as the "soil and atmospheric structure, plant and animal biomass, etc., that, taken together, forms the basis of all ecosystems" (Akerman 2003). In the same way that social capital is defined as accessing resources through social relationships in humans, we can think of natural capital as accessing resources through the ecological relationships in a network. Correspondingly, the quantitative methods for identifying social capital and its effects through the processes of a human network may help with the quantification of natural capital (Akerman [2003] notes that there are important challenges to the quantification of natural capital). By better understanding the role of natural capital within ecological networks, we believe that human networks can begin to interact with ecological networks more effectively and efficiently.

We can draw on the rich history of the study of networks in ecosystems (e.g., Hannon 1973) to inform the types of studies of human networks advocated here. First, calls by those who study food webs for comprehensive, disaggregated data (Cohen *et al.* 1993; Gaedke 1995) should be applied to human social systems as well. The more information we have about exactly who interacts with whom, the more we will be able to understand how, where, and why resources flow in social systems. Network researchers in other domains, such as among bankers or teachers, are increasingly approaching this ideal (Uzzi 1996, 1999; Frank and Yasumoto 1998; Frank *et al.* 2004). Second, longstanding ecological theory suggests that ecosystems are more robust to external disturbances when made up of compartments (May 1973; McCann *et al.* 1998; Jordan and Molnar 1999; Krause *et al.* 2003). The theory of resilience and efficiency informs the presence and value of the social equivalent to compartments, cohesive subgroups. Third, those who study ecological systems are recognized as expert in characterizing both the internal dynamics of such complex systems as well as how such systems react to external disturbances through intrasystem interactions (e.g., Zimmerman *et al.* 2003). These same principles could be used to explore the internal dynamics of human systems and how human systems react to external disturbances, for example, by exploring how different sets of fishers increase or decrease their interactions under economic scarcity.

Like many applications of networks and related theories, our approach in this chapter is necessarily interdisciplinary (cf. Pickett *et al.* 1999; Heemskerk *et al.* 2003; Wali *et al.* 2003). Descriptions of global trade are generated by economists, terms of global trade and regulation are studied by political scientists and lawyers, and aquatic ecosystems are studied by ecologists. Sandwiched between these is a sociological understanding of the resources that flow through people's networks and how those resources affect action. The sociological approach has proven critical to understanding exchange and resource sharing between people that cannot be completely reduced to purely economic terms (Granovetter 1985), and thus we believe it will be fruitful when applied to fishers and fisheries managers. Some would say the macroeconomic forces eventually drive all change. Others might argue that change can be navigated primarily through greater understanding of the ecosystem and how to manipulate it. But if fisheries managers ultimately must engage and rely on human action, then fisheries researchers must increase their understanding of the networks in which those actions are embedded.

References

Abrahamson, E. and Rosenkopf, L. 1997. Social network effects on the extent of innovation diffusion: a computer simulation. *Organization Science* **8**: 298–309.

Acheson, J. M. 1988. *The Lobster Gangs of Maine*. Hanover, CT: University Press of New England.

Acheson, J. M. 2003. *Capturing the Commons*. Hanover, CT: University Press of New England.

Acheson, J. M. and Gardner, R. 2004. The origins of territoriality: the case of the Maine lobster fishery. *American Anthropologist* **106**: 296–307.

Acheson, J. M., Wilson, J. A., and Steneck, R. S. 1998. Managing chaotic fisheries. In *Linking Social and Ecological Systems: Management Practices and Social Mechanisms for Building Resilience*, eds. F. Berkes, and C. Folke. Cambridge, UK: Cambridge University Press, pp. 390–413.

Agresti, A. 1984. *Analysis of Categorical Data*. New York: John Wiley.

Akerman, M. 2003. What does "natural capital" do? The role of metaphor in economic understanding of the environment. *Environmental Values* **12**: 431–48.

Anonymous. 1998. The Haribon netsman training program. *SPC Live Reef Fish Information Bulletin* **4**: 7-12.

Baerveldt, C., Van Duijn, M. A. J., and Van Hemert, D. A. 2004. Ethnic boundaries and personal choice: assessing the influence of individual inclinations to choose intraethnic relationships on pupils networks. *Social Networks* **26**: 55-74.

Bendix, R. and Fisher, L. 1961. The perspectives of Elton Mayo. In *Complex Organizations*, ed. A. Etzioni. New York: Holt, Rinehart & Winston, pp. 113-126.

Berkes, F. 1998. Indigenous knowledge and resource management systems in the Canadian sub-arctic. In *Linking Social and Ecological Systems: Management Practices and Social Mechanisms for Building Resilience*, eds. F. Berkes and C. Folke. Cambridge, UK: Cambridge University Press, pp. 98-128.

Berkes, F. and Folke, C. 1998. Linking social and ecological systems for resilience and sustainability. In *Linking Social and Ecological Systems: Management Practices and Social Mechanisms for Building Resilience*, eds. F. Berkes and C. Folke. Cambridge, UK: Cambridge University Press, pp. 1-26.

Bidwell, C. and Yasumoto, J. 1999. The collegial focus: teaching fields: collegial relationships and the instructional practice in American high schools. *Sociology of Education* **72**: 234-256.

Blau, P. M. 1977. *Inequality and Heterogeneity*. New York: Macmillan.

Blumenthal, D. and Jannink, J.-L. 2000. A classification of collaborative management methods. *Conservation Ecology* **4**(2). Available online at www.consecol.org/vol4/iss2/art13/

Bourdieu, P. 1986. The forms of capital. In *The Handbook of Theory: Reseach for the Sociology of Education*, ed. J. G. Richardson. New York: Greenwood Press, pp. 241-258.

Breiger, R. 1981. Structure of economic interdependence among nations. In *Continuities in Structural Inquiry*, eds. P. M. Blau and R. K. Merton. San Francisco, CA: Sage Publications, pp. 353-380.

Burger, J., Ostrom, E., Norgaard, R. B., Policansky, D., and Goldstein, B. D. 2001. *Protecting the Commons: A Framework for Resource Management in the Americas*. Washington, DC: Island Press.

Burt, R. S. 1992. *Structural Holes*. Cambridge, MA: Harvard University Press.

Burt, R. S. 2000. The network structure of social capital. In *Research in Organization Behavior*, eds. R. I. Sutton and M. Staw Barry. Greenwich, CT: Elsevier Science, pp. 345-423.

Chase-Dunn, C. and Grimes, P. 1995. World system analysis. *Annual Review of Sociology* **21**: 387-417.

Christie, P. and White, A. T. 1997. Trends in development of coastal area management in tropical countries: from central to community orientation. *Coastal Management* **25**: 155-181.

Christie, P., White, A. T, and Buhat, D. 1994. Community-based coral reef management on San Salvador Island, the Philippines. *Society and Natural Resources* **7**: 103-117.

Cohen, J. E., Beaver, R. A., and Cousins, S. H. 1993. Improving food webs. *Ecology* **74**: 252-258.

Coleman, J. S. 1988. Social capital in the creation of human capital. *American Journal of Sociology* **94**: 95-120.

Coleman, J. S. 1990. *Foundations of Social Theory*. Cambridge, MA: Harvard University Press.

Conway, F. D. L., Gilden, J., and Zvonkovic, A. 2002. Changing communication and roles: innovations in Oregon's fishing families, communities and management. *Fisheries* **27**(10): 20-29.

Coralfilm. 2003. *Coral Reef Adventure*. Available online at www.coralfilm.com/f_story.html

Dietz, T., Ostrom, E., and Stern, P. C. 2003. The struggle to govern the commons. *Science* **302**: 1907–1912.

Doreian, P. 2000. Causality in social network analysis. *Sociological Methods and Research* **30**: 81–114.

Dowding, K. 2004. The theory of policy communities and policy networks. *Contemporary Political Studies* **5**: 59–78.

Duke, J. B. 1993. Estimation of the network effects model in a large data set. *Sociological Methods and Research* **21**: 465–481.

Durkheim, E. 1933. *Division of Labor in Society*. New York: Macmillan.

Edwards, A. J. and Shepherd, A. D. 1992. Environmental implications of aquarium-fish collection in the Maldives, with proposals for regulation. *Environmental Conservation* **19**: 61–72.

Faris, J. C. 1972. *Cat Harbour: A Newfoundland Fishing Settlement*. Toronto, Ontario: University of Toronto Press.

Feld, S. L. 1981. The focused organization of social ties. *American Journal of Sociology* **86**: 1015–1035.

Festinger, L., Schachter, S., and Back, K. 1950. *Social Pressures in Informal Groups*. Stanford, CA: Stanford University Press.

Finlayson, A. C. and McCay, B. J. 1998. Crossing the threshold of ecosystem resilience: the commercial extinction of the northern cod. In *Linking Social and Ecological Systems: Management Practices and Social Mechanisms for Building Resilience*, eds. F. Berkes and C. Folke. Cambridge, UK: Cambridge University Press, pp. 311–338.

Frank, K. A. 1998. The social context of schooling: quantitative methods. *Review of Research in Education* **23**: 171–216.

Frank, K. A. 2002. *The Dynamics of Social Capital*. New Orleans, LA: International Social Networks Association.

Frank, K. A. and Fahrbach, K. 1999. Organization culture as a complex system: balance and information in models of influence and selection. *Organization Science* **10**: 253–277.

Frank, K. A. and Yasumoto, J. Y. 1998. Linking action to social structure within a system: social capital within and between subgroups. *American Journal of Sociology* **104**: 642–686.

Frank, K. A., Zhao, Y., and Borman, K. 2004. Social capital and the diffusion of innovations within organizations: application to the implementation of computer technology in schools. *Sociology of Education* **77**: 148–171.

Frank, O. and Strauss, D. 1986. Markov graphs. *Journal of the American Statistical Association* **81**: 832–842.

Freeman, L. 1992. Filling in the blanks: a theory of cognitive categories and the structure of social affiliation. *Social Psychology Quarterly* **55**: 118–127.

Gabbay, S. M. and Leenders, R. T. A. J. 1999. CSC: the structure of advantage and disadvantage. In *Corporate Social Capital and Liability*, eds. R. T. A. J. Leenders and S. Gabbay. London: Kluwer, pp. 1–16.

Gadgil, M., Natabar, S. H., and Mohan Reddy, B. 1998. People, refugia and resilience. In *Linking Social and Ecological Systems: Management Practices and Social Mechanisms for Building Resilience*, eds. F. Berkes and C. Folke. Cambridge, UK: Cambridge University Press, pp. 30–47.

Gaedke, U. 1995. A comparison of whole-community and ecosystem approaches (biomass, size distributions, food web analysis, network analysis, simulation models) to study the structure, function and regulation of pelagic food webs. *Journal of Plankton Research* **17**: 1273–1305.

Gilden, J. 1999. *Oregon's Changing Coastal Fishing Communities.* Corvallis, OR: Oregon State University Press.

Gittell, R. J. and Vidal, A. 1998. *Community Organizing: Building Social Capital as a Development Strategy.* Thousand Oaks, CA: Sage Publications.

Granovetter, M. 1973. The strength of weak ties: network theory revisited. *American Journal of Sociology* **18**: 279–288.

Granovetter, M. 1985. Economic action and social structure: the problem of embeddedness. *American Journal of Sociology* **91**: 481–510.

Hannon, B. 1973. The structure of ecosystems. *Journal of Theoretical Biology* **41**: 535–546.

Hardin, G. 1968. The tragedy of the commons. *Science* **162**: 1243–1248.

Hargreaves, A. 1993. Individualism and individuality: reinterpreting the teacher culture. In *Teachers' Work: Individuals, Colleagues, and Contexts*, eds. J. W. Little and M. W. McLauglin. New York: Teachers College Press, pp. 51–76.

Harrison, A. 1996. Globalization, trade and income. *Canadian Journal of Economics* **48**: 417–447.

Heemskerk, M., Wilson, K., and Pavao-Zuckerman, M. 2003. Conceptual models as tools for communication across disciplines. *Conservation Ecology* **7**(3). Available online at www.consecol.org/vol7/iss3/art8/

Hirsh, P. M. and Levin, D. Z. 1999. Umbrella advocates versus alidity police: a life-cycle model. *Organization Science* **10**: 199–212.

Hoff, P. D. 2005. Bilinear mixed-effects models for dyadic data. *Journal of the American Statistical Association* **100**: 286–295.

Holling, C. S. 1995. What barriers? What bridges? In *Barriers and Bridges to Renewal of Ecosystems and Institutions*, eds. L. H. Gunderson, C. S. Holling, and S. Light. New York: Columbia University Press, pp. 14–16.

Hollingshead, A. B. 1940. Human ecology and human society. *Ecological Monographs* **10**: 354–366.

Holthus, P. 1999. The Marine Aquarium Council, certifying quality and sustainability in the marine aquarium industry. *SPC Live Reef Fish Information Bulletin* **5**: 34–35.

Homans, G. C. 1950. *The Human Group.* New York: Harcourt Brace.

Hughes, T. P., Baird, A. H., Bellwood, D. R., *et al.* 2003. Climate change, human impacts, and the resilience of coral reefs. *Science* **301**: 929–933.

Idoine, J. 2004. *American Lobster.* Woodshole, MA: National Marine Fisheries service. Available online at www.nefsc.noaa.gov/sos/spsyn/iv/lobster.html

Jordan, F. and Molnar, I. 1999. Reliable flows and preferred patterns in food webs. *Evolutionary Ecology Research* **1**: 591–609.

Kapferer, B. 1973. Social network and conjugal role in urban Zambia: towards a reformulation of the Bott hypothesis. In *Network Analysis: Studies in Human Interaction*, eds. J. Boissevain and J. C. Mitchell. New York Mouton, pp. 67–82.

Kim, S. and Shin, E.-H. 2002. A longitudinal analysis of globalization and regionalization in international trade: a social network approach. *Social Forces* **81**: 445–471.

Krause, A., Frank, K. A., Mason, D. M., Ulanowicz, R. E., and Taylor, W. W. 2003. Compartments exposed in food-web structure. *Nature* **426**: 282–285.

Krause, A. E., Taylor, W. W., and Mason, D. M. 2004. Potential overfishing impact on the southeastern Lake Michigan food-web, poster presented at the *Quantitative Ecosystem Indicators for Fisheries Management International Symposium*. March 31–April 3, 2004, Paris, France.

Krempel, L. and Plumper, T. 2002. Exploring the dynamics of international trade by combining the comparative advantages of multivariate statistics and network visualizations. *Journal of Social Structure* **4**(1). Available online at www.cmu.edu/joss/content/articles/volume4/KrempelPlumper.html

Lal, P., Lim-Applegate, H., and Scoccimarro, M.. 2001. The adaptive decision-making process as a tool for integrated natural resource management: focus, attitudes and approach. *Conservation Ecology* **5**(2). Available online at www.consecol.org/vol5/iss2/art11/

Lawler, E.J. and Yoon, J. 1998. Network structure and emotion in exchange relations. *American Sociological Review* **63**: 871–894.

Lazega, E. and Van Duijn, M. 1997. Position in formal structure, personal characteristics and choices of advisors in a law firm: a logistic regression model for dyadic network data. *Social Networks* **19**: 375–397.

Lee, K.N. 1993. *Compass and Gyroscope: Integrating Science and Politics for the New Environment.* Washington, DC: Island Press.

Lee, K.N. 1999. Appraising adaptive management. *Conservation Ecology* **3**(2) Available online at www.consecol.org/vol3/iss2/art3/

Lin, N. 1999. Sunbelt keynote address. *Connections* **22**: 28–51.

Lin, N. 2001. *Social Capital: A Theory of Social Structure and Action.* New York: Cambridge University Press.

Los Angeles Times. 2002. Swimming against the tide of overfishing. *Los Angeles Times*, Los Angeles, December 29, 2002.

Lynch, K.D. 2001. Formation and implications of interorganizational networks among fisheries stakeholder organizations in Michigans's Pere Marquette River watershed. Ph.D. thesis Michigan State University, East Lansing, MI.

Maiolo, J.R. and Johnson, J. 1992. Determining and utilizing communication networks in marine fisheries: a management tool. *Proceedings of the Gulf and Caribbean Fisheries Institute,* **1992**: 274–296.

Marsden, P.V. and Friedkin, N.E. 1994. Network studies of social influence. *Sociological Methods and Research* **22**: 127–151.

May, R.M. 1973. *Stability and Complexity in Model Ecosystems.* Princeton, NJ: Princeton University Press.

Mayo, R. and O' Brien, L. 2000. *Atlantic Cod.* Woodshole, MA: National Marine Fisheries Service. Available online at www.nefsc.noaa.gov/sos/spsyn/pg/cod/

McCann, K., Hastings, A., and Huxel, G.R. 1998. Weak trophic interactions and the balance of nature. *Nature* **395**: 794–798.

McCay, B.J. 2001. Community-based and cooperative fisheries: solutions to fisher's problems. In *Protecting the Commons: A Framework for Resource Management in the Americas*, eds. J. Burger, E. Ostrom, R.B. Norgaard, D. Policansky, and B.D Goldstein. Washington, DC: Island Press, pp. 175–194.

McClanahan, T., Polunin, N., and Done, T. 2002. Ecological states and the resilience of coral reefs. *Ecology and Society* **6**(2): 18.

McDonough, M.H., Cobb, M., and Holecek, D.F. 1987. Role of communication science in social valuation of fisheries. *Transactions of the American Fisheries Society* **116**: 519–524.

Mueller, K.B. 2004. The role of a social network in the functioning of the Grand Haven Charter Boat Fishery, Lake Michigan. M.S. thesis, Michigan State University, East Lansing, MI.

Nadel, S.F. 1957. *The Theory of Social Structure.* London: Cohen & West.

Office of Technology Assessment. 1977. *Establishing a 200-Mile Fisheries Zone*, NTIS Order #PB-273578. Washington, DC: U.S. Government Office of Technology Assessment.

One World 2007. *The OneWorld Network.* Available online at http://us.oneworld.net/

Palsson, G. 1998. Learning by fishing: practical engagement and environmental concerns. In *Linking Social and Ecological Systems: Management Practices and*

Social Mechanisms for Building Resilience, eds. F. Berkes and C. Folke. Cambridge, UK: Cambridge University Press, pp. 48–66.

Pattison, P. and Robbins, G. 2002. Neighborhood-based models for social networks. *Sociological Methodology* **32**: 301–337.

Pauly, D., Christensen, V., Dalsgaard, J., Froese, R., and Torres, F. 1998. Fishing down marine food webs. *Science* **279**: 860–863.

Pickett, S. T. A., Burch, W. R., and Grove, J. M. 1999. Interdisciplinary research: maintaining the constructive impulses in a culture of criticism. *Ecosystems* **2**: 302–307.

Portes, A. 1998. Social capital. *Annual Review of Sociology* **24**: 1–24.

Portes, A. 2000. The hidden abode: sociology as analysis of the unexpected. *American Sociological Review* **65**: 1–18.

Portes, A. and Sensenbrenner, J. 1993. Embeddedness and immigration: notes on the social determinants of economic action. *American Journal of Sociology* **98**: 1320–1350.

Pretty, J. 2003. Social capital and the collective management of resources. *Science* **302**: 1912–1914.

Putnam, R. D. 2000. *Bowling Alone: The Collapse and Revival of American Community*. New York: Simon & Schuster.

Raub, W. and Weesie, J. 1990. Reputation and efficiency in social interactions: an example of network effects. *American Journal of Sociology* **96**: 626–654.

Ricciardi, A. and Rasmussen, J. B. 1998. Predicting the identity and impact of future biological invaders: a priority for aquatic resource management. *Canadian Journal of Fisheries and Aquatic Science* **55**: 1759–1765.

Richardson, J. 2003. With record catches, why are prices so high? *Portland Press Herald*, Portland, OR, July 6, 2003.

Roethlisberger, F. J. and Dickson, W. J. 1941. *Management and the Worker*. Cambridge, MA: Harvard University Press.

Rubec, P. J., Cruz, F., Pratt, V., Oellers, R., and Lallo, F. 2000. Cyanide-free, net-caught fish for the marine aquarium trade. *SPC Live Reef Fish Information Bulletin* **7**: 28–34.

Safina, C.. 1998. *Song for the Blue Ocean: Encounters along the World's Coasts and Beneath the Seas*. New York: Henry Holt.

Schechter, M. G. 1999. Globalization and civil society. In *The Revival of Civil Society: Global and Comparative Perspectives*, ed. M. G. Schechter. New York: St. Martin's Press, pp. 62–80.

Simmel, G. 1955. *Conflict and the Web of Group Affiliations*, trans. K. H. Wolff and R. Bendix. New York: Free Press.

Sporrong, U. 1998. Delecarlia in central Sweden before 1800: a society of social stability and ecological resilience. In *Linking Social and Ecological Systems: Management Practices and Social Mechanisms for Building Resilience*, eds. F. Berkes and C. Folke. Cambridge, UK: Cambridge University Press, pp. 67–92.

Steins, N. 2001. New directions in natural resource management: the offer of actor-network theory. *IDS Bulletin* **32**(4): 18–25.

Stepp, J. R., Jones, E. C., Pavao-Zuckerman, M., Casagrande, D., and Zarger, R. 2003. Remarkable properties of human ecosystems. *Conservation Ecology* **7**(3). Available online at www.consecol.org/vol7/iss3/art11/

Stovel, K., Savage, M., and Bearman, P. 1996. Ascription into achievement: models of career systems at Lloyds Bank. *American Journal of Sociology* **102**: 358–399.

Strauss, D. and Ikeda, M. 1990. Pseudolikelihood estimation for social networks. *Journal of the American Statistical Association* **85**: 204–212.

United Nations, 1992. *United Nations Framework Convention on Climate Change*. Available online at http://unfccc.int/cop4/ftconv.html

United Nations, 1997. Kyoto Protocol to the United Nations Framework Convention on Climate Change. Available online at http://unfccc.int/files/essential_background/kyoto_protocol/application/pdf/07a01.pdf

USAID Philippines. 2006. *Transforming the Marine Aquarium Trade in the Philippines*. Manila: Marine Aquarium Council. Available online at http://philippines.usaid.gov/oee_envgov_marine_tmat.php

Uzzi, B. 1996. The sources and consequences of embeddedness for the economic performance of organizations: the network effect. *American Sociological Review* **61**: 674–698.

Uzzi, B. 1999. Embeddedness in the making of financial capital: how social relations and networks benefit firms seeking financing. *American Sociological Review* **64**: 481–505.

Vanderploeg, H. A., Nalepa, T. F., Jude, D. J., et al. 2002. Dispersal of emerging ecological impacts of Ponto-Caspian species in the Laurentian Great Lakes. *Canadian Journal of Fisheries and Aquatic Science* **59**: 1209–1228.

Wali, A., Darlow, G., Fialkowski, C., et al. 2003. New methodologies for interdisciplinary research and action in an urban ecosystem in Chicago. *Conservation Ecology* **7**(3). Available online at www.consecol.org/vol7/iss3/art2/

Walker, B. and Steffen, W. 1997. An overview of the implications of global change for natural and managed terrestrial ecosystems. *Conservation Ecology* **1**(2). Available online at www.consecol.org/vol1/iss2/art2/

Wasserman, S. and Pattison, P. 1996. Logit models and logistic regressions for univariate and bivariate social networks. I. An introduction to Markov graphs. *Psychometrika* **61**: 401–426.

Weber, M. 1947. *The Theory of Social and Economic Organization*, eds. A. M. Henderson and T. Parsons. New York: Macmillan.

Weber, M. 1958. *Essays in Sociology*, eds. H. H. Gerth and C. W. Mills New York: Oxford University Press.

Wellman, B. and Wortley, S. 1990. Different strokes from different folks: community ties and social support. *American Journal of Sociology* **96**: 558–588.

Wood, E. M. 2001. *Collection of Coral Reef Fish for Aquaria: Global Trade, Conservation Issues, and Management Strategies*. Ross-on-Wye, UK: Marine Conservation Society.

Woodward, J. 1965. *Industrial Organization: Theory and Practice*. New York: Oxford University Press.

Woolcock, M. 1998. Social capital and economic development. *Theory and Society* **27**: 151–208.

Zhao, Y. and Frank, K. A. 2003. An ecological analysis of factors affecting technology use in schools. *American Educational Research Journal* **40**: 807–840.

Zezima, K. 2003. Tracking the route of your lobster dinner. *New York Times* July 24, 2003.

Zimmerman, C. R., Fukami, T., and Drake, J. A. 2003. An experimentally derived map of community assembly space. In *Unifying Themes in Complex System II: Proceedings of the 2nd International Conference on Complex Systems*, eds. Y. Bar-Yam and A. Minai. Cambridge, MA: Perseus Publishing, pp. 427–436.

Appendix: Estimation of social network models

One's first inclination to estimate social network models such as in (16.2) might be to use maximum likelihood techniques such as those that are available to estimate the parameters in standard logit models (e.g., Agresti 1984). But it is difficult to define the likelihood for the data given the parameters in model (16.2) because the observations are not independent. For example, the relation between i and i' is not independent of the relation between i' and i.

There are currently two new approaches for accounting for dependencies in social network models such as (16.2). First, the p^* approach, developed by Frank and Strauss (Frank and Strauss 1986; Strauss and Ikeda 1990) and described by Wasserman and Pattison (1996) shows that estimates from the standard logit model as in (16.2) can be described as based on a pseudo-likelihood if one conditions on key relations between other pairs in the network. That is, an explicit set of covariates is entered into the model to control for structural dependencies. For example, whether i' informs i might be a function of the number of informants they have in common, the tendency of i' to inform others, etc. In a key point, Strauss and Ikeda argue that a Markov assumption implies that one need only account for relations that involve either i or i' – the "stars" around the actors involved (although this has been extended to account for less direct ties defining neighborhoods [Pattison and Robbins 2002]).

In another estimation alternative, one can account for the nesting of pairs within nominators (i) and nominees (i') using an application of multilevel models with cross-nested effects (Lazega and Van Duijn 1997; Baerveldt et al. 2004; Hoff 2005). These are called p_2 models because they estimate and control for the variances of actors' tendencies to send and receive nominations. One advantage of the p_2 approach over p^* is that effects of people can be modeled and tested at a separate level than those of the pair, without attributing most effects to characteristics of the network structure as represented by the p^* covariates. Frank (2002) applied this type of p_2 model to estimate the extent to which teachers were more likely to help close colleagues than other members of their schools.

H. CHRISTOPHER PETERSON AND KARL FRONC

17

Fishing for consumers: market-driven factors affecting the sustainability of the fish and seafood supply chain

INTRODUCTION

The fish and seafood industry successfully satisfies the annual consumption demands of millions of consumers worldwide. The industry's ability to continue this record of success is open to question. Rising incomes and trends in consumer tastes and preferences indicate a potentially dramatic increase in future demand; the future supply side of the industry equation is far more uncertain. The realities of fish and seafood production and the resulting threat to wild stocks of fish and seafood species will require careful management to assure both a sustainable supply of wild catch and sustainable growth in aquaculture to meet demand.

The uncertainty regarding the survival and renewal of wild stocks of fish and shellfish and their habitat is especially challenging. The fish and seafood industry shares similar characteristics with other industries – e.g., mining or timber – utilizing renewable resources. Some species can be readily managed with more sustainable practices; others, such as long-living, late-maturing fish, may not be renewable in our lifetime. The tremendous growth in aquaculture attests to its potential to counterbalance declines in wild harvest. Aquaculture, however, has its own sustainability issues that also need to be resolved, including water quality concerns, invasive species release, disease, use of biotechnology, and drug use for enhanced growth or control of disease.

Even if the industry responds with sustainable management practices, consumers or those who supply consumers may or may not be willing to pay for the change. The grocery retail and hospitality industries directly respond to consumer demands daily. In general,

Globalization: Effects on Fisheries Resources, ed. William W. Taylor, Michael G. Schechter, and Lois G. Wolfson. Published by Cambridge University Press. © Cambridge University Press 2007.

these retailers' knowledge of and relationship with consumers can allow them to predict demand and then coordinate with other supply-chain members (distributors, processors, wild harvesters, and aquaculture producers) to successfully balance demand with supply. The retailers also have the potential to utilize price, product differentiation, and marketing communication to reach and influence consumers about fish and seafood products that have been produced with sustainable methods, e.g., restaurant menu labeling of any sustainability certification recognized. In turn, they could signal the other supply-chain participants of consumers' demand for fish and seafood products sustainably produced. A question that remains open is whether the current supply chain is adequately coordinated to assure an effective balance of basic demand and supply, let alone communicate additional signals about consumer concerns for sustainability.

These introductory remarks lay the foundation for the topic of this chapter – an exploration of the interrelationships between consumer demand, the fisheries supply chain, and sustainability. The chapter is organized accordingly. First, we examine general trends and influences on consumer demand for fish and seafood. Second, we look at the various economic actors in the fish and seafood supply chain. Issues of sustainability and the interaction with consumer demand and supply-chain functioning follow in the third major section. Marine stewardship, eco-labeling and certification, government policy, and the role of aquaculture are also dealt with in this section. Finally, key conclusions are presented, along with the need for more extensive research to support the resolution of concerns and incomplete knowledge.

To the extent possible, this chapter is written from a global perspective. However, the available information about U.S. consumers, markets, and supply-chain participants is far more extensive than it is for many other places in the world. Although the developing regions of the world are mentioned, the authors' knowledge base about fish and seafood demand, supply, and sustainability in these regions is limited. Much work remains to be done in the research arena to understand fully the impacts on these developing regions of the issues discussed. The U.S. experience is thus used as a case study because of the existing information base and its parallels with much of what occurs in other parts of the developed world. Where the U.S. experience diverges, the authors have attempted to point out any known distinctions.

CONSUMER DEMAND

The place to begin a discussion of fish and seafood sustainability is with demand, its likely growth, and the factors that will drive that growth. If demand is expected to drop or even stabilize, sustainability may be less of an issue going forward. However, just the opposite is likely – i.e., demand will likely continue strong growth.

Globally, human consumption of fish and seafood doubled in volume to over 90 million tonnes from the early 1970s to the late 1990s (FAO 1997:24–27). This represents a compound growth rate of about 2.8 percent per year. Again based on FAO data, for the 5 years ending in 2001, the growth rate per year was about 2.45 percent with total consumption just under 100 million tonnes. At this rate, 2010 consumption could easily reach 124 million tonnes. When this number is added to non-food uses of approximately 30 million tonnes (in which growth is relatively flat), total world demand would be predicted to reach 154 million tonnes in 2010. This section discusses reasons why this growth rate would be expected to continue, if not accelerate.

Factors affecting demand in the U.S. and the developed world

The factors affecting U.S. demand are generally reflective of those in many other developed world countries, although the specifics – e.g., value of demand, total weight consumed, and rank of varieties preferred – do vary by nation. Using U.S. demand as a proxy for the developed world allows for useful expanded treatment of the analysis of demand.

Several broad observations help frame the discussion of consumption and demand patterns. In the United States, fish and seafood consumption was valued at $17.7 billion in 2002 and amounted to approximately 15 pounds (7 kg) per person. According to Mintel (2002a), 89 percent of the U.S. population consumes fish and seafood, compared with 96 percent for poultry, 93 percent for beef, and 84 percent for pork. Thus, fish and seafood are less frequently consumed than the two primary meat sources of protein. The top ten most popular fish and seafood items make up almost 90 percent of all seafood consumption in the United States. These items (listed in order by annual weight consumed per capita) are shrimp, tuna, salmon, Alaska pollock, catfish, cod, crabs, clams, flatfish, and tilapia (NMFS 2004). Given this concentration on a relatively few species, consumption trends probably heighten the sustainability issues for these species. The majority of consumption is either fresh or frozen (67 percent), canned seafood is on a downward trend at 28 percent, with cured at 5 percent. Consumers exhibit strong preference for fresh fish and seafood at both restaurants and retail, probably because of the increased perception of quality with fresh product (Bisaillon-Cary and Meser 2004). This preference is also consistent with general consumer preference for freshness across many produce and protein food categories.

Factors shaping fish and seafood demand include income, age, geographical location, taste, eating away from home, convenience and experience with preparation, dietary and health trends, and quality perceptions. Environmental consciousness, certification, labeling, and origin of fish are also important factors, but they are included in the sustainability discussion later in the chapter.

Income

In general, higher income allows the consumer to purchase more basic staple foods until a point is reached where more appealing non-staple foods such as meat protein are preferred. Typically, households in developed nations with the highest incomes are most likely to choose seafood. Consumption is aligned with middle to older age groups who have more disposable income and a greater interest in a healthier lifestyle. Additionally, higher-income consumers tend to eat out more, which means their higher-cost fish and seafood are often consumed as fresh prepared in a restaurant. According to Mintel's findings (2002a), those with lower incomes who like fish or seafood still eat it but less often, and select less expensive fish and seafood options than those with the highest incomes. As incomes rise and more middle-income consumers emerge in the developing nations of the world, the income effect would likely be strongly positive on demand there as well as in the developed world.

Age

As age increases, so does attention to diet and health. Accordingly, Mintel (2002a) surveys reveal that consumers in their twenties say that lack of preparation time and skill both limit their consumption of fresh and frozen seafood. Consumers under the age of 35 are the least likely to choose seafood over other meat options. In the same survey, the "baby boom" generation households (consumers aged 40 to 58 in 2004) are most likely to consume seafood, while retiree households were identified as the least likely of any group to choose fresh or frozen products, probably because of preparation capability. Overall, increasing age tends to correlate with increasing consumption of fish and seafood. The aging of the developed world's population will likely increase the rate of growth in fish and seafood demand.

Geographical location

Respondents in the U.S. South are the most likely to choose seafood or fish over any meat other than poultry. This is likely a result of the abundant availability of fresh fish from the Gulf of Mexico and the Atlantic Ocean as well as availability of the established catfish aquaculture industry (Mintel 2002a). Seafood available in Florida, Louisiana, and the Carolinas will be fresh and somewhat less expensive than seafood at inland locations that incur shipping and refrigeration costs. The northeastern, southern, and western household regions were more likely to use fresh fish, probably because of the large number of fisheries and small fishing operations along coastlines. Households in the north central region were the only ones more likely to use frozen fish than fresh, which points to limited availability and higher prices in their region (Mintel 2002a). Geographic areas with inland coastlines, such as the Great Lakes, and along ocean coastlines allow for extensive wild fishing which somewhat skews actual fish consumption for the region. It would be expected globally that nearness to coasts and inland fisheries would have positive upward pressure on seafood demand. However, well-developed processing and distribution for fish and seafood products probably tend to offset the distance disadvantage in developed countries.

Taste trends

A National Fisheries Institute study found growing preference for broiled, baked, and grilled fish and seafood with sauces and seasonings, and less for fried fish and seafood items (Mintel 2002a). Increasing diversity of available fish and serving suggestions contribute to a more diverse diet. For example, sushi and sashimi bars are increasingly popular with Americans and consumers in other developed countries. In line with these trends, fish and seafood are being selected for menus by more hospitality entities, including campus contractors, medical centers, and retirement communities. Aquaculture, less expensive imports, and higher-quality frozen and highly processed products make it possible to increase fish and seafood presence on menus everywhere. Food service operators offer cost-effective dishes supplied not only to the hospitality industry but at an increasing rate to supermarkets. Contrary to what consumers say about preferring fresh products, many food operators increasingly use frozen, value-added, and convenience seafood products. Such products include thaw-and-bake gourmet appetizers, consistently sized meaty fillets for grilling or broiling, stuffed fish roulades, and fish croquettes. New methods of marinating and brining make products look and taste desirable while meeting low-cost objectives (Silver 2003). The bottom line is that taste preference, coupled with evolution in product form, favors increased demand for fish and seafood.

Eating away from home

With the strong eating-out trend in the United States, it comes as no surprise that 40 percent of all meals are eaten away from home. Fish and seafood meals are even more likely to be eaten outside the home. This trend favorably supports growth of the hospitality industry, which already satisfies over 70 percent of all fish and seafood demand in the United States (Mintel 2002a). Though 25 percent of respondents consume seafood two to three times a week (either at home or away from home), 20 percent consume seafood at restaurants at least once a week. Almost one-third of respondents ordered seafood at a restaurant once a month, and one-fifth twice a month (Mintel 2002b). Consumers generally prefer a full-service restaurant for variety and quality reasons, and their consumption is often driven by situational and emotional factors such as celebration of special occasions. Growth in fish and seafood consumption is thus further reinforced by the growing trend to consume food outside the home.

Convenience and experience with preparation

The trend toward eating out is strong, but home meals are also changing. Meals prepared at home consist of fewer dishes and take less time to prepare. In addition to this broad trend, many consumers are reluctant to deal with raw fish and seafood at home because of food safety issues and lack of knowledge in proper preparation. Despite these concerns, Simmons surveys published in Mintel (2002a) reveal that consumers still greatly favor fresh fish over frozen. In dollar sales, fresh fish has nearly 80 percent of the total market. Though sales of shelf-stable seafood are decreasing, canned tuna is still very popular and present in 90 percent of all households.

As fish processors and retailers increase efforts to address consumer concerns, the segments involving frozen and refrigerated pre-seasoned fish and seafood are expected to grow more quickly than other segments of the fish and seafood market. In 2002, frozen and refrigerated segments were small in comparison with the general fresh seafood market, but Mintel (2002b) estimates that, by 2010, they could together equal 20 percent or more as consumers who are time-pressed and avoiding delicate cooking choices will find seafood to be easier and quicker to prepare than they thought and less expensive. In other words, even though consumers prefer freshness, some will substitute convenience for freshness in home preparation. Again, evolution in product forms for home use favors increasing demand.

Dietary, health, and quality perceptions

Current trends in diet, health claims, and safety scares together shape consumer preferences for seafood. Diets low in carbohydrates and rich in protein and beneficial omega-3 fats encourage fish and seafood consumption. A National Fisheries Institute survey revealed health, diet, and calorie content as important reasons for fish and seafood consumption and that nine out of ten respondents perceived fish and seafood as healthier than beef and pork (Mintel 2002a). Physicians and medical organizations such as the American Heart Association have actively promoted fish and seafood as an integral part of a healthy lifestyle. Mintel (2002a) predicts that "baby boomers" reaching their fifties and sixties will further increase consumption of seafood. Nearly three out of four respondents agree on seafood's positioning as a healthy meal choice.

Countering these positive messages about fish and seafood in one's diet, the news media regularly bring attention to issues such as contamination with heavy metals, polychlorinated biphenyls (PCBs), antibiotic residue and bacteria content, and soon to be commercialized genetically modified organism (GMO) fish products. These conflicting claims about the health benefits and risks of fish and seafood consumption may be a source of confusion for consumers, and thereby limit demand growth. As evidence of the confusion, Mintel (2004) found that 28 percent of consumers surveyed think that fish and seafood are most likely to be tainted in comparison with other meats, and 15 percent have cut back on the amount of fish eaten because of concerns about mercury levels. On the other hand, 42 percent have increased the amount of fish and seafood consumed because they have heard about the health benefits of these products. In balance, if safety concerns can be managed by suppliers, the general health trends greatly favor increased demand.

Factors affecting developing world demand

Fish and seafood already represent an important source of protein for the developing world as a whole. Developing populations derive 20 percent of their overall protein intake from the category, which it delivers to only 13 percent in developed populations (Delgado *et al.* 2003.). Given that proximity to supply (coastal and inland waterway regions) is especially crucial to the developing world, those countries that have access to fish and seafood probably gain even more than 20 percent of their protein from this source and less well-endowed nations fall below this level. Generally speaking, successful development efforts would be expected to spur fish and seafood demand even further. Increasing population generally, rising incomes, development of food processing and distribution industries and channels, and the likely shifting of consumer preferences in favor of processed and protein rich foods all favor expansion of fish and seafood demand. Much research remains to be done to track and to understand how developing population demand will specifically change in the fish and seafood arena. However, the general trends would all argue for expanding, not contracting, consumption.

Conclusions about consumer demand

In the developing world, population and income growth are fueling significant increases in demand for fish and seafood as a protein source. In the developed world, trends in diet and health, convenience, income, age, and taste preference favor increased demand for fish and seafood products even as the population remains relatively stable. The lone concerns arise from consumers' lack of preparation knowledge for home consumption, broad safety concerns about fish and seafood contamination, and environmental concerns related to what fish are consumed. In total, the consumer demand picture suggests that past compound growth rates in excess of 2.5 percent per year are likely to continue and might even accelerate. If there are limits to this demand growth rate, they will probably come from the supply side and not from the preferences of consumers.

FISH AND SEAFOOD SUPPLY

Given the demand analysis, two issues appear particularly relevant to the ultimate discussion of fisheries sustainability. First, what are the expectations for growth in supply? Demand pressures could be largely mitigated if supply growth can keep pace. Second, what is the nature of the fish and seafood supply chain and its ability to transmit signals about demand? If the supply chain is not well coordinated from the consumer backward to basic inputs, then demand signals may not get transmitted properly, most especially signals that might provide the basis for consumer-responsive strategies supporting sustainability. Possible consumer-responsive strategies are discussed in the following section. The analysis, then, needs to address questions of what signals now travel through the supply chain and to be based on what each participant in the chain needs for economic survival and business health.

Expected growth in supply

Global harvest of wild fish stock declined slightly from 93.5 million tonnes in 1996 to 91.3 million tonnes in 2001. In the same period, aquaculture rose from 26.7 million tonnes to 37.5 million tonnes (FAO 1997). Only about 70 percent of the world's harvest of fish and fishery products is consumed as food; the rest is either discarded as bycatch, lost because of perishability, or processed for alternative use such as aquaculture and pet feed. Fish and shellfish products represent about 16 percent of the animal protein supply and 6 percent of total global protein supply. World trade in seafood is estimated at more than $100 billion per year. The major markets are those of Japan, the United States, and the European Union, which depend on imports for 30 to 60 percent of their consumption (Aquaculture Production Technology 2006).

Global production is projected to rise by an average 1.5 percent annually through 2020. Two-thirds of this growth will come from aquaculture, raising the share of aquaculture in total output from 30 percent to about 40 percent. Developing countries, led by China, now provide 70 percent of global production and will be responsible for the majority of the growth of supply (Delgado *et al.* 2003). The disparity between the supply growth projection of 1.5 percent and the projected demand growth rate of 2.5 percent estimated in the prior section already signals challenges for fish and seafood sustainability. These challenges are analyzed in the following section of the chapter.

The U.S. supply situation mostly mirrors that of the world. The United States is the world's fifth largest seafood harvester (NOAA/NMFS 1998). In 2003, the total commercial catch was 9.4 billion pounds (4.26×10^9 kg), down from 9.6 billion pounds (4.35×10^9 kg) in 1996 (Johnson 2002). Total U.S. aquaculture production grew most significantly between 1982 and 1992. Since then, the total U.S. aquaculture annual increase in production has slowed, and total U.S. aquaculture production was 898 million pounds (4.07×10^8 kg) in 1998 (Western Regional Aquaculture Center 1998). Aquaculture has expanded steadily, with a farm gate value of $751.1 million in 1994 and $978 million in 1998 (Tiu n.d.). Aquaculture production in the United States was mixed for 2002, with catfish production up 33 million pounds (1.50×10^7 kg) (to 630 million pounds [2.87×10^8 kg]) and trout, tilapia, and salmon declining (Johnson 2002); U.S. aquaculture has 4000 farms engaged in production and accounts for approximately 18 0000 jobs (Tiu n.d.). As in the broader global situation, U.S. supply growth will likely have to come from increases in aquaculture or imports if projected demand is to be met.

The supply chain

The fish and seafood supply chain (Fig. 17.1) extends from consumers back to basic resource providers. Retailers, both grocery and hospitality, are the suppliers to end consumers. Retailers are supplied by various types of processors and distributors who are, in turn, supplied by harvesters of wild catch and producers of aquaculture. Harvesters rely heavily on fishing rights and vessel ownership; aquaculture producers depend on inputs of water, land, and equipment. Starting from the retail end, we will describe each major set of participants in the supply chain with an emphasis on what motivates the economic behavior of each and what signals each appears to send to the remainder of the chain. Little research has been done to document the nature of the fish and seafood supply chain in either developed or developing nations. This analysis relies on the U.S. supply chain as a representative case. It parallels the supply chains found in most developed economies. The basic elements of the chain are also similar to those found in the developing world. The major difference is likely to be less product diversity (species and product form) because of limited technology for packaging, storage, and distribution in the developing context.

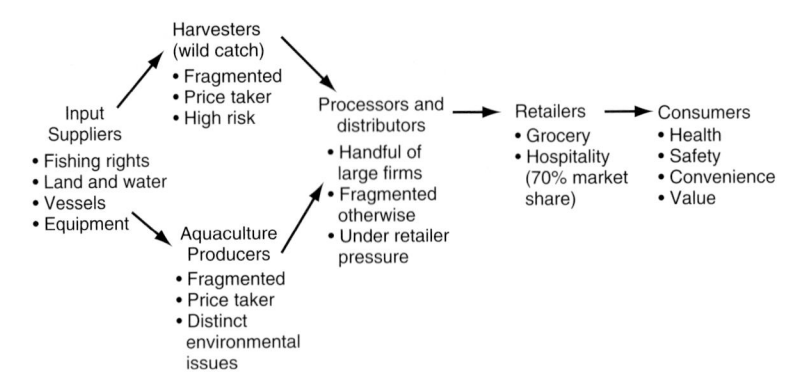

Figure 17.1 Fish and seafood supply chain.

Retailers: grocery

Fish and seafood products move through two distinct channels to end consumers: commercial retail stores, including super- and hypermarkets, local fish markets, and other specialty retail outlets; and commercial/non-commercial hospitality entities, including fast-food restaurants and full-service restaurants – e.g., casual or fine dining – and institutional providers, both public and private – e.g., schools, prisons, and medical care facilities. The first of these channels is most often referred to as the grocery channel, and it is analyzed first.

Total retail grocery sales of *all fish and seafood* in the United States are predicted to increase 25 percent from 2002 to 2007 (Mintel 2002a). Products move through grocery retailers in four product forms: fresh, frozen, refrigerated, and shelf-stable.

Fresh fish will continue to dominate the market, growing at a slightly faster rate – 26 percent – than the total retail market for fish and seafood. Retail prices for fresh fish will have a great impact on the overall value of the market and its actual growth rate. Some supermarkets have already sought to cut costs in the category by eliminating or reducing service at the seafood counter (Mintel 2002a).

Growth will be even higher for sales of *frozen and refrigerated products*. These products will be particularly attractive to consumers seeking convenience through new packaging designs, fully prepared and seasoned products, and other high-value attributes (Mintel 2002a). More specifically, total retail sales of *frozen* fish and seafood are predicted to grow 45 percent from 2002 to 2007. Frozen products that incorporate cooking instructions and nutritional information in their packaging should appeal to health-conscious consumers as well as those with limited cooking skills or time (Mintel 2002a). Likewise, total retail sales of *refrigerated* fish and seafood are predicted to grow rapidly, up 36 percent from 2002 to 2007. Supermarket chains will gradually shift away from full-service seafood departments and toward items that are delivered to the store prepackaged or packaged on site and placed in self-service coolers. These products can offer the convenience of individual packaging and preseasoning, allowing the consumer to prepare a healthy meal quickly (Mintel 2002a).

Total retail sales of *shelf-stable/canned* fish and seafood are predicted to remain constant (Mintel 2002a). This segment consists mainly of canned tuna, which will likely lose sales as a wider variety of frozen and refrigerated fish becomes available.

The retail grocery channel is increasingly dominated by a few large chains. Consistency in supply, quality, and price are critical to supplier success in serving this channel. Each chain attempts to differentiate its offerings to consumers, in part through private labels that serve to create the chain's own brand image. In addition, these chains prefer to handle their own distribution functions, purchasing as near to directly from primary supply as possible.

Generally, grocery retailers track consumer trends quite closely and attempt to respond to their particular customer needs as a matter of business strategy. As a result, the signals that they send to their processors and distributors tend to focus on these needs. Consistency in supply, quality, and price are most often at the heart of these signals. Sustainability of supply will likely be part of the quality signals if it serves a chain's strategy to differentiate itself from competition. For example, Whole Foods, with its primary strategic emphasis on organic and wellness attributes for its products, would be far more likely to signal sustainability as an attribute for supply than a general grocery retailer.

Retailers: hospitality

The hospitality channel for fish and seafood retail is particularly important – it delivers over 70 percent of all fish and seafood in the United States. By the definition of Huffman (2003) and Smet (2003), the hospitality industry is made up of businesses that practice the act of being hospitable, such as lodging, food service, travel and tourism, meeting and convention planning, and institutional catering (including government, prisons, charities, schools, and hospitals). Various hospitality retailers are taking different approaches toward their product offerings.

An emphasis on a varied menu and an attempt to attract a younger demographic characterize the efforts of the leading *fast-food seafood restaurants* in reviving a stagnant segment. Common reasons for less frequent consumption of seafood in quick-service restaurants include limited preparation options, concerns regarding freshness, and a preference for burgers and chicken when eating fast food. Most fast-food restaurants have expanded menus to the point that a consumer can find more than one type of fare, satisfying the varying needs of a group of people and making it unnecessary for a customer to leave a fish restaurant because he or she wanted something else (Mintel 2002b).

Consumers tend to opt for seafood at *full-service restaurants* for variety and quality (Mills 2001). Though common seafood choices such as salmon and tuna remain popular center-of-the-plate options, today's hospitality consumers may choose from new trendy dishes based on common fish prepared in new ways or dishes consisting of less common fish and seafood species. For example, restaurant chains, such as O'Charley's, have introduced new limited-time seafood promotions such as Cajun gumbo made with blue crabmeat, barbecue shrimp sautéed in chipotle barbecue sauce, and yellowfin tuna steak. Chevy's Mexican Restaurants is testing new seafood options, including a ceviche appetizer made with orange roughy and an ahi tuna entrée. If successful, these items will be incorporated into the chain's menu nationwide. Salads showcasing seafood as the main protein are also gaining in popularity (Silver 2003).

Bars and restaurants are adding raw fish (sushi and sashimi) to their menus thanks to the perception that these are trendy and a healthful alternative to such staples as potatoes and pizza. Some campus contractors, including Philadelphia-based Aramark, are adopting a national strategy to make sushi available to their campus accounts (Silver 2003).

For some *non-commercial operations* (such as medical centers), simple seafood items are enough to update the menu plan. Seafood is also in increasing demand at retirement homes. The Guest Services Corporation, whose food service operations are found in several prestigious facilities, relies strongly on farm-raised catfish because of its neutral flavor and easy incorporation into about any kind of cuisine (Silver 2003).

Some seafood is becoming more economical for the non-commercial operator (e.g., hospitals, prisons, colleges) to purchase because of the availability of significantly cheaper Asian imports. Shrimp owes its popularity also to aquaculture and higher yields that have resulted in more favorable market prices and better product consistency. Now shrimp can be found through the entire retail and hospitality industry (Silver 2003).

Many operators use frozen, value-added, and convenience seafood products to round out their seafood selections. Advanced methods of processing, both at sea and in large facilities located a short distance offshore, have led to improved fresh frozen seafood products. Not only are these items relatively inexpensive, they are also available year round and can be used in a variety of menu applications. Marinated and brined fresh frozen fish has entered hospitality kitchens. Besides improved taste and texture variables, convenience is the key factor for restaurants' adoption to save labor costs (Silver 2003).

Across its various outlets, the hospitality industry in the developed world has thus found a way of increasing per capita consumption of fish and seafood by offering greater variety of species, flavors, and preparation styles. Indeed, improved harvesting and sustainable fishing practices, the growth of aquaculture, and higher-quality frozen and value-added products make it easier than ever to include fish and seafood of all forms on restaurant menus (Silver 2003).

Chefs are often the source of information educating their suppliers about sustainable seafood issues. Careful selection of seafood suppliers allows chefs to respond to the growing demand for seafood by offering menu items that taste good and are more sensible for the environment (Seafood Choices Alliance 2002a). Examples of such sustainable seafood choices include Pacific albacore/tombo tuna, catfish, Alaska halibut, Alaska salmon, and mahi-mahi. According to a Seafood Choices Alliance nationwide study, 62 percent of restaurants would like to find and buy from environmentally responsible suppliers, 60 percent of seafood retailers agree, 46 percent of chefs and 34 percent of retailers would look for fish with little to no bycatch, 41 percent of chefs and 29 percent of retailers would look for non-overfished fish. In return, wholesalers confirmed they would be willing to supply to meet those preferences and needs (Seafood Choices Alliance 2002b).

At the same time, retailers try to make it easier for consumers to eat more fish and seafood. The result has been increased cross-fertilization between restaurant and retail operations, derived in large part from vendors that ship to both restaurants and food stores. In fact, seafood suppliers often develop a product for food service operations and then modify the more popular items for retail (Silver 2003).

As with grocery retailers, hospitality retailers also track consumer trends closely to vary menus and attract their customer base. The signals that they send to suppliers are first about consistency in supply, quality, and price – all parallel signals to the grocers. For those retail organizations that rely on chefs for critical menu decisions or that seek key sources of differentiation beyond these basics, additional attributes can and do matter. This influence has already been noted. In these instances, signals about sustainability as a supply attribute would more likely be sent to processors and distributors.

Processing and distribution

Processing and distribution run the gamut of firm types and sizes from small local processors to large branded seafood manufacturers. In branded seafood, dominant processor/manufacturers include StarKist, Bumble Bee, Chicken of the Sea, Gorton, and Aurora Foods. Branded seafood consists mostly of shelf-stable and frozen products. The growing market for private label products is supplied by many of the brand name manufacturers, but 44 percent of private label sales in 2001 came from small, likely regional manufacturers (Mintel 2002a).

Expansion of supermarkets and independent distributors utilizing alliances and contractual agreements with suppliers is changing relationships in the supply chain. In the past, large seafood processing and marketing companies used control over assets such as distribution systems and marketing as a competitive advantage while influencing markets through pull and push marketing strategies such as, respectively, consumer media advertising and trade promotion. Currently, the consistency in quality, freshness, and quantity demanded encourages direct coordination between supermarkets and primary processors, bypassing other links. This mutually beneficial relationship, however, causes dependency on limited suppliers, somewhat decreases the retailers' bargaining market power, and benefits processors with higher margins. For suppliers, this implies the need for balancing processing on demand, so much dependent on uncertain fish stock harvest, and the ability to add desired value (Trondsen 1997).

Processors are motivated to utilize economies of scale, lower cost through location selection, and differentiate product based on retailer demands to increase their relevance to retail customers, either in grocery or in hospitality. Cost containment strategies include, among others, horizontal cooperation and integration of firms, and use of low-cost plants abroad (Trondsen 1997). Quality and consistency enhancement strategies include controlling supply from several harvesters and locating near harvest to provide the freshest possible quality product.

However, advantages from strategies that favor proximity to buyer/consumer markets must be weighed against potential high transportation costs and quality loss during transit. To fulfill agreements with buyers, processors decrease uncertainty by managing harvest stock inventory through upstream coordination with fisheries (Trondsen 1997).

Given the fragmented nature of much of the processor/distributor link in the supply chain, signals from retailers are often seen as being about broad issues that allow economies in operation – ability to deliver needed volume, consistency, general quality, and price. Sustainability attributes may often be lost unless the processor/distributor is using these attributes as a critical element of differentiation or serves a retailer who uses this strategy in its promotion to consumers.

Harvesters and producers

This segment of the supply chain is highly fragmented, with many and varied competitors. For wild catch, harvesters include fishing companies that range in size from fishing fleets to one fishing rig (Mintel 2002a). In aquaculture, numerous farm producers exist, ranging from multiple acres and species to a single-tank, one-species operation. The vast majority of supply does move through processors, but some companies, such as Fresh Choice Seafood, are attempting to market direct to consumers through the Internet and direct delivery channels.

Harvester or aquaculture producers are typically selling for offered price with few opportunities to differentiate themselves from one another. As a result, they must manage for efficiency and minimize the production risks that they face. Harvesters in particular are exposed to some significant production uncertainties caused by natural fish stock fluctuations and weather. Harvesters can pursue fish stock management practices designed to save and regrow fish populations. Improving management of wild fisheries in both quantity and quality can be achieved and overfishing reduced through such strategies as bycatch reducing devices, modern harvesting, and satellite sensing and vessel tracking technologies. However, these means alone cannot solve production and sustainability issues without policy changes fishery access rights and incentives for public/private coordination among user groups and increased responsibility by all users of a given fishery (Delgado *et al.* 2003).

Changes in harvesting combined with processing are also emerging. Factory trawlers with on-board facilities process raw fish and seafood into high-quality packs. Experiments are under way in which seasonally available and captured fish and seafood are kept alive through aquaculture technology until sold (Delgado et al. 2003). Trufresh of Connecticut began freezing lobsters with a technique it used for years on salmon. Trufresh is using the technique as part of a strategy to expand its product line as it launches a retail business on the Internet. The possibility of lobsters to live after being deep frozen allows for flexibility in delivery while preserving fresh-like attributes of lobster (Lindsay 2004).

In contrast to wild harvesters, aquaculture producers are faced with a different set of constraints (to be discussed in the next section), even though these constraints result in much the same business behavior driven by the need to minimize cost in response to price pressure from buyers. The aquaculture producer has some advantages over the wild harvester. Supply can be more readily expanded in response to demand, and it can be managed for consistency given the control that the farm setting allows versus the wild environment. Nonetheless, key limits exist in this ability to control. More on this point is presented in the next section.

The predominant signal from the rest of the supply chain that reaches the harvester or the aquaculture producer is often simply about price coupled with highly generalized concerns about consistency in supply and quality. Evidence of supply relationships for the harvester or the aquaculture producer relying on sustainability attributes tends to be anecdotal or driven by regulation – e.g., dolphin-safe tuna – and is further addressed in the next section.

Conclusions about supply

Globally, aquaculture provides the growth potential for the fish and seafood industry. Many product categories will likely become dominated by aquaculture supply. Wild harvest has likely peaked, with many species in limited supply. Comparing the expected 2.5 percent annual growth rate of demand with the expected 1.5 percent rate for supply suggests that demand pressure will create more challenges for fisheries sustainability in the future.

In the supply chain, retailers are the key link for coordinating market signals about consumer demand. The grocery retail channel calls for increasing consistency of quantity, quality, and price. The hospitality industry moves 70 percent of fish and seafood to consumers and stimulates much of the innovation in product. Retail efforts to increase the variety and availability of fish and seafood products will increase the needed supply volume of preferred species, with additional demand created for those not traditionally preferred. Brand manufacturers have some power in the chain but depend on the retailers' market signals. Small processors and harvester/producers are fragmented and will likely respond to signals from the large players in the chain. Management practices throughout the chain are likely dominated by price-driven efficiency concerns as well as delivering on specific consistency demands for quality and quantity. The needs of retailers for fresh, continuously available, quality product will likely create more need for tight coordination of the supply chain and innovation in production and processing methods. One implication of the retailers' needs is that aquaculture will become more important because it can produce a more standardized and readily available product. At best, supply chain signals consistent with sustainability attributes would seem driven by the particular differentiation approach of a key retailer rather than travel generally or easily across the chain. Many of these conclusions about the supply chain must be made tentatively, given that much of the fish and seafood chain has not been systematically studied.

SUSTAINABILITY

The issues

The case has so far been made that demand for fish and seafood products will likely continue to expand, perhaps dramatically, in many parts of the world. On the supply side, wild catch in total is not likely to grow, meaning that most growth in supply will have to come from aquaculture. Further, the existing supply chain is largely but loosely coordinated around generalized signals about consistency in quantity and quality of supply and about cost efficiency to keep price in line with buyer and consumer expectations. What impacts will these key conclusions about supply and demand have on fish and seafood sustainability?

The Food and Agriculture Organization (FAO) warns that rising demand puts ever more pressure on threatened stocks of wild fish and seafood. Most of the global wild fisheries are being exploited. Seventy percent of the world's fish stocks are exposed to intensive fishing, and many stocks are close to collapse; U.S. waters appear to be overfished and habitat degraded. Though increasing demand for seafood is being satisfied by traditional wild stock harvest and aquaculture, both of these production approaches can be practiced in ways that raise questions about their sustainability – i.e., meeting today's needs without degrading the possibility of meeting future needs (Haland 2002). More specifically, aquaculture, while preserving wild stock, creates its own impacts on the environment through changes in land and water use for alternative activities including recreation, potential water pollution, spread of diseases, and increased demand for wild fish species used as feed for farmed fish.

In the light of these threats to sustainability, it is rather alarming that nearly one-third of wild fish harvest is reduced to oils and feeds for terrestrial animals, pets, and farmed fish. Moreover, nearly one-quarter of all wild fish harvested, about 20 million tonnes, is discarded as non-target species (so called bycatch) (Delgado *et al*. 2003). Finding alternative economic uses for bycatch may actually discourage adoption of technologies and harvest management techniques designed to reduce this waste. Apart from bycatch, fishing techniques themselves, such as bottom-trawling, and blast and poison fishing, also contribute to disruption of whole ecosystems, including seabirds and marine mammals (Delgado *et al*. 2003).

Limited capacity to expand wild catch does exist in approximately one-quarter of the world's marine wild fisheries located in the tropics. These areas are not yet fully exploited and can provide growth opportunities to harvesters with local fishing rights. On the one hand, this new output is going to be, in many cases, of low quality or value and thus used to satisfy increasing domestic demand. On the other hand, when this new output is of high quality or value, it will likely be exported to wealthy nations. In the developed world where little opportunity exists for expansion of wild catch, harvesters will gradually either leave the sector or integrate with aquaculture as the developed nations establish friendly import regimes and defend their waters and wild stock from further exploitation (Delgado *et al*. 2003). For example, recently popularized orange roughy is being harvested farther from the shore and deeper than ever before, causing its local extinction.

The solutions to these sustainability issues will likely take a number of forms, including government policy intervention and various industry strategies that use consumer responsiveness to enhance sustainability. The possible consumer-responsive strategies are of particular interest given the themes of this chapter, and they come in two forms: eco-labeling and aquaculture.

Government policy

Declines in wild fish and seafood populations may have the perverse effect of heightening the exploitation of the dwindling resource. Declining catch results in high prices for what is caught. The higher prices, coupled with the reduced long-term potential, may in turn limit the incentives for private firms to invest in innovation, new products, and environmentally friendly technologies (Trondsen 1997). Ultimately, government policy intervention will likely be essential if sustainability is to be achieved in the current economic climate. Such intervention will likely have to encompass changes in trade policy, harvest regulations, and fishing rights. Any one of these changes is complex in its own right to negotiate, implement, and then monitor compliance. When changes in all three areas must be addressed, the challenges become ever more daunting to a policy solution.

Just one of many possible examples of the policy issues involved provides evidence of the challenges. Government subsidies can distort global trade flows and increase threats to sustainable fisheries in that subsidies can result in fishing operations taking place when they would otherwise not be profitable to pursue. As a result, the subsidies potentially provide incentives to overfish or engage in other unsustainable fishing practices. As reported by Mintel (2004), Japan, China, Korea, Taiwan, and the European Union (EU) provide annual subsidizes of $15 billion to $20 billion to their commercial fishing fleets. Friends of Fish – which includes the United States, Australia, Chile, Iceland, New Zealand, the Philippines, Ecuador, and Peru – has forced negotiations through the World Trade Organization (WTO) to curtail these subsidies. The outcome of these negotiations is not yet clear, but the extent (quantity and value) of fisheries trade involved is highly significant.

Creating policy solutions demands significant public and industry support. For example, most Americans favor international treaties to protect the oceans and say they are willing to eat less of those fish and seafood categories threatened by overfishing (McClure 2004). Seventy-two percent of survey respondents agreed that protecting oceans is best done globally through treaties. Support for global treaties is critical because policies pursued by individual countries may not be effective when done alone. Sixty percent said they were willing to eat less of certain kinds of fish to help improve ocean health. Fifty-six percent agreed the government should spend money on research to reduce pollution. On the other hand, only 47 percent favored government regulation restricting use of the seashore, and only 46 percent supported local efforts to reduce business and economic development of coastal areas (McClure 2004). Getting public support for a full range of sustainability policy incentives would thus seem to be problematic. Industry support will also be questionable unless policy helps support the needed economic investments and process transitions to move from current methods of production to more sustainable ones.

Consumer responsiveness: eco-labeling and traceability certifications

The only alternative to government-mandated solutions is some form of market-oriented and thus consumer-responsive solution. Such a solution requires a market or consumer-demand motivation, proper signaling of the demanded product or service attributes throughout the supply chain, and a coordinated response by suppliers to deliver the attributes demanded. The demand motivation for a solution is relatively clear from the prior analysis in the chapter. For example, American consumers are concerned about the health of their diet and the health of the environment. More than 100 U.S. fish stocks are suffering from overfishing. Consumers shopping for seafood caught or farmed in an environmentally sustainable manner can make a difference to them personally and to the environmental impact on fisheries (Environmental Defense 2002). Business strategies for retailers (grocer and hospitality), processors, and harvesters/producers could potentially be effective in meeting these consumers' needs.

Eco-labeling is one specific strategy for meeting consumer demands in this arena. An eco-label becomes the signal around which the supply chain can be coordinated to deliver the sustainability attribute for the end consumer. Once the label is defined with proper monitoring for compliance, each firm in the supply chain can participate in the harvesting, farming, or processing activities implicit in receiving the label. A survey by the National Fisheries Institute found that one-third of consumers are familiar with and support eco-labeling (Mintel 2002a). The purpose of eco-labeling is to utilize consumer market power to provide incentives for better management practices of fisheries. Dolphin-safe tuna is a positive example of a successful management practice adapted by fisheries to differentiate their product from foreign competition and to please the growing concern of consumers for wildlife preservation and non-targeted species (bycatch) disposal.

In addition to environmental issues, consumer food safety concerns have highlighted the importance of traceability and the certification of the supply chain that a product has moved through from catch to plate. Such certification is already well established in the EU. The MSC certificate on products that move through the chain of the Marine Stewardship Council (MSC) custody documentations assures U.S. consumers about the safety, authenticity, and wholesomeness of the fish and seafood products (Johnson 2002). The MSC certificate has been the only international seafood label that provides traceability back to the source. The Alaska salmon fishery was the first one to earn such a certificate for salmon, and now about 100 Alaska salmon products are carrying the MSC eco-label. Whole Foods became the first U.S. retailer to promote the label, and Xantera Parks and Resorts became the first food service operation promoting eco-labeled Alaska salmon on its menu. Now Vital Choice Seafoods, Thompson Outfitters, SeaBear Smokehouse, and others have chain of custody or traceability certificates (Marine Stewardship Council 2002).

Consumer response to eco-labeling depends on the market credibility of the entity issuing such a label. Consumers in the United States tend to trust governmental certificates; in some countries (such as Denmark), Marine Stewardship Council and World Wide Fund for Nature certificates are preferred (Pickering *et al.* 2002); U.S. consumers are found to be less likely to choose a certified product than Norwegian consumers when on a low budget, when considering salmon or when female; U.S. consumers who are educated, belong to environmental organizations, and are facing a high price for a certified product are more likely to choose certified products than Norwegians; U.S. consumers with high incomes and environmental consciousness are equally likely as Norwegians to choose certified product (Donath 2000).

A study by Roheim and Donath (2003) also found that eco-labeled fresh seafood is favored over non-labeled even with a price premium. However, the study does not conclude any intent of consumers to cross between species and switch from non-labeled favorite fish to less popular eco-labeled alternatives. This finding may be due to the fact that taste buds matter more to consumers than eco-labeling, and it raises a critical question about how far eco-labeling can go to increase sustainable consumption.

A far weaker form of eco-labeling uses Internet-based lists or ratings of fish and seafood products that either meet or do not meet certain sustainability standards. The Audubon Society has established the Audubon Fish Scale, which makes it easy to distinguish which seafood choices are abundant, and well managed and have few problems associated with fishing methods (Seafood Choices Alliance 2002c). The Seafood Choices Alliance also has established a website called *The Fish List* that indicates species (either wild or farmed) in combination with harvesting/raising practices and rate them "enjoy" or "avoid" (Seafood Choices Alliance 2002d). The impact of such ratings is not clear. Members of either the Audubon Society or the Seafood Choices Alliance would have strong reason to follow the ratings, but whether retailers, processors, and harvesters/producers find this approach compelling or economically viable for supply chain relationships cannot be known without more study.

Country-of-origin labeling (COOL) is another form of traceability certification. Like the Internet lists just mentioned, COOL is a weaker form of eco-labeling. The buyer must understand the probable correlation between a particular source country and the likelihood of sustainable fishing practices. COOL may be more useful to predict safety attributes of products than sustainability attributes. The United States requires COOL for fish and seafood products. In addition, the label must distinguish between wild catch and farm raised. The EU also requires COOL for seafood products.

Perhaps the most authoritative work to date from the university research community on the topic of information economics in the market for seafood attributes is by Wessells (2002). She presents an appropriate conceptual framework for studying the issues and reports on the state of past empirical research on the topic. She concludes the following:

> In the empirical studies discussed above, there is a demand for information on product quality, seafood safety, environmental friendliness, and even the name of broker of the individual bluefin tunas sold in Tsukiji market. The analyses indicate that these attributes are valuable to the consumer, and they are willing to pay for them. However, demand is subject to the nature of the attribute. In other words, there are some attributes consumers may attach value to, and some they may not . . .
>
> Although not included in the list of empirical studies, the supply side of the market for attributes also exists, but there are no empirical studies to date examining this portion of the seafood market. We know from the economic framework that producers may supply attributes and inform consumers of them only when the private marginal costs are less than the marginal benefits. (Wessells 2002:161)

In other words, the potential to develop, implement, and succeed at consumer-responsive strategies that utilize incorporation and promotion of sustainability attributes for fish and seafood makes such strategies appear promising, but little is known about how effective these strategies are in practice. Most especially, the gap in knowing how and why the supply chain responds is a critical gap indeed. More research into this entire area is much needed.

Consumer responsiveness: aquaculture as the sustainable alternative

As presented through statistics earlier in the chapter, aquaculture has already become the alternative to wild catch. Its growth rate as a source of supply has been dramatic. Aquaculture has relieved pressure on threatened wild stocks while providing a ready source of product alternatives to meet growing consumer demands. In addition to matching species grown to demand, aquaculture allows for just-in-time slaughter and processing that heightens consumer responsiveness for attributes such as freshness. Aquaculture was the fastest growing U.S. agricultural subsector during the 1980s, with a 265 percent increase in production (Chopak 1992). Aquaculture is currently the fastest growing method of food production in the world (Haland 2002).

Aquaculture's growth will be further enhanced by the rising prices for the dwindling supply of wild catch. Wild fish and seafood are very likely to be more costly than other food products. In this economic environment, wild harvesters should focus on maximizing the long-term value of the wild catch, while aquaculture producers focus on meeting the demand growth that would otherwise go unfulfilled. Because of the production economics and harvest regulations, in the long run the competition between wild catch and aquaculture could lead to wild stock recovery. Tighter coordination of production could even develop between the wild harvest and aquaculture through joint production strategies, e.g., hold previously harvested wild catch in aquaculture facilities until ready for sale.

As more types of fish become farmed and regulatory standards are further developed, the total supply of fish and seafood will become more readily available and stable. As a result, prices may not fluctuate quite as wildly as in the past, though it will probably be after 2010 before aquaculture has enough dominance as a portion of supply to make prices more easily predicted. As seafood prices begin to stabilize, a wider variety of products becomes available, and food safety is assured, consumers will grow comfortable purchasing products that they may have seen as luxury items before but now accept as occasional dinner options at home (Mintel 2002b).

Two limits apply to the role that aquaculture might play in fish and seafood sustainability. The first limit is consumer acceptance of farmed fish versus wild catch. If the consumer has a strong preference for wild caught, aquaculture will have much less impact on the sustainability of the wild catch. The results of consumer surveys on consumer acceptance are mixed. A National Fisheries Survey revealed that 34 percent of respondents disagreed with the statement that no difference exists between wild and farm-raised salmon, and 45 percent did not know there is a difference (Mintel 2002a). According to the same survey, 26 percent preferred naturally caught fish and 25 percent preferred farm-raised fish. Nearly half agreed that aquaculture is a good alternative to wild catch; only 4 percent disagreed. More than half did not know if aquaculture contributes to ocean pollution, and only 20 percent agreed that it does (Johnson 2002). In another survey (Bureau of Seafood and Aquaculture n.d.), Hispanics strongly preferred wild (47 percent) over farmed (4 percent) with 31 percent being neutral; 24 percent strongly preferred U.S.-caught and 53 percent did not care about country of origin. Consumers in the United Kingdom and Denmark have a general preference for wild-caught over farmed fish and hold negative attitudes toward farmed fish as being less natural. This perception hinders willingness to label products as farmed and alternative names are used (Pickering *et al.* 2002).

The second limit to aquaculture relates to whether aquaculture is itself sustainable as a production approach. Expanding aquaculture will face competition for land and water resources used for other activities. Increasingly scarce sources of fresh water will make aquaculture expansion more challenging. Consumers are also aware of issues of heavy metals, antibiotic residue, and bacteria content in aquaculture. For example, some farmed salmon have been shown to provide ten times more contaminants such as dioxins, PCBs, and other toxins as wild salmon as a result of differences in diet (Seafood Choices Alliance 2002a).

Other constraints include disease, contributions to water and land pollution, lack of fish feed derived from wild fish, and concerns over escaped modified or exotic fish. One of the first studies of its kind tested the ecological impact of modified fish escaping from an aquaculture farm. This preliminary study found that all the salmon (wild and modified) thrived as long as there was enough food to go around, but faced with food shortages, modified (by growth hormone in this case) individuals in the mixed group outcompeted their wild mates (Schirber 2004).

If aquaculture is to be *the* sustainable alternative to wild catch, then its farming methods must prove to be sustainable in their own right. Selective breeding, appropriate genetic modification, alternative feeds, farming the most desired yet endangered species, water and population control, improved feed conversion efficiency, and use of bycatch will all need to be practiced to assure the sustainability of aquaculture (Delgado et al. 2003). One might anticipate that an organic production standard for aquaculture could be readily developed and used to enhance sustainability. This strategy would also heighten product differentiation for fish and seafood so produced.

CONCLUSIONS

The interrelationship between fish and seafood demand, supply, and sustainability is a complex story complete with a number of opportunities and challenges. Trends in consumer demand are clearly positive for the fish and seafood industry. Globally, demand will rise in the developing world and in the developed world. Fish and seafood products provide diets with basic protein, very healthful alternatives to other protein sources, and significant variety. The retail and hospitality industries, acting out of self-interest, have every incentive to make fish and seafood products more convenient, tastier, more appealing to prepare, and more prevalent.

As consumers come to expect consistency, availability, quality, and reasonable price, more pressure will be put on the fish and seafood supply chain to be more consumer responsive, with all the implications for production, processing, and supply chain relationships and alliances. In fact, with the peaking of wild catch, meeting the demands of consumers will be increasingly difficult without the presence of a viable aquaculture industry throughout the world. In response to this difficulty, incentives may exist to threaten wild catch sustainability even further, pulling even more volume from the seas or altering local developing country diets in less favorable directions to support exports to developed nations. Food safety and pollution concerns may be heightened as further expansion of supply results in production from more marginal areas or by more threatening methods.

Government regulation will no doubt be needed to stem some of the more abusive practices on the production side. However, the very consumer demand that causes some significant part of the problem may be at the heart of the solution. Consumers have desires for healthy, safe, and environmentally friendly sources of food. Eco-labeling and a sustainable aquaculture can result in enhanced wild populations of fish and seafood. Market opportunities clearly exist for those firms that want to pursue either sustainable wild product or sustainably farmed product. In either case, coordination throughout the supply chain will be essential to delivering the product that consumers want with the certifications needed to prove the desired attributes are present. If the fish and seafood industry goes fishing for its consumers in the most effective way, then the industry has great potential to find compatibility between its economic needs and sustainability of wild fisheries. However, the sustainable outcome desired appears to be only a potential for now.

This chapter has attempted to collect and synthesize the information available to explore its theme. The work of preparing the chapter has convinced the authors that a need exists for improved worldwide data collection of fish and seafood statistics as well as a need for more research in a number of areas. First, the depth and breadth of markets for product produced in a sustainable manner are not well known. Many consumers may not be well informed at this point, and the value trade-off between sustainability attributes and the price consumers are willing to pay is not well defined. Both need sound market research. Research into sustainable production practices for both wild catch and aquaculture must be pursued as a foundation for creating best practices. Finally, issues of social equity have not been addressed here. The differential impacts of altering fish and seafood production and marketing strategies on lower-income consumers, either in developed nations or in developing ones, are not well known but should be studied. Generally speaking, the articulation of a comprehensive research agenda and the pursuit of specific research projects under this agenda are much needed in the area of consumer-responsive strategies consistent with fish and seafood sustainability.

References

Aquaculture Production Technology. 2006. *World Trade and Future Consumption: Seafood and Aquaculture Production*. Available online at www.aquaculture.co.il/Markets/markets.html

Bisaillon-Cary, J. and Meser, J.R. 2004. *A Seafood Resource Guide for Exports to the E.U.* Portland, ME: Maine International Trade Center. Available online at www.seafood-norway.com/?obj=123&title=Import%20/%20export%20guide&lang=en

Bureau of Seafood and Aquaculture. n.d. *National Market Analysis of Hispanic Consumer Attitudes towards Seafood and Aquaculture Products*. Available online at www.fl-seafood.com/industry/research_reports.htm

Chopak, C. 1992. *What Consumers Want: Advice for Food Fish Growers*, MSU Extension Bulletin No. E-2410. East Lansing, MI: Michigan State University.

Delgado, C., Wada, N., Rosegrant, M., Meijer, S., and Ahmed, M. 2003. *Fish to 2020: Supply and Demand in Changing Global Markets*. Washington, DC: International Food Policy Research Institute (IFPRI) and World Fish Center publication. Available online at www.ifpri.org/pubs/books/fish2020/oc44front.pdf

Donath, H. 2000. Consumer preferences for eco-labeled seafood in the United States. *Proceedings of the Meeting of the International Institute of Fisheries Economics and Trade*, July 10–14, 2002, Corvallis, OR. Available online at http://oregonstate.edu/dept/IIFET/2000/papers/donath.pdf

Environmental Defense. 2002. *Seafood Selector: A New Interactive Tool*. Available online at www.environmentaldefense.org/article.cfm?contentid=1822

FAO (Food and Agriculture Organization). 1997. *The State of World Fisheries and Aquaculture*. Rome: Food and Agriculture Organization of the United Nations.

Haland, B. 2002. *Clean Conscience Consumption of Seafood? Eco-labeling*. Oslo: World Wildlife Fund. Available online at www.wwf.no/pdf/ecolabels_for_seafood.pdf

Huffman, L. 2003. *Introduction to Hospitality: Restaurant, Hotel and Institutional Management*. Available online at www.hs.ttu.edu/RHIM2210/html/introduction/tsld002.htm

Johnson, H. 2002. *Market Outlook in the International Fish and Seafood Sector: Alternative Products/Uses and Food Safety Issues*. Available online at www.salmonfarmers.org/resources/Studies11.pdf

Lindsay, J. 2004. Some frozen lobsters return to life. *Associated Press* March 14, 2004. Available online at www.xradiograph.com/wp/index.php?p=3

Marine Stewardship Council. 2002. *MSC Eco-label Helps Consumers Identify Certified Wild Alaska Salmon: Traceability*. Available online at www.msc.org/html/ni_108.htm

McClure, R. 2004. Action to save oceans backed. *Seattle Post-Intelligencer* February 16, 2004. Available online at http://seattlepi.nwsource.com/local/160810_oceans16.html?searchpagefrom=1&sepercent0Aarchdiff=2

Mills, S. 2001. Let them eat seafood. In *Restaurants USA*. Available online at www.restaurant.org/rrusa/magArticle.cfm?ArticleID=15

Mintel. 2002a. *Fish and Seafood U.S. Report*. Chicago, IL: Mintel International Group.

Mintel. 2002b. *Dining out*, vol. 1, *QSR – U.S. Report*. Chicago, IL: Mintel International Group.

Mintel. 2004. *Fish and Seafood U.S. Report*. Chicago, IL: Mintel International Group.

NMFS (National Marine Fisheries Service). 2004. *Annual Report to Congress on the State of U.S. Fisheries – 2003*. Silver Spring, MD: National Marine Fisheries Service, U.S. Department of Commerce. Available online at www.nmfs.noaa.gov/sfa/statusoffisheries/statusostocks03/Report_Text.pdf

NOAA/NMFS (National Oceanic and Atmospheric Administration/National Marine Fisheries Service). 1998. *Ocean Facts on Sustaining Marine Fisheries*. Available online at www.yoto98.noaa.gov/facts/resource.htm

Pickering, H, Jaffry, S., Whitmarsh, D., *et al.* 2002. Seafood labeling trends. *Proceedings of Aquachallenge Workshop*, April 27–30, Beijing. Available online at www.aquachallenge.org/programme/programme_menu.html

Roheim, C. and Donath, H. 2003. *Battle of Taste Buds and Environmental Convictions: Which One Wins?* Kingston, RI: University of Rhode Island. Available online at www.farmfoundation.org/projects/documents/Roheim.pdf

Schirber, M. 2004. GM salmon muscle in on wild fish when food is scarce. *Scientific American*. Available online at www.scientificamerican.com/article.cfm?chanID=sa003&articleID=000A483E-ECB8-10C4-ACB883414B7F0000

Seafood Choices Alliance. 2002a. Chefs demand a smarter supply. *Afishianado, Seafood Choices Alliance Newsletter*. Available online at www.seafoodchoices.com/resources/afishionado_pdfs/Afishionado%20Fall%202002.pdf

Seafood Choices Alliance. 2002b. Fish market trends to watch: sourcing ocean-friendly seafood. *Afishianado, Seafood Choices Alliance Newsletter*. Available online at www.seafoodchoices.com/resources/afishionado_pdfs/Afishionado%20Fall%202002.pdf

Seafood Choices Alliance. 2002c. Conservation corner: in-depth profile – Audubon's living oceans. *Afishianado, Seafood Choices Alliance Newsletter*. Available online at www.seafoodchoices.com/resources/afishionado_pdfs/Afishionado%20Fall%202002.pdf

Seafood Choices Alliance. 2002d. *The Fish List*. Available online at www.seafoodchoices.com

Silver, D. 2003. Food service focus: ocean of options. *Food Product Design*. Available online at www.foodproductdesign.com/archive/2003/0703FFOC.html

Smet, T. 2003. *The Different Definition(s) of Hospitality and Tourism in Relation to Economic Studies*. Available online at www.timsmet.com/

Tiu, L. n.d. *Aquaculture White Paper*, Ohio State University Aquaculture Program. Available online at http://southcenters.osu.edu/aqua/white.htm

Trondsen, T. 1997. Value-added fresh seafood: barriers to growth. *Journal of International Food and Agribusiness Marketing* **8**(4): 55–78.

Wessells, C.R. 2002. The economics of information: markets for seafood attributes. *Marine Resource Economics* **17**: 153–162.

WRAC (Western Regional Aquaculture Center). 1998. *Western Region Aquaculture Industry Situation and Outlook Report*, vol. 6. Available online at www.fish.washington.edu/wrac/images/PART4.PDF

WILLIAM KNUDSON AND H. CHRISTOPHER PETERSON

18

Globalization and worth of fishery resources in an integrated market-based system

INTRODUCTION

Fish and seafood products are a significant part of the global agrifood sector. Fishery products are an important source of protein, especially for low-income food-deficit countries (LIFDCs) (FAO 2003a). Furthermore, world trade in fishery products continues to increase. Much of this trade is north–south, providing an important source of foreign exchange for low-income countries.

Several developing trends will affect the global fishery industry. Though capture fisheries (using fishing techniques to harvest wild fish and seafood products) remain responsible for the majority of fish and seafood product output, aquaculture is a large and growing part of the market. One of the factors contributing to the growth of aquaculture is the growth of supermarkets throughout the world. These markets require a standard product and a stable supply. Another factor affecting the seafood industry is the growth of eating away from home. In the United States, more than half the seafood eaten is consumed in restaurants, and restaurants also depend on a consistent product and a stable supply (Mintel 2002). Aquaculture is well suited to meet these requirements.

Because it appears unlikely that additional output from capture fisheries is possible, the potential for increased international trade in aquaculture products is great provided that the threats facing aquaculture can be addressed. As the aquaculture industry matures, it is increasingly taking on the same characteristics as terrestrial agriculture, with the attendant subsidies and potential trade restrictions designed to protect domestic producers. This situation presents a potential threat to increased international trade.

Globalization: Effects on Fisheries Resources, ed. William W. Taylor, Michael G. Schechter, and Lois G. Wolfson. Published by Cambridge University Press. © Cambridge University Press 2007.

This chapter will provide a general overview of the world fishery industry with respect to the capture fish industry. Three major forces for increased globalization and integration will also be discussed: the growing role of aquaculture, the globalization of the retailing sector, and the growing role of international trade. Future trends will also be discussed.

THE CAPTURE FISH INDUSTRY

Production

The capture fish industry remains an important economic activity throughout the world. Table 18.1 outlines the global level of fish production from 1996 to 2002. During this period, capture fishery production ranged from 88.1 million tonnes in 1998 to 96.8 million tonnes in 2000. Total world production ranged from 118.7 million tonnes in 1998 to 134.4 million tonnes in 2002. From 1996 to 2002, capture fishery production increased by 0.9 percent and aquaculture production increased by 49.6 percent.

These figures include production from China. Excluding China, total capture output was 77.7 million tonnes in 2002, and total aquaculture production was 22.9 million tonnes for a total of 100.6 million tonnes (FAO 2002b). It is conceded that China is the world's largest producer of seafood, but there has been some question about the accuracy of its statistics (*The Economist* 2003). It appears that the total amount of world fishery production may be overstated.

Though capture fish production is flat, and may be at its full potential level of output (Valdimarsson 1998), it is still the major source of seafood production. In 2001, it accounted for 70.9 percent of all seafood production. Inland fisheries accounted for 9.6 percent of capture fishery production in 2001, and ocean capture accounted for 90.4 percent.

Table 18.1 *World fishery production, 1996–2002 (million tonnes)*

	1996	1997	1998	1999	2000	2001	2002
Total capture	93.7	95.7	88.1	95.0	96.8	94.2	94.6
Total aquaculture	26.7	28.7	30.6	33.4	35.5	37.8	39.8
Total	120.4	124.4	118.7	128.4	132.3	132.0	134.4

Source: FAO (2003a).

Table 18.2 *Icelandic catches of non-traditional species (tons)*

Species	1989	1992	1996
Grenadiers	2	210	808
Starry ray	99	317	1493
Greater silver smelt	8	657	808
Deep-sea rosefish	1374	13 845	52 994
American plaice	565	1468	7027
Dab	2233	3044	7954

Source: Valdimarsson (1998).

One reason that the capture fish catch has been relatively stable is the fact that fishers have been substituting less desirable species for increasingly scarce and more desirable species. For example, shark is becoming more common as a seafood item as a result of the reduction in harvests for species such as cod (Seafood Choices 2005). Iceland, a major seafood-producing country, has also seen an increase in the harvest of non-traditional species, as Table 18.2 illustrates.

The substitution of non-traditional species for traditional species brings up two points. First, what was once bycatch (species that were caught in the process of capturing valuable species and traditionally thrown away) is increasingly becoming commercially viable. This situation has the benefit of increasing the efficiency of capture fishing activities. Second, simply looking at total output may be misleading; that is, total capture fishery output may be stable, but the population of the most commercially desirable species may continue to decline. It has been estimated that 47 percent of the main fish species are fully utilized (catches at or near their sustainable limits), and another 18 percent are overexploited (catches greater than their sustainable limits) (FAO 2002a), meaning that the populations of these species will decline in the future.

Another stabilizing factor in the capture industry is the widespread adoption of the 200-mile (320-km) limit. This limit has allowed individual nations to control the size and distribution of the fish catch. This, in turn, has helped maintain the fish catch at more sustainable levels. However, the development of large fishing vessels capable of processing fish while at sea is putting pressure on smaller, less efficient fishers. Fish in those countries that attempt to maintain a traditional fishing industry will be at a cost disadvantage to those nations that allow their fishers to use the latest technology and fully exploit economies of scale. Even if the capture fish industry is able to maintain catches at their current level, fewer people will be employed in the industry in the future.

Consumption

Seafood products remain an important source of protein for the world's population. Globally, seafood accounted for approximately 16 percent of animal protein intake and likely exceeds 20 percent for some LIFDCs (FAO 2002a). Seafood has the advantage of being the most natural source of protein available because wild fish select their own food (Valdimarsson 1998). Food animals raised in captivity eat what their owner makes available to them. Table 18.3 shows the world utilization of seafood products from 1996 to 2001. The figures in this table include China and, as previously discussed, may be overstated.

This table shows two major aspects of the global fishery industry. The first is increasing levels of human consumption. Seafood products used for human consumption increased by 13 percent from 1996 to 2001. This increase can be explained by two factors: increasing incomes raise the demand for seafood products in industrialized societies (particularly for restaurant-prepared meals), and increasing world population increases the demand in developing countries.

Table 18.3 also points out that a fairly large portion of the total production of fishery products is devoted to non-human consumption. From 1996 to 2001, the use of seafood products for non-food uses varied from 26.8 percent in 1996 to 21.3 percent in 1998. This variation in non-food use is likely due to variations in the anchovy catch (Vannuccini 2003). As a general rule, approximately 25 percent of the world's fishery catch is used for non-food consumption. The primary non-food use

Table 18.3 *World utilization of seafood products 1996–2001 (million tonnes)*

Utilization	1996	1997	1998	1999	2000	2001
Human consumption	88.0	90.8	92.7	94.4	96.7	99.4
Non-food uses	32.2	31.7	25.1	32.2	33.7	29.4
Total	120.2	122.5	117.8	126.6	130.4	128.8

Source: FAO (2002a).

is fishmeal used in terrestrial animal feed and as an aquacultural feed for species such as salmon and trout.

FORCE ONE: THE ROLE OF AQUACULTURE

The growth of aquaculture

In 2002, aquaculture accounted for 29.6 percent of all fishery production. In 1970, aquaculture accounted for only 5.3 percent of all production (FAO 2003a). The rate of growth from 1970 to 2000 averaged 8.9 percent per year. During the same time period, capture fisheries increased by 1.4 percent per year, and terrestrial meat production increased by 2.8 percent per year (FAO 2003a). Some estimate that aquaculture will become the largest source of seafood by 2030; this situation may already be the case in the United States (*The Economist* 2003).

Aquaculture has also become a major industry. In 2002, the size of the industry worldwide was $60 billion. Table 18.4 breaks down the value of world aquaculture by category. Finfish are the dominant category, accounting for 59.5 percent of all aquaculture sales. Crustaceans, particularly shrimp, account for 20.1 percent of all aquaculture sales. Shrimp is now the largest seafood species in the U.S. in dollar terms (Pritchard 2003). Increasingly, U.S. consumers obtain their shrimp from farm-raised sources, particularly from southern Asian countries.

A definite differential exists between those nations that use marine areas for aquaculture production and those that use inland waterways. Developed nations tend to specialize in marine aquaculture; developing countries tend to specialize in inland aquaculture

Table 18.4 *Value of aquaculture production 2002 (U.S.$ billions)*

Category	Value
Finfish	32.0
Mollusks	10.5
Crustaceans	10.8
Others	0.5
Total	53.8

Source: FAO (2003a).

Table 18.5 *Aquaculture fishery production 2002*

Marine areas		Inland fishing areas	
Country	Production (tonnes)	Country	Production (tonnes)
China	560 404	China	16 370 361
Norway	551 332	India	2 046 234
Chile	479 132	Indonesia	731 979
Japan	268 396	Bangladesh	721 025
United Kingdom	146 930	Vietnam	390 000
Canada	127 621	Philippines	359 322
Greece	63 059	Egypt	340 556
Taiwan	52 880	United States	330 869
Faeroe Islands	50 946	Thailand	295 064
South Korea	48 073	Taiwan	190 013
Total	2 348 773	Total	21 775 423
Share of world total	90.2%	Share of world total	94.2%

Source: FAO (2003a).

production (as also does the United States). Table 18.5 gives the top ten producing nations for both marine and inland aquaculture areas in 2001. The table shows China's dominance in both types of aquaculture. This dominance holds true even if the Chinese figures are overstated.

For the most part, major producers of inland fisheries are developing countries. The most obvious exception is the United States with its well-developed catfish industry. Furthermore, 93 percent of finfish aquaculture production in developing countries consists of omnivorous/herbivorous and filter-feeding fish species (FAO 2003a). Some groups and organizations believe that this type of aquaculture is more sustainable and environmentally responsible than the production of carnivorous species.

Developed economies are much more involved in marine aquaculture. Japan, Norway, Spain, and the United Kingdom are major producers and consumers of seafood products. Much of inland aquaculture is carried out by small enterprises, but much of marine aquaculture is carried out by large firms that use capital-intensive, labor-saving technologies. Another major difference between developed and developing countries is the fact that 73.8 percent of total aquaculture finfish production in developed countries consists of carnivorous varieties such as salmon and trout (FAO 2003a). The production of this type of species has been criticized for adverse effects on the environment.

The increased dependence on aquaculture may reduce pressure on wild fish populations. Increased aquaculture production could put sufficient downward pressure on prices to render traditional commercial marine fishing inefficient and lead to an eventual restoration of some species that are currently overfished.

Threats to aquaculture growth

Because of its ability to deliver a consistent level of supply and consistent quality, the retail/supply chain system favors aquaculture. Many threats to the aquaculture industry may need to be addressed, however. The first of these is environmental. One issue is the destruction of habitat from waste and management practices. Shrimp and salmon farming are frequently cited as industries that have poor environmental records (*The Economist* 2003). Another issue is genetically modified fish escaping into the wild. It is feared that the cross-breeding of genetically modified fish with wild fish will adversely affect wild fish stocks. Additional safeguards and improved management techniques are likely to be required if the industry is to grow, especially in developed countries that have effective environmental regulations.

Another threat is the possible effects on human health. Many aquaculturlists use antibiotics as a prophylactic measure to prevent disease (FAO 2003a). The effects of antibiotic residue are an important issue that is just beginning to be addressed by the aquacultural community. Some fear that antibiotic residues will reduce the long-term effectiveness of these antibiotics. Additional health concerns are polychlorinated biphenyls (PCBs) and other toxins in fish. It appears that the level of PCBs is higher in farm-raised salmon than in wild salmon (Hites *et al.* 2004). However, other researchers believe that the health benefits of eating salmon outweigh the risks, and that current levels of contaminants are below levels established by the U.S. Food and Drug Administration (FDA) and the World Health Organization (WHO) (Santerre 2004).

A threat related to the environment is the sustainability of raising carnivorous fish species such as salmon. These fish require more protein than do herbivorous fish such as catfish and tilapia. This requirement raises the question of the long-term viability of feed species such as anchovies and sardines. However, the issue may be addressed by improving breeding techniques and so improving feed conversion rates, transferring fishmeal from hog and poultry feed to fish feed, using bycatch as fish feed, and increasing use of plant protein as fish feed.

These issues are a source of some debate in the scientific community. Some environmentalists believe these are serious issues; some scientists and others believe the dangers are overstated. Effective adoption of technology coupled with improved management techniques should allow the industry to address most of these threats in a way that allows the industry to develop. If not, the industry will likely face further regulation and the increased use of health warnings by various health and environmental organizations.

The changing structure of aquaculture

There is strong evidence to suggest that aquaculture is beginning to develop some of the characteristics of terrestrial agriculture. That is, fewer larger farms are beginning to dominate. The primary reason for this dominance is the existence of economies of scale. Given that fish and seafood products are commodities, each producer's output is a perfect or nearly perfect substitute for another producer's. Those firms that produce at the lowest cost will be successful at the expense of other firms. Aquaculturalists have control over breeding, feeding, and size of the fishing stock that capture fishers do not. This control means that aquaculturalists have the means to reduce their costs. For example, it has been estimated that the operating costs per kilogram of salmon in Norway declined by two-thirds in real terms over a 15-year period (FAO 2002a).

This trend also appears to be true for the U.S. aquaculture industry. Table 18.6 provides the breakdown by sales class for catfish and trout. Similar figures exist for other types of aquacultural output. It should be noted that, to be considered a farm by the U.S. Department of Agriculture, an enterprise must generate at least $1000 in revenues.

In catfish production, the largest 103 farms account for 59 percent of sales, and those farms that have sales at or above $500 000 account for 78 percent of all sales. The average sale per farm for those operations with $1 million or more in sales is $2.6 million. The situation for trout is similar. Trout farms with more than $500 000 in sales account for 63 percent of total trout sales.

The trend also exists in the salmon industry. Firms adopting cost-reducing technologies have increased production, leading to a

Table 18.6 Structure of U.S. catfish and trout aquaculture operations, 1998

	Sales class					
	$1000 to $24999	$25 000 to $49 999	$50 000 to $99 999	$100 000 to $499 999	$500 000 to $999 999	More than $1 000 000	Total

	$1000 to $24999	$25 000 to $49 999	$50 000 to $99 999	$100 000 to $499 999	$500 000 to $999 999	More than $1 000 000	Total
Catfish							
Number of farms	515	112	165	354	121	103	1370
Total sales	$3 550 000	$3 916 000	$11 827 000	$78 779 000	$84 128 000	$268 510 000	$450 710 000
Percent of total	1	1	3	17	19	59	100
Trout							
Number of farms	333	56	64	82	17	9	561
Total sales	$2 673 000	$2 000 000	$4 731 000	$16 701 000	$12 775 000	$33 594 000	$72 474 000
Percent of total	4	3	7	23	18	45	100

Source: U.S. Department of Agriculture (1998).

reduction in prices. Smaller firms will likely leave the industry, and larger firms will continue to consolidate (*The Economist* 2004).

FORCE TWO: THE CHANGING RETAIL SECTOR

One industry change that is altering aquaculture and the fishery industry is the increasing penetration of large supermarket chains throughout the world. The enhanced stability of supply and the consistency of product that aquaculture provides make stocking seafood products more desirable to the large supermarket chains that increasingly dominate the food retail industry. These supermarkets tend to buy their product from large commercial producers (Boselie *et al.* 2003).

The trend for large supermarket chains is becoming a worldwide phenomenon. In the United States, Wal-Mart became the largest food retailer in the space of a few years. The activities of Wal-Mart and the use of mergers, including purchases of U.S. firms by foreign firms, are driving this trend toward a national retail market in the United States (Cotterill 1999). This transformation has already occurred in Europe, where supermarket concentration is much higher than it is in the United States. These large food-retailing chains are increasingly becoming involved in seafood retailing. For example, Costco, a relative newcomer to seafood retailing, now sells 15 000 tonnes of farm-raised salmon a year (*The Economist* 2004).

Though far behind Europe and the United States, food retailing in Africa, Asia, and Latin America is increasingly affected by large supermarkets. In Latin America, supermarkets accounted for 50 to 60 percent of food retail sales. In Mexico, Wal-Mart has a 30 percent market share of all food purchases (Reardon *et al.* 2003). The level of retail concentration in Asia varies from country to country. In the more advanced nations of East Asia, supermarket sales account for approximately two-thirds of retail packaged and processed food sales. The figures for the rest of Asia are lower, but the trend is toward an increasing role of supermarkets (Reardon *et al.* 2003:1142). According to a referee who reviewed an earlier draft of this chapter, Tesco Lotus, Carrefour, and others are beginning to dominate Southeast Asian food distribution. The trend is similar in Africa, although supermarkets currently play a much smaller role in this continent (Reardon *et al.* 2003).

The causes of the growth of supermarket retailing in developing regions are familiar: increased urbanization, the increase in the number of women in the workforce, and supermarkets' ability to sell processed products at lower prices. This trend has been coupled with increased foreign investment from U.S. and European retailers (Reardon *et al.* 2003). The ability to produce a consistent product and a reliable supply allows aquacultural farms and distributors to meet the requirements established by large food retailers.

The need to provide a reliable supply and a consistent product also provides an opportunity for aquacultural producers to enhance their profitability through product differentiation. Aquacultural producers may be able to earn additional profits if they are able to meet consumer preferences for taste and other attributes, such as hormone-free, antibiotic-free, and other food safety and environmental attributes that some consumers prefer and are willing to pay for. Branded products have now become common in the chicken section of retail establishments and are becoming increasingly common in the beef and pork sections as well. It is likely that this trend will spread to the seafood sections of supermarkets.

FORCE THREE: INTERNATIONAL TRADE

Size of international trade

Another aspect of the global integration of the world fish and seafood industry is the growth of international trade. International trade is an important aspect of the world's seafood industry. It is particularly important for developing countries. In 2001, total trade in fish and fishery products was approximately $59.4 billion (Vannuccini 2003). From 1992 to 2001, exports as a percentage of production increased from 34.1 percent to 38 percent (FAO 2002b). Developing countries account for 52 percent of all exports, the European Union for 21 percent, the United States for 6 percent, and others for 22 percent. The European Union is the major importing region, representing 33 percent of all imports. The other major importers are Japan, with 26 percent of all imports, and the United States, with 17 percent. Developing countries represent 17 percent of all imports (FAO 2003b).

The United States is a good example of the increasing role of international trade in seafood products. In 2002, the United States imported $19.7 billion in edible and non-edible fish products. The United States also exported $11.7 billion in these products in 2002 (Pritchard 2003). The trend of increased seafood imports is likely to continue. Salmon imports have increased by more than 100 percent, imports of tilapia have increased by 230 percent, and imports of shrimp have increased by 41 percent over the past 5 years (Harvey 2003).

One reason that the role of trade in fish products is increasing is the relatively high income elasticity of demand for seafood products. Income elasticity of demand is a measure of change in demand for a product resulting from a change in income. As incomes and the standard of living increase, the demand for seafood products increases faster than the demand for other food products. Over the past 50 years, per capita consumption of fish has almost doubled (*The Economist* 2003). The demand for seafood products will continue to increase as standards of living increase, especially for popular species such as salmon, shrimp, tilapia, and cod.

Related to income and the demand for seafood, especially, is the increase in the number of meals eaten away from home. The trend to eat more and more meals away from home will likely boost the level of international trade in seafood products. More than half the money spent on food in the United States is spent in restaurants. According to one survey, 70 percent of the amount that U.S. consumers spent on seafood was spent at restaurants in 2001 (Mintel 2002). Many U.S. consumers believe that seafood is something to be purchased and consumed away from home because of the difficulty of cooking seafood (Mintel 2002). This trend could spread to other countries as standards of living increase and two-income households become more common around the globe.

Trade issues

Though the level of international trade in fish and fishery products has increased, several things may keep the level of international trade in seafood from reaching its potential. As previously noted, the aquaculture industry is becoming more and more important as a source of seafood globally. Also, aquaculture production in the developed countries is increasingly dominated by fewer and larger farm operations. As commercial aquaculture increasingly resembles commercial agriculture, the potential for trade disputes increases.

Fish and fishery products are not covered by the World Trade Organization (WTO) Agreement on Agriculture; they are covered by Market Access for Non-Agricultural Products. Currently, import tariffs on fish products in developed countries are about 4.5 percent. However, non-tariff barriers and tariffs on processed products continue to be a source of disagreement (FAO 2003b). Non-tariff barriers and tariffs on processed products reduce the amount of international trade in fish and fishery products. Fish and fishery products are part of the Doha Round WTO negotiations.

A typical example is illustrated in the growing dispute between Norway and Great Britian and Ireland in the salmon industry. At the time this chapter was written, the European Union had agreed to open an investigation into Norwegian salmon imports into the European Union at the request of the United Kingdom and Irish governments (*Aftenposten* 2004). Large increases in the supply of farm-raised salmon have put downward pressure on prices, and Norwegian exports have increased sharply. If the British and Irish are successful, salmon imports into the European Union (including those from Chile and the Faeroe Islands as well as Norway) would be curtailed through the use of quotas. Given the fact that aquaculture is still a developing industry in many parts of the world, the potential for trade disputes is relatively high as the industry grows and matures.

CONCLUSION AND FUTURE TRENDS

Though the capture fish and fish product industry is the primary source of seafood and will continue to be so for the next 10 to 20 or more years, aquaculture will continue to grow in size and importance. The demand for seafood continues to increase, and the level of fish and fish products generated by the capture fish industry is flat. The threats to the aquaculture industry, such as threats to the environment and human health concerns, may slow the growth but probably will not stop it. Improved genetics and management practices should allow the industry to address many of these concerns. Furthermore, production of species such as tilapia and catfish is likely to be as environmentally sound, if not more so, than the production of terrestrial protein sources.

Another major trend affecting the seafood industry is the alteration of the marketing chain. The fact that supermarkets are in the ascendancy the world over also improves the relative position of the aquaculture industry. Supermarkets want a steady stream of product of a consistent quality. Supermarkets are the dominant food retailer in North America and Europe and are growing in importance in the rest of the world. Aquaculture is well suited to meet these requirements, particularly those producers that are large and those firms that are vertically integrated.

The income elasticity of demand for seafood is relatively large compared with that for other food products. As incomes and standards of living increase, so will the demand for seafood. This trend will lead to an increase in the amount of international trade in seafood products, provided that individual countries and trading blocs do not attempt to protect their own producers through the use of trade barriers. Instituting and enforcing trade barriers would hamper trade in seafood products as is currently the case with agricultural commodities.

The trend towards the domination of aquaculture by fewer and larger firms will also increase. Although some potential for product differentiation exists, fish and seafood products are currently sold as commodities, and those producers that can take advantage of economies of scale will be successful in the future. This is particularly true in developed countries. The increased efficiencies derived from aquaculture will put downward pressure on the prices of some species. This downward pressure, in turn, will lead adversely affected fish farmers to a call for trade protection.

References

Aftenposten. 2004. EU opens trade probe into salmon imports. March 5, 2004.

Boselie, D., Henson, S., and Weatherspoon, D. 2003. Supermarket procurement practices in developing countries: redefining the roles and public and private sectors. *American Journal of Agricultural Economics* **85**: 1155–1161.

Cotterill, R. 1999. *Continuing Concentration in the U.S.: Strategic Challenges to an Unstable Status Quo,* Food Marketing Policy Center Research Report No. 48. Storrs, CT: University of Connecticut Department of Agricultural and Resource Economics.

FAO (Food and Agriculture Organization). 2002a. *The State of World Fisheries and Aquaculture 2002.* Rome: Food and Agriculture Organization of the United Nations.

FAO. 2002b. *FAO Yearbook of Fishery Statistics.* Rome: Food and Agriculture Organization of the United Nations.

FAO. 2003a. *Review of the State of World Aquaculture.* Rome: Food and Agriculture Organization of the United Nations.

FAO. 2003b. *Fisheries Trade Issues in the WTO,* AD/I/Y4852E/1/9.03/500. Rome: Food and Agriculture Organization of the United Nations.

Harvey, D.J. 2003. *Aquaculture Outlook,* LDP-AQS-18. Washington, DC: U.S. Department of Agriculture, Economic Research Service.

Hites, R.A., Foran, J.A., Carpenter, D.O., et al. 2004. Global assessment of organic contaminants in farmed salmon. *Science* **303**: 226–229.

Mintel. 2002. *The Fish and Seafood Market.* Chicago, IL: Mintel International Group.

Pritchard, E. 2003. *Fisheries of the United States.* Silver Spring, MD: U.S. Department of Commerce, National Marine Fisheries Service.

Reardon, T.C., Timmer, P., Barrett, C.B., and Berdegue, J. 2003. The rise of supermarkets in Africa, Asia and Latin America. *American Journal of Agricultural Economics* **85**: 1140–1146.

Santerre, C.R. 2004. Review of "Global assessment of organic contaminants in farmed salmon." *Purdue News,* January 8, 2004.

Seafood Choices. 2005. *Shark*. Available online at www.seafoodchoices.org/
The Economist. 2003. The promise of a blue revolution. August 9, 19–21.
The Economist. 2004, Salmon a-slumping. January 17, 49.
U.S. Department of Agriculture. 1998. *Census of Aquaculture*. Washington, DC: National Agriculture Statistics Service.
Valdimarsson, G. 1998. Developments in fish food technology: implications for capture fisheries. *Journal of Northwest Atlantic Fishery Science* **23**: 233–249.
Vannuccini, S. 2003. *Overview of Fish Production, Utilization, Consumption and Trade*. Rome: Food and Agriculture Organization of the United Nations.

19

Can transgenic fish save fisheries?

INTRODUCTION

Fisheries are in trouble. For decades, there have been warnings that fish harvests have reached or exceeded sustainable limits and that collapse of capture fisheries might be imminent. Recent evidence has overwhelmingly confirmed these dire predictions (Pauly et al. 1997; Pauly and Maclean 2003). Despite increased fishing effort and more effective equipment, total catch levels have remained stable or decreased every year since the mid 1990s (Vannuccini 2003). Dismal as this total catch statistic might be, it unfortunately paints a deceptively rosy picture. Ever-increasing inputs of money and technology are required to merely tread water – a constant total catch under these circumstances means a diminished return per unit of fishing effort. Moreover, looking only at tonnes of fish caught (the typical representation of total catch) masks the dramatic shifts that have taken place in the species making up that total catch (Garcia and Newton 1997). Increasing catches of low-value species (so-called "trash fish") obscures the decline in almost every high-value demersal fishery and the profound impact that changing fish populations have had on the aquatic food web.

At the same time that fishers are expending more effort to catch fewer and less valuable fish, demand for fish is increasing at a rapid pace. The human population grows year by year, and food security continues to lag behind. Although Nobel Prize-winning economist Amartya Sen makes a persuasive argument that distribution issues, rather than actual food shortages, can explain much human hunger (Sen 1999), at some level more mouths to feed mean an increased demand for food. Because fish are a cheap and accessible source of protein for the Earth's poorest inhabitants, increasing populations

Globalization: Effects on Fisheries Resources, ed. William W. Taylor, Michael G. Schechter, and Lois G. Wolfson. Published by Cambridge University Press. © Cambridge University Press 2007.

in developing countries put fisheries under pressure. Paralleling this upsurge in demand from the world's poor has been an increased demand from the wealthiest nations, where fish is both a status symbol and part of a growing shift toward health-based diets. Globalization has facilitated these trends – the spread of high-tech equipment has meant that fish can be caught more efficiently, and improved transportation and trade links permit easier access to the world's most affluent markets.

Unfortunately, this increased demand for fish could not have come at a worse time for the fisheries themselves. Overfishing, a growing problem for decades, has reached crisis proportions under these new pressures. In 2002, the United Nations Food and Agriculture Organization (FAO) reported that 75 percent of the world's fisheries were overfished, threatened, or fully exploited (FAO 2002a). Prospects for expansion or increased production from these already stressed stocks are indeed slim. Much more likely is the prospect that these stocks will further decline unless remedial actions are taken to reduce overfishing.

In this world of downward trends and depressing statistics, aquaculture, or fish farming, was heralded as the great hope. Aquaculture was to relieve pressure on deteriorating wild fish stocks by replacing capture fisheries with fish farming. This shift to fish husbandry would allow wild populations to recover because reduced fishing pressures would mean greater cohort survival and reproduction. At the same time, aquaculture production could expand almost infinitely to satisfy the world's ever-increasing demand for fish.

Nice story. Too bad it did not work out that way. The fastest-growing sector of aquaculture involves raising carnivorous fish for Western markets. Even advocates of fish farming concede that these farmed predators do not find their way to developing or poor countries but instead grace the tables of the world's wealthy (*The Economist* 2003).[1] Fish are big business, and in an increasingly globalized market, demand is measured in dollars, not in hungry mouths. Moreover, aquaculture

---

[1] Official statistics indicate that most aquaculture takes place in China and serves local needs, but recent evidence suggests that China's fishery statistics have been unreliable for at least the past decade (*The Economist* 2003). As a result, the FAO now calculates total catch and aquaculture yield without reference to those numbers (FAO 2002a). Thus figures for carp cultivation, which purport to be some ten times those for salmon cultivation, are suspect. To avoid being mired in questionable statistics, this chapter focuses on salmon farming, which grows fish primarily for the North American market.

did not develop in a void. The resources now devoted to aquaculture had competing uses and users. Shifts in allocating these resources have had profound impacts on marine ecosystems.

The rise of carnivorous fish aquaculture has had significant negative impacts on wild fisheries and has created a host of other environmental problems. As a consequence, many of aquaculture's most vocal boosters are now promoting transgenic (or genetically modified) fish as the new and improved great hope – the way to achieve the earlier promises on which aquaculture has failed to deliver (Aleström 1996). Unfortunately, aquaculture of transgenic fish has all of the risks and drawbacks of existing aquaculture practices, plus an additional layer of hazards created by the genetic modifications being proposed. In short, transgenic fish will not save aquaculture, and aquaculture as it is currently practiced cannot save fisheries.

AQUACULTURE

In the past few decades, aquaculture has contributed an ever-increasing share of global fish supplies. More than 220 species of finfish and shellfish are currently farmed around the world (Naylor et al. 2000), and the business of aquaculture is booming. From 1970 through 2000, the percentage of global fish production attributable to aquaculture increased almost nine-fold – from 3.9 percent of total production in 1970 to 27.3 percent in 2000 (FAO 2002a). This amounts to an increase of 9.2 percent a year since 1970 – much faster growth than capture fisheries or, indeed, land-based meat production (FAO 2002a). Today, aquaculture supplies more than one-fourth of all fish that humans eat (Naylor et al. 2000), with low-income countries most dependent on fish. Globally, aquaculture is still predominantly small-scale and rural, producing species low in the food chain that require few or no inputs or capital investment (over 80 percent of total global finfish production is cyprinid fishes) (FAO 2002b).

Industrial aquaculture, by contrast, focuses on top-carnivore species for sale in a globalized fish market. Developing countries produce mainly omnivorous, herbivorous, and filter-feeding fish species, developed countries produce mainly higher-value carnivorous fish species. Although these carnivorous fish represented only 12.7 percent of total global finfish production by weight in 1999, they accounted for 34.7 percent of total production by value (FAO 2002b). The main fish species traded internationally are Atlantic and coho salmon, shrimp, and tuna (Nambiar 1999). Japan is the top importer of fish and fishery products, and together with the United States, France, Spain, Germany, and Italy, accounts for two-thirds of global fish imports (Nambiar 1999). Not only has this globalized trade in fish shaped fishing behaviors – resource investment and preferred catch – but it has also channeled commercial aquaculture toward the high-value carnivorous species that consumers prefer in affluent societies (Allsopp 1997; Ackefors 1999).[2] This preference has fueled an export-driven trend toward farming of predatory fish such as salmon, one of the most heavily cultivated fish species (Sandnes and Ervick 1999).

Dwindling wild stocks and high market prices have made salmon a particularly attractive candidate for aquaculture. By the mid 1980s, salmon aquaculture production exceeded worldwide harvests of wild salmon (Anderson 1997), and by the late 1990s, salmon aquaculture was producing more than 900 000 tonnes of fish per year (Sandnes and Ervick 1999). The vast majority of the aquaculture enterprises in developed countries are thus devoted to carnivorous species destined for the international fish market (FAO 2002a). The U.S. Department of Commerce has publicly committed itself to building domestic aquaculture into a $5-billion industry by 2025 – a five-fold increase from 2001 levels (DOC 1999). On August 5, 2005, the Bush administration proposed national offshore aquaculture legislation to Congress. If enacted, this legislation would grant the Secretary of Commerce broad discretion to issue permits for marine aquaculture in the 3.4 million square miles (8.8 million km^2) of the United States exclusive economic zone (EEZ) (Senate Bill 2005). As of 2006, Senate hearings have continued on implementation of the National Offshore Aquaculture Act of 2005.

In the United States alone, sale of carnivorous aquaculture products, largely salmon, grew from $45 million in 1974 to over $1.1 billion in 2000 (Price Waterhouse Coopers 2001) and is a €200-million business in the European Union (Central Statistics Office 2003). Farm-raised salmon has not only brought down the price to consumers (farmed salmon typically sells for $4 to $5 per pound [$8 to $10 per kilogram], as opposed to $15 per pound [$30 per kilogram] for wild salmon) but

---

[2] In developing countries, aquaculture tends to focus on omnivorous/herbivorous fish or filter-feeding species. Most of that production is consumed domestically and thus never enters international trade. Though intensive farming of omnivorous/herbivorous fish raises some of the same issues explored in this chapter, this discussion will focus on industrial and commercial aquaculture of predatory species.

has smoothed out the hithero cyclical availability of the fish (*Seattle Times* 2004). Steady supplies are a critical precondition to global trade, and just as farming has enabled a globalized salmon market, so, too, the globalized market demands a continuous supply of salmon and, therefore, continued salmon aquaculture. Neither the increased aquaculture production nor the decrease in price seems to have resulted in reduced fishing effort or lower catches of wild salmon (Naylor *et al.* 2000). If these trends were to continue, there would be little net environmental benefit in the form of reduced fishing pressure, a situation that would undercut one of the primary justifications for significant public investment in aquaculture. Indeed, as aquaculture production continues to expand and intensify, both its reliance and its impact on ocean fisheries are likely to increase. Minimizing aquaculture's impacts on wild fisheries will become an ever more pressing concern.

Environmental problems associated with industrial aquaculture

Unfortunately, raising these carnivorous species in aquaculture requires large quantities of wild-caught fish for use as feed. Indeed, the normal input is 2 to 5 kg of wild fish biomass for every kilogram of these high-market-value species produced (Naylor *et al.* 2000; *Issues in Ecology* 2001; Weber 2003). Inefficient conversion rates mean that two to five times more food is fed to the farmed fish than is ultimately produced by aquaculture (Pauly *et al.* 2003). As a result, 1 tonne of industrial farmed Atlantic salmon has an ecological footprint approximately twice that of commercially captured sockeye, chum, or pink salmon – the aquatic equivalent of "robbing Peter to pay Paul" (Belton *et al.* 2004).

Between 1987 and 1997, as global aquaculture production more than doubled (Vannuccini 2003), the demand for feed exploded. That feed has to come from somewhere, and, unfortunately, capture fisheries are the main source. By 1997, four of the top five and eight of the top 20 capture species were destined for use as feed for aquaculture and livestock consumption (Naylor *et al.* 2000). Rather than sparing fisheries from further pressures, aquaculture, as currently practiced, has placed new and damaging stresses on already threatened fisheries.

There is now heavy fishing pressure on small pelagic fish such as mackerel, anchovies, and sardines, all of which are used as fish feed. This fishing not only depletes food stocks relied upon by wild fish, marine mammals, and seabirds but also pits the needs of poor developing countries that rely on these pelagics for food against the U.S. love affair with salmon. Increased aquaculture production thus intensifies rather than diminishes the pressures that capture fisheries place on aquatic ecosystems.

Increased fishing pressure is not the only threat that aquaculture poses to the survival and rehabilitation of threatened fish stocks. Aquaculture also modifies marine and coastal habitat to the detriment of wild species. Many wild salmon populations are threatened or endangered and, therefore, fall under the protection of the U.S. Endangered Species Act (ESA). Among the populations of salmonids that have been listed as threatened or endangered are populations of Atlantic salmon (FWS 2000), various populations of coho salmon (NOAA 1996, 1997, 1998), and populations of chinook salmon, sockeye salmon, and steelhead trout (NMFS 2004). This is not an exhaustive list. For many of these listed species, habitat protection measures have been required under the ESA. To the extent that aquaculture affects the habitat of these listed species, it not only violates the ESA but also further jeopardizes the already precarious survival of wild populations. The ESA status of these fish populations may be in flux. In 2001, a federal district court in Oregon decided that the National Marine Fisheries Service (NMFS) had inappropriately distinguished between hatchery salmon and wild salmon when determining their listing status under the ESA (Alsea Valley Alliance 2001). The court remanded the matter to NMFS for further action. The Bush administration elected not to appeal this decision, and an attempt by environmental groups to appeal this decision was dismissed. In April 2004, the Bush administration announced that, contrary to the advice of its scientific advisory council, it intended to count hatchery salmon when calculating wild populations of these fish for ESA purposes (Harden 2004; Myers 2004). Thus, the future of ESA protection of these species is unclear as of this writing.

Shrimp farming is notorious for the attendant destruction of coastal zones, particularly mangrove swamps (Kaiser 2001). This destruction of mangrove swamps and forests in the United States may have significant regulatory implications under the wetlands provisions of the Clean Water Act. Despite the regulatory and statutory regime intended to protect these ecologically valuable wetlands, mangrove swamp losses have continued to mount in the United States. Florida, for example has lost at least 80 percent of its mangroves (Florida Department of Environmental Protection 2004), and overall U.S. mangrove losses have been significant (Wilkie and Fortuna 2003). Outside the United States, mangrove protections are even sketchier.

The ecological threats from uncontrolled shrimp farming are well known; the impacts that netpen fisheries have on coastal ecosystems are less appreciated. Many of the farming practices associated with netpen aquaculture pose significant environmental risks that directly affect wild species. These risks include displacing wild populations through escape and outcompetition or interbreeding; creating reservoirs for parasites and diseases; and introducing organic wastes, chemicals, and pollutants into the marine environment.

Impacts of escaped fish on wild populations

To be profitable, a fish-farming operation must keep possession of its fish stocks. Fish farming can be conducted in either land-based or sea-based facilities, but containment is far more reliable and successful in land-based facilities. Because sea-based facilities use pens that are vulnerable to accidents that can release large numbers of farmed fish, sea pens pose a much greater risk of escape than do land-based aquaculture facilities. Despite this serious risk of escape, most aquaculture is conducted in sea pens, which require much smaller initial capital investments, are less expensive to operate, and are, therefore, much more profitable. In most jurisdictions, including the United States, aquaculture is only lightly regulated, and few if any penalties are imposed for releases of farmed fish stocks (Naylor *et al.* 2003). Even the government agencies charged with regulating aquaculture describe escapes as "routine" (CEQ/OSTP 2001). With no regulatory penalties that might force aquaculture operations to internalize this environmental cost, escaped fish are generally seen as a cost of doing business. Containment standards have thus been viewed primarily through a commercial rather than an environmental lens.

When commercial sea pens fail, an event that happens with unfortunate frequency, large numbers of farmed fish can escape. It is virtually impossible to recapture these escaped fish. Most of these escapes are attributable to storm damage, accidental mechanical damage from boats and harvesting equipment, predator damage, or human error and malfeasance (Devlin and Donaldson 1992). Damage can be difficult to discover and repair. For example, in 2002, a fish farm in British Columbia discovered its net had ruptured only when commercial fishing boats contacted it after catching hundreds of Atlantic salmon in one 24-hour period (Morton and Volpe 2002). More than 32 000 Atlantic salmon escaped into the Pacific before the hole could be repaired. Best management practices can minimize the risks from human error (NASCO 2000), but profit concerns mean that few commercial aquaculture operations employ sea pens built to withstand extreme or unpredictable environmental conditions.

The scope of the resulting escape problem can be staggering. For example, 170 000 salmon escaped from a Maine salmon farm during one storm in December 2000 (Atlantic Salmon Federation 2001), and 600 000 from a single incident in the Faeroe Islands (Gardar 2002). In 2002, approximately 2 million Atlantic salmon escaped from sea pens in the North Atlantic (BBC 2003). More than 1 million Atlantic salmon have escaped in the Pacific Northwest since 1991 (Alaska Department of Fish and Game 2001), and a similar number have escaped in Scotland since 1997 (Aitken 2002). The literature abounds with reports of other large escapes from sea-pen aquaculture (*Seattle Times* 1997; McDowell 2002; Reuters 2002). Norway has such a history of mass escapes that up to 90 percent of the returning salmon populations to some rivers are farm escapees (Saegrov *et al.* 1997; Fleming *et al.* 2000). As startling as those numbers are, they are likely a significant underestimate of total escapes because chronic net pen leakage probably permits even more fish to escape than do these large events (Volpe *et al.* 2001).

Unlike traditional livestock species, farmed salmon are still very similar to their wild progenitors, making them better adapted to survival in the wild. There is a growing body of evidence that these escaped fish are fully capable of establishing themselves in the environments to which they have escaped. This phenomenon already poses ecological risks to native salmon stocks (NRC 2002). Recent studies indicate that 30 to 40 percent of Atlantic salmon caught in the Northern Atlantic Ocean are of farmed fish origins (Hansen *et al.* 1993; Jönsson *et al.* 1996). In some parts of Norway, fish of farmed origins are the majority of animals captured (Saegrov *et al.* 1997). On the east coast of North America, escaped farm salmon outnumber wild fish by as much as ten to one in some rivers (Atlantic Salmon Federation 2002). These escapees can threaten wild populations through competition and interbreeding. This is a particular concern for a species such as salmon because the farmed populations outnumber wild populations by orders of magnitude. Escapees can, therefore, overwhelm the wild populations. For this reason, in *United States Public Interest Group v. Salmon of Maine*, the U.S. District Court of Maine banned use of non-native salmon hybrids in Maine waters (USPIRG 2003).[3]

---

[3] The First Circuit upheld this decision on appeal in August 2003 (*United States Public Interest Group v. Atlantic Salmon of Maine*, 339 F.3d 23 [1st Cir. 2003]).

Farmed fish tend to be more homogeneous than wild populations. Introgression can, therefore, reduce the genetic variability of the wild fish populations and thus reduce the population's ability to adapt to changed environmental circumstances (Clifford et al. 1998). Moreover, because they have been bred to human tastes and needs rather than evolved to successfully occupy ecological niches, escaped farm fish are frequently less fit overall than are their wild counterparts (McGinnity et al. 2003; Ocean Studies Board 2004). Although this reduced fitness translates into reduced breeding success (Fleming et al. 2000), escaped farm salmon do breed in the wild and do hybridize with wild fish. Through sheer numbers, farm escapees can swamp the wild populations even if the farmed fish are at a selective disadvantage. Similarly, repeated annual escapes, even of small numbers, hinder the ability of natural selection to purge disadvantageous genes (Fleming et al. 2000). These effects on the wild population can be significant and deleterious because the ratio of more fit wild fish to less fit escaped farmed fish will likely be low. Under such circumstances, significant modifications of the wild populations and of the ecosystem as a whole may be unavoidable. Aquaculture can thus reduce fitness and diminish genetic diversity of wild fish populations because the genetically distinct escapees will interbreed with wild populations.

Even when escapees have no wild relative with which to crossbreed, escaped farmed fish can disrupt wild fish populations and, indeed entire ecosystems, by competing with wild populations for scarce habitat and food (Fleming et al. 2000). For example, 80 percent of the farmed salmon in British Columbia are Atlantic salmon *Salmo salar*. Rampant escapes from fish farms have given Atlantic salmon the opportunity to establish naturalized populations in the Pacific Northwest (McKinnell and Thompson 1997). Free-ranging Atlantic salmon have been captured in British Columbian waters since the mid 1980s and have been routinely encountered since the late 1990s (Volpe et al. 2001). There is clear evidence that Atlantic salmon are successfully spawning in British Columbian waters and thus have become an invasive species (Volpe et al. 2000). Because Atlantic salmon display significant niche overlap with juvenile steelhead–rainbow trout and Pacific salmon, the escapees are likely to compete with the native species for food. Many populations of steelhead are already at high risk (meaning that, over the past decade, population estimates have been below 20 percent of the long-term average). The escaped Atlantic salmon, which often have a size-at-age advantage and outcompete the native species under many conditions, may threaten the long-term survival of the native populations. The native wild salmon populations are already highly threatened, so escaping farmed fish could jeopardize the survival of these species.

Reservoirs for infection and disease

Aquaculture crowds carnivorous fish into densities not found in the wild (USPIRG 2003). A typical sea pen can contain anywhere from 5000 to 16 000 fish, and a single farm can stock as many as 250 000 fish (USPIRG 2003). This crowding facilitates the spread of disease among the farmed fish. Because the sea pens are open to the environment, pathogens can be dispersed by tidal currents or from feces and urine of infected salmon. The population density within a sea pen allows the amplification of pathogen and parasite loads for subsequent transmission back to wild populations (Saunders 1991). Farmed fish thus serve as a reservoir for diseases that can be spread to wild fish (Todd *et al.* 1997; Tully *et al.* 1999). For example, sea lice infestations are endemic in most areas with intensive salmon culture (Watershed Watch 2001). Salmon farms have been correlated with a more than three-fold increase in abundance of lice infestations of wild fish (Tully *et al.* 1999). When salmon farms are situated along salmon migration routes or in wild salmon habitats, the results can be devastating to already endangered wild populations. For example, major sea lice infestations in British Columbia have been correlated with significant decreases in numbers of fish returning to spawn (Naylor *et al.* 2003) and are believed to be responsible for the catastrophic collapse of the wild sea trout population (Pearson and Black 2001). Bacterial and viral diseases such as infectious salmon anemia also run rampant in fish farms and can infect wild populations (Ocean Studies Board 2004).

Introduction of wastes and pollutants

Aquaculture stresses and degrades the marine ecosystem in other ways as well. Sea pens are open systems, and salmon feces, fish feed, and other organic wastes are freely discharged into the aquatic environment. This typically results in excess nitrogen and phosphorus loads in the immediate vicinity of the sea pens. This nutrient overloading causes eutrophication problems (De Silva 1999), and underneath every fish pen is a footprint or "dead zone" – a shadow of oxygen depleted and contaminated sediment (Pearson and Black 2001; USPIRG 2003). Nutrient loading is, of course, a significant and widespread problem attributable to many causes in addition to aquaculture. Nevertheless, nutrient loading from aquaculture can have significant local impacts. Proper rotation and fallow periods can minimize these effects over the long term. Unfortunately, the industry's track record with rotation and fallow periods is not very good (USPIRG 2003). There is even less regulation of the organic wastes produced from shellfish aquaculture. At least one federal court has concluded that organic wastes from these operations were not "the type of materials the drafters of the Act would classify as 'pollutants.'" The court reasoned that one of the purposes of the Clean Water Act was "the protection and propagation of shellfish," and, therefore, a large-scale commercial mollusk farm in the Puget Sound did not produce wastes within the meaning of the act (Association to Protect Hammersly 2002). By reading the Clean Water Act narrowly (and circularly), the Ninth Circuit exempted an entire aquaculture industry from environmental scrutiny.

In addition to organic wastes that may or may not be subject to regulation, fish farms also release a wide range of chemical pollutants, including pesticides, antifoulants, and antibiotics. The Environmental Protection Agency (EPA) has only recently begun to consider regulating these discharges under the Clean Water Act (EPA 2002). In the absence of regulation, industry has been free to discharge chemical contaminants used in aquaculture. The use of parasiticide drugs such as cypermethrin to control sea lice infestations is particularly problematic. Cypermethrin is applied as a bath. That means the parasiticide is diluted with water and poured directly into a sea pen enclosed in a tarpaulin. After the treatment, the tarpaulin is removed and the cypermethrin is released directly into the surrounding waters (USPIRG 2003). At least one study has demonstrated that cypermethrin plumes can persist in marine waters for significant periods of time and can spread over fairly large distances (Ernst *et al.* 2001). Because cypermethrin is highly toxic to many marine organisms, this is a serious concern.

Antifoulants, antibiotics, and dyes are also discharged into the marine environment from net pen aquaculture. Antifoulants containing copper are typically used to retard growth of organisms on the sea pen nets. This copper leaches into the marine environment, where it can be toxic to wild populations (USPIRG 2003). Antibiotics are routinely administered to fish through feed based formulations. Estimates of the quantities of antibiotics used in aquaculture range from 70 000 (MacMillan 2001) to 433 000 pounds (35 to 215 kg) per year (Benbrook 2002). Few data are available upon which to build these estimates of antibiotic use. Across all kinds of animal husbandry, the use of antibiotics has become extremely controversial in light of the potential to create resistant bacteria (Mellon *et al.* 2001). Some have suggested that aquaculture practices raise these same concerns (Benbrook 2002). Because antibiotics are administered to fish in their feed, which is dispersed in the water, use of antibiotics in aquaculture directly doses the environment. Resistant bacteria have been identified not only in the farmed fish, but also in wild fish and the sediment beneath net pens (Kerry *et al.* 1994). The development of resistant bacteria due to antibiotic use in aquaculture may have human health repercussions (CDC 1999). Interactions between humans, antibiotics, fish, bacteria, and aquatic environments are still poorly understood, and conclusions are few and far between (National Aquaculture Association 2004). What is clear is that the aquaculture industry's antibiotic use practices are based on ignorance of likely or possible environmental or human health effects, not on peer-reviewed experimental evidence that such antibiotic use is safe.

Requiring properly run land-based facilities could go a long way toward eliminating many of the environmental risks associated with netpen aquaculture.[4] As early as 1990, the U.S. Department of Agriculture recognized that rigorous design standards, constant monitoring systems, and emergency plans would be necessary to prevent escapes of fish from land-based aquaculture facilities (*Federal Register* 1990). Unfortunately, the United States has only rudimentary and ad-hoc standards for land-based aquaculture. A uniform, industry-wide set of standards would not only encourage compliance with environmental standards but might also promote environmentally sound innovation. Weak as these existing land-based environmental protections are, however, the state of netpen marine aquaculture is even worse. To date there are few federal or state legal requirements

---

[4] In addition to these environmental risks, there are also a host of human health concerns swirling around aquaculture. One recent study concluded that farmed salmon contain significantly higher concentrations of polychlorinated biphenyls (PCBs), dioxins, and other organic pollutants than wild fish (Hites *et al.* 2004). Canthaxanthin and astaxanthin, dyes derived from petroleum by-products, are used to give farmed salmon flesh the same pink color as wild salmon. Hoffmann-La Roche provides the SalmoFan – a color chart with assorted shades of pink – to help salmon farmers create the color they think their customers want. However, these dyes may cause retinal damage, and their use has been curtailed throughout the European Union (European Commission 2002). After three class action lawsuits were filed in 2003, most American supermarkets now notify their customers that farmed salmon contain dyes. The complaints in those lawsuits are available at www.smithandlowney.com/salmon/complaints/

that marine net pen aquaculture facilities must meet. The requirements that do exist are scattered among the U.S. Department of Agriculture, the National Marine Fisheries Service, the Food and Drug Administration, the Environmental Protection Agency, various state agencies, and the Army Corps of Engineers, virtually ensuring that no comprehensive or coordinated oversight will occur.

TRANSGENIC FISH

In just over a decade, genetic engineering (genetic modification or biotechnology) has emerged as a powerful tool for agricultural production. By transferring genetic material from organism to organism, researchers can create wholly new transgenic organisms. The Cartagena Protocol on Biosafety provides a useful definitional starting point for a discussion of transgenic organisms (though for purposes of the protocol, the equivalent term "living modified organism" is used). Under the protocol, a living modified organism is "any living organism that possesses a novel combination of genetic material obtained through the use of modern biotechnology" (Cartagena Protocol 2000). These techniques typically include microinjection, electroporation, use of microprojectiles, and liposome-mediated transformation. For fish, microinjection has been the preferred technique, with a success rate of about 10 percent (meaning that out of every 100 eggs injected, about 10 will integrate and express the transgene). A much smaller percentage (about 1 percent) of these transgenic individuals will pass the transgene on to their offspring (Beardmore and Porte 2003).

Although these percentages are comparable to the success rates for mammalian transformations, fish reproduction mechanisms – they produce eggs in large quantities, and those eggs develop outside the fish's body – make fish a particularly attractive candidate for genetic modification. The first reports of the application of genetic engineering to fish appeared in the 1980s (Maclean and Talmar 1984; Zhu et al. 1985). Since then, a burst of genetic modification activity has occurred in aquaculture research and development. Indeed, by 1990, 13 species of transgenic fish had been produced in laboratories around the world (Kapuscinski and Halleman 1991), and in 2003, the FAO reported 23 aquatic transgenic species (Beardmore and Porte 2003). Other aspects of fish biology, however, have prompted the National Research Council (NRC) to call for caution (NRC 2002). In particular, the NRC identified the unresolved containment problems of aquaculture as posing serious environmental issues and cited the many critical unknowns surrounding these genetically modified animals that prevent any informed judgement about whether or how to proceed with commercialization of transgenic fish. A host of ethical questions also surrounds use of the technology that should be addressed before large-scale commercial use of transgenic animals or fish is permitted.

Unlike domestic farm-based animals, farmed fish easily become feral and compete with indigenous populations. Because of their novel characteristics, transgenic escapees could pose even greater threats to wild populations than do conventional farmed fish. As described above, the minimal environmental regulations imposed on aquaculture have been wholly unable to successfully resolve the escape problem, and fish farmers seem to treat escaped fish as a cost of doing business. Given the increasing demand for fish and the lack of an adequate regulatory structure, it is perhaps not surprising that, despite these risks, aquaculture biotechnology research has pressed on full bore. This expansion will almost certainly result in the escape of transgenic fish unless immediate steps are taken to prevent such an occurrence.

The majority of the transgenic fish research and development efforts to date have focused on improving fish growth rates or efficiency of food conversion (Beardmore and Porte 2003). Creating transgenic fish with increased cold tolerance is a second area of significant research, though to date this research has not been as successful. Increased growth means reaching marketable size sooner and, therefore, reducing overhead costs for fish farmers. Researchers have used molecular biology techniques to modify at least 14 fish species – including varieties of carp, trout, salmon, and channel catfish – so they will grow two to 11 times faster than their non-modified counterparts (Dunham 2003). Through insertion of additional copies of fish growth hormone (GH) genes, coupled with mammalian growth promoters, researchers have been able to accelerate fish growth rates from 10 percent up to a 30-fold increase compared with non-transgenic fish (Rahman *et al.* 1998, 2001; Rahman and Maclean 1999). The economic attraction of these genetic modifications is obvious.

Increased size in response to GH transgenes varied by species, but overall those fish strains that had already been subject to extensive conventional breeding to enhance the desired phenotype showed the smallest increases (Devlin *et al.* 2001). Different transgene constructs also produced different rates of growth (Beardmore and Porte 2003). For example, use of an all-fish GH gene construct to make transgenic Atlantic salmon has produced a two-fold increase in the transgenic fish growth rate (Du *et al.* 1992), while use of ocean pout antifreeze promoter and salmon GH cDNA elevated circulating GH levels in coho salmon up to 40 times the levels found in unmodified fish (Devlin *et al.* 1994). This hormonal change equated to a five-to 30-fold increase in weight after 1 year of growth (Du *et al.* 1992; Devlin *et al.* 1994, 1995a, b, 2001). Under laboratory conditions, this increased growth rate has been correlated with a significant increased efficiency in feed conversion (NRC 2002). Thus, if aquaculture of high-trophic level-carnivorous fish is to continue, transgenic fish might provide a means to reduce the pressures on wild fish stocks.

Unanswered challenges to regulating transgenic fish

Although the potential upside from applications of molecular biology to fisheries is enormous, so are the risks. The very factors that make transgenic fish an attractive commercial prospect might also pose serious risks once these fish escape into the wild. And the negative consequences could be devastating. In the mid 1980s, federal policy declared that biotechnology products would be evaluated under the same laws and processes used to review products produced without biotechnology. However, many genetically engineered organisms confound conventional regulatory categories, forcing regulators to rely on increasingly creative interpretations of the existing laws to respond to *sui generis* challenges posed by these GMOs. Not only are the lines of authority for regulating animal biotechnology unclear, but there are serious questions about the legal and technical capabilities of the agencies to address potential hazards posed by the technology (NRC 2002).

In the United States, the Food and Drug Administration (FDA) claims primary regulatory authority over transgenic animals, including fish, by virtue of its new animal drug authority under the Food Drug and Cosmetics Act (CEQ/OSTP 2001). The relevant provisions define a new animal drug as "any drug intended for use in animals other than man [sic] including any drug intended for use in animal feed" (FDCA 2005). The FDA has interpreted this authority to extend to transgenic fish, reasoning that the transgene and the protein for which it codes would be new animal drugs. The Food Drug and Cosmetics Act gives the FDA legal authority to regulate the food safety aspects of these transgenic salmon, but the emerging consensus is that the bigger risk is that transgenic fish will find their way into the wild and pose a significant environmental threat (NRC 2002).

The federal government claims that, as part of its safety assessment for a new animal drug, the FDA considers "environmental effects that directly or indirectly affect the health of humans or animals." However, the government explicitly concedes that the FDA's authority does not extend to all environmental impacts (CEQ/OSTP 2001). The limits on the FDA's authority raise real questions about whether the FDA has enough flexibility and expertise to address the environmental and ecological issues unique to genetically modified fish.

Although no transgenic fish have yet been grown or marketed in the United States as food, the first transgenic fish went on sale in the United States on January 5, 2004. The transgenic "GloFish®" is an aquarium zebra fish genetically engineered to glow in the dark through expression of various fluorescent pigments. These fish are currently available in Taiwan and are, or soon will be, marketed in every state of the United States except California, where they are banned.[5]

Despite the federal government's sweeping claims that the FDA's authority to regulate transgenic ornamental fish as well as transgenic food fish is adequate to prevent environmental harms (CEQ/OSTP 2001), the agency announced in 2003 that it would not regulate GloFish®. This decision not to regulate rested on a three-sentence official statement in which the FDA announced that "because tropical fish are not used for food purposes, they pose no threat to the food supply. There is no evidence that these genetically modified fish pose any more threat to the environment than their unmodified counterparts which have been widely sold in the United States" (FDA 2003). This decision was not based on *any* environmental risk assessment process, nor did the agency conduct an environmental assessment under the National Environmental Policy Act (NEPA) and its implementing regulations at 40 CFR Parts 1500–1508 and 10 CFR Part 1021.

Instead, the FDA seems to have merely assumed that transgenic fish are the "substantial equivalent" of conventional fish. This assumption flies in the face of a significant body of scientific scholarship detailing the various behavioral and survival differences between con-

---

[5] These fish are marketed under the name "Night Pearl GloFish®" by Taikong Corporation of Taiwan, www.azoo.com.tw/select.html, and under the name "GloFish®" by Yorktown Technologies in the United States. California law prohibits the import or sale of transgenic fish without a permit or an exemption under California Code of Regulations Title XIV, Section 671. Yorktown Technologies requested an exemption which the California Fish and Game Commission denied on December 3, 2003 (California Fish and Game Commission 2003).

ventional fish and their genetically altered counterparts. It also ignores the significant regulatory concerns identified by the NRC (2002) and by the agency itself when it initially asserted this authority (CEQ/OSTP 2001). A coalition of consumer groups sued the FDA on January 14, 2004, to challenge this decision as a failure to regulate (International Center for Technology Assessment 2004). The government filed a motion to dismiss on April 19, 2004, and the litigation is slowly wending its way through the court system.

The FDA's decision not to regulate GloFish® highlights the inadequacy of the existing U.S. regulatory structures. Even though release of these fish into the wild might well have significant ecological impacts, no federal agency has evaluated these impacts. A transgenic, highly mobile organism has thus been loosed into the commerce stream without any sense of the likely or possible environmental repercussions. Neither the FDA nor any other agency has taken steps to satisfy NEPA's requirements designed to protect the environment. This failure is particularly troubling because commercial production of these fish will inevitably lead to release of some proportion of these fish into the wild (USGS 2001). Indeed, at least 185 exotic fishes have been caught in U.S. waters, and 75 of these are known to have established breeding populations. Over half of these introductions are due to the release or escape of aquarium fishes (USGS n.d.; Zhuikov 2004). Similar patterns occur elsewhere (Western Australia Department of Fisheries 2006). There is no reason to assume that these GloFish® will suffer a different fate, and it will certainly be impossible to monitor thousands or millions of households purchasing these fish to ensure proper confinement. Responsible regulatory oversight must consider this set of questions *before* trangenic ornamental fish are sold in the aquarium trade.

Aqua Bounty's permit application

The FDA is currently considering an application from Aqua Bounty Farms[6] for what would be the first permit to grow transgenic salmon commercially for food under the same new animal drug authority it putatively exercised in deciding not to regulate GloFish®. The Trade Secrets Act requires that the FDA keep secret *all* of the investigations and pre-market notifications that precede the release of a new animal drug, including whether any such petition exists. In the absence of

---

[6] Now known as Aqua Bounty Technologies, Inc.

Aqua Bounty's public disclosure of its application, the Trade Secrets Act would have prevented any public participation in the FDA decision-making process, despite explicit NEPA statutory provisions requiring transparency and public participation in the environmental assessment process. The Trade Secrets Act prevents the FDA from discussing whether any other applications have been filed for approval of other transgenic fish. The FDA acknowledges that this duty of secrecy creates a clear conflict with NEPA – a conflict, moreover, that prevents the agency from fulfilling its duties under NEPA to ensure a public airing of significant environmental impacts (CEQ/OSTP 2001). The GloFish® precedent calls into question both the scope of the FDA's authority to consider wholly ecological impacts and its willingness to exert whatever authority the agency might possess. There is a significant possibility that important environmental concerns will not find their way into the regulatory decision-making process, particularly as commercial pressures on the FDA to approve GM fish are mounting. The populations of wild Atlantic salmon are severely depleted, and farmed Atlantic salmon are threatening native salmon populations around the world. Transgenic salmon would pose a new threat to the continued survival of these wild populations.

Aqua Bounty genetically modifies its salmon by microinjecting a transgene construct consisting of an ocean pout antifreeze protein (AFP) promoter linked to the chinook salmon GH cDNA (Fletcher *et al.* 2000). This transgene construct enables the fish to produce growth hormone year round rather than only during the spring and summer. As a result, Aqua Bounty's transgenic fish grow up to six times faster than non-transgenic farmed salmon (Stokstad 2002). The company acknowledges that its transgenic fish pose significant risks for wild populations and that sea pen aquaculture is associated with negative environmental consequences (Fletcher *et al.* 2000). According to the NRC, Aqua Bounty has requested that the FDA authorize it to sell transgenic fry to industrial salmon farms that will raise the transgenic fish in sea cages (NRC 2004). Because of the lack of transparency in the FDA drug approval process, it is unclear whether Aqua Bounty has collected and submitted the body of data needed before the FDA can responsibly assess whether this proposed distribution of transgenic salmon poses a heightened threat to the fitness of wild salmon populations (NRC 2002; Pew Initiative on Food and Biotechnology 2003). It is clear, however that neither Aqua Bounty, nor anyone else for that matter, has published the relevant information in any peer-reviewed scientific journals. The catalogue of scientific uncertainties indicates that more research is needed on these questions before transgenic fish can be safely commercialized.

Additional risks posed by aquaculture of transgenic fish

The environmental effects of transgenic fish will to some extent depend on aquaculture practices. One of the most highly touted economic benefits of transgenic fish is that they are purported to reach market weight more rapidly than conventional farmed stock. If farms do not alter their annual level of production, the local environment would clearly benefit. The fish would grow for 18 months, rather than 24 to 30, leaving a 6-month to 1-year fallow period in which the sea pen site could recover. If, however, fish farmers took advantage of the faster growth to run more fish production cycles, the environmental load on the site would increase accordingly. Evidence from current aquaculture practices suggests that farmers are already disregarding growing cycle restrictions and stocking limits imposed to preserve the marine environment (USPIRG 2003). Thus it seems unrealistic to assume that those same farmers would use transgenic fish as an opportunity to decrease environmental impacts rather than to increase profits.

Transgenic fish as invasive species

The possible impact of escaped transgenic fish on wild populations is probably the greatest science-based concern raised by the new technology. We already know from experience with conventional aquaculture that physical containment measures fail with disturbing frequency. On the basis of what is currently known about transgenic salmon, it is impossible to predict adequately the environmental outcomes should these fish escape or be released to the wild.

Conventional farmed salmon are an environmental nuisance upon escape. Transgenic fish that escaped into natural ecosystems could pose a much bigger environmental threat. This danger mainly arises for those transgenic fish endowed with new genes that improve fitness traits such as mating success or the ability to withstand harsh conditions. There is little published information about whether adult transgenic fish are larger than their conventional counterparts (a variable that tends to relate directly to mating success), but at least one study has shown that transgenic fish modified to produce higher levels of GH not only grow more rapidly but also grow larger (de la Fuente *et al.* 1999). The establishment of a thriving transgenic fish population in an ecosystem where it never existed could crowd out native fish populations. These dangers are only poorly understood and have yet to be thoroughly considered by any of the regulatory agencies charged with protecting and preserving the marine environment. There simply is not yet enough information to predict when and where transgenic fish would be likely to become an invasive species.

Trojan gene scenario

Beyond these near-term ecological effects, there are also real concerns about the effects of transgenic fish interbreeding with wild populations. A transgenic fish that has a survival advantage in the wild could outcompete its wild relatives. For example, some experimental evidence suggests that transgenic coho salmon modified to express high levels of GH will be able to outcompete wild coho salmon for food (Devlin *et al.* 1999). Changes in the genetic makeup of well-adapted wild populations may ultimately affect their abilities to withstand environmental change.

Even if they are not well adapted for survival in the wild, transgenic animals may have detrimental impacts on the genetic structure of wild populations by allowing the introgression of "exotic" genes into natural gene pools. Of particular concern is the so-called "Trojan gene" effect, whereby transgenic animals that are poorly adapted for survival in the wild exhibit traits that give them a mating advantage (Muir and Howard 1999). Many transgenic fish have been modified to generate faster growth and/or larger size, traits typically associated with male mating success (Howard *et al.* 1998). These positive fitness traits are balanced by other characteristics, such as reduced swimming speed (Farrell *et al.* 1997) and aggressive food pursuit (Jönsson *et al.* 1996), that suggest the transgenic fish may have a viability disadvantage. This matrix of favorable reproductive traits and maladaptive pleiotropic traits raises concerns that transgenic fish may introduce Trojan genes to their wild relatives – genes that increase mating success but decrease ultimate viability. Such genes would reduce the mean fitness of the populations exposed to them and, in extreme cases, might drive populations to extinction by reducing the fitness of the progeny of transgenic fish breeding with wild individuals (Muir and Howard 1999, 2001; Hedrick 2001).

At this point, it is not clear whether GH transgenic fish will possess either the mating advantage or the viability disadvantage central to the Trojan gene scenario. The Trojan gene possibility is largely based on computer simulations of non-salmonid reproduction and on extrapolations from behavioral studies. These data are necessarily preliminary, but there is evidence that non-transgenic farmed salmon exhibit characteristics that predispose them to such Trojan gene effects, such as reduced survival of progeny from matings between farmed and wild salmon (Fleming et al. 1996, 2000; McGinnity et al. 1997, 2003).

The ultimate physical containment system for growing transgenic fish of course, would be closed-system land-based facilities. Because transgenic fish may more efficiently convert feed, it might be economically feasible to raise these fish in land-based aquaculture facilities. Requiring that transgenic fish be grown in closed system land-based facilities might eliminate much of the risk to wild fish associated with raising transgenic fish.

Biological containment

In a landmark settlement of a Clean Water Act lawsuit brought by a coalition of public interest organizations, one fish-farming company agreed to a ban on the company's growing genetically engineered salmon strains in Maine (Environmental Law Center 2002; USPIRG 2002). The same plaintiffs brought another federal lawsuit against other Maine aquaculture companies and obtained an injunction banning transgenic fish from Maine waters pending further safety research (USPIRG 2003). In particular, the court ordered that biological containment mechanisms be explored.

Biological containment can reduce the risks to wild fish from escapees. In the context of aquaculture, biological containment typically means raising sterile triploid fish or sterile transgenic fish carrying antifertility genes tailored into their genomes (Aleström et al. 1992; Donaldson et al. 1993). Sterilization techniques are relatively easy and inexpensive, but success rates are highly variable. There is an overwhelming consensus, even among advocates of this technology, that neither perfect containment nor 100 percent sterilization of transgenic fish will be possible (Maclean and Laight 2000; Dunham 2003). Given the huge numbers of fish in commercial aquaculture operations, typically hundreds of thousands per pen, and the concomitant large numbers of escapees, even a small percentage of residually fertile transgenic fish might be enough to pose a threat of cross-breeding (Kapuscinski and Brister 2001).

In addition, even effective sterilization will not necessarily neutralize the risks to wild populations. Escaped sterile fish might still engage in courtship and spawning behavior, disrupting breeding in wild populations and decreasing overall reproductive success. Even without reproducing, waves of escaped sterile fish could also create ecological disruptions by competing with wild fish. If transgenic fish have a competitive advantage, wild fish will be overwhelmed as each sterile escapee cohort is replaced by another equally strong cohort. Transgenic fish that do not have a competitive advantage would still stress fragile marine ecosystems through their sheer numbers.

Enhanced ability to transfer disease

Genetic engineering has also focused on increasing resistance of fish to pathogens (Traxler *et al.* 1999; Melamed *et al.* 2002). The possibility of increased resistance is of obvious commercial interest. However, it does raise an additional environmental concern. Transgenic fish might act as reservoirs for diseases and parasites to which they are resistant, thereby increasing the risk of transferring diseases and/or parasites to wild populations (Tully *et al.* 1999). Aquaculture already creates disease reservoirs, but in conventional aquaculture the nature of this risk is necessarily limited by the possibility that the disease will kill its host fish. Creating transgenic fish immune to the disease would increase the risk dramatically because infected fish could serve as hosts for the infectious agent without expressing any of the negative manifestations of the disease. Infected transgenic fish could persist for long periods of time, thus spreading the infection or disease.

Economic consequences

Biotechnology and transgenic fish are being touted as having the potential to revolutionize aquaculture. These predictions should be viewed skeptically. Although direct extrapolations from agronomy to aquaculture may be suspect, lessons drawn from the marketing of genetically modified field crops, which have been on the market for 5 years, can be instructive about the direction that any market for transgenic fish is likely to take. The lessons are not salutary.

Were agricultural biotechnology's potential oriented toward meeting the needs of the world's poor, genetic engineering might already be a means to provide more and better food for the growing and undernourished human population while, at the same time, decreasing agriculture's devastating ecological impacts. To date, however, biotechnology has not been directed toward those ends. Instead, major agricultural conglomerates have used biotechnology to consolidate their hold on agricultural markets and have used patented seeds as a means to further expand their reach (Monsanto 2002; Fernandez-Cornejo 2004). The genetically modified crops currently on the market (corn, soybean, cotton, and canola) have been modified to withstand spray with patented herbicides or to endogenously produced pesticidal proteins. These modifications have made the crops easier and more profitable to grow in Iowa but have offered little benefit to the world's poor, and they provide consumers no nutritional, environmental, or financial benefits.

Over the past decade, globalization has led to tremendous consolidation in the aquaculture industry (Naylor *et al.* 2003). Large, vertically integrated conglomerates have created a competitive advantage out of economies of scale and changes in technology. As a result, a handful of multinational firms now dominate global aquaculture production. Given this industry structure, there is little reason to believe that aquaculture will deviate from the path blazed by biotechnology in the heavily consolidated agricultural industry. The focus of the developing aquaculture industry is likely to be on benefits to producers rather than to consumers or the environment.

Another product of globalization has been expanding intellectual property claims to fish broodstocks. These intellectual property claims are likely to have significant negative impacts on the poorest communities – those most in need of the increased fish supplies for basic survival. These effects will further compound the already skewed impacts of aquaculture itself, which transforms a common property and multiuse resource into a privately owned, single-use resource.

CONCLUSION

Aquaculture has not been the panacea it was touted to be. Current industrial aquaculture practices, which focus on commercially valuable carnivorous species, have led to decisions based on economic rather than environmental considerations. Without clear regulation designed to consciously tie aquaculture development to healthy natural ecosystems, the aquaculture industry is unlikely to develop to its full potential or continue to supplement ocean fisheries.

Much more can be done to make aquaculture a net environmental positive rather than an additional drag on vulnerable fish populations and ecosystems. For that to happen, governments must take the lead in protecting coastal ecosystems and in requiring responsible aquaculture practices. First, government policy should encourage aquaculture of native herbivorous and omnivorous fish rather than exotic carnivorous fish. To protect aquatic ecosystems, governments should regulate sea pens by setting maximum stocking limits to minimize eutrophication and dead zones; prohibiting pesticide use, except for short-term emergency treatment; and setting stringent waste management standards to prevent release of toxic substances into the environment. In addition, there must be rigorous design standards for pens and cages to reduce the risks of escape.

If public and private interests act jointly to reduce the environmental costs generated by fish farming, present unsustainable trends can be reversed, and aquaculture can make an increasingly positive contribution to global fish supplies.

With regard to transgenic fish, too many critical unknowns complicate risk assessment and management decisions. Without more information, it is not possible to make informed judgments about whether or how the technology can be safely exploited in an open-water commercial setting. According to the FAO Code of Conduct for Responsible Fisheries (FAO 1995), Article 7.5.1, nations should apply the precautionary approach widely to conservation, management, and utilization of living resources to protect them and conserve the aquatic environment. Under the precautionary approach, the absence of adequate scientific information is not a reason for postponing or failing to take conservation and management measures. Precaution dictates that any exploitation of transgenic fish be limited to land-based facilities for the foreseeable future.

References

Ackefors, H. E. 1999. Environmental impact of different farming technologies. In *Sustainable Aquaculture: Food for the Future*, eds. N. Svennevig, H. Reinertsen, and M. New. Rotterdam, The Netherlands: A. A. Balkema, pp. 145–169.

Aitken, M. 2002. Staggering extent of fish farm escapes: over a million salmon lost from cages threaten Scotland's wild stock with extinction. *The Mail on Sunday* April 14, 2002.

Alaska Department of Fish and Game. 2001. *Reported Escapes and Recoveries of Atlantic Salmon in Washington State, British Columbia, and Alaska.* Available online at www.adfg.state.ak.us/special/as/docs/esc_rec87-01.pdf

Aleström, P. 1996. *Genetically Modified Fish in Future Aquaculture: Technical, Environmental and Management Considerations.* Available online at www.agbios.com/docroot/articles/02-254-005.pdf

Aleström, P., Kisen, G., Klungland, H., and Andersen, Ø. 1992. Fish gonadotropin-releasing hormone gene and molecular approaches for control of sexual maturation: development of a transgenic fish model. *Molecular Marine Biology and Biotechnology* **1**: 376–379.

Allsopp, W. H. L. 1997. Aquaculture performance and perspectives. In *Global Trends: Fisheries Management*, eds. E. L. Pikitch, D. D. Huppert, and M. P. Sissenwine. Bethesda, MD: American Fisheries Society, pp. 153–165.

Alsea Valley Alliance. 2001. *Alsea Valley Alliance v. Evans*, 161 F. Supp. 2d 1154.

Anderson, J. 1997. The growth of salmon aquaculture and the emerging new world order of the salmon industry. In *Global Trends: Fisheries Management*, eds. E. L. Pikitch, D. D. Huppert, and M. P. Sissenwine. Bethesda, MD: American Fisheries Society, pp. 175–184.

Association to Protect Hammersly. 2002. *Association to Protect Hammersly v. Taylor Resources*, 299 F.3d 1007 (9th Cir. 2002).

Atlantic Salmon Federation. 2001. Catastrophic salmon escape prompts calls for moratorium on the aquaculture industry: largest documented escape ever in US or Atlantic Canada. Available online at www.asf.ca/Communications/2001/feb/catastrophe.html

Atlantic Salmon Federation. 2002. *Atlantic Salmon Aquaculture: A Primer*. Available online at www.asf.ca/backgrounder/asfaquacbackgrounder.pdf

BBC. 2003. Farm threat to wild salmon. October 20, 2003.

Beardmore, J. A. and Porte, J. S. 2003. *Genetically Modified Organisms and Aquaculture*, FAO Fisheries Circular No. 989. Rome: Food and Agriculture Organization of the United Nations. Available online at http://200.198.202.145/seap/pesquisa/pdf/Aquicultura/Tecnologia/2.pdf

Belton, B., Brown, J., Hunter, L., Letterman, T, Mosness, A., and Skledany, M. 2004. *Open Ocean Aquaculture*. Available online at www.mindfully.org/Water/2004/Aquaculture-Open-Ocean-IATP10Feb04.htm

Benbrook, C. 2002. *Antibiotic Drug Use in U.S. Aquaculture*. Available online at www.iatp.org/fish/library/antibiotics

California Fish and Game Commission. 2003. *December Meeting Summary*. Available online at www.dfg.ca.gov/fg_comm/2003/12-03-03summary.html

Cartagena Protocol. 2000. *Cartagena Protocol on Biosafety*. Available online at www.biodiv.org/biosafety/protocol.asp

CDC (Center for Disease Control). 1999. *Center for Disease Control Memo to Record, Use of Antimicrobial Agents in Aquaculture: Potential for Public Health Impact*. Available online at www.nationalaquaculture.org/pdf/CDC%20Memo%20to%20the%20Record.pdf

Central Statistics Office. 2003. *Fisheries Statistics 2002*. Available online at www.cso.ie/publications/agriculture/fishery.pdf

CEQ /OSTP. 2001. *Assessment: Case Studies of Environment Regulation for Biotechnology*, Case Study No. 1, *Growth Enhanced Salmon*. Available online at www.ostp.gov/html/ceq_ostp_study2.pdf

Clifford, S. L., McGinnity, P., and Ferguson, A. 1998. Genetic changes in Atlantic salmon (*Salmo salar*) populations of northwest Irish rivers resulting from escapes of adult farm salmon, *Canadian Journal of Fisheries and Aquatic Sciences* **55**: 358–363.

de la Fuente, J., Guillen, I., Martinez, R., and Estrada, M. P. 1999. Growth regulation and enhancement in tilapia: basic research findings and their applications. *Genetic Analysis* **15**: 85–90.

De Silva, S. S. 1999. Feed resources, usage and sustainability. In *Sustainable Aquaculture: Food for the Future, Proceedings of the 2nd International Symposium on Sustainable Aquaculture*, Oslo, eds. N. Svennevig, H. Reinertsen, and M. New. Rotterdam, The Netherlands: A. A. Balkema, pp. 221–244.

Devlin, R. H. and E. M. Donaldson. 1992. Containment of genetically altered fish. In *Transgenic Fish*, eds. C. L. Hew and G. L. Fletcher. Singapore: World Scientific Press, pp. 229–265.

Devlin, R. H., Yesaki, T. Y., Biagi, C. A., *et al.* 1994. Brief communication: Extraordinary salmon growth. *Nature* **371**: 209–210.

Devlin, R. H., Yesaki, T. Y., Donaldson, E. M., Du, S.-J., and Hew, C. L. 1995a. Production of germline transgenic Pacific salmonids with dramatically increased growth performance. *Canadian Journal of Fisheries and Aquatic Sciences* **52**: 1376–1384.

Devlin, R. H., Yesaki, T. Y., Donaldson, E. M., and Hew, C. L. 1995b. Transmission and phenotypic effects of an antifreeze/GH gene construct in coho salmon (*Oncorhynchus kisutch*). *Aquaculture* **137**: 161–169.

Devlin, R. H., Johnsson, J. I., Smailus, D. E., *et al.* 1999. Increased ability to compete for food by growth hormone transgenic coho salmon (*Oncorhynchus kisutch* Walbaum). *Aquaculture Research* **30**: 479–482.

Devlin, R. H., Biagi, C. A., Yesaki, T. Y., Smailus, D. E., and Bryatt, J. C. 2001. Brief communication: Growth of domesticated transgenic fish. *Nature* **409**: 781–782.

DOC (Department of Commerce). 1999. *U.S. Department of Commerce Aquaculture Policy*. Available online at www.nmfs.noaa.gov/trade/newgrant.htm#Top.

Donaldson, E. M., Devlin, R. H., Soler, I. I., and Piferrer, F. 1993. The reproductive containment of genetically altered salmonids. In *Genetic Conservation of Salmonid Fishes*, eds. J. G. Cloud and G. H. Torgaard. New York: Plenum Press, pp. 113–129.

Du, S. J., Gong, Z., Fletcher, G. L., *et al.* 1992. Growth enhancement in transgenic Atlantic salmon by the use of an "all fish" chimeric growth hormone gene construct. *Bio/Technology* **10**: 176–181.

Dunham, R. A. 2003. *Status of Genetically Modified (Transgenic) Fish: Research and Application*, FAO/WHO Expert Consultation on Safety Assessment of Foods Derived from Genetically Modified Animals including Fish. Available online at www.fao.org/biotech/index.asp?lang'en

Environmental Law Center. 2002. Judge approves landmark settlement of clean water act lawsuit against Heritage Salmon, Inc. Press release, July 29, 2002.

EPA (Environmental Protection Agency). 2002. *Effluent Limitations Guidelines and New Source Performance Standards for the Concentrated Aquatic Animal Production Point Source Category*, Proposed Rule. 67 Fed. Reg. 57871-57928 (September 12, 2002).

Ernst, W., Jackman, P., Doe, K., *et al.* 2001. Dispersion and toxicity to non-target aquatic organisms of pesticides used to treat sea lice on salmon in net pen enclosures. *Marine Pollution Bulletin* **42**: 433–444.

European Commission 2002. *Opinion of the Scientific Committee on Animal Nutrition on the Use of Canthaxanthin in Feedingstuffs for Salmon and Trout, Laying Hens, and Other Poultry*. Available online at http://europa.eu.int/comm/food/fs/sc/scan/out81_en.pdf.

FAO (Food and Agriculture Organization). 1995. *FAO Code of Conduct for Responsible Fisheries* Rome: Food and Agriculture Organization of the United Nations.

FAO. 2002a. *The State of World Fisheries and Aquaculture*. Rome: Food and Agriculture Organization of the United Nations. Available online at www.fao.org/docrep/005/y7300e/y7300e00.htm

FAO. 2002b. *Aquaculture Development and Management: Status, Issues and Prospects*. Rome: Food and Agriculture Organization of the United Nations. Available online at www.fao.org/docrep/meeting/004/y3277E.htm

Farrell, A. P., Bennett, W., and Devlin, R. H. 1997. Growth enhanced transgenic salmon can be inferior swimmers. *Canadian Journal of Zoology* **75**: 335–337

FDA (Food and Drug Administration). 2003. FDA Statement regarding GloFish, December 9. Available online at www.fda.gov/bbs/topics/NEWS/2003/NEW00994.html

FDCA. 2005. Food, Drug and Cosmetics Act, 21 U.S.C. § 321(v).
Federal Register. 1990. Research proposal on transgenic fish: publication of environmental assessment. 55 *Fed. Reg.* 5752-5757
Fernandez-Cornejo, J. 2004. *The Seed Industry in U.S. Agriculture: An Exploration of Data and Information on Crop Seed Markets*, Regulation, Industry Structure, and Research and Development, Agriculture Information Bulletin No. AIB786. Available online at www.ers.usda.gov/publications/aib786/
Fleming, I. A., Jonsson, B., Gross, M. R., and Lamberg, A. 1996. An experimental study of the reproductive behavior and success of farmed and wild Atlantic salmon (*Salmo salar*). *Journal of Applied Ecology* **33**: 893-905.
Fleming, I. A., Hindar, K., Mjølnerød, I. B., *et al.* 2000. Lifetime success and interactions of farm salmon invading a native population, *Proceedings of the Royal Society of London B* **267**: 1517-1523.
Fletcher, G. L., Goddard, S. V., and Hew, C. L. 2000. Current status of transgenic Atlantic salmon for aquaculture. In *Proceedings of the 6th International Symposium on the Biosafety of Genetically Modified Organisms*, eds. C. Fairbairn, G. Scoles, and A. McHughen, July 2000, Saskatoon, pp. 179-189.
Florida Department of Environmental Protection. 2004. *Mangroves: "Walking Trees."* Available online at http://wwdep.state.fl.us/coastal/habitats/mangroves.htm
FWS. 2000. *Endangered and Threatened Species: Final Endangered Status for a Distinct Population Segment of Anadromous Atlantic Salmon* (Salmo salar) *in the Gulf of Maine*. 65 *Fed. Reg.* 69459-01, November 17, 2000.
Garcia, S. and Newton, C. 1997. Current situation, trends and prospects in world capture fisheries. In *Global Trends: Fisheries Management*, eds. E. L. Pikitch, D. D. Huppert, and M. P. Sissenwine. Bethesda, MD: American Fisheries Society, pp. 2-27.
Gardar, J. 2002. 600 000 Faroese salmon on the run after storms. *Intrafish* February 28, 2002.
Hansen, P., Jacobsen, J. A., and Lund, R. A. 1993. High numbers of farmed Atlantic salmon *Salmo salar*, observed in oceanic waters north of the Faroe Islands. *Aquaculture Fisheries Management* **24**: 777-781.
Harden, B. 2004. Hatchery salmon to count as wildlife. *Washington Post* April 29, 2004.
Hedrick, P. W. 2001. Invasion of transgenes from salmon or other genetically modified organisms into natural populations. *Canadian Journal of Fisheries and Aquatic Sciences* **58**: 841-844.
Hites, R. A., Foran, J. A., Carpenter, D. O., *et al.* 2004. Global assessment of organic contaminants in farmed salmon. *Science* **303**: 226-229.
Howard, R. D., Martens, R. S., Innes, S. A., Drnevitch, J. M., and Hale, J. 1998. Mate choice and mate competition influence male body size in Japanese medaka. *Animal Behavior* **55**: 1151-1163.
International Center for Technology Assessment. 2004. *International Center for Technology Assessment. v. Thompson*, Docket No. 1:04-CV-00062-RMU, Complaint filed January 14, 2004.
Issues in Ecology. 2001. No. 8: *Effects of Aquaculture on World Fish Supplies*. Available online at www.esa.org/sbi/sbi_issues/issues_pdfs/issue8.pdf
Jönsson, E., Johnsson, J. I., and Björnsson, B. T. 1996. Growth hormone increases predation exposure of rainbow trout. *Proceedings of the Royal Society of London B* **263**: 647-651.
Kaiser, M. J. 2001. Ecological effects of shellfish cultivation. In *Environmental Impacts of Aquaculture*, ed. K. D. Black. Sheffield, UK: Sheffield Academic Press, pp. 51-75.

Kapuscinski, A. R. and Brister, D. J. 2001. Genetic impacts of aquaculture. In *Environmental Impacts of Aquaculture*, ed. K. D. Black. Sheffield, UK: Sheffield Academic Press, pp. 128-153.

Kapuscinski, A. R. and Hallerman, E. M. 1991. Implications of introduction of transgenic fish into natural ecosystems. *Canadian Journal of Fisheries and Aquatic Sciences* **48**: 99-107.

Kerry, J., Hiney, M., Coyne, R., et al. 1994. Frequency and distribution of resistance to oxytetracycline in microorganisms isolated from marine fish farm sediments following therapeutic use of oxytetracycline. *Aquaculture* **123**: 43-54.

Maclean, N. and Laight, R. J. 2000. Transgenic fish: an evaluation of benefits and risks. *Fish and Fisheries* **1**: 146-172.

Maclean, N. and Talmar, S. 1984. Injection of cloned genes with rainbow trout eggs. *Journal of Embryology and Experimental Morphology* **82**: 187.

MacMillan, J. R. 2001. Aquaculture and antibiotic resistance: a negligible public health risk? *World Aquaculture* **32**(2): 49-51, 68.

McDowell, N. 2002. Stream of escaped farm fish raises fears for wild salmon. *Nature* **416**: 571.

McGinnity, P., Stone, C., Taggart, J. B., et al. 1997. Genetic impact of escaped farmed Atlantic salmon (*Salmo salar* L.) on native populations: use of DNA profiling to assess freshwater performance of wild, farmed and hybrid progeny in a natural river environment, *ICES Journal of Marine Science* **54**: 998-1008.

McGinnity, P., Prodöhl, P., Ferguson, A., et al. 2003. Fitness reduction and potential extinction of wild populations of Atlantic salmon, *Salmo salar*, as a result of interactions with escaped farm salmon. *Proceedings of the Royal Society of London B* **270**: 2443-2450.

McKinnell, S. and Thompson, A. J. 1997. Recent events concerning Atlantic salmon escapees in the Pacific. *ICES Journal of Marine Science* **54**: 1221-1225.

Melamed, P., Gong, Z., Fletcher, G., and Hew, C. L. 2002. The potential impact of modern biotechnology on fish aquaculture. *Aquaculture* **204**: 255-269.

Mellon, M., Benbrook, C., and Benbrook, K. 2001. *Hogging It: Estimates of Antimicrobial Abuse in Livestock*. Cambridge, MA: Union of Concerned Scientists.

Monsanto. 2002. *Monsanto Canada Inc. v. Schmeiser*, September 4, 2002, Canada Federal Court of Appeals, 2. F.C. 165.

Morton, A. and Volpe, J. 2002. A description of escaped farmed Atlantic salmon *Salmo salar* captures and their characteristics in one Pacific salmon fishery area in British Columbia, Canada in 2000. *Alaska Fisheries Research Bulletin* **9**: 102-110.

Muir, W. M. and Howard, R. D. 1999. Possible ecological risks of transgenic organism release when transgenes affect mating success: sexual selection and the Trojan gene hypothesis *Proceedings of the National Academy of Sciences of the United States of America* **96**: 13853-13856.

Muir, W. M. and Howard, R. D. 2001. Fitness components and ecological risk of transgenic release: a model using Japanese medaka (*Oryzias latipes*). *American Naturalist* **158**: 1-16.

Myers, R. A., Levin, S. A., Lande, R., et al. 2004. Policy forum: hatcheries and endangered salmon. *Science* **303**: 1980.

Nambiar, K. P. 1999. Global market for fish and fishery products. In *Sustainable Aquaculture: Food for the Future, Proceedings of the 2nd International Symposium on Sustainable Aquaculture*, Oslo, eds. N. Svennevig, H. Reinertsen, and M. New. Rotterdam, The Netherlands: A. A. Balkema, pp. 246-262.

NASCO (North Atlantic Salmon Conservation Organization). 2000. *Guidelines on Containment of Farm Salmon*. Available online at www.nasco.int/pdf/nasco_res_slgguidecontain.pdf

National Aquaculture Association. 2004. *Drugs Used in U.S. Aquaculture Industry*. Available online at www.natlaquaculture.org/pdf/Drugs%20and%20Chemicals%20in%20US%20Aquaculture%2011.10.pdf

Naylor, R. L., Goldburg, R. J., Primavera, J. H., *et al.* 2000. Effect of aquaculture on world fish supplies. *Nature* **405**: 1017–1024.

Naylor, R. L., Eagle J., and Smith, W. L. 2003. Salmon aquaculture in the Pacific Northwest: a global industry. *Environment* **45**: 18–39.

NMFS. (National Marine Fisheries Service). 2004. *Enumeration of Threatened Marine and Anadromous Species*. 50 C.F.R. 223.102, Code of Federal Regulations. Washington, DC: U.S. Government Printing Office.

NOAA (National Oceanic and Atmospheric Administration). 1996. *Endangered and Threatened Species: Threatened Status for Central California Coast Coho Salmon Evolutionarily Significant Unit (ESU)*. 61 Fed. Reg. 56,138-56,149, October 31, 1996.

NOAA. 1997. *Endangered and Threatened Species: Threatened Status for Southern Oregon/Northern California Coast Evolutionarily Significant Unit (ESU) of Coho Salmon*. 62 Fed. Reg. 24588, May 6, 1997.

NOAA. 1998. *Endangered and Threatened Species: Threatened Status for the Oregon Coast Evolutionarily Significant Unit of Coho Salmon*. 63 Fed. Reg. 42587, August 10, 1998.

NRC (National Research Council). 2002. *Animal Biotechnology: Science-Based Concerns*. Washington, DC: National Academies Press. Available online at www.mindfully.org/GE/GE4/Animal-Biotechnology-Concerns-NRC-Aug02.htm

NRC. 2004. *Biological Containment of Genetically Modified Organisms*. Washington, DC: National Academies Press.

Ocean Studies Board. 2004. *Atlantic Salmon in Maine, Committee on Atlantic Salmon in Maine, National Research Council*. Washington, DC: National Academies Press.

Pauly, D. and Maclean, J. 2003. *In a Perfect Ocean*. Washington, DC: Island Press.

Pauly, D., Christensen, V., Dalsgaard J., Froese, R., and Torres, F. 1997. Fishing down marine food webs. *Science* **279**: 860–863.

Pauly, D., Bennett, E., Christensen, V., Tyedmers, P., and Watson, R. 2003. The future for fisheries. *Science* **302**: 1359–1361.

Pearson, T. H. and Black, K. D. 2001. The environmental impacts of marine fish cage culture. In *Environmental Impacts of Aquaculture*, ed. K. D. Black. Sheffield, UK: Sheffield Academic Press, pp. 1–31.

Pew Initiative on Food and Biotechnology. 2003. *Safety of Transgenic Animals Raises Questions*. Available online at http://pewagbiotech.org/newsroom/summaries/

Price Waterhouse Coopers. 2001. *Northern Aquaculture Buyers Guide*.

Rahman, M. A. and Maclean, N. 1999. Growth performance of transgenic tilapia containing an exogenous piscine growth hormone gene. *Aquaculture* **173**: 333–346.

Rahman, M. A., Mak, R., Ayad, H., Smith, A., and Maclean, N. 1998. Expression of a novel piscine growth hormone gene results in growth enhancement in transgenic tilapia (*Oreochromis niloticus*). *Transgenic Research* **7**: 357–369.

Rahman, M. A., Ronyai, A., Engidaw, B. Z., *et al.* 2001. Growth performance of transgenic tilapia containing an exogenous piscine growth hormone gene. *Journal of Fish Biology* **59**: 62–78.

Reuters. 2002. Caged Scottish salmon escape, threaten wild cousin. April 3, 2002.

Sandnes, K. and Ervick A. 1999. Industrial marine fish farming. In *Sustainable Aquaculture: Food for the Future, Proceedings of the 2nd International Symposium on Sustainable Aquaculture,* Oslo, eds. N. Svennevig, H. Reinertsen, and M. New. Rotterdam, The Netherlands: A. A. Balkema, pp. 97–107.

Saegrov, H., Hindar, K., Kalus, S. and Lura, H. 1997. Escaped farm salmon replace the original salmon stocks in the River Vosso, Western Norway, *ICES Journal of Marine Science* **54**: 1166–1172.

Saunders, R. 1991. Potential interaction between cultured and wild Atlantic salmon. *Aquaculture* **98**: 51–61.

Seattle Times 1997. Farm salmon escape pens. News Service, July 21, 1997.

Seattle Times 2004. Study finds higher level of toxins in farmed salmon. News Service, January 9, 2004.

Sen, A. 1999. *Development as Freedom.* New York: First Anchor Books.

Senate Bill. 2005. *National Offshore Aquaculture Act of 2005,* Senate Bill 1195. Available online at http://thomas.loc.gov

Stokstad, Eric. 2002. News focus: Engineered fish–friend or foe of the environment? *Science* **297**: 1797–1798

The Economist. 2003. The promise of a blue revolution. August 9, 2003.

Todd, C. D., Walker, A. M, Wolff, L., *et al.* 1997. Genetic differentiation of populations of the copepod sea louse *Lepeophtheirus salmonis* (Krøyer) ectoparasite on wild and farmed salmonids around the coast of Scotland: evidence from RAPD markers, *Journal of Experimental Marine Biology and Ecology* **210**: 251–274.

Traxter, G. S., Anderson, E., *et al.* 1999. Naked DNA vaccination of Atlantic salmon *Salmo salar* against IHNV. *Diseases of Aquatic Organisms* **38**: 183–190.

Tully, O., Gargan, P., Poole, W. R., and Whelan, K. F. 1999. Spatial and temporal variations in the infestation of sea trout (*Salmo trutta* L.) by caliged copepod *Lepeophtheirus salmonis* (Krøyer) in relation to sources of infection in Ireland. *Parasitology* **119**: 41–51.

USGS (U.S. Geological Survey). 2001. Got Fish? Already tired of that holiday gift aquarium? Think before you dump and create an even bigger problem. Available online at www.msgs.gov/newsroom/article.asp?ID=517

USGS. n.d. *Problems with the Release of Exotic Fish.* Available online at http://nas.er.usgs.gov/fishes/dont_rel.html

USPIRG (U.S. Public Interest Research Group). 2002. *United States Public Interest Research Group v. Heritage Salmon, Inc.,* (2002) D. C. Maine CIV-00-150-B-C. Available online at www.med.uscourts.gov/opinions/kravchuk/2002/MJK_02192002_1-00cv150_USPISG_v_Heritage.pdf

USPIRG. 2003. *United States Public Interest Research Group v. Atlantic Salmon of Maine,* (2003) D. C. Maine CIV. 00-151-B-C, CIV-00-149-B-C. Available online at www.med.uscourts.gov/opinions/carter/2003/GC_05282003_1-00cv151_USPIRG_v_AtlanticSal.pdf

Vannuccini, S. 2003. *Overview of Fish Production, Utilization, Consumption and Trade.* Rome: Food and Agriculture Organization of the United Nations. Available online at ftp://ftp.fao.org/fi/document/trends/overview/2001/commodit/2001fishery-overview.pdf

Volpe, J. P., Taylor, E. G., Rimmer, D. W., and Glickman, B. G. 2000. Evidence of natural reproduction of aquaculture escaped Atlantic salmon (*Salmo salar*) in a coastal British Columbia river. *Conservation Biology* **14**: 899–903.

Volpe, J. P., Anholt, B. R., and Glickman, B. G. 2001. Competition among juvenile Atlantic salmon (*Salmo salar*) and steelhead (*Oncorhynchus mykiss*): relevance to the invasion potential in British Columbia. *Canadian Journal of Fisheries and Aquatic Sciences* **58**: 197–207.

Watershed Watch. 2001. *Salmon Farms, Sea Lice and Wild Salmon*. Conquitlam, British Columbia: Wateshed Watch Salmon Society.

Weber, M. L. 2003. What price farmed fish. In *Title of Book*. Silver Spring, MD: SeaWeb Aquaculture Cleaninghouse, pp. 24–26.

Western Australia Department of Fisheries. 2006. *Invasive Aquarium Species*. Available online at www.fish.wa.gov.au/hab/broc/invasivespecies/aquarium/index.html

Wilkie, M. L. and Fortuna, S. 2003. *Status and Trends in Mangrove Area Extent Worldwide*, FAO Forest Resources Assessment Working Paper No. 63, Appendix I. Rome: Food and Agriculture Organization of the United Nations. Available online at www.fao.org/documents/show_cdr.asp?url_file=/DOCREP/007/J1533E/J1533E00.htm.

Zhu, Z. Y., Li, G., He, L., and Chen, S. 1985. Novel gene transfer into the fertilized eggs of the goldfish *Carassius acuratus* L. 1758. *Journal of Applied Ichthyology* **1**: 31–34.

Zhuikov, M. 2004. Dumping of aquarium fish causing trouble in Duluth or something's fishy in Rock Pond. Available online at www.greatlakesdirectory.org/mn/051204_great_lakes.htm

TRACY DOBSON AND HENRY A. REGIER

20

Contributing to fisheries sustainability through the adoption of a broader ethical approach

Fishing is a human right for the many, not for the few.[1]

INTRODUCTION

We begin with a background discussion highlighting the global fisheries crisis that implies need for expanded application of ethical consideration to fisheries. A general discussion of ethics follows to provide some context for particular applications to fisheries governance. We outline principles in law and policy that may reverse the present fisheries downward spiral. Discussed are the role of science and risk assessment, the precautionary principle, the public trust doctrine, an effective female work model, effective commons management, and the Food and Agriculture Organization (FAO) *Code of Conduct for Responsible Fisheries* (FAO 1995). Some recommendations are based on proven results; others rest on speculation, as one might expect within a regime of adaptive management. In each instance, we endorse changes in management behavior on the part of all participants that should result in sustainable and equitably shared fisheries.

Modern technology, that sometimes useful, sometimes dreadful set of human inventions, has facilitated the emergence of environmental crises around the globe. Among all species, we are the one that has found the most effective ways to escape natural constraints, for a period of time, by employing technology to satisfy our myriad, insatiable needs and desires. We know that early humans using simple techniques such as spears extinguished species in North America and

---

[1] Kurt Christensen, quoted in "Fisherman seeks to harvest ailing Baltic, gently." *New York Times*, July 26, 2003, p. A4.

Globalization: Effects on Fisheries Resources, ed. William W. Taylor, Michael G. Schechter, and Lois G. Wolfson. Published by Cambridge University Press. © Cambridge University Press 2007.

Australia. But now we have positioned ourselves to extinguish not just a species here and there, but thousands of species and vast ecosystems in a very short time-frame. Human activities are driving forces behind these mostly negative changes, which may be manifested as a cascade of ecosystem transformations resulting in degraded and unsustainable fisheries, among many other losses.

The globalization ideology now encompasses the world's fisheries. Industrial fishing operations catch and preserve fish for air delivery to a host of wealthy destinations. For example, diners in East Lansing, Michigan, may choose Caribbean grouper, Chilean sea bass, Atlantic salmon, or Pacific tuna from the array of offerings at Mitchell's Fish Market. The taste for fish for dining or for feed or a variety of other purposes has brought about the near demise of many species of importance to humans. A coalescence of spectacularly efficient fishing equipment and fish processing gear and local fish depletions has resulted in fishing fleets relentlessly combing the oceans' both coastal/national and international commons, leaving waters emptied of valued fish in their wake, until major commercially significant stocks (and their ecological associates) have lost health and viability (Ellis 2003; Myers and Worm 2003; Pauly and Maclean 2003; Pew Oceans Commission 2003; Murphy 1994).

We also argue that a significant contributor to the global fisheries crisis is the dominant economic practice: the currently prevalent version of state-subsidized capitalism. Though it has spurred technological and economic growth, its ethical tenets also result in negative consequences that are economic (Stiglitz 2002; Chua 2003) but spread inevitably to the social and ecological domains (O'Connor 1994; Bakan 2004). The history of the decline of the world's magnificent fisheries reveals commercialization and profit-taking as driving forces (Bakan 2004). In analyzing the question "Is capitalism sustainable?" O'Connor (1994) theorizes that it is a system where capitalists withdraw value and shift the costs to labor and the state. A good specific example is the story of the cod fisheries. As eloquently described by Kurlansky (1997), the cod built the early New England economy. Because these fish were so extraordinarily abundant, it didn't take humans long to see and seize the economic potential for a cod industry. History further shows that surpassing economic growth unfolded, with riches for the few and dangerous work for fishers viewed as labor inputs. In the end of the story, this originally rich species was driven to near extirpation by human exploitation in most of its range (Pauly and Maclean 2003). Hegemonic capitalism not only drives resource depletion, it also destroys the sense of community that may be necessary for long-term survival and certainly is necessary for peace and harmony to prevail (Bakan 2004). In fisheries, this is evidenced by the inability of fishers, managers, and scientists to agree to much-needed fisheries closures until too late. Another example of the societal damage that can be inflicted is the increasing gulf between rich and poor in the United States as compared with more egalitarian institutions and policies in some other democracies (Smeeding 2004). Indeed, the United States is becoming more and more like so-called Third World nations in this respect (Chua 2003).

We may decide to continue to walk this path of environmental destruction until our own species hits the wall, or we may choose to change our ways so that our kind and others may continue to live on this planet for the foreseeable future. We may choose to climb out of our denial and take individual and collective action based on the obvious: we cannot continue to expand economic growth and use of material resources without end. We are the one species with the knowledge and compassion to make this choice, to preserve ourselves and other creatures through love and reason (Sanders 1998; Morito 2002). It is in this spirit that we explore in this chapter the case for intensifying our profession's commitment to the ethics of stewardship of global fisheries. Through our analysis and recommendations, we hope to contribute to thought and discussion in the fisheries management literature about these critical issues.

We will begin with a discussion of ethics generally as background to a consideration of applications to fisheries governance. We outline the adoption in law and policy of principles that will forward the shift from the present fisheries downward spiral. Our recommendations include changes in management behavior that have proven effective in other situations.

ETHICS BACKGROUND

We begin by acknowledging that a focus on sustainability implies an anthropocentric perspective, and our analysis seeks to achieve what will sustain our species in the long run. (We do not limit our ethics to a posteriori or consequentialist considerations but also emphasize a priori or deontic considerations [see below].) As it happens, this sustainability approach will necessarily involve giving high value to other species, in this case most notably fish, upon which humans depend in a variety of important ways. It also involves developing knowledge through various mechanisms such as scientific experimentation and shared experience, as joined together in adaptive management, to create a base of facts and inferences. Further, we need to decide what should be done with this accumulated knowledge. The "shoulds" are developed through a human process, because humans have the requisite mental capacity to do this, that employs values to develop guidance about moral thought and action (Morito 2002). Thus systems of ethics are devised to protect critical human values through norms, policies, laws, religious doctrines, and so forth, and, ideally, ethics steer human behavior to desirable practice.

By "ethics" we mean moral principles that are converted into pragmatic guidelines specifying "good behavior." Fisheries governance and participation should be guided by ethical principles that will ultimately result in the continuation of our complex earthly biological system because we have identified it as a critically important value. Two large classes of ethical principles come together in a commitment to sustainability. One class relates to considerations of duty that extend beyond direct human self-interest and to which we commit before we act; these may be termed a priori deontic considerations. Another class relates to considerations that relate to human material interests that may extend beyond immediate concerns and indefinitely into the future; these may be termed a posteriori, consequentialist, and/or utilitarian considerations. In an ecosystem approach, as now commonly invoked with respect to fisheries, both deontic and consequentialist considerations are emphasized.

The combined moral principle that undergirds our thinking in this respect holds that a web of interconnected beings/organisms, as it has emerged naturally, is a good that should be sustained. Behaving in ways that facilitate the support and maintenance of a healthy web of living things across the planet should result in achieving a dynamic (recognizing dynamism as a key ecological principle) sustainability over the long term, or as far as we can see. In contrast, behavior that forecloses future choices is probably a bad thing because it cuts strings from the web of life. Removing top predators from the Great Lakes Basin, for example, produces ripples far into the future. One of the most painful and visible effects of loss of lake trout in Lake Michigan in the 1960s (from overfishing, exotic sea lamprey predation, chemical contamination, etc.) was the annual spring mortality of exotic alewife which then accumulated in nauseating drifts on beaches near Chicago (Loftus and Regier 1972). An example from elsewhere, the near extinction (through human exploitation) of North American sea otters in the early 1900s caused the collapse of populations of bald eagles and seals. The sea otters eat sea urchins, which if left unchecked destroy the habitat for seal and eagle prey species (Roush 1989). Thus, we argue for a fisheries ethics to provide general guidance that will need to be interpreted sensitively in many different contexts. Agreeing with Morito (2002), we hold that all species are "loci of valuational activity" and are "part of the network of loci that constitute the moral community." Because of humans' particular characteristics, they are in position to engage in thinking and rendering judgments unlike other species (at least, as far as we know). Ethical decision-making includes democratic and inclusive processes, reliance on the precautionary principle (see below), and seeking to avoid doing harm.

A caveat: the intellectual domain of ethics may be delineated narrowly so as to exclude specifically esthetic and/or spiritual considerations. Here spiritual considerations may include those that may be termed "numinous," i.e., referring to a spirit or divinity that may reside in a special place or object. Our intention is not to marginalize esthetic and numenic considerations but rather to emphasize ethical considerations that are consistent with widely appreciated esthetic qualities and deeply felt numinous responsibilities. These, too, are consistent with an ecosystem approach.

ROLE OF SCIENCE

Having sketched the relevant ethical framework, we move on to consider the important place held by science and its associated disciplines in the present era. Unquestionably, good science is a meaningful ingredient in policy development and governance. It is necessary but not sufficient, however. Its data-based theories may be useful in forecasting some features of the future, but it sometimes blinds us from seeing the bigger picture as we become mesmerized by details. With its emphasis on replicability and objectivity, science has a strong tendency to exclude from consideration non-quantitative information that, along with the data gathered and analyzed by scientists, is essential to reaching the understanding needed for decisions. Perspectives from other cultures that rely on different epistemologies may thus be eliminated from policy-making (Morito 2002).

A better meta-paradigm, termed an ecosystem approach, emerged in the Great Lakes Basin several decades ago with strong leadership by fisheries experts. It has come to be applied elsewhere with variations appropriate to different contexts, as with respect to oceanic fisheries, for which Busch *et al.* (2003) identify three dimensions:

- include stakeholders' perspectives and human goals
- consider the health and vitality of ecosystems into the indefinite future
- include the larger landscape and connections among other landscapes.

This broadened vision, if widely employed, should prove beneficial.

We should note that even the very best science may not be incorporated in decision-making. Indeed, particular scientific findings are all too often ignored because of political considerations (Cairns 2001), failure to present findings in a user-friendly format, or scientific arrogance (Dodd 2000). It is not our purpose to provide a detailed critique of data-driven science here, and we restrict our additional comments to the following section on risk assessment because it stands as an example of a helpful tool that may be misused to the detriment of fisheries governance.

RISK ASSESSMENT: SCIENCE VS. ETHICS?

Overreliance on a posteriori utilitarian ethics and related algorithms, e.g., as embraced in risk assessment, indirectly contributes to losses of environmental values and to increased levels of insecurity (O'Brien 2000). We may have entered a period in history where important risks are unquantifiable (Coxe 2003). This presents a particularly daunting challenge within the vast, watery commons of our planet. Moreover, in our rush to exploit, we may have replaced ethics with risk assessment. In some instances, it provides seeming legitimacy to actions that, when we scratch beneath the surface, are in fact, based on politics, economics, or some other non-scientific basis, as a "scientific" cover for the true ambitions and objectives. In this connection, greater commitment to a priori deontic or duty-related ethics should help reduce pressure on stocks, slow habitat destruction, and prevent some exotic introductions.

The future reality of interest to investors implicitly includes ecological issues like those mentioned in our introduction. Western governments have been moving toward "managing" environmental issues, for example, using algorithms of quantified risks for purposes of prioritizing particular protective and corrective actions. From Coxe's perspective, such quantification may become progressively less feasible in our new period of history.

His notion is not new to well-informed and intelligent experts in risk assessment. With respect to environmental issues in Western cultures, Ravetz and others have been documenting a similar notion for several decades (see Ravetz 1999). In his view, "Now we face the paradox that while our knowledge continues to increase exponentially, our relevant ignorance does so, even more rapidly. And this is ignorance generated by science!" In support of that judgment call, Ravetz (1986) used the issue of nuclear energy. Regier (1988) used the issue of contaminants in the Great Lakes to illustrate a similar inference. The increasing difficulty of quantification is yet another indication of the danger of overreliance on risk assessments (O'Brien 2000).

We describe the risk assessment process to better illustrate our argument. According to Morito (2002), a risk assessment process should proceed along the following lines. Construction of a pollutant-emitting facility on the lake is proposed. The permitting authority begins to collect information about the degree of threat from the proposal to human health, to other animals and plants, and to the ecosystem generally. In addition to other sources of information, the authority seeks a risk assessment. Science contributes through expert analyses which, insofar as possible, identify and quantify risks. In the second stage of a proper risk assessment process, an assessment is created that brings in social acceptability through consultation with affected individuals and groups. Unfortunately, in some cases, the outcome of the first stage may articulate a clarity that ignores the degree to which components of risks may be unknown or unquantifiable, and the second stage may rely on expert opinion rather than that of affected communities and persons. Such risk assessments are inherently flawed and unethical. Those who will be affected must be involved in making the decision about acceptability of risk. Excluding them or providing incomplete or incompetent analysis is in direct conflict with the ethical standards set out earlier. Compounding the problem, some decision-makers may weigh the risk assessment as the decisive factor rather than viewing it as an additional source of information, giving it more weight that it deserves in view of its inherent limitations, even under the best circumstances.

With widespread awareness that at least some important environmental and other risks are unquantifiable and may well remain so, why are pro-business governments implementing, embracing, and giving significant weight to quantitative risk assessments in environmental governance processes? This may be part of a strategy of crippling environment-related stewardship within governments because "excessive red tape" (environmental protection) slows down or, in some cases, halts economic development. It may also provide a decision-maker faced with significant uncertainty and overwhelming political pressure a basis for making a decision or, cynically, a fig leaf of scientistic legitimacy for decisions really made on other, seemingly less publicly acceptable grounds by a bureaucrat who is expected to act definitively as a technocrat on issues that may well transcend that person's technical competence.

Another way to understand this dilemma is with the help of a schema by B. Wynne, as cited by Healy (1999), that distinguishes between various notions related to "risk." In Wynne's schema:

> Risk – system behavior is known, and outcomes can be assigned probabilistic distributions.
> Uncertainty – important system parameters are known, but not the probability distributions.
> Ignorance – what is not known is not known.
> Indeterminacy – causal chains, networks, or processes are open and thus defy prediction.

Through this framework, we can point out yet another way in which risk assessment, like cost–benefit analysis, can be used to provide biased, inaccurate, incomplete information or to reach predetermined ends. In effect, governmental embrace of a risk assessment approach may result in duly accredited experts fitting any and all aspects of a problem that might fall into one of the second and third classes into the first class, resulting in an assessment that substantially masks its true basis. So the risk assessment approach, as it is currently being implemented with respect to environmental issues in the United States and Canada, may be more political than scientific in essence. To the extent that the process is scientific it may be more adversarial than objective. We argue that for the risk assessment tool to provide useful insight and contribute to a balanced and fair decision, it must be used in a transparent decision-making process and must not be permitted to dictate outcomes.

The formal legal system as an adversarial process puts great emphasis on due process in which the rhetoric of contending parties can be assessed fairly, say, by an expert judge or by a jury of citizen peers. Conventions that are roughly comparable to formal legal procedures in this respect are coming to be developed with respect to risk assessment on environmental issues, considered in a broad sense (Ravetz 1999), resulting in winners and losers and outcomes that may be less than optimal. While the process works well in the legal system, its use in scientific contexts is likely less advisable. Instead, peer review should remain a central strength of scientific knowledge creation.

We argue that, in the context of infusing science with an ethics that values creating and maintaining sustainable fisheries, approaches in discovering and using knowledge needed in the effort should be chosen to expand perspectives and invite debate rather than settling on closed and rigid visions of acceptability.

In this context, O'Connor's (1999) contrasts of the epistemological and ethical stances of control-oriented technocrats with reciprocity-oriented democrats according to the following dualisms strengthen our point. Within each dualism, the first relates to the control mode and the second to the reciprocity mode.

- Laplacian reconciliation – all knowledge shall, ideally, be integrated within a single and internally consistent conceptual framework vs. dialogical reconciliation – a diversity of perspectives and modes of understanding coexist in irreducible plurality.
- The Cartesian epistemology – privileges "objective" description (leading to universal knowledge) and explanation based on axiomatic formulations of categories for system description and behavior vs. the complexity epistemology – postulates an irreducible plurality of pertinent analytical perspectives for a situation of enquiry.
- Domination ethic – knowledge is conceived in instrumental terms, allowing the knowing subject to act upon and control the interaction with the object; calculation, prediction, and contractual certainty are privileged vs. hospitality ethic – knowledge is pursued and exploited based on forms of courtesy and dialog; tolerance of tensions, and admission of (legitimate) antagonisms that may imply (mortal) combats; dignity is important.

With respect to fisheries and related environmental issues in our Great Laurentian Basin, there had been a concerted push for several decades toward reciprocal as opposed to control approaches in governance programs. During the past decade, however, political strategists seeking to weaken and shrink formal government institutions to the advantage of free-enterprise business interests may have seen an opportunity to subvert effective governance by urging a return to emphasizing "sound science" but manipulating the science behind the scenes to achieve desired political ends. The Bush administration's refusal to participate in the Kyoto Protocol stands as a good example of such governance. In that case, it was reported that despite an international scientific consensus, the administration insisted that more research was needed to determine whether climate change was occurring and, if it was, what caused it. At the same time, findings of senior Environmental Protection Agency (EPA) scientists that contradicted this position were deleted from a government report. Additional evidence of such a trend in the U.S. federal bureaucracy is presented by Urstadt (2003), who documents what may be a flagrant case of such subversion within the U.S. federal government. He concludes that the Data Quality Act as an extension of the Paperwork Reduction Act is intended not to improve data but, through constant contention, to suppress it, with production of reams of pointless extra paperwork.

We turn now to laying out some approaches that we believe will address global fisheries crises ethically and pragmatically.

PRECAUTIONARY PRINCIPLE

The inclination to discount information that lacks precision or full coverage, in other words, where uncertainty or ignorance or indeterminacy exists, seems at least as common as the proclivity to choose the explanation and data that lead to the desired outcome in the face of seemingly overwhelming evidence to the contrary on, say, fish stock abundance in the North Atlantic (Pauly and Maclean 2003). In a similar vein, Cairns (2001) opines that human hubris and "exuberant optimism for economic growth" in effect result in denial of ecological facts. These tendencies demonstrate a dangerous propensity to choose opportunistic fantasy over disinterested fact and, for example, to cling to an unrealistic faith in human ability to "control nature" or blind faith in nature's "unlimited" fecundity. It is the "live for today" mentality of the grasshopper in the parable involving the industrious ant. This behavior further connects with the economic "rational man" who always selfishly chooses that action that will most directly benefit him without regard for how it harms his neighbors, his community, or his environment (see, e.g., Russell [2001] and Wikipedia [2004]). While some look to the distance and envision the future and how actions in the present may shape it, others, who seem to hold a large percentage of the power positions in the corporate sector and government, wear glasses that see only 12 to 18 months out. Unfortunately, incentives within the political and financial systems cause these individuals to employ short-term thinking that focuses typically on building the "bottom line," in the case of business, and winning re-election, in the case of government. In the U.S. House of Representatives, the election cycle interval is 2 years, so campaigning for the next election begins the day after an election, and campaigning and fund-raising never end.

Happily for the environment, for water and fisheries, countervailing forces advocate adoption of the precautionary principle as one way to respond to our knowledge shortcomings so that valuable species and ecosystems are not lost. Following the principle dictates that we make no rush to judgment, no snap decisions where gaps or fuzziness in knowledge exist. Moreover, it reverses a commonly employed regulatory approach of responding to problems after they occur and waiting until the scientific evidence is irrefutable. As greater and greater numbers of humans have come to recognize more clearly the fallibility of human knowledge and the knowledge production system (i.e., science) on virtually every subject, we have moved closer to accepting the precautionary principle. It is incorporated in the Rio Declaration (1992), the Treaty of the European Union (EEA 2001) and FAO policy statements, including the 1995 Code of Conduct for Responsible Fisheries (see below). The European Community is incorporating the precautionary principle in its Common Fisheries Policy (EC 2004). At this point in our discussion, providing specific, fisheries-relevant content to the principle will clarify our argument. One useful definition is found in the Gilchrest–Farr Fisheries Recovery Act (2000): "exercising additional caution in favor of conservation in any case in which information is absent, uncertain, unreliable, or inadequate as to the effects of any existing or proposed action on fish, essential fish habitat, other marine species, and the marine ecosystem in which a fishery occurs." Cairns' (2001) formulation of the precautionary principle finds seven themes:

> A willingness to take action in advance of formal justification of proof.
> Proportionality of response.
> A preparedness to provide ecological space and margins for error.
> A recognition of the well-being and interests of non-human entities.
> A shift in the onus of proof onto those who propose change.
> A greater concern for impacts on future generations.
> A recognition of the need to address ecological debts.

If major political actors such as governments and transnational corporations can be persuaded to fold these themes into policy and action, the survival rate of species and ecosystems should be significantly enhanced.

As appropriate and reasonable as the adoption of a precautionary approach might seem, acceptance is far from widespread, as evidenced by the ever-declining quality of many ecosystems due to human activities. We argue here that increasing evidence of environmental decline in fish stock quality and abundance should be taken as an alarm bell telling us that the time has long since come for fishers, fishing industry companies, fisheries managers, and government, working with the public, to adopt the precautionary principle as a basic tenet of sustainable fisheries governance. Scientific management under "maximum sustainable yield" (MSY) has proven itself an abysmal failure. As noted by Francis, we find ourselves in a new situation where the expected time-frame for feedback from our actions and the pace at which it is actually occurring no longer match. Feedbacks are no longer immediate, so that our overfishing actions may not create serious problems until our grandchildren are starving (Francis 2002). We need better, surer guiding principles, and the precautionary principle shines as one with much to offer.

The precautionary principle is not beholden to any particular philosophy. We could hold that duty to future generations requires that we assure that our actions do not eliminate fisheries. From a simplistic utilitarian perspective, it can be instrumental in preserving the greatest good for the greatest number. Relying again on Francis, it supports the notion that "the most important target is the long-term health of the interaction between nature, the economy, and the legal system." Closing fishing areas for periods of time, creating marine protected areas or aquatic reserves, and creating marine conservation trusts (Bratspies 2003) will reduce short-run fish catch. Continued fishing pressure in many fisheries, however, may produce fish today but is likely to lead to extirpation or extinction in the near or medium term. What does application of the precautionary principle suggest in this situation? It prescribes that conservation measures be applied in circumstances where declines are precipitous and steady, as they are in many areas currently. Reducing fishing pressure for a period of time now will give us the opportunity to learn more about species, habitats, and ecosystems so that guidelines may be developed that will allow some exploitation but at sustainable levels. Use of the precautionary principle also would suggest that whatever exploitation system is devised after a recovery period is allowed for fish stocks, we should reduce the suggested exploitation level to provide a greater margin of safety. We have learned that safety margins are needed with dynamic systems. In this way, we are more likely to be able to sustain a greater degree of stability in exploitation level rather than simply driving into the next crash, necessitating long-term fishing area closures again, further damaging livelihoods and economies (not to mention risking the loss of species and habitats).

Integrating the precautionary principle in local, national, regional, international, and corporate policy will also lead to greater levels of interdisciplinary collaboration in knowledge production as we attempt to develop the most complete understanding possible, viewing from multiple disciplinary perspectives. Increasingly, calls are made for drawing on collaboration among many disciplines to address conservation issues. Biological data are useful, but decision-makers need also to consider history, culture, politics, psychology, law, and other disciplines to formulate a properly holistic and successful paradigm (International Summit on Science and the Precautionary Principle 1998; Ewel 2001; Ehrlich 2002). Members of the Society for Conservation Biology are launching one such effort. Beginning in 2001, a working group of experts coming from many disciplines lumped into "social science" formed to better promote focused discussion about cross-disciplinary work. Furthermore, the organization's board maintains designated positions for scholars from social science and the humanities.

PUBLIC TRUST DOCTRINE REVIVAL

An additional valuable support for achieving fisheries sustainability is the public trust doctrine. Fundamentally rooted in a utilitarian stance, the public trust doctrine, long buried in law books, is experiencing a resurrection that could also be drawn upon to conserve fisheries ecosystems. Codified by Roman emperor Justinian in 529 (*Codex Justinianus* 529), who stated, "By the law of nature these things are common to all mankind, the air, running water, the sea and consequently the shores of the sea." In less distant history, the public trust doctrine was embraced by the English (1225, 1647), French (1000), and Spanish (1200) and carried into their colonies, including the United States, where it resides as part of the common law of the 50 states. Its use and significance ebbed and flowed until 1970, when Professor Joseph Sax, seeking to motivate and strengthen the growing body of environmental protection legislation, moved the public trust doctrine front and center in his seminal explication in the *Michigan Law Review*, one of the country's most prestigious legal journals (Sax 1970). Historically looked to primarily by those seeking support for protection of or public access to aquatic resources, it has won judicial backing for protecting watersheds and ecosystems. In 1983, the California Supreme Court (majority members later recalled for this decision) halted Los Angeles' plan to drain ecologically rich but non-navigable Mono Lake (California Supreme Court 1983). Previous decisions had focused on *navigable* waters. In essence, the public trust doctrine charges the government to act as a trustee for the public, to protect the public's continuing right to natural resources (if we broaden from the narrow, traditional focus on navigable waters).

The principle that the people have a right to access and exploit the commons as well as to expect governmental protection of the commons, though being challenged currently by privatization (an offshoot of capitalism) and other forces, can be found in most if not all countries. Some with environmental frameworks of recent vintage, such as Malawi, have enshrined the public trust doctrine in national constitutions or statutes. Wherever it resides in countries' legal frameworks, it adds another important principle to undergird actions to preserve fisheries or to stand in defense against destructive actions. Giving full weight to the doctrine would require that countries protect the interests of the public to healthy and sustainable fisheries over those of private entrepreneurs or their own revenue enhancement through the sale of public rights (Illinois Central Railroad v. Illinois 1892). For instance, if the public trust doctrine were the controlling principle, essential mangrove wetlands would be held a critical public resource in India which must be protected rather than permitting private parties to remove them to introduce shrimp aquaculture. Such a holding rooted in public trust is essential to a variety of fish species whose young reside in the mangroves during certain life stages.

Though the concept and application of the public trust doctrine is well known in legal and some other circles, it needs to be promoted throughout the relevant domains of fisheries management and science. Its potential power as a tool in the hands of fisheries managers and the public is such that substantial attention should be devoted to raising awareness about its existence and force. It seems that common law concepts fall to the background and typically surface only in infrequently read court decisions. Publicizing the law of public trust will empower the public to demand more long-term vision in management and will empower managers to withstand the politics of special interests. Our new fisheries ethics requires that long-term ecosystem integrity and equity steer our governance course for the foreseeable future. The holism implied in these ethics is also located in the public trust doctrine as it encompasses a substantial component of the non-human environment. Through such decisions government would secure the public interest in fisheries and plant its flag in the heart of commons governance over individualistic and fisheries-destroying private profit-taking.

A WORD ON COMMONS GOVERNANCE

The problem of fish stock depletions, extirpations, and extinctions plays out in areas in which numerous fishers, from the small-scale/artisanal to the industrial factory trawler fleet, have rights to exploit stocks. This raises questions about access and governance. The nature of governance will be briefly explored here, drawing on Dobson *et al.* (2002) on managing migratory species in the Great Lakes Basin, especially Lake Erie.

Processes of allocation of goods and bads among stakeholder humans and other creatures permeate all of nature and culture and must be addressed systemically. No technical legal fixes or silver bullets will suffice in specific cases, let alone in generic cases. And no universalistic systemic solutions to this issue will be found, if reality is perceived as a complex of evolving living things in a four-dimensional, nested spatio-temporal mosaic. But a set of partial guidelines to a balance of rights and responsibilities can perhaps be inferred for various classes of ecosystemic phenomena. In practice, such guidelines depend on both a priori deontological and a posteriori consequentialist ethical principles, as well as on esthetic and numinous or spiritual considerations. Such considerations are seldom teased apart in the course of decision-making; instead, a kind of decent pragmatism subsumes them tacitly. We proceed here in such a pragmatic, if optimistic, way.

With respect to the management of human uses of natural resources, a simplistic and partial understanding of frequently experienced difficulties was exaggerated as a *tragedy of the commons* by Hardin in 1968. Using many historical and contemporary case studies, socio-ecologists then showed that a *tragedy of the commons* can occur when a resource (e.g., harvestable yield of goods, assimilative capacity for bads) is freely accessible and open to use by anyone. But such extreme openness was not a common occurrence historically in natural/cultural ecosystems anywhere in the world. The socioecologists corrected some of the shortcomings of Hardin's rather generalized description, etiology, and corrective treatment, and provided cases of a *comedy of the commons*. But the socioecologists' contributions have been less appealing to neo-liberal conservatives with an urge to privatize than was Hardin's original piece. So a tragedy of Hardin's tragedy continues as misinformation or disinformation in some political circles. Ill-informed privatization may entrain unbearably high transaction costs, which a private property owner may try to externalize unfairly to others. Ostrom and others now include consideration of both tragedies and comedies in their study of commons-oriented governance regimes and refer to a balanced approach as *drama of the commons* (Ostrom *et al.* 2002).

In many areas of the planet, natural resource governance now proceeds as a combination of governments, non-governmental organizations (NGOs), and interested individuals cooperating in some kind of partnership. Many mechanisms are employed to facilitate this new, more inclusive and, it is hoped, more successful approach to resource conservation. Such mechanisms are normally devised and employed in response to a crisis situation involving a common property/pool resource.

Countless versions of an implicit common property resource approach to governance have emerged among humans over thousands of years. From an ecological perspective, this construct overlaps with such notions as niche differentiation among organisms of different species and the complementarity of selfishness and selflessness within successful selective processes within evolution, etc.

Until recently, the formal studies within this approach to governance within the common property resource research community seem to have been limited mostly to small socioecological systems or small cultural–natural ecosystems in which the immediate users of a common property resource are empowered to participate actively in the relevant governance process. In cases where the local cultural–natural ecosystem is nested and intermeshed within a larger regional one, a co-management form of common property resource governance may emerge. Where such co-management forges an explicit link between local governance and regional or national governance, the adjective "cross-scale" may be added to the co-management term (Berkes 2002).

Apparently the common property resource approach has, as yet, seldom been extended to governance of fisheries in which different sectors of a fishery place quite different demands on the available resources, e.g., ceremonial, artisanal, recreational, extensive capture commercial, and intensive aquaculture commercial fishers. Also, the common property resource approach as such may have seldom been applied to cases of migratory or straddling stocks of fish that are subject to harvest by a complex of fishers on each side of a jurisdictional boundary that plays a prominent role in the interjurisdictional governance of the relevant cultural–natural ecosystem, but this situation is changing.

Through "stakeholder analysis," the common property resource approach has been recently extended to governance of fisheries in which different sectors of a fishery place quite different demands on the available resources. The International Development Research Centre in Canada and the International Union for the Conservation of Nature and Natural Resources have been using such stakeholder analyses, according to Berkes (2002). With fisheries, direct stakeholders could include ceremonial, artisanal, recreational, extensive capture commercial, and intensive aquaculture commercial fishers.

Also according to Berkes (2002), there is a growing literature on international and global common property resource regimes. But it has been difficult to apply it to cases of migratory or straddling stocks of fish because the exclusion and subtractability aspects cannot be addressed directly. With Oran Young's (2002) approach to governance regime formation, interjurisdictional arrangements like co-management can be crafted to include cross-scale interactions (Ostrom *et al.* 2002). The current global fisheries crises necessitate a focus on governance questions, on how we can better link (and build where needed) local, regional, national, and international governance institutions to collaborate in remediation and conservation to achieve sustainable fisheries development. These common property resource approaches are encompassed within our proposed ethics rooted in collaboration, ecosystem integrity, and equity.

INFLUENCE OF GENDER: THE FEMALE APPROACH

As we seek models, frameworks, and strategies to induce change in fisheries governance and fishing behavior, we find that women environmental protection advocates provide a useful example of effective conservation. Thus, we suggest that a "female collaborative model" should be employed. In brief, women provide diverse role models of highly effective action, including an ability to target, organize, and act decisively and successfully in the face of crises. By highlighting the strengths of women's efforts in conservation, we do not mean to demean those of men. We must be honest in revealing our opinion, however, that the prevailing impoverished social paradigm of male dominance has been less effective than one in which women and men work together in a cross-gender partnership where preconceived notions do not dictate roles based on gender characteristics, and a social shift in this direction is occurring (Regier and Kay 2001). One of the myriad benefits of this shift, at least as demonstrated in human responses to date, is that women who are educated and valued outside of the household tend to have much smaller families (Sen 1997). Among the impacts of this outcome is a smaller human population that will reduce demand for fish!

Until quite recently, it was assumed that protection of the environment was primarily a male pursuit. Research on the history of environmentalism in the United States reveals, however, that women were deeply involved from its beginning (Gottlieb 1993; Taylor 1997). Sometimes led by credentialed women, most often conservation crusades were (and are) led by middle- and working-class housewives who simply came to the task when someone was needed (see, e.g., Joseph 2004). Along with the hundreds of specific environmental protection campaigns led primarily at the grassroots level, women such as Rachel Carson and Sandra Steingraber wrote sweeping indictments of business and government regulation-as-usual that permitted and even facilitated enormous environmental destruction. Indeed, Carson (a scientist and a writer), who wrote first in *The Sea around Us* (1950) of the wonder of the world's ocean ecosystems, is credited with launching the modern environmental movement in 1962 with her *Silent Spring*, which eloquently and cogently detailed the horrific damage caused by herbicides and pesticides. *Living Downstream* (Steingraber 1999) continues the tale with reports of the impacts of ubiquitous and toxic synthetic chemicals in the United States.

What motivates women's activism, and why is it often successful? With respect to the first question, many answers have been offered. To generalize and simplify, it is argued by some scholars that women have a strong sense of the future as it could and should be lived through their children (Shiva 1988). They are, in effect, a living embodiment of the precautionary principle because they are ready to take action to preserve a healthy future for their issue. Harkening back to our earlier discussion on risk assessment, we could say they are typically highly risk averse. Indeed, attitude surveys generally indicate that women are most concerned about local environmental degradation because of concerns about family, while men more often focus concern on more distant or global issues such as climate change and ozone layer depletion (Solomon *et al.* 1989; Mohai 1992; Stern *et al.* 1993). Other scholars suggest that women are inherently closer to nature (which is frequently depicted as a nurturing mother). Another perspective argues that, like nature, women are marginalized ("virgin" land/resources and women to be exploited for male purposes) and so stand with it in solidarity as they might with other oppressed human groups (indigenous peoples, people of color, the disabled). They were so directly dependent on nature previously (and still are in developing countries) that they were driven to act to protect it because they were protecting their very lives (Shiva 1988; Merchant 1996). Salleh (1994) argues that it is particularly in moments of crisis that women glimpse and act on hidden political potentials as a consequence of their current position "inside/outside relations of production." This position makes them more likely and able to see and to act on environmental threats.

To explain the high degree of success of women-led conservation, we look to the literature that compares women's and men's work styles as averaged statistically and not as exhibited by every individual. Women tend to be persevering, goal-oriented workers who often adopt a collaborative, team approach. Their emphasis typically is on problem-solving rather than personal gain from successful or visible outcomes (Kanter 1993). This approach causes them to focus on tasks and to involve all whose interests and expertise are pertinent to the cause, from the grassroots to the head of state, from affected community members to scientists and politicians. Women's usual strategy is to consult widely and to try to keep all participants well informed, resulting in stronger commitment and continuing participation and trust among team members (Eagly and Johannesen-Schmidt 2001). Employing broad and continuing consultation is also a political strategy that demonstrates their political savvy. The success of the Love Canal campaign led by Lois Gibbs, which concluded in 1978 with President Jimmy Carter's order for the government to buy out the residents, rested on effective politics as much as anything else (Gibbs 1997). Some biologists contend that male competitive behavior is hardwired, as is women's cooperative behavior (Wilson 1975, 1978; Blum 1997).

The female model we propose is part and parcel of the ethical components set out above that focus on collaboration, inclusiveness, and respect for others unlike ourselves. It is a model that will best achieve reduced fishing pressure and fish habitat destruction. Effective leaders will draw participants from all interested and affected groups and maintain open lines of communication. Something like this has happened in recent decades in some of the remedial action plans under the Great Lakes Water Quality Agreement of 1978 as amended in 1987, and in some of the adaptive management processes under the aegis of the Great Lakes Fisheries Commission, but more is needed. This approach can also be found embedded in the FAO's impressively comprehensive Code of Conduct for Responsible Fisheries.

FAO CODE OF CONDUCT FOR RESPONSIBLE FISHERIES

An outgrowth of the Law of the Sea Convention and discussions in Rio de Janeiro at the 1992 Earth Summit, among others, the 1995 FAO Code of Conduct for Responsible Fisheries (CCRF) seeks to provide an all-encompassing policy which, if followed, would likely rehabilitate fisheries worldwide. Carefully and thoughtfully crafted, it identifies virtually every imaginable source of fisheries degradation, from the obvious overfishing to nations' failure to regulate fishing behavior of vessels flying their flag to unselective gear and poor monitoring. It is an admirable effort that has the potential for real impact if the peoples of the world who are connected in any way with fishing enact it through a concerted and ethical desire for fisheries equity today as well as in the future. It recognizes the importance of fishing for human nutrition, employment, and trade, as well as the need for research and monitoring of fishing activities and aquatic ecosystems. Though voluntary, it seeks adoption and implementation of its principles and fisheries management strategies by FAO member and non-member states alike. In other words, its goal is to achieve adherence to its principles throughout the fisheries of Planet Earth. It recognizes that, to achieve the code's goals, governments must work not only with one another but also with all organizations in the field, including regional and global fishing organizations, NGOs, and local fishers. Moreover, the code specifically states that a "precautionary approach" and the "best scientific evidence available" should be employed with a view to long term sustainability of fisheries.

Is the crisis sufficiently apparent within the fishing industry and its governance and politics that the majority who care about fisheries will promote and adopt the CCRF? Knowing that, no matter how great the degradation, some desperate or dastardly individuals and businesses will continue fishing depleted stocks, destroying habitat, and introducing exotics, will the concerned majority collaborate with like-minded local, regional, national, and international bodies to track down and punish transgressors so that implementation will have a chance to work? Bratspies (2003) reminds us that, for this project to achieve success, voluntary compliance is essential. The combined available enforcement resources of all nations of the world are insufficient for enforcement if violation is widespread. Thus, we need a strong majority of fisheries participants who are knowledgeable and committed to equity and long-term sustainability of the fisheries.

The important role that voluntary participation must play in developing sustainable fisheries leads us to consider the fundamental nature of humans, to speculate on whether, in the fisheries context, humans are capable of exercising the necessary self-restraint to achieve sustainability. Fehr and Rockenbach (2003) reported in *Nature* that "Altruistic cooperators are willing to cooperate ... although cheating would be economically beneficial for them." Seemingly flying in the face of economics' "rational man" theory, this conclusion holds even when the altruism of others will occur not simultaneously with their own but at some future time. Indeed, Fehr and Rockenbach conclude that "altruistic cooperation is an important behavioral force."

Concomitantly, their work and research on willingness to adopt the precautionary principle suggests that willingness to be altruistic or to abide by this principle will depend to a significant degree on the amount of trust that exists among the participants. Establishing trust between collaborators is fundamental to making progress. The transaction costs of political and other interactive processes spiral out of control in the absence of a sufficient measure of trust among the actors. The *drama of the commons* approach (Ostrom et al. 2002) is rooted in trust between stakeholders. Building a trustworthy governance process is also embedded in the CCRF, and it calls for processes that are transparent and timely. High levels of trust between participants and in scientific findings result in a significantly increased willingness to live by the precautionary principle as well as much more altruistic behavior (Fehr and Rockenbach 2003).

If this is an accurate assessment of the power of altruism, we may have a powerful, if unwieldy, force to apply to the global fisheries crises. According to a 2000 FAO progress report, CCRF implementation has been steady but slow (Doulman and Willmann 2006). Dissemination, coordination, and monitoring constitute a massive project that is taking years to unfold. At the international level, four specific plans of action have been drafted to spell out in detail practices to address particular code provisions. Addressed to date are: fishing capacity, sharks, seabirds, and illegal, unregulated, and unreported fishing (FAO 2006). Replies to an FAO survey from 69 countries, however, indicated that 70 percent of those who should be concerned with the CCRF didn't know about it (Doulman and Willmann 2006). This is clearly not a task for the faint of heart! Diligence and perseverance, with an eye to the benefits for global ecosystems and our great-grandchildren, must stimulate continuing commitment and effort. Working from the bottom up and the top down, connecting through the mid-levels of governance as advocated by Dobson et al. (2002) for Great Lakes Basin fisheries governance, guided by the ecosystem approach, will create the multilevel interconnected partnerships needed to conserve fisheries globally.

CONCLUSION

Globally, virtually every important fishery is in trouble. We suggest that hegemonic and state-subsidized capitalism, dated technology, "scientific" risk assessment, and lack of governance coordination are at least partially responsible for this environmental disaster. We have argued that, as ethical beings concerned about the health and viability of ecosystems the world over, we should adopt responsible behaviors that are more likely than not to lead to conservation and sustainability of fisheries. Greater commitment to duty-related ethics as well as long-term utility-related ethics, in combination with evidence-driven science, should help reduce pressure on stocks, slow habitat destruction, and prevent some exotic introductions.

Specifically, we urge that scientific organizations and their members should review annual meeting programs and scholarly journal emphases to make environment and resource ethics a core issue. Our undergraduate and graduate programs associated with environmental concerns should contain components that include the study of environmental ethics. Once more thoroughly infused in education, it should become a normal part of the concerns taken up in all aspects of natural resources and fisheries management.

Broadly inclusive processes should be employed at all levels of governance, from the local to the international and global, to reinvigorate existing strategies and create and deploy new strategies to conserve our enormous fisheries wealth so that it sustains us today but is also available for future generations. Fisheries are critical components of the web of life. We also believe that fisheries merit continued existence whether or not they directly benefit humans. Respect for all beings, even in the absence of evidence of a beneficial role for particular species, accords with deepening respect among the diversity of *Homo sapiens* and results in greater earthly harmony. It is also a pretty smart hedge against our limited knowledge!

References

Bakan, J. 2004. *The Corporation: The Pathological Pursuit of Profit and Power*. Toronto, Ontario: Viking Canada.
Berkes, F. 2002. Cross-scale institutional linkages: perspectives from the bottom up. In *The Drama of the Commons*, eds. E. Ostrom, T. Dietz, N. Dolsak, *et al.* Washington, DC: National Academy Press, pp. 293–321.
Blum, D. 1997. *Sex on the Brain: The Biological Differences between Men and Women*. New York: Penguin Books.
Bratspies, R. 2003. Finessing King Neptune: fisheries management and the limits of international law. *Harvard Environmental Law Review* **25**: 213–258.
Busch, W.-D. N., Brown, B. L., and Mayer, G. F. (eds.) 2003. *Strategic Guidance for Implementing an Ecosystem-Based Approach to Fisheries Management*. Silver Springs, MD: National Marine Fisheries Service, National Oceanic and Atmospheric Administration, U.S. Department of Commerce.
Cairns, J., Jr. 2001. Exuberant optimism vs the precautionary principle. *Ethics in Science and Environmental Politics* **2001**: 46–50.
California Supreme Court. 1983. Audubon Society *v.* Los Angeles Department of Water and Power, 464 U.S. 977; 104 S. Ct. 413; 78L. Ed. 2d 351.
Carson, R. 1950. *The Sea around Us*. New York: Oxford University Press.
Carson, R. 1962. *Silent Spring*. New York: Houghton Mifflin.
Chua, A. 2003. *World on Fire: How Exporting Free Market Democracy Breeds Ethnic Hatred and Global Instability*. New York: Random House.
Coxe, D. E. 2003. *The New Reality of Wall Street*. Toronto, Ontario: McGraw-Hill.
Dobson, T., Regier, H. A., and Taylor, W. W. 2002. Governing human interactions with migratory animals, with a focus on humans interacting with fish in Lake Erie: then, now, and in the future. *Canada–U.S. Law Journal* **28**: 389–446.
Dodd, D. 2000. *The Great Gulf: Fishermen, Scientists, and the Struggle to Revive the World's Greatest Fishery*. Washington, DC: Island Press.
Doulman, D. J. and Willmann, R. 2006. *The Future of the Code of Conduct*. Rome: Fisheries Department, Food and Agriculture Organization of the United Nations.
EEA. 2001. *Consolidated Version of the Treaty Establishing the European Community*, Title XIX, *Environment*, Article 172.2. Brussels: EEA.
Eagly, A. and Johannesen-Schmidt, M. 2001. Leadership styles of women and men. *Journal of Social Issues* **57**: 781–797.
EC (European Community). 2004. *Common Fisheries Policy*. Available online at www.jncc.gov.uk/marine/fisheries/reports/rpt_precautionary.html
Ehrlich, P. 2002. Human natures, nature conservation, and environmental ethics. *BioScience* **52**: 31–43.
Ellis, R. 2003. *The Empty Ocean*. Washington, DC: Island Press.
Ewel, K. C. 2001. Natural resource management: the need for interdisciplinary collaboration. *Ecosystems* **4**: 716–722.
FAO (Food and Agriculture Organization). 1995. *Code of Conduct for Responsible Fisheries*. Rome: Food and Agriculture Organization of the United Nations.
FAO. 2006. Implementation of the 1995 FAO Code of Conduct for Responsible Fisheries, International Institutions and Liaison Service. Available online at www.fao.org/fi/ipa/ipae.asp

Fehr, E. and Rockenbach, B. 2003. Detrimental effects of sanctions on human altruism. *Nature* **442**: 137–140.

Fisheries Recovery Act. 2000. HB 4046, 106th Congress, 2000.

Francis, R. C. 2002. Some thoughts on sustainability and marine conservation. *Fisheries* **27**(1): 18–22.

Gibbs, L. 1997. *Dying from Dioxin: A Citizen's Guide to Reclaiming our Health and Rebuilding Democracy.* Montreal, Quebec: Black Rose Books.

Gottlieb, R. 1993. *Forcing the Spring: The Transformation of the American Environmental Movement.* Washington, DC: Island Press.

Hardin, G. 1968. The tragedy of the commons. *Science* **162**: 1243–1248.

Healy, S. 1999. Extended peer communities and the ascendance of post-normal politics. *Futures* **31**: 655–669.

Illinois Central Railroad v. Illinois. 1892. 146 U.S. 387, 13 S. Ct. 110, 36 L. Ed. 1018.

International Summit on Science and the Precautionary Principle. 1998. *Summit Report.* Available online at www.uml.edu/centers/LCSP/precaution/back.over.html

Joseph, S. 2004. Women advocates in Great Lakes conservation. Master's thesis, Department of Fisheries and Wildlife, Michigan State University, East Lansing, MI.

Kanter, R. 1993. *Men and Women of the Corporation.* New York: HarperCollins.

Kurlansky, M. 1997. *Cod: A Biography of the Fish that Changed the World.* New York: Putnam.

Loftus, K. H. and Regier H. A. (eds.) 1972. Proceedings of the Salmonid Communities in Oligotrophic Lakes (SCOL) Symposium, 1971. *Journal of the Fisheries Research Board of Canada* **29**: 613–986.

Merchant, C. 1996. *Earthcare: Women and the Environment.* New York: Routledge.

Mohai, P. 1992. Men, women, and the environment: an examination of the gender gap in environmental concern and activism. *Society and Natural Resources* **5**: 1–9.

Morito, B. 2002. *Thinking Ecologically: Environmental Thought, Values, and Policy.* Black Point, Nova Scotia: Fernwood Publishing.

Murphy, R. 1994. *Rationality and Nature: A Sociological Inquiry into a Changing Relationship.* Boulder, CO: Westview Press.

Myers, R. A. and Worm, B. 2003. Rapid worldwide depletion of predatory fish communities. *Nature* **423**: 280–283.

O'Brien, M. 2000. *Making Better Environmental Decisions: An Alternative to Risk Assessment.* Cambridge, MA: MIT Press.

O'Connor, M. 1994. Introduction: Liberate, accumulate – and bust? In *Is Capitalism Sustainable: Political Economy and the Politics of Ecology*, ed. M. O'Connor. New York: Guilford Press, pp. 1–22.

O'Connor, M. 1999. Dialogue and debate in a post-normal practice of science: a reflexion. *Futures* **31**: 671–687.

Ostrom, E., Dietz, T., Dolsak, T. N., et al. (eds.) 2002. *The Drama of the Commons.* Washington, DC: National Academy Press.

Pauly, D. and Maclean, J. 2003. *In A Perfect Ocean: The State of Fisheries and Ecosystems in the North Atlantic Ocean.* Washington, DC: Island Press.

Pew Oceans Commission. 2003. *America's Living Oceans: Charting a Course for Sea Change.* Philadelphia, PA: Pew Charitable Trust.

Ravetz, J. R. 1986. Usable knowledge, usable ignorance: incomplete science with policy implications. In *Sustainable Development of the Biosphere*, eds. W. C. Clark and R. E. Munn. Cambridge, UK: Cambridge University Press, pp. 415–432.

Ravetz, J. R. (ed.) 1999. Special issue: Post-normal science. *Futures* **31**.
Regier, H. A. 1988. Will we ever get ahead of the problem? In *Aquatic Toxicology and Water Quality Management*, Vol. 1, eds. J. O. Nriagu and J. S. S. Lakshminarayana. New York: John Wiley, pp. 1–5.
Regier, H. A. and Kay, J. J. 2001. Phase shifts and flip flops. In *Encyclopedia of Global Environmental Change*, vol. 5, ed. R. E. Munn. Chichester, UK: John Wiley. pp. 422–429.
Roush, G. J. 1989. The disintegrating web: the causes and consequences of extinction. *Nature Conservancy Magazine* **39**: 6.
Russell, C. S. 2001. *Applying Economics to the Environment*. New York: Oxford University Press.
Salleh, A. 1994. Nature, woman, labor, capital: living the deepest contradiction. In *Is Capitalism Sustainable: Political Economy and the Politics of Ecology*, ed. M. O'Connor. New York: Guilford Press, pp. 106–124.
Sanders, S. R. 1998. The stuff of life. *Audubon* July/August, 1998.
Sax, J. 1970. The public trust doctrine in natural resource law: effective judicial intervention. *Michigan Law Review* **68**: 471–566.
Sen, A. 1997. Population: delusion and reality. In *The Gender Sexuality Reader: Culture, History, Political Economy*, eds. R. Lancaster and M. Di Leonardo. New York: Routledge, pp. 89–106.
Shiva, V. 1988. *Staying Alive: Women, Ecology and Development*. London: Zed Books.
Smeeding, T. M. 2004. *Public Policy and Economic Inequality: the United States in Comparative Perspective*, Working Paper No. 367, Luxembourg Income Study Working Paper Series. Syracuse, NY: Maxwell School of Citizenship and Public Affairs, Syracuse University.
Solomon, L. S., Tomaskovic-Devey, D., and Risman, B. J. 1989. The gender gap and nuclear power: attitudes in a politicized environment. *Sex Roles* **21**: 401–414.
Steingraber, S. 1999. *Living Downstream: An Ecologist Looks at Cancer and the Environment*. Reading, MA: Addison Wesley.
Stern, P. C., Dietz, T., and Kalof, L. 1993. Value orientations, gender and environmental concern. *Environment and Behavior* **25**: 322–348.
Stiglitz, J. 2002. *Globalization and Its Discontents*. New York: W. W. Norton.
Taylor, D. 1997. American environmentalism: the role of race, class and gender in shaping activism 1820–1995. *Race, Gender and Class* **5**: 16–59.
Urstadt, B. 2003. One-act farce: deregulation by disputation. *Harper's Magazine* June 2003: 52–53.
Wikipedia. 2004. http://en.wikipedia.org/wiki/Homo_economicus.
Wilson, E. O. 1975. *Sociobiology: The New Synthesis*. Cambridge, MA: Harvard University Press.
Wilson, E. O. 1978. *On Human Nature*. Cambridge, MA: Harvard University Press.
Young, O. 2002. Institutional interplay: the environmental consequences of cross-scale interactions. In *The Drama of the Commons*, eds. E. Ostrom. Washington, DC: National Academy Press, pp. 259–291

Part V Conclusions and recommendations

TRACY L. KOLB AND WILLIAM W. TAYLOR

21

Globalization and fisheries: recommendations for policy and management

INTRODUCTION

Fisheries have an inherent ecological and sociological nature (Moyle and Cech 1982), so optimally, fisheries are managed to sustain both components. Finding this balance becomes increasingly complex in a globalized world. Specifically, a globalized world is one in which information, communication, and transportation innovations are combined with loss of cultural boundaries, as well as aggressive trade liberalization, growth in transjurisdictional investment, deepening integration of capital markets, and accelerated speed of capital relocation (Conca 2001). Globalization makes the scope of complexity for environmental regulation increasingly ineffective (Friedman 2000). This book covers the current debate about the effect of globalization on world fisheries, delivers a review and synthesis of research in the field, and, most importantly, illustrates how globalization has become one of the primary drivers influencing the sustainability of the world's fisheries resources.

However, global fisheries are also an industry and every industry can be broken down into its supply chain components. A supply chain is the sequence of steps, often completed by different firms and locations, that produce a final good from primary factors. The supply chain starts with raw materials, continues with production, and ends with final assembly, distribution, and wholesale (Deardorff 2005). Global fish production constitutes a supply chain, delivering fish and fish products or recreation to customers throughout the world (Fig. 21.1).

Because of the complexity of the global fisheries industry, we believe that it is useful to organize our summary through a supply chain analysis. This approach allows us to assess the impacts of

Globalization: Effects on Fisheries Resources, ed. William W. Taylor, Michael G. Schechter, and Lois G. Wolfson. Published by Cambridge University Press. © Cambridge University Press 2007.

```
Raw material  →  Production  →  Manufacturing Distributing  →  Wholesaler  →  Customer/Consumer
```

Raw material:
Biological inputs
Political, social, and economic infrastructure

Production:
Fish production wild/farmed

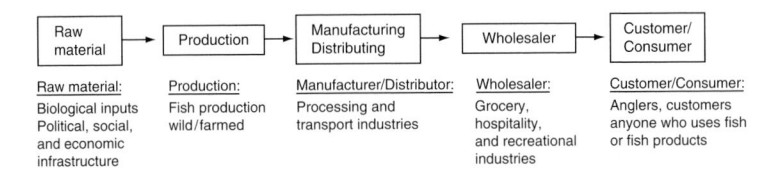

Manufacturer/Distributor:
Processing and transport industries

Wholesaler:
Grocery, hospitality, and recreational industries

Customer/Consumer:
Anglers, customers anyone who uses fish or fish products

Figure 21.1 The fisheries supply chain.

globalization on the global fisheries supply chain, and, using information provided by the authors of this book's chapters, identify intervention points for improving fisheries sustainability (Tables 21.1 and 21.2).

GLOBALIZATION'S EFFECTS ON THE FISHERIES SUPPLY CHAIN

Fisheries supply chain: raw materials

We define "raw material" in the fisheries supply chain as ecosystem productivity, which is a function of quantity and quality of habitat. For the purpose of this synthesis, raw material also includes existing political, economic, technological, and social structures that are currently in place for harvesting and managing fish and fish products. Globalization has positively affected the supply chain's impacts on fisheries sustainability through global awareness of fish declines and support of projects to protect and improve these vulnerable stocks. Globalization has negatively affected the supply chain's impacts on fisheries sustainability by facilitating a reduction in marine ecosystem productivity, while concurrently influencing governments to wholesale their fisheries resources to outside commercial operations whose interests are often not aligned with local fisheries sustainability.

Several chapters detail the positive effects that globalization can have on the raw materials of the fisheries supply chain. For example, globalization has allowed for the rapid spread of remediation techniques for aquatic ecosystems through the interchange of important tools for sampling, analyzing, monitoring, reporting, and enforcing fisheries regulations while also allowing scientists and managers to share research information (see Taylor *et al.*, Chapter 1).

In most instances the effects of globalization are not as clear-cut. More often than not globalization can accelerate the remediation, as well as decline in fish production. For instance, advances in global communications, capital transactions, and transportation, while allowing money and materials designated for conservation efforts to move

Table 21.1 *Positive impacts of globalization on the fisheries supply chain and fisheries sustainability*

Effect	Supply chain location	Chapters
Global awareness of fish declines and support of projects to protect and preserve vulnerable stocks	Input	Taylor et al. (Chapter 1), Vincent et al. (Chapter 7)
Rapid spread and interchange of remediation, monitoring, sampling, analyzing, reporting, and enforcement technologies	Input	Taylor et al. (Chapter 1)
Large- and small-scale fishers can work together to decrease destructive fishing practices	Production	Pollnac (Chapter 9), Frank et al. (Chapter 16)
Increases access to aquaculture technologies providing a source of sustainable and safe protein, alleviating pressure on overfished stocks, and allowing for rehabilitation of diminished species	Production	Taylor et al. (Chapter 1), Alder and Watson (Chapter 2); Molnar and Daniels (Chapter 11), Taylor and Leonard (Chapter 12), Hegarty (Chapter 15), Peterson and Fronc (Chapter 17), Knudson and Peterson (Chapter 18), Bratspies (Chapter 19),
Access to technology increases aquaculture of farmed fish that do not require fish feed.	Manufacturing	Naylor et al. (Chapter 10), Molnar and Daniels (Chapter 11)
Homogenization of seafood processing standards while decreasing seafood discharge/waste	Manufacturing	Hegarty (Chapter 15)
New labeling regulations that inform retailers the commercial designation, production method, and catch area	Manufacturing, Wholesale	Hegarty (Chapter 15), Peterson and Fronc (Chapter 17)

Table 21.1 (cont.)

Effect	Supply chain location	Chapters
Increases demand for stable fish supply, favoring aquaculture	Wholesale	Taylor et al. (Chapter 1), Alder and Watson (Chapter 2), Molnar and Daniels (Chapter 11), Taylor and Leonard (Chapter 12), Hegarty (Chapter 15), Peterson and Fronc (Chapter 17), Knudson and Peterson (Chapter 18), Bratspies (Chapter 19)
Information available to customers allowing them to purchase sustainably	Customer	Seares et al. (Chapter 3), Peterson and Fronc (Chapter 17), Knudson and Peterson (Chapter 18), Dobson and Regier (Chapter 20)
Communication and technological advances provide public with the tools to organize through non-governmental organizations and conservation groups	Customer	Taylor et al. (Chapter 1), Taylor and Leonard (Chapter 12), Bavington and Kay (Chapter 14), Frank et al. (Chapter 16), Dobson and Regier (Chapter 20)
Dispersal of power to edges of supply chain, allowing customers to drive supply chain practices	Input, Production, Manufacturing, Wholesale Customer	Kolb and Taylor (Chapter 21)

Table 21.2 Negative impacts of globalization on the fisheries supply chain and fisheries sustainability

Effect	Supply chain location	Chapters
Advances in global communication, capital transactions, and transportation facilitate business transactions internationally, opening new trade channels for fish products	Input	Alder and Watson (Chapter 2)
Promotes privatization of industries that are not aligned with the goals of fisheries sustainability	Input	Seares et al. (Chapter 3)
Facilitates a reduction in fisheries stocks through efficient fishing, cold-storage, and transportation technologies	Production	Alder and Watson (Chapter 2)
Pressures developing countries to wholesale fishing access rights to commercial interests that aren't aligned with local fisheries sustainability	Input, production	Alder and Watson (Chapter 2), Holeck et al. (Chapter 6), Ruddle (Chapter 8), Naylor et al. (Chapter 10), Hegarty (Chapter 15), Peterson and Fronc (Chapter 17), Knudson and Peterson (Chapter 18)
Pressures the commercial fishing industry to emphasize short-term return on investment	Production	Alder and Watson (Chapter 2), Pollnac (Chapter 9), Naylor et al. (Chapter 10)
Increases market opportunities for bycatch	Production	Vincent et al. (Chapter 7)
Increases access to aquaculture technologies that decrease water quality, increase the potential for disease, and decrease genetic fitness	Production	Taylor et al. (Chapter 1), Alder and Watson (Chapter 2), Rose and Molloy (Chapter 4), Faisal (Chapter 5), Molnar and Daniels (Chapter 11), Peterson and Fronc (Chapter 17), Knudson and Peterson (Chapter 18), Bratspies (Chapter 19)

Table 21.2 (cont.)

Effect	Supply chain location	Chapters
Increase in trade of live species opening new markets for previously under-exploited aquarium species	Production/Manufacturing	Vincent et al. (Chapter 7)
Increased rates of transportation of exotic invasive species	Manufacturing	Taylor et al. (Chapter 1), Alder and Watson (Chapter 2), Rose and Molloy (Chapter 4), Faisal (Chapter 5), Holeck et al. (Chapter 6), Taylor and Leonard (Chapter 12), Folland and Schechter (Chapter 13)
Increases demand for stable fish supply, favoring aquaculture (see above)	Wholesale	Taylor et al. (Chapter 1), Alder and Watson (Chapter 2), Rose and Malloy (Chapter 4), Faisal (Chapter 5), Molnar and Daniels (Chapter 11), Peterson and Fronc Chapter (17), Knudson and Peterson (Chapter 18), Bratspies (Chapter 19)
Encourages wholesale industry to meet customer demands, regardless of whether these demands are sustainable	Wholesale	Peterson and Fronc (Chapter 17), Knudson and Peterson (Chapter 18)
Increasing the spatial and social complexity between production and consumption along the supply chain, insulating customers from the negative impacts of their consumption choices	Input, Production Manufacturing, Wholesale, Customer	Kolb and Taylor (Chapter 21)

Restructuring of the traditional corporate paradigm, centralization of capital and emphasis on outsourcing production, manufacturing and distribution to places with least restrictive environmental and labor regulations	Input, Production Manufacturing, Wholesale, Customer	Kolb and Taylor (Chapter 21)
Dispersal of power to edges of supply chain, allowing industries to drive supply chain practices	Input, Production, Manufacturing, Wholesale, Customer	Kolb and Taylor (Chapter 21)

more freely and quickly (see Vincent *et al.*, Chapter 7), also make it much easier to conduct business transactions internationally, accelerating and opening new trade channels for vulnerable fish and fish products (Arbo and Hersoug 1997; Alder and Watson, Chapter 2).

Globalization also influences economies in developing countries which are responsible for management of diverse aquatic ecosystems and fish stocks. For instance, more than three-quarters of World Trade Organization (WTO) members are developing countries and countries in transition to market economies. Both the WTO and the International Monetary Fund are internationally powerful political and economic entities that encourage countries to move towards a free-market economy, based on evidence that freer trade equals economic growth (WTO 2006). Freer trade includes actions such as removal or reduction of tariffs on imported goods, elimination of restrictions on foreign investment, increasing exports, decreasing subsidies, opening new domestic industries, and privatization of state-run industry, all of which can have tremendous consequences for ecosystem productivity and fisheries sustainability.

Two examples of these consequences are detailed in the chapter by Seares *et al.* (Chapter 3), The first example discusses how reform and privatization of the electricity sector in some tropical countries rich in freshwater resources has resulted in an increase in the number of hydropower dams. Generally dams fragment aquatic habitat and alter habitat conditions downstream of the dam (Hayes *et al.*, in press). The second example examines how removal of subsidies is shifting cotton farming from the United States to developing countries. Cotton is very vulnerable to pests, requiring multiple yearly applications of insecticides, herbicides, and defoliants prior to harvest (Seares *et al.*, Chapter 3). A shift in cotton production location from the United States to a developing county results in greater impact to freshwater ecosystems because environmental regulations in developing countries are less stringent.

For developing countries, establishing a free-market economy also generates foreign exchange to supplement the national budget or service the national debt (Alder and Watson, Chapter 2; Bavington and Kay, Chapter 14; Peterson and Fronc, Chapter 17; Knudson and Peterson, Chapter 18). However, developing countries can have a difficult time capitalizing on fisheries export markets because international protocols can be impediments, due to the high cost of technology, facilities or expertise needed to comply with international regulations, e.g., Hazard Analysis and Critical Control Points (HACCP) (see Alder and Watson, Chapter 2; Hegarty, Chapter 15). The fishing sector can therefore become highly valuable, and, unable to be fully utilized, encourages the sale of access rights to foreign fishing fleets.

The by-products of this arrangement are potentially destructive for both of the development goals: the growth and stabilization of the national economy and the sustainable harvest of the fishery resource (see Alder and Watson, Chapter 2; Holeck et al., Chapter 6; Ruddle, Chapter 8; Naylor et al., Chapter 10; Hegarty, Chapter 15; Peterson and Fronc, Chapter 17; Knudson and Peterson, Chapter 18; Bratspies, Chapter 19). Rather than working with countries to establish joint ventures and develop fishing industries within the host country, developed countries purchase fishing access rights outright, removing a significant source of long-term economic assets that could help a developing nation build and stabilize its financial infrastructure. Most critical for fisheries, commercial industry is extracting a resource that it has little responsibility or obligation to sustain for the future.

Fisheries supply chain: production

We define the production level for the fisheries supply chain as the harvest of wild and capture fish stocks. Globalization has affected production's impacts on fisheries sustainability negatively by accelerating overfishing through the spread of more efficient fishing technologies, emphasizing goals that conflict with sustainability and increasing the market for bycatch. Globalization of production has impacted fisheries sustainability with mixed results by facilitating the development of the global aquaculture industry.

Global capture fisheries production has decreased by some 3 million tonnes between 2000 and 2002 and continues to decrease (FAO 2004). Because of these downtrends, wild capture fisheries cannot keep expanding in terms of numbers of fishers and vessels, and the production industry has had to consolidate as opportunities diminish (FAO 2004). At the same time, globalization is facilitating the communication and dissemination of inexpensive technologies for vessels, gear, and navigational equipment that make it possible for fishing vessels to increase harvest through more effective fishing gear (Martinez 1995; Alder and Watson, Chapter 2).

The rise of a global economy has sharpened competition and squeezed profit margins in many industries, so that to be competitive, a company *must* sell to the biggest market in order to recompense in volume for dwindling profit margins. This is exemplified by the commercial fishing industry which has seen tremendous consolidation and vertical integration over the last 20 years (Alder and Watson, Chapter 2; Pollnac, Chapter 9; Naylor et al., Chapter 10). When companies become vertically integrated the structure of their supply chains is fundamentally altered. Resources are managed further away from where they are exploited so companies can shift production to locations where operational costs are lower. Globalization increases the speed with which capital can relocate, forcing companies to increase short-term return on investment in order to retain shareholders. The combination of these two effects leads to numerous incentives to overfish stocks in the short term. However, this is not always the case. In a few instances, where responsibility is recognized, large- and small-scale fishers are able work together, decreasing destructive fishing practices such as blasting, poisoning, and overfishing (Pollnac, Chapter 9; Frank et al., Chapter 16).

Because of these economic pressures, the fisheries supply chain is currently in crises as many fish stocks including Patagonian toothfish, tunas, coral-reef fish, whitefish, snapper, orange roughy, and Pacific salmon are currently being overfished (Alder and Watson, Chapter 2; Taylor and Leonard, Chapter 12; Bavington and Kay, Chapter 14; Knudson and Peterson, Chapter 18). Overfishing is a particularly pernicious problem because it ultimately diminishes the sustainable production of fish for food and recreation, thus limiting the economic productivity of stocks for years to come, restricting subsistence and recreational uses, and reducing genetic diversity and ecological resilience (Botsford 1997; Pauly et al. 1998).

Another dangerous consequence of globalization is the increase in market opportunities that utilize bycatch (Vincent et al., Chapter 7). Demand for previously unused bycatch has promoted additional sorting to extract these fishes. This is opening up new markets for previously unused fish. For instance, new markets have fostered continued global expansion of trawling for sea horses, because there are new incentives to promote trawling operations for sea horses into areas where target fishing would, on its own, be unprofitable (Vincent et al., Chapter 7).

Aquaculture is a central component of production along the FSC. There has been a constant downward trend since 1974 in the proportion of wild fish stocks offering potential for expansion. In contrast global production from aquaculture continues to grow in terms of both quantity and its relative contribution to the world's supply of fish for direct human consumption (FAO 2004).

The drawbacks that aquaculture can have on fisheries sustainability are numerous, including water quality concerns, invasive species release, disease, use of biotechnology, and drug use for enhanced growth (Taylor *et al.*, Chapter 1; Alder and Watson, Chapter 2; Rose and Molloy, Chapter 4; Faisal, Chapter 5; Molnar and Daniels, Chapter 11; Peterson and Fronc, Chapter 17; Knudson and Peterson, Chapter 18; Bratspies, Chapter 19). However, globalization via the transfer of inexpensive technologies is allowing for the growth of an internationally significant capture fish industry. This industry, if carefully regulated, has the potential to provide a source of relatively inexpensive, sustainable, and safe protein to an ever-increasing human population, alleviate pressure on overfished marine and freshwater fish stocks, and allow for rehabilitation of diminished, threatened, and endangered species (Taylor *et al.*, Chapter 1; Alder and Watson, Chapter 2; Molnar and Daniels, Chapter 11; Taylor and Leonard, Chapter 12; Hegarty, Chapter 15; Peterson and Fronc, Chapter 17; Knudson and Peterson, Chapter 18; Bratspies, Chapter 19).

Fisheries supply chain: manufacturing/distributing

The manufacturing component of the fisheries supply chain includes processors of fish or fish products and the transport–shipping industry. Globalization of processing and distribution positively impacts fisheries sustainably by providing the impetus for seafood safety standards that seek to minimize seafood waste. Globalization of production has mixed effects on fisheries sustainability by facilitating the growth of aquaculture, which has potentially mitigating effects on wild fish protein supply. Globalization of distribution has negatively impacted production effects on sustainable fisheries through the role of global shipping and transportation in accelerating the spread of invasive species.

At the manufacturing level of the fisheries supply chain, globalization is affecting both the fish that are processed, and the processing itself. For example, as mentioned above, the spread of technology used in aquaculture facilities is allowing for an increase in farmed fish such as tilapia, mussels, oysters, and catfish. These fish are now becoming a staple in the retail industry (Naylor *et al.*, Chapter 10; Molnar and Daniels, Chapter 11) and can be grown with feed containing little or no fishmeal, which is a clear advantage over carnivorous mariculture species that require high amounts of wild caught fish protein (Naylor *et al.*, Chapter 10; Molnar and Daniels, Chapter 11; Bratspies, Chapter 19).

By providing the technological infrastructure, political will, and global customer base, globalization of manufacturing is allowing for the homogenization of processing standards and standardization of seafood safety regulations (Hegarty, Chapter 15). When large quantities of food pass through many waypoints, extending the processing, storage, and distribution chain, control is more difficult. There is a greater risk of exposing food to contamination (Hegarty, Chapter 15). One effect of standardization of seafood safety requirements is pressuring companies to consolidate processing plants, thereby decreasing seafood discharge and waste, while allowing for greater utilization of fish products that otherwise would have had to be discarded.

The global economy has increased demand for and transport of goods. This demand has and continues to result in the transfer (both intentional and unintentional) of exotic species from one ecosystem to another on unprecedented scales. Exotic species are a critical impediment to sustainability of native species, especially in freshwater environments (Taylor et al., Chapter 1; Alder and Watson, Chapter 2; Rose and Molloy, Chapter 4; Holeck et al., Chapter 6; Taylor and Leonard, Chapter 12; Folland and Schechter, Chapter 13). Exotic species affect the health and productivity of fisheries and their ecosystems (Faisal, Chapter 5; Holeck et al., Chapter 6; Folland and Schechter, Chapter 13). It's also worth noting that globalization of international trade and technology is one of the foremost contributors to the increase in the exploitation of marine ornamental species due to improved ability to transport these fish alive (Vincent et al., Chapter 7).

Fisheries supply chain: wholesale

The wholesale level of the fisheries supply chain includes the retail fisheries industries, such as grocery, hospitality, and recreation. Globalization positively affects the impacts that the wholesale industry has on sustainability by facilitating the growth of the aquaculture, which is taking pressure off wild-caught fish stocks as well as becoming the leading supplier for the supermarket industry. By driving seafood labeling regulations, globalization of wholesaling is also having positive impacts for threatened and endangered fish stocks because consumers have the choice to consume seafood products that are sustainably fished. Globalization of wholesaling is having mixed impacts by facilitating an increase in overall supply and demand for seafood products, potentially putting pressure on global fish stocks.

Growing numbers of consumers are eating out (Peterson and Fronc, Chapter 17; Knudson and Peterson, Chapter 18). In the United States, more than half the seafood eaten is consumed in restaurants (Peterson and Fronc, Chapter 17). This trend to eat more and more meals away from home is boosting levels in global trade of farmed seafood products because retail – both grocery and hospitality – calls for increasing consistency of quantity, quality, and price. Potentially, because of its stability, aquaculture will supply the bulk of retail seafood needs, thereby alleviating some of the burden on wild stocks (Peterson and Fronc, Chapter 17).

Another substantial impact at the wholesale level of the fisheries supply chain is an increase in the homogenization of seafood safety standards. The standards are driving new labeling regulations for retailers in certain countries (Hegarty, Chapter 15). These labels inform retailers of "the commercial designation of the species, production method and catch area" (Food Standards Agency 2003). Retailers will thus be able to inform consumers which fish and seafood products have been produced with sustainable methods sending feedbacks to the other supply chain participants notifying them of consumers' demand for fish and seafood products that are sustainably produced (Peterson and Fronc, Chapter 17).

However, there is ambiguity to whether the current fisheries supply chain is adequately synchronized to assure an effective balance of supply and demand. In fact declines in fish and seafood populations may have the opposite effect of heightening the exploitation of wild-catch fisheries in order to keep up with customer demand (Peterson and Fronc, Chapter 17). Because supply is decreasing while demand is increasing, there will be incentives to overharvest fisheries even more, as further expansion of supply results in production from more marginal areas or by using more damaging methods (Peterson and Fronc, Chapter 17).

Fisheries supply chain: customer

The customer can be anyone who uses fish or fish products. Globalization has positive effects on fisheries supply chain consumer impacts by allowing information to reach customers so they can make informed choices about their consumption habits. However, globalization has negative effects on fisheries supply chain consumer impacts by increasing the spatial and social distance of transactions, insulating customers from the direct impacts of their own consumption choices.

Globalization of communication and information provides the public with tools to more efficiently organize through the development of non-governmental organizations (NGOs) and conservation groups (Taylor et al., Chapter 1; Taylor and Leonard, Chapter 12; Bavington and Kay, Chapter 14; Frank et al., Chapter 16; Dobson and Regier, Chapter 20). Like the retail industry for seafood products, labeling is also becoming more common in developed countries for consumers who are concerned with conservation (Hagerty, Chapter 15). In a globalized world, customers have the unique ability to influence the fisheries supply chain if they are committed to purchase products that support healthy environments and freshwater systems (Seares et al., Chapter 3; Peterson and Fronc, Chapter 17; Knudson and Peterson, Chapter 18; Dobson and Regier, Chapter 20).

However, even if the industry can respond with sustainable management practices, consumers or those who supply consumers may or may not be willing to pay for the change (Peterson and Fronc, Chapter 17). This is partly because the global supply chain disrupts the traditional local feedback mechanisms by increasing the spatial and social distance between production systems and consumers. In a globalized world both the fishing industry and its customers are no longer able to see the direct relationship between their own consumption patterns and corresponding ecological effects, which leads them to support ecological, social, and economic policies that are unsustainable.

FISHERIES AND SUSTAINABILITY: A MUTUAL COEXISTENCE

Globalization of communication and trade has promoted an exponential increase in the diversity of the world's markets and provided multiple entry points for individuals to influence fisheries sustainability. This means that no longer are the only financial titans investment companies, banks, and governments. Today, everyday, global citizens can use the Internet to manage and trade their financial resources in all of the markets of the world. This coalition of customers is growing, expanding, and becoming a significant arm of power and change (Friedman 2000). It is this emerging individual power that guides our final recommendations for governments, managers, industry, and customers as they face the challenge of creating sustainable fisheries for the future.

This book covers three major consequences of globalization. The first consequence is a sole international focus on short-term economic gain that will accelerate the depletion of natural resources. This depletion will be exacerbated by insufficient monitoring devices, ineffective multijurisdictional governance, and poorly defined property rights (Vincent et al., Chapter 7). The second consequence is that countries will lower environmental standards to attract investment, leading to a "race to the bottom" where all regard for environmental issues is lost in the competition with one another (Martinez 1995). The third consequence is a perpetuation of our current economic system which insulates producers and traders from the full environmental costs of their actions. These costs, such as loss of future production, when taken into account, would compel companies to adopt sustainable practices (Daly 1993; Goldsmith 1996).

These arguments encapsulate many of the negative effects of globalization on fisheries sustainability, and have been discussed at length throughout the book, with suggestions for prevention and mitigation. However, there are three consequences of globalization's impacts on the fisheries supply chain that have profound impacts on societies and environment that require additional discussion. These are: increasing complexity with greater spatial and social distance between production and consumption along a given supply chain, restructuring of the traditional corporation paradigm, and dispersal of power to the edges of the supply chain (Conca 2001). Understanding the dynamics associated with these changes will provide fisheries managers, policy-makers, and customers with potential entry points for mitigation along the fisheries supply chain.

The first of these consequences is the increase in the dispersion and complexity along the fisheries supply chain. As an example, it's not unlikely for a customer to eat surimi that was caught by a Korean company, using a fishing vessel constructed in Panama, in Russian waters, processed in China, and shipped to the United States or Europe. This is an instance where the global fisheries supply chain has disrupted any local feedback mechanism that may be in place by increasing the spatial and social distance between production systems and consumers. The surimi-eating customer has no idea how his or her consumption choice affects fisheries sustainability, because those effects are occurring in Korea, Panama, Russia, and China.

The same case also exemplifies the second consequence. This consequence is the breakdown of traditional corporation paradigms known as Fordist. "Fordism" simply refers to the system of mass production that was popularized by the Ford Motor Co., and became popular in industrialized countries after World War II. The term "post-Fordism" refers to shifts from a dominantly Fordist system of production to a system that is more decentralized (Conca 2001). As these companies make the shift towards post-Fordism, they become little more than roving centralized capital sources. They outsource production, manufacturing, and distribution to places with the least restrictive environmental and labor regulations, but at significant social and environmental cost. In this situation, trade is not only between input and customer, but occurs at every level along the fisheries supply chain, having enormous ramifications for responsibility, monitoring, and governance structures that are designed to protect fisheries sustainability.

Does the complexity of the fisheries supply chain and the breakdown of international regulatory systems imply that the future of world fisheries is hopeless? We believe that it does not. We think that the greatest potential obstacle to sustainable fisheries, as well as the greatest potential champion for ecological progress, is the customer.

Conca (2001) discusses this situation as a dispersal of power to the edges of the supply chain. In one sense this is a dispersal of power "from the factory to the board room" as multinational corporations grow and condense. However there is also a great concentration of power allotted to the individual as customer. Because globalization has uprooted most of the obstacles that once limited the movement and reach of people, while concurrently allowing everyday citizens to be connected to the global market, there is more individual power to influence markets and states than at any other time in history. For the manager, policy-maker, or customer who places importance on fisheries sustainability, globalization's real strength lies in its ability to break down barriers to communication and facilitate the spread and exchange of ideas, beliefs, and knowledge worldwide.

Short of government-mandated solutions for fisheries sustainability coupled with effective international management and enforcement for transboundary stocks, one solution to address the problem of increasing fragmentation of the fisheries supply chain and post-Fordist business models is a market-based, consumer-responsive tactic. Such a solution requires adequate market or consumer demand motivation, proper signaling of the demanded product or service attributes throughout the supply chain, and a coordinated response by suppliers to deliver the attributes demanded. Eco-labeling is one strategy that can be used to achieve this goal. The eco-label can become the site at which the fisheries supply chain can be coordinated to deliver the sustainability attribute for the customer. The label can list the monitoring effort, as well as harvesting, farming, or processing activities associated with the product (Peterson and Fronc, Chapter 17).

Another consumer-responsive strategy is the development of a Web-based, relational database that allows customers to research and trace the spatial evolution of the products they purchase. This product database would inform customers which products were produced using sustainable methods along the fisheries supply chain. Customers could then choose to purchase products from companies that were not in violation of environmental statutes themselves or were not contracting with companies who were. This database would allow customers to make the crucial connection with their behaviors and consumptive effects and those consequences that they don't necessarily experience. It is important to note that government or NGO certification of ecologically fished products, from a publicly *trusted* source, is essential for eco-labeling or consumer databases to be consistent and effective. In this globalized world, technological and communication infrastructures are in place to facilitate more effective monitoring and governance for fish stocks than at any other time in history; a feature which should be utilized to further enhance our sustainable responses.

We also agree with Lynch *et al.* (2002) that information and awareness campaigns from fisheries managers are needed to increase awareness and public support for global fisheries. This is because, if sustainable fisheries initiatives are to grow and continue to exist, then managers must rely on positive attitudes from the public in order to support their management decisions.

Although there needs to be a component of responsibility at the customer level, there is also a need for responsibility at the resource extraction level. Because globalization is causing the fisheries supply chain to become more fragmented there is an opportunity for the findings of social networks to take a leading role in sustainable fisheries extraction, as evidenced in chapters by both Pollnac (Chapter 9) and Frank *et al.* (Chapter 16). In each chapter, members of a community access resources within the context of their community relationships. These members identify not only in their relationships to others in their social networks, but with the community in which networking takes place. Thus, they are more disposed to extract natural resources with a responsibility for the value of the resource for the future of the community.

Eventually, government policy and public sentiment as expressed by consumer preference must align with the goal of sustainability in order to coexist successfully with a globalized economy. Changes in global trade, harvest regulations, and fishing rights will have to be made in this context. Ultimately, the world's fisheries are not just a business, but a public trust, and only via multilevel governance as empowered by the world's citizens can we hope to enforce and regulate globally shared and common resources for sustainable futures.

References

Arbo, P. and Hersoug, B. 1997. The globalization of the fishing industry and the case of Finnmark. *Marine Policy* **21**: 121-142.

Botsford, L. W., Castilla, J. C., and Peterson, C. H. 1997 The management of fisheries and marine ecosystems. *Science* **277**: 509-515.

Conca, K. 2001. Consumption and environment in a global economy. *Global Environmental Politics* **1**(3): 53-71.

Daly, H. 1993. The perils of free trade. *Scientific American.* **290**: 50-57.

Deardorff, A. 2005. *Deardorff's Glossary of International Economics.* Available online at www-personal.umich.edu/~alandear/glossary/

FAO (Food and Agriculture Organization). 2004. *State of the World Fisheries and Aquaculture.* Rome: Food and Agriculture Organization of the United Nations.

Food Standards Agency. 2003. *Fish Labelling (England) Regulations 2003.* Available online at www.foodstandards.gov.uk/foodindustry/regulations/ria/fishlabelriafinal

Friedman, T. L. 2000. *The Lexus and the Olive Tree.* New York: Anchor Books.

Goldsmith, E. 1996. Global trade and the environment. In *The Case against the Global Economy and for a Turn toward the Local,* eds. J. Mander and E. Goldsmith. San Francisco, CA: Sierra Club, pp. 78-91.

Hayes, D. B., Dodd, H. R. and Lessard, J. L. In press. Conservation considerations for small dams on cold-water streams. *Proceedings of the 4th World Fisheries Congress,* May 2-6, 2004, Vancouver, British Columbia.

Lynch, K. D., Jones, M. L., and Taylor, W. W. 2002. Sustaining salmon fisheries in Canada and the United States: recommendations for policy and management. In *Sustaining North American Salmon: Perspectives across Regions and Disciplines,* Bethesda, MD: American Fisheries Society, pp. 379-395

Martinez, A. 1995. Fishing out aquatic diversity. *Seedling* July: 2-13. Available online at www.grain.org/publications/jul951-en.cfm

Moyle, P. B. and Cech, J. J., Jr. 1982. *Fishes: An Introduction to Ichthyology.* Englewood Cliffs, NJ: Prentice Hall.

Pauly, D., Christensen, V., Dalsgaard, J., Froese, R., and Torres, F., Jr. 1998. Fishing down marine food webs. *Science* **279**: 860-863.

WTO (World Trade Organization). 2006. Free trade cuts the cost of living. Available online at www.wto.org/english/thewto_e/whatis_e/10ben_e/10b04_e.htm

Index

Acipenser see lake sturgeon
   A. fulvescens 31, 173
Aeromonas salmonicida 127
Agreement on Straddling Fish Stocks and Highly Migratory Fish Stocks 207, 295
agriculture 26, 77-80, 95, 103, 246
   expansion 78
aldicarb 79
alewife 28, 29, 159, 173
Alosa
   A. pseudoharengus 28, 159
amphipod 173
anglers 24-26, 317
Anguilla anguilla 125
Anguillicola crassus 125
antifoulants 478
Aphanomyces astaci 126
Aqua Bounty permit application 484
aquaculture 8, 23, 31-32, 123-125, 128, 147, 198, 244, 245, 247, 249, 282, 297, 298, 424-449, 453-465, 469, 470-472, 535-538
aquarium trade 29, 64, 123, 126, 146, 191, 195, 203, 472-480
aquatic animals 120-149
Aquatic Animal Health Standards Commission 132
Aquatic Ornamental Trade Association (AOTA) 379
Aral Sea 27, 83, 85
artisanal fisheries 198, 207
Asia 8
Australian Nature Conservation Agency (NCA) 146
Australian Wildlife Protection Act 191

bacteria 111
bacterial kidney disease 122, 124, 127
bait 29, 126, 196-197
Balanus improvisus 157
ballast 27, 104, 123, 160, 161, 163
Bank for International Settlements 4
barnacle 157
Bentenan 233-240
billfish 56-58
biodiversity 313-315
biomass 291
bivalve 157
Bretton Woods 7, 76
Bonamia ostreae 125
Boundary Waters Treaty 29, 110
bounded solidarity 398
Bosmina coregoni maritima 173
brackish water 49
brood stock 31, 146
bycatch 187, 189, 196, 198, 230-231, 351, 442, 455, 457, 536

Canada 32
Canadian Department of Fisheries and Oceans (DFO) 344
capital
   markets 527
   natural 415
   relocation 527
capitalism 500-501
capture fisheries 49-66, 454-456
carp
   Asian 28
   common 28, 159
Cartesian epistemology 507
Cerocopagis pengoi 157, 164, 169, 173
certification 205, 207
China 5, 9, 454
Cichla ocellaris 156
cladoceran 157
Clean Water Act 30, 77, 473, 478
climate 96, 107
   climate change 293, 508
Clostridium botulinum 174
cod 32, 129, 344-350

545

Code of Conduct for Responsible
    Fisheries (CCRF) 518-520
codes of practice 149
Codex Committee on Fish and Fishery
    Products (CCFFP) 373, 375
cold water disease 124
co-management 226, 403, 514
comedy of the commons 513
communication 2, 3
commercial fisheries 244-249
    creation of 126
    effects on 75, 291
    fishing 47
    fleets 47-62
    industry 25, 293, 298
    ship design 47-48
commissions 36
Common Fisheries Policy 37
common property/pool resource
    514-515
compartments 415
conservation 9, 77, 85, 202-204
containment 474
Convention on International Trade in
    Endangered Species (CITES) 10,
    185, 205, 207, 208
coral reef 64, 66, 400-413
Cordylophora caspia 157
cotton 79
country-of-origin labeling (COOL) 446
Cox, Robert W. 15
Crassostrea angulata 125
    C. gigas 125
crayfish 125
cross-scale 514
Cryptosporidium 104
ctenophore 157
cultural–natural ecosystem (CNE)
    514, 515
Cyprinus carpio 28, 159

dams 81-83, 293, 534
degradation
    aquatic ecosystem 23, 26-29, 84
    fisheries 188
demand 8, 187, 192, 292, 296, 298,
    426-430, 464, 469
demersal fish products 52-53
Des Voeux 222-223
developed countries 47-70, 540
developing countries 6, 47-70, 82,
    279, 431, 459, 464, 470, 472
Dietary Supplement Health and
    Education Act (DSHEA) 374
diseases 22, 138-140
    aquaculture 474, 477
    climate impacts on 107
    vectors 144

    waterborne 94, 96
    zoning 141-144
Diseases of Fish Act 127, 131
distribution 537-538
domination ethic 507
drama of the commons 514, 519
Dreissena
    D. polymorpha 157, 159
    D. rostriformis 165
dreissenids 164, 172
Dutch East India Company 7

Echinogammarus ischnus 164,
    169-172, 173
eco-labeling 12, 425-447, 542
economic
    development 7, 292, 297
    impact 122
    integration 7-9
ecosystem
    approach 70, 312-313, 503
    management 312-313, 318,
    332-357
EEZ *see* Exclusive Economic Zone
El Niño 55, 293
Endangered Species Act 30, 77, 292, 473
enforcement 38
Enterococcus 110, 111, 112
Enterocytozoon salmonis 128
enforceable trust 392
Environment Canada 313
Environmental Protection Agency
    (EPA) 313, 480
epidemiological tools 129
epistemic communities 316
epizootic ulcerative syndrome 124
erythrocyte inclusion body
    syndrome 124
Escherichia coli 110, 111, 112
    E. coli O157 104
ethics 502
Eurasian ruffe 159
Europe 8
European Inland Fishery Advisory
    Commission (EIFAC) 144, 147
European Union 3, 12, 191
eutrophication 78, 106, 477
Exclusive Economic Zone (EEZ) 6, 32,
    55, 160, 295, 471
exotic species 22, 23, 26, 27, 29, 122,
    144, 538
exports 51-68
    large pelagic 56
    small pelagic 55-56

fair trade 364, 372
FAO *see* Food and Agriculture
    Organization

FDA *see* Food and Drug Administration
female collaborative model 515
Fiji 216–224
fish aggregating devices (FADs)
    231–240
fisheries
    blast 234
    boutique 64
    commissions 13
    export-oriented 59–62
    industry 538–539
    large-scale 229–241
    management 6, 28, 202, 207, 413, 518
    poison 234
    small-scale 229–241, 332–342, 470
    sustainable 332, 443, 507, 519
Fisheries Amendment Act 219, 220
Fisheries Conservation Act 219
Fisheries Ordinance of 1922
    (New Guinea) 224
Fisheries Products International
    (FPI) 349
Fisheries Recovery Act 509
Fishing pressure 24–26, 293
fishmeal 55–56, 455, 457
Fish Protection Act 219
Flavobacterium psychrophilum
Food and Agriculture Organization
    (FAO) 144, 147, 469
    Aquaculture Production database 50
    Code of Conduct 518–520
    Fisheries Commodities Production
        and Trade database 49–51
Food and Drug Administration
    (FDA) 365
food security 47–70, 271, 279
Fordism 541
foreign exchange 62–64
foreign species 6
freshwater
    fishing sector 49
    systems 75–86
furunculosis 127

Gadus morhua 344
gaffkemia 125
Gammarus fasciatus 173
General Agreements on Tariffs and
    Trade (GATT) 7, 39, 76, 187, 297
genetically modified organisms
    (GMOs) 379
geographic information system (GIS) 38
Giardia 104
gill net 234
global commodity 270
global governance 6, 10, 13, 305–310,
    319, 322, 326
global positioning system (GPS) 38

globalization
    definition 2, 21, 75, 95, 270, 385, 469
    rate of 215–216
GMOs *see* genetically modified
    organisms
governance 507, 514–515
Great Lakes 10, 27, 28, 29, 31, 83, 84,
    94, 107, 131, 157–161, 305, 306,
    386, 503, 513
Great Lakes Fisheries Commission 34,
    159, 308
Great Lakes Water Quality
    Agreement 29
groundwater contamination 102–103
growth hormone (GH) genes 481
Gymnocephalus cernuus 159
Gyrodactylus salaris 127

Haplosporidium nelsoni 140
    see also MSX
harmful algal blooms 99, 106
Hazard Analysis and Critical Control
    Points (HACCP) 60, 368, 369, 372
health
    certification 141
    fish 282
Hippocampus spp. 188, 201
Homarus
    H. americanus 125
    H. gammarus 125
hospitality industry 424–438
hydropower *see* dams
hydrozoan 157

Ichthyophthirius multifiliis 156
ignorance 506
indeterminacy 506
India 4
individual transferable quotas (ITQs) 65
Indonesia 231–233
infectious hematopoietic necrosis
    virus (IHNV) 124
infectious hypodermal and
    hematopoietic necrosis virus
    (IHHNV) 124
infectious pancreatic necrosis virus
    (IPNV) 124
Institute for Agriculture and Trade
    Policy 85
Integration, vertical 64–65, 349
International Aquatic Animal Health
    Code 132, 133, 138
International Commission for the
    Northwest Atlantic Fisheries
    (ICNAF) 344
International Council for the
    Exploration of the Seas (ICES)
    144, 147

International Joint Commission (IJC) 29, 160, 308
International Monetary Fund (IMF) 7, 75
international norms 311
International Standards Organization (ISO) 5
international treaties 443
International Whaling Commission 5
Internet 14, 15, 22, 24, 35, 199
invasive
　biological 157
　invasional meltdown 161-162
　movements 122
　species 105, 476, 486
　vectors 156-164
　see also exotic species; non-indigenous species
invertebrates 58-59

Joint Expert Committee on Food Additives (JECFA) 375
Jubi 234
jurisdictional stress 323

Kimberley 222-223

Labrador 335
Lake Ontario 28
lake sturgeon 31, 173
landing data 50-55
land wars 219
Laplacian reconciliation 507
Larceny Act 219
large-scale fisheries 229-241
Lates niloticus 156
law
　customary 218
　hard 309, 310
　soft 309
leaching 80
Lernaea cyprinacea 124
lobster 125
Londe 233-236
low income food deficit countries (LIFDCs) 453
Lythrum salicaria 158

Malaysia 4
management 202
　community-based 217-218
　decisions 338-356
　interjurisdictional 37
　managers 412
mangroves 63, 67, 473, 512
Maori 216-227
Manual of Diagnostic Tests for Aquatic Animals 132, 141

marine ecosystems 337
　fishing sector 49-70
　property right 218-227
Marine Aquarium Council 205
Marine Mammal Protection Act 39
Marine Stewardship Council 205, 445
market 292, 469
market access for non-agricultural products 464
Marteilia refringens 125
media 202
Michigan United Conservation Clubs (MUCC) 316-318
Minihasa, Maluku coast 233-240
Mnemiopsis leidyi 157
mollusk species 158
morbillivirus 121
MSX 140
multijurisdictional
　governance 34, 541
　management 34
multinationals 7, 10, 12, 297
mutual coercion 391
Mysis relicta 173
Myxobolus cerebralis 126

NAFTA *see* North American Free Trade Agreement
NAFO *see* North Atlantic Fisheries Organisation
Native tribes 296, 299
Neogobius melanostomus 29
network effects model 409
Newfoundland 344-350
New Zealand 216-227
Nile perch 156
Nile tilapia 123
NOBOB *see* ballast
non-indigenous species 156
non-governmental organizations (NGOs) 5, 10, 12, 13, 84, 315
non-traditional species 455, 456-463
North American Free Trade Agreement (NAFTA) 187, 297
North Atlantic Fisheries Organization (NAFO) 344

Oceania 216
OECD *see* Organisation for Economic Cooperation and Development
OIE *see* World Organization of Animal Health
omega-3 fats 430
Oncorhynchus
　O. clarki pleuriticus 28
　O. keta 292
　O. kisutch 28
　O. mykiss 156, 292

O. nerka 292
O. tshawytscha 28, 128, 292
Oreochromis niloticus 123
Organisation for Economic Co-operation and Development (OECD) 75
ornamental fish 126, 278, 379, 400–406, 483–484
ostracization 396
Ostrea edulis 125
overexploitation 185
overfishing 9, 185
oyster 125, 140
Oyster Fisheries Act 219

Pacifastacus leniusculus 125
Pacific Salmon Commission 295
Pacific Salmon Treaty 295
Pajeko 234–238
Pakistan 4
Papua New Guinea 217–226
Parvoviridae 124
pathogen 113, 121–149, 477
peacock bass 156
Pearl Shell and Bêche-de-mer Fishery Ordinance of 1894 (Papua) 225
Pearl, Pearl-Shell and Bêche-de-mer Ordinance 1911-1932 (Papua) 225
Pelang 233–237
Petromyzon marinus 159
piggybacking 200
pilchard 121
policy 216–219, 225, 255–263, 306, 315–322, 385, 391, 501, 503, 527
  ecosystem approach 312–313
  government 443–444, 490
  international 208
political integration 7
pollutants 56, 281, 293, 478
pollution 94–96, 110, 220, 293–294, 442–444
Ponto-Caspian region 157, 162, 163, 172, 174
precautionary principle 12, 332, 503, 509, 510–511, 516, 519
privatization 84, 512, 534
Procambarus clarkii 126
processing 537–538
  frozen trout 126
property rights 65, 69–70, 208
public trust doctrine (PTD) 511–513
purple loosestrife 158
purse seine 235

quagga mussel 164, 165, 172
quarantine 145

rainbow smelt 173
reciprocity transactions 393
recreational
  fisheries 24, 25, 126, 293
  water protection 111
Renibacterium salmoninarum 124
resource rights 222
restaurant 453–464
riparian zones 78
risk 474, 506
  analysis 145, 370
  assessment 144, 504–505, 506, 520
Rivers and Streams Ordinance 224
round goby 29, 164, 167–169, 172

safety standards, seafood 429–431, 539–544
salmo
  S. trutta 28, 156
  S. salar 28, 476
salmon 8, 127, 128, 291, 292–299, 471
  Atlantic 28, 247, 470, 472, 476
  chinook 28, 292, 473
  coho 28, 292, 470, 473
  Pacific 128, 476
  sockeye 292, 473
  steelhead 292, 473, 476
Salmonella 104, 365, 371
Salvelinus namaycush 31, 173
Sanitary and Phytosanitary Measures Agreement (SPS Agreement) 60–62, 132, 146, 379
scarcity 75, 84, 94
Sea Around Us project database 50–52
seafood 221, 348, 365, 369
  contamination 350, 430
  industry 424–431, 438–439, 455, 456–463
  products 434, 435
sea horses 8, 9, 187–209
sea lamprey 29, 159, 308
sewage 112–113
  sanitary sewer overflows (SSO) 102
sheep 158
shrimp 39, 124, 126, 129, 173, 196, 206, 470, 473
small-scale fisheries 229–241, 332–342, 470
social
  capital 390–391, 393, 398, 400, 404–405, 406, 410, 411
  cohesive subgroups 391, 406–408, 415
  enforceable trust 406
  impacts 65, 69–70
  network 388–389, 402–404, 406, 410
  relationships 391
sociogram 406
SOHO *see* systems

spatial feedback 352
spring viremia of carp virus (SVCV) 126
stakeholders 25, 34, 82, 202, 203, 294, 298, 299, 412, 515
stewardship, environmental 310
St. Lawrence Seaway 159
Streptococcus iniae 284
subsidies 443-447, 534
subsistence 199, 293
supermarkets 350, 424-435, 453-463
supply chain 432-441, 527
   customer 539-544
   production 535-537
   manufacturing/distributing 537-538
   raw materials 528-535
   wholesale 538-539
sushi 428-429
sustainability 187, 298
systems
   Bertalanffy's general systems theory 335
   Koestler's notion of holons and holarchy 337
   self-organizing, holarchic, open systems (SOHO) 335-357

tariffs 7, 464
taura syndrome 124
Technical Barriers to Trade Committee (TBT) 379
technology 5, 9, 120, 196, 199, 281, 282, 283, 399, 468, 499
   internet 9
   fishing gear 9
temporal feedback 352
Thailand 4
Thurston 223-224
tilapia 13, 269
Tombak 234
Torres Strait Islands 225
total coliform bacteria 109
tourism 95, 202
toxaphene 80
traceability 372
trade 3-5, 47, 52
   aquatic animals 127, 130, 188-209
   barriers 7
   fair 364, 372
   free 77, 186
   international 120, 121, 125-126, 131, 144, 463, 472
   liberalization 47-64, 527
   live fish 64
traditional medicine 185, 188-197
Tragedy of the Commons 34, 513
transgenic fish 274, 377, 468, 470, 480
transjurisdictional investment 527
transportation

transoceanic shipping 157, 159
technology 2, 199
trawling 189, 196, 206
trends 297
Trojan gene scenario 487
trout
   brook 127
   brown 28, 127, 156, 159
   cutthroat 28
   lake 31, 173
   rainbow 127, 156
Trout Unlimited (TU) 318-319
tuna
   dolphin-safe 12, 14, 35, 444
   product 56-58, 470
turtle
   excluder devices 4, 9, 39
   shrimp-turtle 4

uncertainty 332-340, 506
United Nations 3, 294, 324
   Convention on the Law of the Sea (UNCLOS) 47
U.S. Fish Commission 159
U.S. Department of Agriculture 479

Vibrio 105, 122
Victoria, Queen 222-223

Waitangi, Treaty of 216
Ward, Barbara, "Spaceship Earth" 9
water
   indicators of globalization 96
   pollution 94-96
   quality 95, 109
   resources 92
   scarcity 94
   use 92
   waterborne agents 100, 103, 104
Welland Canal 159
WHO *see* World Health Organization
wild stocks 123, 424-440, 473
WIPO *see* World Intellectual Property Organization
World Bank 7, 75
World Health Organization (WHO) 377
World Intellectual Property Organization (WIPO) 5
World Organization of Animal Health (OIE) 131-138, 147
World Trade Organization (WTO) 4, 8, 12, 13, 76, 77, 84, 132, 186, 187, 206, 376, 464, 534
World War II 3, 7
WTO *see* World Trade Organization

zebra mussels 29, 159, 164, 172
zooplankton 294